BREATHING WATER IN A WARMING WORLD

sidestonepress

BREATHING WATER IN A WARMING WORLD

Principles and Applications of the Gill-Oxygen Limitation Theory

Daniel **Pauly** & Johannes **Müller**

Published by Sidestone Press, Leiden
www.sidestone.com
E-mail: info @ sidestone.nl
Phone: (+31)(0)71-7370131

Imprint: Sidestone Press Academics
This book has been peer-reviewed. For more information see www.sidestone.com

Lay-out & cover design: Sidestone Press
Photograph cover: Fish artwork | kichigin19 | stock.adobe.com

ISBN 978-94-6427-147-8 (softcover)
ISBN 978-94-6427-148-5 (hardcover)
ISBN 978-94-6427-149-2 (PDF e-book)

DOI: 10.59641/k3n9h0i1j2

Contents

Foreword

Our evolutionary trappings have not prepared us for this book. As terrestrial, air-breathing creatures that have only traipsed the Earth for a few million years – an evolutionary blink of an eye – we humans struggle to comprehend the notion that oxygen, such an essential element to our existence, could possibly be limiting. We are comfortable inhaling air infused with plenty of oxygen, about 21% by volume to be exact. Unless we scale a tall mountain, where oxygen concentrations might be half or a third of that at sea level, its availability just doesn't concern us. Yet, over deep time, oxygen has been a defining commodity for terrestrial animals because the Earth's atmospheric oxygen levels have fluctuated markedly over millions of years, from roughly half to nearly double that of current levels. But most importantly for this book, dissolved oxygen concentrations in water – in lakes, rivers, and the sea – both now and in the past are only a tiny fraction of those in the atmosphere. Water-breathing animals struggle to get enough of the stuff.

None of this registered with me until a few years ago when, while working at my desk at Florida International University, an email message from Dr. Daniel Pauly popped up on my computer screen. Any respectable marine scientist knows of the famous fishery biologist Daniel Pauly who has published more consequential scientific works than any marine biologist in history. Dr. Pauly has gone toe-to-toe with the world's industrialized fishing interests to expose their destructive practices and introduced several important ecological concepts now common in textbooks, "*shifting baselines*" and "*fishing down the food web*" among them. He wrote me to discuss sponges and…oxygen.

By that time, I had conducted research in tropical marine ecosystems for nearly four decades, primarily on the ecology of lobsters and, yes, sponges. But oxygen? I was certainly aware of local episodic events in aquatic ecosystems in which dissolved oxygen levels plummeted for one reason or another, asphyxiating fishes and other animals and resulting in massive, smelly die-offs. Though dramatic, I considered those infrequent events. But Daniel's Gill-Oxygen Limitation Theory (GOLT)? Never heard of it. So, Daniel sent me some papers on O_2-related phenomena, including his original paper introducing the theory derived from his doctoral dissertation, penned in Germany 45 years ago. It was a revolutionary idea. Through subsequent emails, teleconferences, and animated discussions during his visits with my research team in the Florida Keys, I found Daniel's scientific arguments for the GOLT persuasive. I became an acolyte of the GOLT. If correct, this theory fundamentally changes marine biology, underpinning the evolution and biology of water-breathing animals with far-reaching consequences for fisheries management across the globe.

I am deeply honored that my friend Daniel Pauly and co-author Johannes Müller invited me to offer this foreword. I have never met Johannes Müller, which is unfortunate for me as an amateur historian because Dr. Müller is the real deal. An interdisciplinary scholar, his writings on the interconnections between the humanities and the natural sciences are fascinating, and who would not be compelled to read some of his purely historical works with titles such as "*A good guy with a sword*" or "*Dracula and the printing press.*" Like me, Johannes spiraled into Daniel's formidable gravity a few years ago, culminating in several co-authored publications on the physiology of fishes. I'm sure there is a story to be told about that transition. Together, Drs. Pauly and Müller have scripted in this book a detailed account of why Earth's rapidly warming waters will fundamentally change the lives of water-breathing animals and, in turn, humanity's interactions with them.

This book is a deep dive into the physics and physiology behind the construct of the GOLT, deftly explaining its significance in a warming world, and offering examples of the ways in which oxygen limitation has driven and still drives the physiology and behavior of a panoply of water-breathing animals, well beyond Daniels' beloved fishes. As the best scientists do, Drs. Pauly and Müller also entertain criticisms of the GOLT and consider alternative hypotheses for its predictions. Indeed, the GOLT has spawned a flurry of research by teams from around the world, including my own. We are putting this provocative theory to the 'acid test.' It is that important.

Dr. Mark J. Butler IV
Professor, Eminent Scholar, & the Walter & Rosalie
Goldberg Chair of Tropical Ecology
Institute of Environment
Department of Biological Sciences,
Florida International University
Miami, Florida, USA

Preface and Acknowledgments

We live in a time when the effects of global warming are unmistakable: forest fires, heatwaves, droughts, melting glaciers, and increasingly unpredictable weather events already give us a glimpse of what can be expected in the coming decades, even if we succeed in reducing our global emissions of greenhouse gases. While the effects of climate change on terrestrial environments attract considerable attention and can be easily observed, the world's oceans and freshwater ecosystems are also massively affected by warming and deoxygenation, but we do not perceive their intensity.

The effects of climate change on animals that inhabit these ecosystems are different and, in some respects, more drastic than the challenges faced by terrestrial animals. This is especially true for all those species that breathe water instead of air, which includes the vast majority of fishes and marine and freshwater invertebrates. This even affects marine mammals, seabirds, and not to forget, humans, for whom fish provide an essential source of nutrition.

Breathing water presents energetic challenges that are hard to imagine for mammals like us. When temperatures rise, water contains less dissolved oxygen, but more importantly, water-breathing animals require more oxygen to maintain their metabolic processes, and their oxygen demand often reaches levels that exhaust the uptake capacities of the fish, crab, or squid in question. Simply put: water-breathing animals in warmer water are short of breath – a physiological state we experience only when we attempt to run a marathon, or ascend Mount Everest without supplementary oxygen.

This book explains the effects of climate warming on water-breathing animals, but its scope is more comprehensive than that. Addressing the current constraints and the impact of rising temperatures on random species of water breathers is not enough; instead, this requires a theoretical foundation that allows for a more comprehensive understanding of the eco-physiological processes that shape underwater life. To do so, this book presents a new interpretation of the mountain of published life-history data on water-breathing animals, and it documents that the oxygen supply to their bodies – which is limited both by the physical properties of water, and by the oxygen supplied by their gills not being able to grow as fast as the oxygen demand of their bodies – makes them extremely sensitive to the warming and deoxygenation of their watery habitats. Indeed, their entire biology is shaped by this extreme sensitivity to the oxygen dissolved in the water they inhabit, as well as its temperature, which directly impacts their oxygen requirements.

Although we advocate, in this book, a theory that has *limitation* in its name, and which is structured around *constraints*, we view the diversity of water-breathing as a triumph of evolution over a medium that presents severe physiological challenges, as it contains 30 times less oxygen than the air we breathe, is 50 times more viscous, and in which molecular diffusion is 10,000 times slower than in air. Indeed, the evolutionary triumph of developing organs that enable life in such a medium allows one to dream – and to suggest in our last chapter – that if life evolved in any of the ancient and deep oceans on the moons of Jupiter and Saturn, it might well be structured by several of the constraints identified in this book.

It is important to provide a thorough mechanistic foundation for the phenomena and trends currently observed in the growth and life-histories of fish and other water-breathing animals, as this helps us avoid fragmented conclusions that differ in each case and each species. Avoiding singular and entirely case-based *ad hoc* solutions is not enough to deal with the problems facing the world's aquatic ecosystems and their inhabitants. Instead, we need a theoretical framework that offers us the tools to understand what we observe, and act accordingly. This book is an attempt to provide such a theory.

This book has numerous antecedents, notably the work of Galileo Galilei, Charles Darwin, D'Arcy Thompson, August Pütter, J.B.S. Haldane, and Ludwig von Bertalanffy. These pioneers are historical figures, and we can thank them only figuratively.

This endeavour would not have been possible without the help and intellectual contributions of colleagues from different fields. Daniel Pauly would like to thank Professor Gotthilf Hempel, his advisor, who, in the late 1970s, gave him the liberty to write an unconventional doctoral thesis. The collaborations he maintained in the intervening decades with various colleagues were also crucial, notably with Rainer Froese, who helped formulate several of the ideas explored in this book. He also would like to thank his wife, Sandra Wade Pauly for her support, and for editing the first draft of this book and checking its references, and Elaine Chu, who not only drafted or adapted its 84 figures – with assistance from Tiffany Mak and Rachel 'Aque' Atanacio – but also helped with some of the computations they required. Johannes Müller would like to thank Noman Saleh, Keywan Sohrabi, Armin Schäfer, Martin Müller, and Matthanja Müller-Pieters, whose ideas, thoughts and observations helped shape this book. We also thank Itsaso Lopez-Ahedo for hunting down multiple typos in our draft and Claudia Moreira Calzadilla for helping with the index.

The first draft of this book was assembled during a 3-month mini-sabbatical that Daniel Pauly spent in early 2024 on the campus of the University of Cape Coast (UCC), Ghana, as a guest of the Africa Center for Excellence in Coastal Resilience Cape (ACECoR); he wishes to thank its Director, Professor Denis Aheto, and his staff and students for their support and friendship, and his colleagues at UCC, notably Dr. Margaret Fafa Akwetey.

Other colleagues we want to thank are Melanie Warren, the first author of the draft that became Chapter 8, 'Bob' Steneck, for his text on kelp also being oxygen-limited, and D.P.'s co-authors of papers whose pre-submission drafts formed the core of Chapters 4, 7, 9, 12, and 14, i.e., Upali Amarasinghe, Nicolas Bailly, Mark Butler, Elaine Chu, Niels Houben, Katia Freire, Mimi Lam, Cui 'Elsa' Liang, Peter Sorensen, Elsa Vasquez, and Weiwei Xian.

Finally, we would like to thank the two reviewers of this book, Drs Mark Costello and Jeppe Kolding, whose numerous suggestions greatly improved its form and content.

1

Breathing Water: Principles and Applications of the Gill-Oxygen Limitation Theory

Synopsis: This chapter provides an overview of the book and outlines the need for a comprehensive theoretical framework to explain altered growth patterns in fish and other water-breathing ectotherms in relation to temperature and oxygen. While the Gill-Oxygen Limitation Theory (GOLT) was not originally developed to address the impact of climate change on fish growth, it offers valuable insights into how rising temperatures affect growth by disrupting the balance between oxygen supply and metabolic demand. The theory suggests that as temperature increases, oxygen availability becomes a limiting factor that constrains growth and shapes central life-history traits. Through this theoretical framework, the book aims to provide a deeper understanding of how climate change influences the biology of aquatic ectotherms.

Why theory?

Life first evolved in water, and aquatic lifestyles represent the ancestral state of all animal life on our planet. Consequently, this is also true for how organisms breathe: gills are much older than lungs and played a vital role in the evolution of vertebrate and invertebrate animals (Gillis and Tidswell 2017). The transition to breathing air – a medium that contains 30 times more oxygen than water – allows for far better diffusion and is 50 times less viscous, offering opportunities for biological life that are seldom fully appreciated.[1] Breathing in water presents physiological and biophysical challenges that inform all aspects of the life-histories and ecologies of aquatic animals. These challenges become all the more visible in times of rapid change, which demands either a high capacity to adapt to new circumstances or a high degree of plasticity, allowing organisms to adjust their physiological organization accordingly (Earhart *et al.* 2022). In our current situation, the most drastic factor of change is climate warming, a phenomenon that affects all life on earth, but impacts aquatic organisms in particular and critical ways (see, e.g., Daufresne *et al.* 2009; Sheridan and Bickford 2011; Cheung *et al.* 2013a, 2013b; 2013c; Dahlke *et al.* 2020).[2] This book presents an overarching theory that allows us to understand these challenges by exploring the fundamental question of what it means to breathe water and how rapid climate change affects organisms that depend on this mode of respiration.

Why do we need such a theory? All forms of scientific knowledge result from questions asked at a specific time and place and the attempts to answer them. Regardless of their size and scope, these questions are posed at historical moments that require solutions to specific research problems that occur at that time. Science and the broad insights it produces transcend this historical situatedness and create frameworks that allow for a better understanding of phenomena that may be entirely unrelated to the context in which the original problem emerged.[3] In times of historical challenges that demand answers and solutions on a large scale and require data from a vast number of scientific disciplines and fields, tying knowledge together and theorizing it is of crucial importance. This book is an attempt to do this: even though it cannot be separated from the specific situation in which it was written – a moment in history at which the effects of global warming are felt more rapidly than expected by many – it reacts to this specific challenge. At the same time, the book presents a theoretical framework with much broader implications and it aims to explain several fundamental ecological, physiological, biochemical, and biomechanical phenomena that shape the life-histories of water-breathing animals, some of which touch upon the very foundation of how we think about biological organisms and their relationships to their environments.

This apparent tension between historical situatedness and maintaining a larger theoretical scope that connects wider bodies of knowledge is not a contradiction. Instead, these two aspects of our endeavor – explaining the effects of climate change on aquatic organisms and developing a broader theoretical framework to understand key features of their ecologies and life-histories in the context of the respiratory medium they inhabit – belong inherently together and depend on each other. The climatic transition through which our planet is currently going has (and will have) consequences we do not yet fully understand.[4] Rapid change is even more difficult to comprehend when its effects will most likely be permanent and irreversible in the foreseeable future. Explanations and interpretations are, therefore, often formulated *ad hoc* and as responses to suddenly occurring and sometimes unexpected phenomena. Such parochial, *ad hoc* answers to new questions may often be unavoidable. Answers of this kind require constant revision and cannot explain more comprehensive sets of phenomena and connect them. Understanding the mechanisms and the impact of quickly accelerating global change requires more than *ad hoc* responses – it requires theories. These theories should allow us to develop an explanatory framework that helps us understand what has happened so far and make reasonable predictions about the effects in the near future.

While the theory presented in this book explains the effects of climate change on the planet's water bodies – oceans, rivers, and lakes – and the animals living therein: fish, squid, crustaceans, and other invertebrates, it is essential to note that this theory was not developed as a response to climate warming or as an attempt to describe its effects. Rather, its theoretical foundation was first developed to describe the growth of tropical fish, a topic that had long been neglected by European and North American fisheries scientists, who had mainly focused on temperate species (Pauly 1979). The theory that emerged from this approach proved capable of explaining several phenomena observed in water-breathing animals, many of which remain poorly understood. It can provide a comprehensive model to understand respiration, metabolism, growth, reproduction, food conversion, adaptability, migration, and other life-history traits in aquatic organisms under changing circumstances, as well as their interactions with their ecosystems and their roles within the food webs of marine and freshwater ecosystems (Pauly 2019, 2021a). When this theory was first developed, the impact of tropical temperatures on the growth and related physiological processes of water-breathing animals was not seen as theoretically important: climate change was not yet a central topic in fisheries science, and its effects were not as immediately felt as they are now. Our explanation of the impacts of climate change on fish is not an isolated hypothesis formulated *post hoc.* Still, while it is based on concepts originally developed for other purposes, it possesses a predictive potential that makes it applicable to diverse phenomena.

The fact that the key elements of the theory elaborated in this book predate the now visible effects of anthropogenic climate warming illustrates once more what theories are and what they are not: theories are neither direct solutions to problems nor are they mere sums of hypotheses (either *ad hoc* or systematized) or sets of answers to specific questions. They are neither accumulations of loosely related statements or propositions nor compilations of observed facts. Their internal coherence does not depend on the ideas of their originators or proponents but on the necessity and specific function of each component within the theory as a whole. The relevance of Charles Darwin's opinions on geology, for example, is limited compared to his views on natural selection, even though it was important that Charles Lyell's 'uniformitarianism' had provided him with the longer timespans in which evolutionary processes could take place.[5] What matters to his theory of evolution is the basic mechanism he identified as its core: biological variation and selection. In other words: theories are not the sum of their internal components but rather their products. Therefore, every element of a theory that does not change the results or outcome of its products can be left out and considered external to it.

For this reason, theories are categorically different from individual hypotheses and are necessary to transcend the level of *ad hoc* observations, explanations, and solutions. In contrast to isolated hypotheses, they also structure how new questions can be asked and how they become meaningful.[6] It is not their task to solve specific problems (for example, temperature-induced size reduction in aquatic animals). When they can achieve this, they do so within a wider explanatory framework and with implications for a whole range of other questions and problems. While many of the basic mechanisms underlying the theory presented in this book may be relatively simple and can be related to widely available empirical data and findings, they have broader implications and an explanatory scope beyond explaining isolated phenomena.

Some of the ideas on which the theory behind this book is built were already developed in the first half of the twentieth century, and some of its core tenets can be traced even further back in the history of science and human knowledge. This also includes the practical and applied knowledge of fishing and aquaculture communities, as well as the rich mathematical and geometrical traditions of learning that preceded and shaped modern science.[7] For earlier authors and people who worked with fish daily, the correlational and causal nexus between temperature, oxygen uptake, and growth in fish and other animals was far less relevant than it is now. In our current situation, the implications of these relationships are of central importance to our understanding of the effects of climate change on the oceans and freshwater bodies of our planet.

Fish are among the groups of animals that may suffer most severely from rising temperatures in the near future: as recent projections suggest, many fish species face extirpation (and even extinction for some) if global warming continues at the rates we see in the present, not to mention the even larger group of water-breathing invertebrates that share similar fates (Dahlke *et al.* 2020). These effects will (and already do) also have a drastic impact on human societies. Fish and aquatic invertebrates supply billions of people with animal proteins and essential micronutrients; they play a crucial role in the planet's ecosystems and may even have an essential role in global carbon storage (Belton and Thilsted 2014; Fisher *et al.* 2010; Hicks *et al.* 2019; Saba *et al.* 2021). Like many effects of global warming, these effects will be distributed asymmetrically, and societies with the lowest 'carbon footprint' will probably feel the consequences of warming and deoxygenating oceans and freshwater bodies much more drastically than those that are the main drivers of global climate change. Ironically, fisheries of countries in the global north may even benefit from ocean warming for at least some decades as fish populations in the tropics and subtropics shift poleward (northward or southward) due to rising temperatures.

Given the severity of these projections, we believe it is necessary to connect the implications of the theory presented here to current developments, even if the scope of that theory is much broader and aims to provide more fundamental explanations for critical biological and ecological traits of fish and other water-breathing animals.

The theory presented here aims to explain how fish and other water-breathing animals, such as molluscs or crustaceans, respond to rising temperatures, i.e., how they grow and reproduce differently and shift their distribution poleward if they can. Additionally, it suggests what current developments will mean for the coming decades and perhaps even wider timeframes. It presents a comprehensive set of explanations for a wide number of biological and physiological traits in water-breathing animals, including larval development, growth, maturation, spawning, mortality, adaptability, migration, and geographical distribution. In the following, we will present the most basic building blocks upon which this theory is based and then outline its general architectural design.

Geometries of life: size, form, and dimension in the biosphere

Although scientific theories in biology are often viewed as categorically different from those in physics, biological processes occur within the physical world. The three spatial dimensions (length, width and height) and the resulting scale and size relationships are fundamental aspects of all biological processes, and the specific geometries of cells, organs, and organisms shape the processes they perform. Physical space is never scale-neutral, and as soon as more than one spatial dimension is involved, a scale change will impact each dimension differently. What follows is that organismic growth inescapably changes the relationships between the dimensions in which size increases. These fundamental geometrical relationships shape the ways in which organisms can function within different size ranges and define how they can regulate and organize basic mechanical and physiological processes. In other words, an increase in size will inevitably meet some form of dimensional constraint, and organisms will need to find strategies to function within these constraints[8].

The theory presented here revolves around two central concepts: *dimensional tension* and *surface allometry*. In the nineteenth and early twentieth centuries, the significance of these concepts was already recognized by biologists and theorists, who discussed their impact on basic functional mechanisms of different organisms, for example, in terms of heat regulation and food conversion in homeotherms or oxygen consumption in water-breathing ectotherms (see, e.g., Bergmann 1847; Spencer 1866; Rubner 1894;

Pütter 1909, 1920; Thompson 1917; Haldane 1926; von Bertalanffy 1951, 1964). The first concept, 'dimensional tension' (Pauly and Cheung 2017) refers to the difference between a 2-dimensional surface and a 3-dimensional volume, a difference around which all physical organisms are organized: while our bodies are three-dimensional volumes, our interactions with the world outside us are essentially two-dimensional: they take place through surfaces, be it as respiratory (lung or gill surface), digestive (intestinal surfaces), thermal (heat exchange via the skin), but also as sensory surfaces. Life on Earth depends on geometrical forms and cannot occur without them; therefore, by definition, it is organized around these dimensional tensions. The tension between the second and the third dimensions applies to all physically organized life forms – from bacteria and mushrooms to elephants and blue whales. Its effects and implications are different for different kinds of organisms. Life has evolved around this tension in such a way that it can function within the constraints imposed upon it. To deal with these constraints and to partly overcome the resulting limitation of resource supply that would otherwise occur, the vital surfaces of biological organisms tend to show allometric growth patterns, i.e., they grow relatively faster than the surfaces of cubes and spheres, for example through the development of folded or filamented structures, as we will discuss in more detail below.

Biological and physiological theories since the nineteenth century began to take surface-volume relations more seriously. Centuries earlier, Renaissance thinkers like Galileo Galilei (1564 – 1642) had already pointed out that life is organized around geometrical relationships and constraints, and explained allometric patterns in and between organisms with references to square-cube laws. A large portion of the geometrical knowledge underlying this kind of explanation predates early modern European science by millennia. Greek geometricians, Indian and Chinese mathematicians, and Egyptian architects had mathematized the consequences of the tension between the second and the third dimension (Needham 1959; McKenzie 2007; Kolachana 2019; Pauly 2019).[9] In ancient China, this resulted, as early as the second millennium BC, in methods for calculating the volumes of highly irregular geometrical objects, and similar endeavors were undertaken in India (Needham 1959). While the description of square-cube laws is typically attributed to Galilei and his *Two New Sciences* ("*Discorsi e dimostrazioni matematiche intorno a due nuove scienze*," 1638), many of his insights were preceded by knowledge that first emerged in East, South, and West Asia, and along Mediterranean shores. Galilei's seminal book applied the geometrical problem of the relationships between surfaces and volumes to mechanical constructions and biological organisms. It presented a masterful dialogue that developed complex mathematical and physical problems in an accessible language.

Galileo demonstrated through several practical examples that allometries of scaling shaped how organisms were organized and how geometrical realities set the conditions under which animals and plants could exist. If an animal grew big, it had to rearrange the organization of all its components in order not to collapse under its own weight: "*Who does not know that a horse falling from a height of three or four cubits will break its bones, while a dog falling from the same height or a cat from a height of eight or ten cubits will suffer no injury? Equally harmless would be the fall of a grasshopper from a tower or the fall of an ant from the distance of the moon. Do not children fall with impunity from heights which would cost their elders a broken leg or perhaps a fractured skull? And just as smaller animals are proportionately stronger and more robust than the larger, so also smaller plants are able to stand up better than larger. I am certain you both know that an oak two hundred cubits high,*

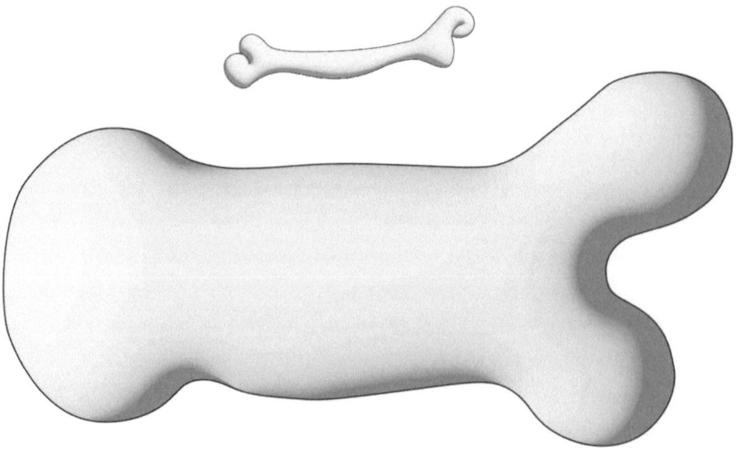

Figure 1.1. Bones of two different sizes in "*Discorsi e dimostrazioni matematiche intorno a due nuove scienze*" (redrawn from Galilei 1638, p. 119).

would not be able to sustain its own branches if they were distributed as in a tree of ordinary size; and that nature cannot produce a horse as large as twenty ordinary horses or a giant ten times taller than an ordinary man unless by miracle or by greatly altering the proportions of his limbs and especially of his bones, which would have to be considerably enlarged over the ordinary" (Galileo 1638).

Organisms must reorganize the relationships between the different components of their bodies to become larger; that is, an increase in size can only be achieved through morphological alterations (**Figure 1.1**).

The second concept, surface allometry, describes a direct consequence of the dimensional tension discussed above, but it refers to a specific consequence of the scaling relationships in organisms. Mechanical constraints resulting from gravity are primarily consequential for the size of terrestrial organisms, but scaling relationships also impact animals and plants in other ways. The relationship between surface area and mass or volume determines the exchange rates of vital resources between organisms and their physical surroundings. Being the site of resource uptake, waste excretion, and thermoregulation (in endothermic taxa), surfaces limit animal metabolic rates. As we will see below, organisms have developed different strategies to increase the area of the respective surfaces through which resource exchange occurs.

While the limiting role of surfaces is a result of the dimensional tension described by square-cube laws, biological surface-volume relationships do not typically manifest themselves in terms of simple spherical or cubic mass-scaling ratios: since organisms depend on adaptive mechanisms that allow them to survive and reproduce in their respective environments, optimization of surface-volume relationships is one of the most fundamental universal evolutionary adaptations in the biosphere, and while even mitochondria need to enlarge their internal surface area to take up resources, the importance of surface-volume optimization increases disproportionally with body and organ size. Allometry, a term first introduced by Huxley and Teissier (1936), from the Greek ἄλλος – 'other' or 'different' – describes the relative increase in size of individual body components with respect to total body size. This has specific implications for different organs and functional traits but, in the case of surface, it often implies deviations from square-cube laws: while the surface of a sphere grows with a power of 2/3 relative to mass, vital surfaces in most biological organisms grow in such a way that their increase follows a different pace than the usual 2/3 surface rule, mostly with positive allometry, i.e., with values larger than 2/3 (but always < 1). In many cases, this can be described as an adaptation to specific energetic demands.

Despite such adaptive strategies that allow organisms to cope with geometrical constraints within the conditions to which they are adapted, organisms remain vulnerable to sudden environmental change: while metabolic rates and available respiratory surface areas necessarily scale very close to each other, only slight changes in one of the vital parameters – for water-breathers mainly temperature or oxygen demand – can have dramatic effects. The physiological traits that allow an animal to survive and reproduce are adjusted to a specific environment and lifestyle; sudden changes to this environment pose challenges to further adaptation.

The theory presented in this book takes these dimensional challenges seriously. It proposes to examine environmental change, in our case, climate change, through the eyes of whole-organism physiology. It is in whole organisms, i.e., bodies, that the effects of global warming are being felt. Rather than focusing on molecules, genes, or cells, we position bodies and entire organisms at the center of our analysis. This is not to say that we regard them as final and irreducible entities but rather as systemic units that can only be understood in their entirety because "*the whole exists for and by means of the parts, and the parts for and by means of the whole,*" as Longo *et al.* (2012) put it.[10]

Organisms act as the interfaces between the environment and the cellular and molecular levels, providing the anatomical and morphological supports for physiological processes. This focus on organisms as a whole will become important later in the book as the most critical constraints can only be understood in the relationships between the different components of an organism to each other, in our case, the relationship between the respiratory organs and the body they have to supply with oxygen.

Geometrical constraints apply to all life on Earth, but this book focuses on one specific constraint, and one that applies to only one group of organisms: water-breathing animals, or water-breathing ectotherms (or 'WBE' in the following). Note, though, that the specification 'ectotherms' may be redundant: the very fact that no homeotherm breathes water (or conversely, that no water-breathing homeotherm exists) speaks for itself, and, as will be shown below, supports the theory presented here. Although some sharks, tuna, and other large active fish have developed limited forms of homeothermy, breathing underwater is so difficult that the extracted oxygen would be insufficient to meet the energetic demands of maintaining a constant body temperature. The energy requirements of mammals and birds, which enable their elevated body temperature, are so high that they depend on oxygen uptake in a medium within which it is much easier to breathe. Water-breathing is only possible for organisms whose metabolic rates are so low that their energy demand can face the specific constraints and limitations of respiration in this medium.[11]

Air and water as respiratory media

Before we discuss the implications of temperature, oxygen, and geometrical constraints for growth and other physiological processes in more detail, we will have to set the central problem of our theory in perspective by taking a short look at the constraints to life in aquatic environments, and especially the difficulties of extracting oxygen from a fluid medium like water.[12] Water constitutes a physical and ecological environment that is, in many ways, fundamentally different from terrestrial habitats, requiring different physiological adaptations (Denny 1993).

In an influential essay on the ecological foundations of animal biogeography from 1913, Richard Hesse (1868 – 1944), professor of zoology in Tübingen, discussed the transition from aquatic to terrestrial lifestyles as a major development in the history of life, an observation he linked to the greater availability of oxygen in the air: "*Living in air presents a number of great benefits to animals. The most important one is the amount of oxygen available. One liter of water contains about 7 cm³ of dissolved oxygen, whereas a liter of air contains 207 cm³ of oxygen. This creates the possibility of a significant increase of combustion processes in the body: aerial animals are therefore generally far more active than water-breathers*" (Hesse 1913). As Hesse further remarked, aquatic animals had made countless attempts to evolve the capacity to breathe air, but only a few of these had resulted in fully terrestrial lifestyles. Those that had been successful have undergone a quick diversification and now occupy a vast array of ecological niches.[13]

Hesse's discussion of the increasing opportunities that came with air-breathing and terrestrial lifestyles was not misguided. As more recent comparative reviews of terrestrial and aquatic biodiversity estimate, 80-85% of all known species on our planet live on land and only 15-20% in water (Grosberg *et al.* 2012; May and Godfrey 1994), although all phyla evolved and, with one exception, are still present in water.[14] The reasons for this disparity may be greater connectivity of marine environments, which could facilitate higher gene flow. Still, it seems counterintuitive that life on Earth began in the sea but then remained less diverse in water than on land. This asymmetry between aquatic and terrestrial biodiversity suggests that life in a fluid medium, such as water, depends on specific body plans that can less easily develop new forms than in air, i.e., on land (May and Godfrey 1994). The physical properties of water require adaptations that are constrained in such a way that they may allow for fewer new developments than would be possible for terrestrial organisms. As mentioned above, water not only contains 30 times less oxygen than air; but it is also 50 times more viscous, and its density is 850 times higher. Moreover, diffusion is 10,000 times slower in water than in air. To use a straightforward comparison, the air at the top of Mount Everest contains more oxygen than the best-oxygenated water layers of the sea or freshwater rivers, streams, and lakes.[15]

The costs of oxygen uptake for water-breathers are very high. Even though there are few existing studies, they suggest that fish invest about 30% of their energy in respiration (**Table 1.1**)[16]. The high costs of breathing also result from the fact that water is far more viscous than air, which makes ventilation of the respiratory organs in water even more difficult. As **Table 1.1** shows, the percentage of energy allocation to breathing has been estimated to be as high as 57% at certain life stages of some fish. The more the organism tries to cover increasing energetic demands with intensified respiration, the more it has to invest in the same activity, and less energy is left for the energetic demand it seeks to cover. The contrast with terrestrial animals is striking, as one recent study sets the differences in perspective: "*air-breathers can increase metabolic rates ca. 280-fold above that of water-breathers for the same ventilation cost*" (García-Gómez *et al.* 2024; see also Makarieva *et al.* 2008).

The data in **Table 1.1.** illustrates that the additional costs of breathing limit energy budgets for activities other than respiration. This means that a sudden higher energy

Table 1.1. Energy for respiration (**R**) in several fish species and age groups (in % of overall metabolism, rounded to the nearest integer); data from Dabrowski (1985); here 'd.s.' refers to estimates from different seasons.

Species	R (%)
Irish pollan (*Coregonus pollan*)	57
Atlantic cod (*Gadus morhua*; 9 years old)	< 49
Vendace (*Coregonus albula*; d.s.)	41 – 59
Largemouth black bass (*Micropterus salmoides*)	40
Atlantic cod (*Gadus morhua*; 3 years old)	< 37
Three-spined stickleback (*Gasterosteus aculeatus*; d.s.)	36 – 52
Northern pike (*Esox lucius*; d.s.)	16 – 50
European eel (*Anguilla anguilla*)	7
Mean	~ 30

demand, for example, through temperature-induced increase of metabolic rates, raises energetic costs disproportionally, which is why fish are so sensitive to rising temperatures, something every aquarist, recreational angler, or professional fisher will be able to confirm. The medium from which this group of animals takes up oxygen sets a constraint on what is possible in their physiology and ecology. Nevertheless, despite the difficulties of extracting oxygen from water and the huge costs of breathing in some species, fish have evolved to thrive within these constraints.

While the theory we present in this book strives for generality and seeks to provide an explanatory framework valid across many species and higher taxa, we do not claim that its explanatory range is universal. Its scope is limited to organisms subjected to the specific constraints imposed by breathing water.[17] The properties of this respiratory medium determine the range of physiological adaptations: extracting oxygen from a fluid and viscous medium requires specialized organs. While small aquatic animals can rely on their body surfaces for respiration, larger fish (if they are not air-breathers) and other WBE rely on gills, i.e., filamentous structures that immensely maximize surface area in relation to body size.

We will refer to the theory presented in this book as the Gill-Oxygen Limitation Theory, or GOLT, which also applies to some WBE that have no gills, such as, e.g., the chaetognaths or arrow worms (see Chapter 14). While the focus on gills may appear oddly specific for a theory that strives for generality, we will show that the structure of gills (or other respiratory surfaces) and the scaling relationship of their surface area relative to body size is of crucial importance to the biology and ecology of this group of animals and their ability to adapt to environmental change.

As mentioned above, the GOLT is a theory of organisms, but to be more specific, it is a theory of individual organisms – as opposed to life-history theories whose scope is at the level of populations rather than individuals. As Morbey and Pauly (2022) have argued, the GOLT is compatible with such life-history approaches and that the two can arrive at similar results. The difference between the two is the mechanistic explicitness of the GOLT, which relates anatomical and morphological factors to observed or calculated growth patterns and therefore allows for statements on the growth of individual organisms and about the heuristics that individual animals use for decision-making on, e.g., reproduction (Chapter 7) or predation (Bakun 2022). While the growth patterns of WBE are also crucial in life-history theories, the GOLT proposes a mechanistic foundation that explains their emergence due to the physical properties of the respiratory medium in which the respective organisms live.

Constraints and optimizations

While biological nature is historical and thereby inevitably shaped by contingent factors, the physiological and anatomical structures of biological organisms follow patterns that suggest wider generalities. The question of how such general patterns occur has puzzled researchers for centuries, especially after the advent of evolutionary theories in the nineteenth century: did such regularities emerge because they were the result of optimal organization, or did they represent the limits of what biological organization could achieve?

Nineteenth-century physiology saw several attempts to formulate generalizations across species and phyla, particularly regarding the role of energy metabolism and the processes that regulate an organism's maintenance, growth, and reproduction. While these processes differed between species and different size classes within species, some patterns could be detected across a wider range of phyla. Many of the early generalizations of metabolic processes revolved around surface-mass relationships. In 1838, Pierre Frédéric Sarrus (1798 – 1861) and Jean-François Rameaux (1805 – 1878) argued that the oxygen consumption of animals was proportional to the surface areas of their bodies. This was explained as an effect of thermoregulation: while smaller animals lost much more heat than larger ones, because their body surfaces were much larger in relation to their bodies, large bodies, by contrast, required far less energy because they could conserve heat much more efficiently (Sarrus and Rameaux 1838). Carl Bergmann (1814 – 1865) later developed a similar explanation to explain the geographical distribution of body size in birds. As he argued, body size correlated with latitude within related families and genera, and the birds from colder latitudes tended to be bigger (Bergmann 1847). The advantage of larger bodies in conserving heat was then used to explain why their greater mass was more advantageous in colder climates.

This reasoning led to an early consensus that metabolism scaled universally with surface area, with scaling exponents of 2/3 relative to body mass, a view for which more support was provided by a set of experiments by Max Rubner (1854 – 1932), who compared metabolic rate and body size in different dog breeds (Rubner 1894). This observation could then be expressed by the equation

$$R = a \cdot W^{2/3} \qquad \text{...(1.1)}$$

where R is the oxygen consumption (or respiratory rate) of an animal, used as a proxy for its metabolic rate (which requires oxygen), W is its body weight (or mass), and '*a*' a coefficient that varies between species, growth stage, and environmental conditions. The exponent 2/3 was based on the notion that (*i*) metabolic rate universally scales proportionally to surface area, and (*ii*) the geometry of the respective organism and the relationship between its mass and the surface that regulated resource uptake did not differ strongly.

The 2/3 rule in metabolic scaling was questioned by new generations of researchers in the early 20th century. As Max Kleiber (1893 – 1976) and Samuel Brody (1890 – 1956) argued, the slope of metabolic rates between species was not proportional to 2/3. Rather, the exponents of basal metabolism relative to body size were closer to 3/4 (Kleiber 1932; Brody and Procter 1932; Brody 1945). The idea of a universal 3/4-power metabolic scaling law gradually replaced the older surface-volume 'law' that assumed the generality of $W^{2/3}$, even though important theorists, such as Ludwig von Bertalanffy (1901 – 1972), insisted on this older model (von Bertalanffy 1951, 1964) for animals whose growth he classified as "*growth type 1*" (see Chapter 2). This insistence was confirmed by his experiments with guppies and brine shrimp, for which 2/3 clearly applied.[18]

Theorists have proposed different mechanistic explanations throughout the twentieth century, but the best-known model today is the one provided by the proponents of the so-called "*Metabolic Theory of Ecology*" or 'MTE' (West *et al.* 1997). These theorists developed the idea that supply networks were space-filling fractal structures that scaled with 3/4. The MTE model is also based on the assumption that physical constraints limit resource uptake through the cardiovascular and respiratory transport networks of organisms: since the transport of nutrients and oxygen in large animals and plants takes more time and has different energetic consequences, their metabolism is slower than in smaller ones. Similar to the older 2/3 model, the MTE model identifies geometrical constraints as the reason behind metabolic scaling relationships.

The regularity in metabolic scaling across a vast number of phyla and size ranges from bacteria to blue whales suggests a more general pattern or even a universal 'law.' As more recent research has shown, the scaling exponents of energy metabolism are not universal across taxa, but vary widely, even though they are consistently < 1 (Glazier 2005; Pauly 2019).[19] This holds true for both inter- and intraspecific scaling relationships, i.e., between different species and within individual taxa – with the exception of larvae and early juveniles (Post and Lee 1996; Wieser 1995). For example, based on their comparative study of data from 25 species, Jerde *et al.* (2019) concluded that they had found "*strong evidence for an intraspecific metabolic scaling coefficient near 0.89 in fish.*"[20] Studies that focus on interspecific relationships also report widely differing patterns: while endotherms (birds and mammals) seem to confirm the 3/4 rule as a whole, metabolic rates scale closer to 2/3 (≈ 0.667) with body mass in Columbiformes (pigeons and doves) as well as in rodents (0.669) (Glazier 2005).[21] The interspecific slopes are generally higher in larger mammals and also in ectotherms.

These differing values raise questions about the assumption of a 'general rule' or a single mechanism that could explain patterns in metabolic scaling from aquatic ectotherms to terrestrial endotherms. As we have seen above, the constraints on growth and size are not the same for all groups of organisms. This suggests that different mechanisms generate the pattern we see between phyla and size classes. There is no universal 'law' but many different constraints resulting in variable values of the scaling exponent. Equation (1.1) can then be written as

$$R = a \cdot W^d \qquad \text{...(1.2)}$$

where the value of 2/3 is replaced by the scaling exponent *d*, and takes values <1 but typically ≥ 0.6.[22] Only in the smallest of all organisms, for example, in prokaryotes and protists, does *d* take values ≈ 1, an observation that will be of interest in Chapter 2.[23]

As a result of this wide distribution of slopes (and of the values that *d* can take), more recent approaches towards explanations of metabolic scaling propose to look beyond the role of physical and geometrical constraints and understand both inter- and intraspecific scaling relationships (thus between and within species) as a result of evolutionary optimization. As White *et al.* (2022) and White and Marshall (2023a, 2023b) have recently argued, metabolic scaling can be explained without reference to physical constraints and as a result of evolved life-history strategies that allow for sufficient somatic growth and then optimal reproduction for a given species in a given environment. Their departure from constraint-based explanations reflects a wider rift between supply-based and demand-based models of metabolic scaling: while supply-based models explain a relative decrease of metabolic rates with increasing body size by referring to constraints and limitations of resource supply, demand-based explanations focus on declining energetic requirements, i.e., they argue that larger organisms become more efficient in their energy use (see, e.g., Glazier 2002).

The argument that evolution acts upon metabolic slopes produces a variety of scaling exponents across the tree of life is valid and important: life on Earth is dynamic, and even though it is subjected to a wide variety of energetic, physical, and geometrical constraints, evolutionary processes find ways to accommodate physiological functions within these constraints. While biological processes occur in the physical world, they constantly challenge their geometrical constraints and are able to overcome strictly determined surface-volume constraints, as a typical example. J.B.S. Haldane (1892 – 1964) has elegantly summarized this in his seminal essay "*On being the right size,*" by stating that "*comparative anatomy is largely the story of the struggle to increase surface in proportion to volume*" (Haldane 1926).

This "*struggle to increase surface in proportion to volume*" is central to the GOLT. While this theory can be categorized as being structured around a constraint- or a supply-based theory of metabolic scaling – in the sense that it explains relationships between body size and metabolic rate by limited supply rather than decreasing demand – it provides a mechanistic and evolutionary explanation for the variability in metabolic slopes in and between species. In contrast to older surface-volume-based explanations of metabolic scaling, which assumed fixed exponents of 2/3 (e.g., Pütter 1920; von Bertalanffy 1951, 1964, i.e., for "*growth type 1*"), the GOLT argues that evolution acts in such a way that metabolic slopes are adjusted to the lifestyles and life-histories of the respective organisms. Curiously, earlier supply-based models that invoked surface-volume relations as the reason for decreasing energy consumption with increasing size seldom indicated which specific surface was the limiting factor.

The GOLT proposes a mechanistic foundation for surface-volume relationships, or metabolic scaling of water-breathing animals, by focusing on their respiratory organs (a factor that does not constrain the growth and metabolic scaling of air-breathers, as we will discuss in more detail in Chapter 11). A robust body of research indicates that the surface area of gills scales proportionally to metabolic rates and matches the observed variety of scaling exponents of *d*, ranging from approximately 0.6 in small fishes (and crustaceans) to 0.9 in very large fish species.[24] The relationship between the slopes of gill surface area and metabolic rates will be discussed in more detail in Chapter 2.

The fact that metabolic scaling does not follow strict 'laws' and that scaling exponents are not universally equal to 2/3 does not invalidate explanations based on surface-volume ratios (or fractal relationships). Slopes of 2/3 do apply to surface-volume ratios of geometrical objects whose shape remains identical when they increase in size, for example, spheres (where surface area increases in proportion to the second power of the radius) or cubes (where the increase is proportional to the second power of the side lengths; see **Figure 1.2A**).

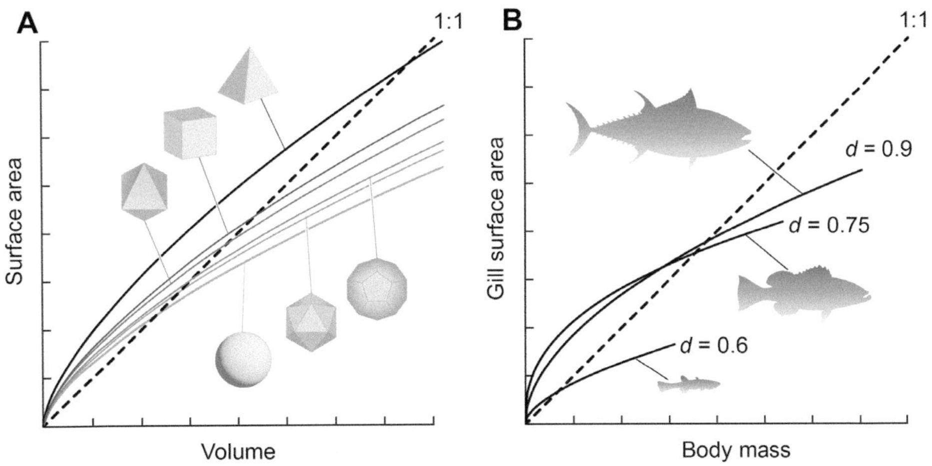

Figure 1.2. Relationships between surface and volume (or mass) in arbitrary units. **A**: surface-volume ratios of the so-called Platonic solids. **B**: Even though gill surface area (*S*) also increases when fish or other WBE grow, it cannot keep up with the subsequent increase in their body weight (*W*) if $d < 1$, as occurs in very small fishes (with $d \approx 0.6$), in medium-size fish ($d \approx 0.75$), or in fishes that can become very large ($d \approx 0.9$). Note that this graph, which illustrates schematically the relationship $S \propto W^d$, does not consider the fact that in reality, $S = \underline{a} \cdot W^d$, with '*a*' being low in fish species that remain small and high in species that attain large sizes, and that values of $d \geq 1$ do not occur in adult fish, except in air-breathing species with very small gill areas (i.e., low values of '*a*').

The 2/3 rule does not frequently apply in biological organisms, whose existence depends on the "*struggle to increase surface in proportion to volume*" (Haldane 1929), and whose organs responsible for the uptake of vital resources, mainly oxygen, develop highly filamentous or folded surfaces whose area can increase hyper-allometrically, with slopes > 2/3 (**Figure 1.2B**).

Consequently, the GOLT acknowledges that metabolic slopes and the scaling factors of gill surface area with whom they correlate are adaptive and the result of evolutionary processes that allow species to grow to a given size in a given environment and at a given time, to reproduce, and to utilize suitable energy resources, e.g., food of size and nutritious value sufficient to sustain all these processes. These 'optimizations' can only occur within geometrical constraints that limit the morphological possibilities of what can further evolve (see Hordijk 2016; McGhee 2015).[25] Unless it changes its shape, no physical object can grow so that its surface area increases at the same rate as its volume.

As we will show in this book, the gills of fish and other water-breathing animals cannot grow at the same pace as the growth of their bodies (thus not with slopes of $d \geq 1$) past its post-larval or early juvenile stages. Only in larvae and early juveniles (whose body mass is very low) can the gills grow with slopes of $d \geq 1$.[26] The life histories of water-breathing animals – from metamorphosis to maturation, first reproduction, and finally, post-reproductive growth – can be explained by the relationship between gill sizes and body size.[27] Even though the GOLT acknowledges the reality of adaptation, i.e., maximizing gill surface area relative to body size, it can still be characterized as a supply-based model of metabolic scaling. We argue that allometric metabolic slopes are not a result of decreasing demand but of limited supply.[28]

Strong support for such supply-based models can be found in studies on the metabolism of cells *in vitro*. Multiple studies have shown that cells extracted from an organism whose metabolism scales relative to body size with an exponent <1 showed higher metabolic rates when cultured *in vitro* (Ahluwalia 2017; see also West *et al.* 2002 and Brown *et al.* 2007). This suggests that it is not reduced demand that shapes metabolic slopes but a whole-organism structure that constrains the supply of oxygen to an organism's different components, organs, and cells. We will discuss the problem of causality in metabolic scaling in more detail in Chapter 16 and address how models of metabolic supply and demand can define these two traits. For now, we will only mention that optimization- and constraint-based models are not opposed to each other, but rather depend on each other. As Maynard Smith (1978) remarked in a review of optimization theories in evolution, optimization can only be meaningfully discussed when it is related to the constraints that define the possibilities of what could actually evolve: "*It is clearly impossible to say what is the 'best' phenotype unless one knows the range of possibilities. If there were no constraints on what is possible, the best phenotype would live forever, would be impregnable to predators, would lay eggs at an infinite rate, and so on. It is therefore necessary to specify the set of possible phenotypes, or in some other way describe the limits on what can evolve.*"

Organisms with such 'best' phenotypes do not exist, and all biological processes are constrained in several ways. Optimization can only meaningfully be understood when it is defined as an optimization concerning these constraining factors: the constraints define what can (or should) be optimized. The relationship between optimizations and constraints will also be important in Chapters 3 and 16, where we will discuss the possibilities of inferring causality in metabolic scaling and the physical constraints on which optimization can act.

Constraints and limitations

The relationship between constraints and optimization is of primary importance in phylogeny, i.e., in the adaptive adjustment of an organism's morphological and physiological traits to its environment in evolutionary time. The constraints within which an organism has evolved can also play a constitutive role in maintaining the organism's functions or even in developing new traits and features. Like physical and chemical processes, biological processes are constrained externally through the availability of resources and the exposure to specific thermal and physical conditions of the system's environment. Unlike chemical and physical processes, biological systems are also self-regulatory, meaning that they operate by constraining themselves. As biological and general systems theorists have long argued, systemic operations depend on internal and 'self-created' constraints that function as closures to the system's operational modes (Montévil and Mossio 2015; Wiener 1948; von Bertalanffy 1952, 1964).[29] The system can collapse whenever such self-controlled constraints are removed, as it can no longer maintain its internal energy balance.[30]

Constraints also have productive functions and are necessary conditions of organic life. They are not just limiting the ability to perform certain operations, but they also constitute these operations. They not only reduce the degrees of freedom of a system but also enable and stabilize biological functions or create new opportunities. A simple example: a heart can only function (by creating pressure) when it acts as a constraint – by constraining the free and uncontrolled flow of blood and pumping it through a closed system (Montévil and Mossio 2015). At the same time, the heart's pumping capacity or the width of arteries and veins are the most crucial and fundamental limiting factors of an organism's ability to perform vital operations. We must be aware of the dual role of constraints in biology when discussing the role of limitations and deficiencies.[31]

While functional constraints are a central tenet of the GOLT, the critical mechanism it addresses is the limitation of oxygen uptake. As we have seen, constraints can play a constitutive role and thus enable certain functions. It is evident that they also play a limiting role by excluding specific options or reducing the degrees of freedom of a given system. But what is the relationship between constraints and limitations? The concept of 'limitation' might be potentially misleading as it suggests that individual organisms are constantly pushed to the brink of their capacity to perform basic vital operations. In that sense, limitations appear to be maladaptations. While examples of such maladaptations can be easily found, they only represent a particular case of limitation. Most forms of limitations, however, are unproblematic and do not endanger the survival and reproduction of an organism or species.[32] Both constraints and limitations can be defined as the reduction of the degrees of freedom, but it is helpful to define these reductions on different levels. A constraint reduces phylogenetic possibilities and can be conceived as having evolved. A limitation, by contrast, is the reduction of some degrees of freedom on an ontogenetic level. Individual organisms experience limitations set by the evolved constraints of their respective body plan (or *Bauplan*). In contrast to phylogenetic constraints, limitations are those reductions of freedom that cannot be overcome by adaptation, and the only possibility to mitigate their effects lies in individual plastic responses.

This distinction between constraints and limitations is conceptually important, particularly in Chapters 4 and 5, where we examine the impact of temperature, growth, and size on fish. When we apply these definitions to the growth patterns of different species in different environments, it would be implausible to explain slow growth or small size to a physiological limitation. The growth of small or slow-growing species can be better understood as a life-history adaptation to specific niches and lifestyles. These adaptations set and determine the limitations that the individual organism will experience, such that its capacity to generate and distribute energy determines its ability to grow, reproduce, and perform basic maintenance activities. In an optimal environment, these limitations may, in fact, never become a problem for an individual organism – for example, when the internally determined growth limit corresponds to the limits that are set by the environment. Speaking of limitations does not imply that an organism is constantly living on the very limits of its existence – it only means that it must function within certain boundaries.

However, the impact of limitations changes rapidly in response to the environmental conditions experienced by the organism. The evolved *Bauplan* of a species allows for a given size at a given temperature and ambient oxygen level. However, this size cannot be reached at higher temperatures or lower oxygen levels. The constraints on body size that evolved for different parameters at certain temperatures will affect the physiological performance of the organism in such a way that limitations occur in different ways and reduce the full growth potential that would otherwise be possible. In short, our stress on gill-oxygen limitation as a mechanism behind reduced growth does not imply that the respiratory system of fishes is a maladaptation. Instead, the GOLT identifies the dimensional tension between water-breathers' respiratory surfaces and their bodies' oxygen demand as the constraint that acts upon the organism's performance. This constraint becomes effective as it limits the range of phenotypic possibilities available under specific environmental conditions, with temperature being the most critical parameter.

Mechanistic theory and the principle of constrained generalization

With the growing importance of science in modern societies and the increasing trust in scientific methods to explain complex problems, the question of what constitutes a valid scientific explanation has become ever more pressing. As a result of the dominance of physics in the modern philosophy of science, the most prominent explanatory models were based on nomothetic accounts that tried to identify general laws of nature that determine the occurrence of individual phenomena. In Carl Gustav Hempel's (1905 – 1997) 'Deductive-Nomological' model, an explanation was valid if an observed phenomenon could be traced back to a more fundamental law that made its occurrence possible (Hempel 1962). This mode of explanation, however, poses a challenge for disciplines in which no such fundamental laws can be identified, most notably the social sciences and the humanities. According to many philosophers of science, the same applies to biology and the life sciences (Glennan 2002; Bechtel and Abrahamsen 2005; Bechtel 2011). In contrast to problems studied in physics and most of its subfields, biological, social, historical, or linguistic phenomena often depend on much more contingent mechanisms and can seldom be reduced to deeper general or even universal laws.

This does not mean, however, that biological phenomena are random or entirely unpredictable (or, to relate this point to our previous section, that their range of biological possibilities is unconstrained), and this book makes precisely the opposite point. Given its stress on geometry and the importance of surface-volume relationships, the GOLT may even appear as a theory that resorts to nomothetic-deductive modes of explanation in the sense of Hempel's model. Indeed, compared to other theories of growth and metabolism, the GOLT pays far more attention to the underlying physical and geometrical factors of biological processes. However, rather than understanding the dimensional tension (Pauly and Cheung 2017) between respiratory surfaces and the bodies they must supply with oxygen as a 'law' that fully determines the outcomes of specific metabolic processes, the GOLT interprets this tension as a constraint on what is biologically possible. As Ross (2023) argues, the surface-cube law should not strictly be understood as a "*law*" but rather as "*a mathematical principle that can be derived without empirical studies.*" The GOLT relies on this mathematical principle in a mechanistic rather than a nomothetic sense and investigates how it acts on the various mechanisms involved in metabolic processes. As we have seen in the previous sections, metabolic slopes cannot be reduced to geometrical properties that universally produce identical values, regardless of whether the exponents are 2/3 or 3/4. Instead, these slopes reflect adaptations to different niches and lifestyles because different life-history strategies result in a diverse range of values.

While biological processes rely on and are constrained by physical principles, they are never entirely determined by physical laws. Hence, biological explanations must identify the (physics-constrained) mechanisms that shape and *cause* the phenomena we observe in the biosphere. In this regard, the GOLT is an explicitly mechanistic theory. This is also important in its use of mathematical models to interpret data. Some of the growth models we present here are also used by proponents of other theoretical approaches, such as the Pütter and von Bertalanffy growth equations, which are the central mathematical underpinning of the GOLT. However, mathematical models themselves do not provide explanations, especially when their parameters are not explicitly related to biological traits. This is often the case in studies that understand their models as merely

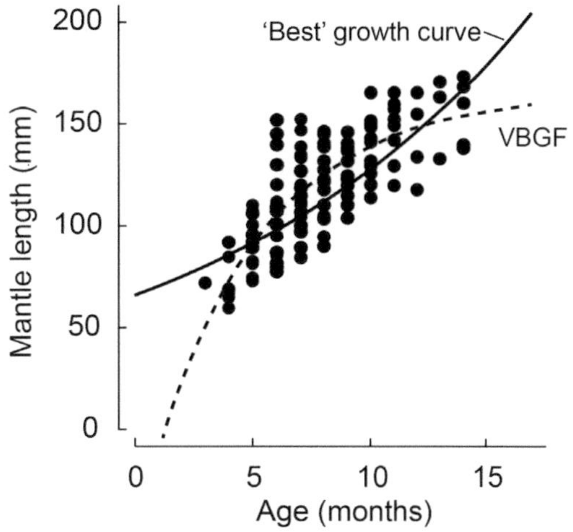

Figure 1.3. Growth curves fitted to age-length data pairs obtained by reading microstructure on the beaks of the African cuttlefish *Sepia bertheloti* (modified from figure 7 in Guerra-Marrero *et al.* 2023 from the coast of Guinea-Bissau). The dashed line is a von Bertalanffy growth function (VBGF), i.e., an asymptotic curve which, as argued in this book, fits size-at-age (or growth) data because it reflects underlying physiological processes. The exponential line meets Akaike's statistical criteria for the 'best fitting growth curve' (both sexes combined), hence this absurd graph.

phenomenological representations of the described datasets, as proposed by White and Marshall (2019), and as practiced by many fishery biologists who use Akaike's 'information criterion' (Akaike 1973) to select the model that best 'fits' their age-at-length data. This practice, which is common among colleagues studying specific groups of WBE[33], leads, in some cases, to grotesque results (**Figure 1.3**).

However, a model requires a mechanistic interpretation that makes its relationship to its object explicit to explain something. If this is the case, a phenomenon can be explained in the sense that a biological mechanism is related to a more general mathematical principle.

Some proponents of the 'new mechanistic philosophy of science' have criticized non-mechanistic modes of explanation – and especially the nomothetic-deductive model – as prone to the pitfalls of what Ludwig Wittgenstein referred to as the human *"craving for generality"* (Wittgenstein 1958; see Glennan 2017). While attempts to over-generalize bear the risk of an explanatory 'overstretch' or of losing sight of specific causal or constitutive mechanisms behind specific phenomena, a scientific explanation always requires a certain degree of generalizability. Explanations that entirely isolate their objects from other phenomena and limit their scope to only individual phenomena lose their explanatory validity. No explanatory argument can be entirely unique, even in extreme mechanistic explanations. The *explanans* must either be able to explain other (often similar) phenomena or show categorical parallels to other *explanantia*. This requirement is primarily formal since the possibility of a genuinely unique causal relationship could exist in theory. However, invoking such causes would inevitably lead to *ad hoc* explanations, where each phenomenon could be explained by a different mechanism, 'law,' or underlying principle.

Both extremes – nomothetic over-generalization and mechanistic over-individualization – fail to formulate a framework for what could count as a valid explanation or to commit themselves to certain constraints on their explanatory scope. Suppose every single phenomenon requires a different explanation. In that case, we cannot formulate general categorical guidelines of explanatory validity, and the same is true for attempts at explanations whose scope is too broad. The GOLT proposes a more balanced approach and commits itself to what we may call the 'principle of constrained generality,' i.e., it strives for generality, but this generality is not limitless; rather, it is limited to specific categories of phenomena and only applicable to specific groups of biological processes and organisms.[34] However, it does not include isolated *ad hoc* explanations since causal or enabling mechanisms are never entirely individual, and their variety is not infinite. Otherwise, strictly speaking, we could not exclude the possibility of miracles.

The rejection of *ad hoc* explanations and the requirement of constrained generality also imply that explanations must always be less complex than the phenomena they should explain. Every described and describable phenomenon and process possesses a certain degree of individuality and particularity, especially in biology. Every fish stock is different, every population is exposed to different thermal and trophic conditions, and every individual of a given population is different to a certain degree. However, even though no scientific or statistical model can account for every single variable, it is still possible to identify several underlying factors and develop generalizing explanations beyond individual variances and differences. In some cases, the validity of the generalizing explanation can be compromised by numerous contingent factors that become more dominant in the specific case or situation, and we will discuss several such examples later in this book. However, singular exceptions can only be used to reject an explanatory model when their persistence suggests an overall revision of the model itself and the search for a new and different set of explanations.

This requirement of constrained generality can be illustrated by the five criteria for acceptable scientific theories defined by Kuhn (1977): accuracy, consistency, scope, simplicity, and fruitfulness. The more generalized an explanation is, the greater its internal consistency and broader scope. The requirement of accuracy constrains explanatory scope. Simplicity, in turn, is a precondition of scope because the ability to generalize is limited by too many sub-clauses and the incorporation of local and specific variables. If we accept that the explanation of a phenomenon must be less complex than the phenomenon itself, it follows that theories should refrain from adding too many layers of complexity and instead, strive for simplicity.[35] Finally, a theory's explanatory fruitfulness for related areas and phenomena is also a result of its ability to generalize within reasonable constraints. We believe the GOLT meets all these criteria, and this book will lay out its explanatory potential.

Breathing water in a warming world: the structure of this book

This book consists of two parts. In the first part, we will lay out the theoretical architecture of the GOLT. The second part will illustrate the explanatory scope of the theory through case studies that discuss applications of the theory to animals across different taxa and/or different processes.

Before we further describe these two parts, we must mention some technical peculiarities of this book. The central tenets of the GOLT have already been discussed in an earlier book titled *Gasping Fish and Panting Squids* (1st edition: Pauly 2010; 2nd: Pauly 2019). One of these particularities is that while this book is certainly not a 3rd edition of the earlier book, some text, equivalent to about ten pages, and several of the figures from its 2nd edition were incorporated herein after slight modifications.[36] The figures, when dealing with only one taxon, will generally include a simple drawing of the animal(s) in question in the hope that this may elicit "*a feeling for the organism*" (Keller 1983), and **Table 1.2** defines the most frequently used variables and acronyms in this book. Another feature of this book is that we do not provide the authority for the scientific names of the fish and other WBE, e.g., we omit 'Linnaeus, 1758' after the scientific name for Atlantic cod (*Gadus morhua*). Instead, we refer the reader to FishBase, the global online encyclopedia on marine and freshwater fishes (Froese and Pauly 2024; www.fishbase.org) for fishes, and to SeaLifeBase (Palomares and Pauly 2024; www.sealifebase.org) for marine non-fish WBE.[37]

These databases provide scientific names with authorities and update them when the taxonomy of various groups is revised, hence the differences between some of the names we use and the references we cite. Additionally, these databases provided the biological traits of multiple species that we used in some of our graphs or to support the statements we make. In the following, these two databases are cited only as 'FishBase' and 'SeaLifeBase.'

After presenting the general scientific framework of the GOLT in the introduction, our second chapter, titled *What is Growth?* will provide the foundation of the GOLT's definition of organismic growth, especially in ectothermic water-breathing animals, and discuss the theory's indebtedness to earlier work, especially the historical contributions by August Pütter and Ludwig von Bertalanffy. Chapter 2 also presents and explains the fundamental mathematical models on which the theory is based. Chapter 3 will outline the geometrical structure of fish and invertebrate gills, as well as the consequences of their shape and growth for the organism's metabolic performance. Chapters 4 and 5 will discuss the impact of temperature on the metabolism and growth of water-breathing animals, focusing first on the biochemical level and then on the ecological consequences of higher temperatures.

The book's first part will conclude with Chapter 6, hopefully clarifying why we use a 'P-diagram' (to be discussed later) to conceptualize fish growth and resolve misunderstandings about what it means and does not.

Chapters 7 to 11 apply the elements of the GOLT elaborated in the first six chapters to the growth and reproduction of what is conventionally viewed as 'fish,' i.e., bony fishes that breathe water, sharks, rays and skates, and chimeras (or chondrichthyans) and air-breathing fish. Two of these chapters will show that an evolutionary ancient heuristic, triggered by the declining oxygen supply of growing fish and other WBE, provides the mechanism that 'tells' them to respond to the environmental stimuli for maturation and spawning that they did not previously perceive as such, while chondrichthyans (and a coelacanth!) provide evidence for this mechanism through a surprising aspect of their biology. The other three chapters in this part of the book present a solution to the old problem of 'compensatory growth,' propose a distinction between two forms of 'gigantism' in fish, of which one is explained by the GOLT, and interpret the high growth performance of air-breathing fish as indirect evidence for the GOLT, which posits that the performance of their gills limits the lesser growth of water-breathing fish and invertebrates.

Chapters 12 to 14 focus on applying the GOLT to non-fish WBE. Chapter 12 deals with water-breathing arthropods and shows not only (*i*) that asymptotic growth, which is necessary for the GOLT, is the rule in Recent species, as it appears to have been in (fossil) Ordovician, Cambrian and Ediacaran species,[38] but (*ii*) that the heuristic helping fish 'decide' when to mature and eventually spawn also occurs in water-breathing arthropods. Chapter 13 applies the GOLT to molluscs, whose hard parts (and their calcification) appear extremely sensitive to the changes in internal (within-body) pH, which shapes various

Table 1.2. Definitions of parameters, symbols and acronyms that occur repeatedly in this book; those that occur in equations are *in italics*.

Item (unit)	Definition
/	This symbol, when occurring between two numbers, letters, symbol or words, means that the first item is divided by the second (exceptions are: 'and/or' and 'L/F' for 'length-frequency data')
Δ	A difference in time, length or weight
a	Intercept in linear regressions; also: multiplicative term of a length-weight relationship, a relationship linking food conversion efficiency (*E*) and weight, or another relationship
b	Slope of a linear regression; the exponent in $W = a \cdot L^b$, *where* usually, $b \approx 3$
°C	Degree Celsius
c.f.	Condition factor, defined by $c.f. = W \cdot 100/L^3$; usually based on TL (cm) and *W* (g)
D	Exponent in a relationship linking gill surface area (*GSA*) and body weight (*W*), i.e., $GSA \propto W^d$, or metabolic rate (*Q*) with weight, i.e., $Q \propto W^d$ (but see d_G and d_Q)
d_G	Exponent in relationship: gill surface area (*GSA*) vs. body weight (*W*), i.e., $GSA \propto W^{dG}$
d_Q	Exponent in a relationship linking metabolic rate (*Q*) with weight, i.e., $Q \propto W^{dQ}$
D	A convenience parameter equal to 3(1-*d*) or *b*(1-*d*), where *b* is from $W = a \cdot L^b$
dW/dt	Growth rate, in weight (or mass)
FCE	Food conversion efficiency, i.e., growth increment/food ingested (in a given time span)
g	Gram
GSA (mm²)	Gill surface area, in same cases replaced by *S*
GSI	Gonadosomatic index, a measure of the weight of ripe gonads relative to that of the body
H	Coefficient of anabolism, i.e., of protein synthesis
k ($year^{-1}$)	Coefficient of the catabolic term in Pütter's equation; note that $k = 3K$
K ($year^{-1}$)	In the VBGF, the rate at which asymptotic size is approached
L (cm)	Any measure of body length, with L_t the length at age *t*
L_∞ (cm)	Asymptotic length in the VBGF, reached after an infinitely long time
L_m (cm)	Mean length at first maturity, i.e., at which 50% of WBE become mature and spawn
L_{max}^D/L_m^D	(Maturation and) spawning threshold, numerically equivalent to Q_m/Q_{maint}
L_{max} (cm)	The largest fish in a species or population, depending on context
m	Weight exponent of the catabolic term in Pütter's equation ($m = 1$, and is mostly ignored)
OCS	Ontogenetic changes of scale, i.e., change in d_G or d_Q
P	Index of growth performance, defined by $P = \log(K \cdot W_\infty)$
Q_{10}	Coefficient expressing the increase in some activity due to a temperature increase of 10 °C.
Q_m	Metabolic rate at first maturity (i.e., L_m)
Q_{maint}	Metabolic rate ensuring the maintenance of all life-supporting activities except growth. Estimated as the metabolic rate of non-growing individuals, i.e., at L_{max} or W_{max}
Q_m/Q_{maint}	Maturation and spawning threshold, numerically equivalent to L_{max}^D/L_m^D
S	A surface; here often a respiratory surface (typically in mm²), e.g. gill surface area (*GSA*)
t (year)	Absolute age (relative age = $t - t_0$)
t_0 (year)	The usually negative 'age' at $L = 0$ predicted by the VBGF
TL (cm)	Total body length; also note FL, i.e., fork length, used for scombroids such as tuna
TSR	Temperature-Size Rule, i.e., the tendency of small (young) WBE to grow faster in warmer temperatures, but to remain smaller than those that grow in lower temperatures
VBGF	Von Bertalanffy Growth Function; different forms exist for growth in length and weight
WBE	Water-Breathing Ectotherms, i.e., animals that breathe water
WBD (μm)	Water-Blood Distance; diffusion distance across a respiratory surface
W (g)	Weight (or mass); W_t is weight at age *t*; here: 'live' or wet weight
W_∞ (g)	Asymptotic weight in the VBGF, reached after an infinitely long time
W_{max} (g)	The heaviest fish in a species or population, depending on context

aspects and predictions of the GOLT. Also, this chapter addresses the widespread notion that cephalopods, besides being smart[39], should be able to escape the physiological constraints that shape the biology of other WBE due to their unique *Bauplan*. Chapter 14 deals with three phyla whose various species have bodies that contain far more water than in fish and lack dedicated gills: sponges (Porifera), medusoid jellyfish and their relatives (Cnidaria), and the enigmatic arrow worms (Chaetognatha), all of which have species with stages that, nevertheless, can be described by asymptotic growth curves, and to which the GOLT appears to apply.

The three closing chapters – 15 to 17 – will bring us back to the more theoretical level and examine two key questions: first, in Chapter 15, we discuss the implications of the GOLT for the evolution of water-breathing ectotherms and address the question of whether gill-oxygen limitation is a constraint that can be 'solved' by natural selection over evolutionary time. The fact that oxygen constraints remain a limiting factor for the metabolism and growth of water-breathing ectotherms has led to the assumption that the GOLT assumes a fundamental maladaptation that lies at the heart of the *Bauplan* of aquatic organisms (Lefevre *et al.* 2018; Scheuffele *et al.* 2021). We address this concern by discussing the criteria of successful adaptation in more detail and connecting the GOLT's central tenets to processes of natural selection and how they act to optimize physiological processes. As we argue, the primary mechanism described by the GOLT is not a maladaptation but a result of evolutionary processes that enable organisms to thrive in a medium that imposes severe constraints. Understanding these challenges and how evolution develops ways to cope with these constraints is crucial in rapid environmental change, which requires different adaptation rates.

The second question about the theoretical foundation is addressed in Chapter 16, where we discuss the methodological problem of how to draw causal inferences from the observations and data we present here. The focus of this chapter will be the phenomena of metabolic scaling, and we will evaluate different optimization- and constraint-based explanations for the fact that metabolic processes decrease in intensity with age and size. As we argue, the GOLT has the potential to bridge the gap between these two modes of explanation and to provide a foundation for future explorations of this topic by offering more unifying and coherent interpretations and explanations of various phenomena.

As already mentioned, the GOLT was not initially developed for this purpose. Rather, it was first designed to explain features of the growth of tropical fishes (Pauly 1979)[40] but has since been extended to water-breathers in general and appears robust enough to suggest a unified theory: "*A theory that exhibits simplicity explains the same facts as rival theories, but with less theoretical content. A unified theory, however, is one that explains more kinds of facts than rival theories with the same amount of theoretical content*" (Keas 2018, citing Thagard 1978). Put differently, scientific advances occur when the theory explaining a class of phenomena can also be used, without substantial modification, to explain another class of phenomena (Keas 2018).

This book aims to provide a unifying understanding of fundamental biological and ecological traits of water-breathing ectotherms by organizing these traits into a coherent theoretical framework. The richness of aquatic life and its features extend beyond what we can present here, and we encourage readers to use the framework presented here as a starting point for further exploration and for relating their observations and data to the models presented in the following chapters. Like any theory that aims for generalization, the GOLT will remain an unfinished project and rely on others to take it further. In this respect, our theory can be seen as a window into the world of aquatic life, and we invite others to look through this window, hoping they will see new phenomena and processes, some of which we could not have envisaged at this point. This invitation is not only addressed to biologists and ecologists but also to researchers in other fields and readers with a more general interest in our topic. While non-biologists may skip some of the more technical details of our argument, we hope that the summaries of these details will be sufficient. Conversely, many biologists might be less interested in the historical background of the theory and how it developed and move on more quickly to the chapters on different taxa and phyla.

The final chapter serves as an epilogue that connects the various topics addressed in this book and identifies underexplored issues that other researchers might wish to try resolving (in addition to some of the anomalies mentioned in other chapters), using the mechanisms presented. We are convinced that the world of aquatic life is still full of surprises and wonder, and we hope the GOLT's window to this fascinating world will open to new and unexpected vistas.

2

What is growth? Pütter, von Bertalanffy, and living systems

"The body and its parts are in a continuous state of dissolution and reconstruction, so they are inevitably undergoing permanent change." Ibn Al-Nafis (1213-1288)[41]

Synopsis: This chapter introduces a general model of organismic growth in fish and other water-breathing ectotherms (WBE), emphasizing its foundation in the balance between anabolism (the build-up of new organismic material) and catabolism (the breakdown of existing material). We explore the historical roots of the GOLT's growth model and acknowledge key contributions by August Pütter, Ludwig von Bertalanffy, and earlier theorists. By integrating these historical, mathematical, and physiological perspectives, this chapter establishes the basis for subsequent discussions on how environmental and physiological constraints shape the growth of aquatic ectotherms.

What slows down indeterminate growth?

Growth and size are essential characteristics of an organism's existence in a three-dimensional world, and they directly impact reproduction, energy uptake, thermal regulation, and many other essential life-historical and ecological traits of living beings.[42] Changes in growth rate can have dramatic consequences, especially in seasonal environments, and both the reduction and maximization of body sizes can push a population out of its ecological niche. On the other hand, different growth trajectories can lead to optimal sizes within a given environment. As mammals, we intuitively view growth as a phenomenon limited to the first stages of life and ends when maturation and reproduction set in. However, this form of determinate growth is typical only for birds and mammals, i.e., for endotherms. In most phyla and species within the animal kingdom, i.e., in ectotherms, growth does not stop in early adulthood but can continue throughout an animal's entire life.[43] Such indeterminate forms of growth can be split into various sub-forms. Still, they all have in common that an increase in metabolically active body mass persists throughout the reproductive life stages and often until death, albeit with a steady decline in growth rates.

In phyla and lower taxa where growth is not genetically fixed, it is inevitable that a mechanism must exist that causes an asymptotic decline in growth rates. As not only early theorists of growth understood, but also farmers and aquaculturists worldwide know, the growth of animals depends on a sufficient supply of resources. Consequently, once the available resources are exhausted, growth should cease. However, it is not evident that individual growth should cease when food resources are abundant; indeed, this may be seen as a paradox.[44]

This problem required theorization and a more general understanding of organismic growth. As Jean-Baptiste Lamarck (1744 – 1829) pointed out, the growth of living and nonliving things is fundamentally different. While non-organic objects can also grow – outwardly similar to organisms or organic matter – this form of growth is different: crystals and minerals accumulate more substances and add them to their structure, but living organisms must also maintain the structures they have already built. As Lamarck explained, *"the body of* [a] *living being is not formed by juxtaposition, as in most mineral substances, that is to say, by the external and successive apposition of particles aggregated* en masse *by attraction, but it is essentially formed by generation, in its principle, it has then grown through intussusception—which means, by the introduction, the transportation, and the internal apposition of molecules borne along and deposited between its parts; that bring about the successive developments of parts which compose the body of this living individual, and from which afterwards also result in the repair activities which preserve it during a limited time*" (Lamarck 1797).[45]

While Lamarck was later criticized for his distinction between organic and non-organic growth (see, e.g., Wiesner 1892), the fact that organisms could not just grow new mass but also had to pay maintenance costs for the mass they had *already* built remained foundational to modern theories of organismic growth. Dead matter can be preserved without maintenance, but living cells require constant energy. This energy is necessary for the specific component's function(s) in the organism and for cell repair activities. According to Johannes Peter Müller (1801 – 1858), one of the pioneers of modern physiology in the nineteenth century, an organism's resource uptake was devoted to both the "*regeneration*" of existing structures and the production of new materials (Müller 1835). This relationship between growth and maintenance was further elaborated in various applied sciences, where growth, or the time required for an organism to reach a specific size, was crucial. Two of these fields in the second half of the nineteenth century were pediatrics and agricultural sciences, and researchers in these disciplines presented new models and data to explore further the energetic investment in growth and other physiological processes, especially the maintenance of cell structure and the replacement of dead cells by new ones. Max Kassowitz (1842 – 1913), one of the most renowned pediatricians of his time, considered the relationship between disintegration and regeneration of cellular substances one of the most critical processes in biology and devoted the entire first volume of his "*General Biology*" to the breakdown and the renewal of cellular protoplasm (Kassowitz 1899).[46]

Dividing available resources between repair (maintenance) and the production of new substances (growth) was already widespread before the twentieth century. However, no mathematical model existed that predicted individual growth curves and the timespan in which final or asymptotic sizes could be approached under specific feeding regimes and/or temperatures. While mathematical growth models for demographic growth or financial interest rate development were already in use, for example, the Gompertz function or Verhulst's logistic growth function, none provided a satisfactory foundation for a mechanistic explanation of individual organismic growth.[47] Such a model requires realistic assumptions about the energy fractions devoted to growth and maintenance. To estimate these fractions, it is necessary to predict the differences in demand and supply during growth, i.e., to explain the energy required during an organism's various growth stages, which is related to the amount of food consumed and processed within a given time.

Pütter's equation and the mathematization of organismic growth

Mechanistic models to determine growth curves and their dependence on the balance between energy uptake and expenditure were produced in the first decades of the twentieth century, a period that marked the development of new theoretical approaches to biology and physiology. Following important breakthroughs in physiology in previous decades, some of which will be described in more detail below and in the next chapter, biologists began to rethink growth in the context of new physiological theories and describe it in mathematical terms.

A pioneer in elaborating such mathematical models was the German physiologist August Pütter, mentioned in Chapter 1. Even though his modeling approach to growth was formative for many growth theories, his work was later seldom read and cited in the original,[48] but was instead mainly disseminated through the work of Ludwig von Bertalanffy (1934, 1951). Between 1917 and 1920, Pütter published a series of four papers on "*Physiological Similarities*" ("*Physiologische Ähnlichkeiten*") and the last one, published in 1920 and titled "*Similarities of Growth*" ("*Wachtumsähnlichkeiten*"), contained an equation that laid the basis for growth models in a wide variety of disciplines, but was typically ascribed to von Bertalanffy by later authors.[49]

The "*Physiological Similarities*" series was not Pütter's first attempt to present a broader generalization of physiological processes. In 1911, he published a book on comparative physiology ("*Vergleichende Physiologie*"), a thoroughly theorized account of physiological phenomena across different phyla and taxa (Pütter 1911). Although this work did not yet contain a mathematical growth model, its explanatory foundations were already in place, including a theory that describes the impact of temperature on metabolism and growth (see Chapter 4).

Pütter's career fell into a period during which physiology was characterized by a renewed quest for theoretical generality and synthesis, especially in Central Europe. Born in Stralsund in 1879, he initially studied zoology in Breslau (now Wrocław, Poland) and earned a doctoral degree with a dissertation on the physiology of vision in aquatic mammals in 1901. In 1903, he went to Göttingen to earn a medical degree under the supervision of the renowned physiologist Max Verworn (1863 – 1921), with whom he collaborated for several years. Despite his prominent status in German physiology at the time, Verworn was a highly idiosyncratic scientist and thinker. He explicitly positioned himself in the tradition of Johannes Peter Müller and published a synthesis of modern physiology in 1894, titled "*General Physiology*" ("*Allgemeine Physiologie*"), but also published on divergent topics, ranging from archeology, philosophy – mainly epistemology –, art, and psychology. (Stang 2023; Verworn 1901).

While Pütter did not engage in direct debates outside his discipline, he was also an unconventional scholar in many respects. As the physiologist Edgar Wöhlisch wrote in his eulogy for Pütter, he "*embodied a type of scholar that is uncommon in the age of biological specialization: a theoretical polymath who strove for deeper natural-philosophical understanding beyond the countless individual findings*" (Wöhlisch 1929). During his lifetime, and as a professor in Kiel and Heidelberg, he was primarily known for his work on endocrinology, the physiology of vision, as well as a controversial theory on the nutritional sources of aquatic animals, which he believed to be largely dependent on dissolved organic matter as a food source (Pütter 1909).[50]

In the final chapter of his "*Comparative Biology*" from 1911, Pütter closed with a discussion of how and where nature showed similarities that could be detected and understood across phyla and taxa. This comparative approach was further developed in the "*Physiological Similarities*" series. As Pütter asserted, identifying similarities was the prerequisite to arrive at any form of generalizing theory in the field of biology: "*Similarity is always 'partial sameness,' or sameness with regard to a specific feature or relationship. Finding these similarities is one of the great tasks of comparative physiology. Every awareness of a similarity in places where we first only saw differences brings us one step further on the way to a general physiology*" (Pütter 1920, p. 298). This quest for generality, understood as the search for "*partial sameness*" was typical for Pütter's understanding of what a genuinely *comparative* physiology should look like.

Despite Pütter's ambition to generalize across phyla, the scope of the 1920 essay was limited to the growth of water-breathing animals. In this model, the energy required for maintenance determines an organism's final size. As Pütter put it, organisms did not stop growing because of the unavailability of additional materials that could be added to the system, "*but because the processes of buildup and breakdown reach an equilibrium*" (Pütter 1920, p. 299). This approach was similar to the conceptualization of growth by his mentor Verworn, who defined growth as the relationship between "*assimilation*" and "*dissimilation*" and stated that growth could only occur if A (assimilation) > D (dissimilation). The point at which growth stopped could be expressed as $A/D = 1$ (Verworn 1901, p. 513).[51]

Pütter reasoned that if growth ceased, "*buildup*" and "*breakdown*" increased at different paces. Based on geometrical considerations, he stated that vital resources could only be absorbed by entering the system through a two-dimensional surface, yet they had to be supplied in a three-dimensional body. Decreasing growth could be explained by decreasing resource supply capacity and a more rapid increase in the negative term with weight. Growth rates could then be expressed as a simple subtraction

$$k \cdot \lambda^2 - k' \cdot \lambda^3 \qquad \text{...(2.1)}$$

Here, λ represents the linear dimension proportional to "*buildup*" and "*breakdown*" raised to the second and the third power, respectively, and k and k' are the corresponding coefficients of synthesis and breakdown. Regardless of the values of the coefficients, the anabolic term ("*buildup*") must increase faster initially (otherwise, the organism would not be able to initiate growth). After a while, however, the catabolic term ("breakdown") catches up, and growth ceases when the two are equal.

Pütter's idea that the relationship between synthesis and breakdown was proportional to surface-volume ratios was not entirely new, and it was already anticipated by Herbert Spencer's "*Principles of Biology*" (1866), which conceptualized growth as the balance between "*surplus assimilation*" and "*expenditure.*" The balance between these two traits depended on the resource supply rate, body size, and their interrelationship. Since resource uptake depended on surfaces, but consumption (or expenditure) was proportional in the organism's entire body, assimilation of surplus resources was easier for smaller organisms but became more difficult in large ones.[52]

This approach to understanding growth was further developed by Patrick Geddes, a biologist and sociologist best known for his work on urban planning, and his colleague John Arthur Thomson. In their book "*The Evolution of Sex*," Geddes and Thomson further developed Spencer's theory and explicitly described growth as the balance between anabolism and catabolism and applied this model to both cells and whole organisms: "*The limit of growth, when waste has overtaken and is beginning to exceed the income or repair, corresponds in the same way to the maximum of katabolic preponderance consistent with life. The limit of growth is the end between anabolism and katabolism, the latter being the winner*" (Geddes and Thomson 1889).[53]

Researchers who followed the reasoning that buildup and breakdown metabolism (however defined) scaled proportionally to surface-volume ratios typically assumed scaling exponents of 2 and 3, including von Bertalanffy (1934, 1951), who found these values confirmed in his experiments on guppies (von Bertalanffy 1938) and brine shrimp (von Bertalanffy and Krywienczyk 1953). However, as Thompson (1917) had already demonstrated, morphological adaptations acted so that organisms could strongly increase surface in relation to volume. Later work showed that the proposed values of the exponents needed revision: while smaller fish and other WBE may indeed show scaling relationships that are similar to the second and the third power (i.e., with $d \approx 2/3$), larger and more active species show very different values (Pauly 1979). The Pütter equation, of the form

$$dW/dt = HL^2 - kL^3 \quad ...(2.2)$$

was revised to

$$dW/dt = HW^d - kW \quad ...(2.3)$$

where dW/dt is the growth rate, W is body weight, and H and k are protein synthesis and breakdown metabolism coefficients, respectively. While Equation (2.2) may be physiologically valid for several species, it only represents one case among many others. In smaller fishes, such as guppies, the values for d assumed by Pütter and von Bertalanffy ($d \approx 0.67$) are close to 2/3. However, the values of d can vary between 0.6 and 0.9 across fish species and other WBE.[54] Equation (2.3) corresponds to the theory presented in this book.

The special and generalized von Bertalanffy growth functions

Equation (2.3) is central to the GOLT and defines its growth concept. It expresses that two factors constrain growth: (*i*) a limiting surface that determines the uptake and utilization of vital substances, and (*ii*) a factor that subtracts a given amount of energetic investment from growth because organic materials need constant maintenance and repair. This equation, but with $d = 2/3$, as in Equation (2.2), was also adopted by von Bertalanffy and entered fisheries science through its incorporation into the now classic fisheries text of Beverton and Holt (1957) "*On the Dynamics of Exploited Fish Populations*." While von Bertalanffy was sometimes accused of incorporating Pütter's growth model without crediting him as the originator, he explicitly acknowledged his indebtedness to Pütter, albeit while criticizing some aspects of the original model.[55]

Von Bertalanffy was an even more eclectic researcher and scholar than Pütter and his predecessors, publishing on a wide range of subjects, from medieval philosophy (as a student) to physiology and mathematical biology. Later in life, he laid the foundation for what became known as General Systems Theory. Whereas Pütter's growth model was explicitly restricted to water-breathing animals, von Bertalanffy attempted to adapt his approach to a wide range of water- and air-breathing animals. To achieve this, he distinguished between three "*growth types*," i.e., (*1*) one in which respiration (and consequently growth itself) was proportional to a surface, (*2*) one in which respiration was proportional to body weight, and (*3*) an "*intermediate*" growth type between surface and weight proportionality (von Bertalanffy 1951, 1964). The first 'type,' which could be described by Equation (2.2) or by Equation (2.3) with $d = 2/3$, was not restricted to water-breathers and – according to von Bertalanffy – also applied to reptiles, mammals, and other air-breathing vertebrates.[56] In this regard, the GOLT deviates from von Bertalanffy's reasoning (and is closer to Pütter), as discussed in this chapter's last section.

If the model in (2.3), which has the form of a *differential* equation, is to serve as the basis for an actual growth curve used to determine length or weight at a given age, it must be integrated. While von Bertalanffy agreed with Pütter on the basic growth model, the two authors integrated this equation differently. Pütter's integrated function for length was given as follows:[57]

$$L_t = L_\infty (1 - ae^{-c \cdot t/L_0}) \qquad \text{...(2.4)}$$

Here, L_t stands for length at age t, L_∞ is the asymptotic length, a is the constant of integration that corresponds to body length when $t = 0$, and c *is* the growth coefficient, expressing the rate at which L_∞ is approached. This equation was slightly modified by von Bertalanffy with k as the growth coefficient and L_0 as the length of the organism at age zero:[58]

$$L_t = L_\infty - (L_\infty - L_0)\, e^{-kt} \qquad \text{...(2.5)}$$

Here, L_t is the length at age t, and k is the growth coefficient. The difference between Equation (2.4) and (2.5), which represent different ways to interpret and integrate Pütter's model, is evident in the definitions of the growth coefficients c and k. Indeed, von Bertalanffy criticized Pütter for an underlying assumption that led to a dimensional mismatch, as it connected final size to the coefficients of the positive and the negative terms of Equation (2.3) in a misleading manner.[59]

The GOLT adopts von Bertalanffy's integration of Pütter's growth model and expresses it as follows for length:

$$L_t = L_\infty (1 - e^{-K(t - t_0)}) \qquad \text{...(2.6)}$$

Here, K is the growth coefficient, and t_0 is the age at which the organism's length would have been zero. This equation will be referred to as the *special* von Bertalanffy Growth Function (or VBGF) for length, which predicts the mean length (L_t) of a given population's WBE at age t. The 3 parameters of the special VBGF are L_∞, the asymptotic length, i.e., the mean length the WBE of a given population would reach after a very long time (actually: infinity, hence the ∞ symbol), K is a coefficient expressing the rate (with time^{-1} as dimension) at which L_∞ is approached (therefore K is *not* a growth rate, contrary to many published statements), and t_0 is the (theoretical) age at which the length of the organisms would have been zero *if* they always grew in the manner predicted by the equation – which they don't (see further below for the reason).

The special VBGF for weight takes the form:

$$W_t = W_\infty \cdot (1 - e^{-K \cdot (t - t_0)})^3 \qquad ...(2.7)$$

where W_∞ is the weight corresponding to L_∞ and all other parameters are as defined above. However, if the length-weight relationship of the WBE in question, of the form $W = a{\cdot}L^b$ has a value of $b \neq 3$, the following version can be used:

$$W_t = W_\infty \cdot (1 - e^{-K \cdot (t - t_0)})^b \qquad ...(2.8)$$

As mentioned above, both Pütter and von Bertalanffy assumed that surface-volume ratios in the metabolic scaling in fish would always lead to exponents of 2 and 3, as in Equation (2.2). However, as we have seen in Chapter 1 and will develop in more detail in Chapter 3, one of the key characteristics of biological organisms is the ability to develop morphological modifications in evolutionary time that rearrange the relationship between the second and third dimensions. Equations (2.6) and (2.7) still assume exponents of 2 and 3 (or 2/3 and 1), i.e., values that only apply to a limited number of species. This is why we refer to these equations as versions of the *special* VBGF.

The growth model that results from the integration of Equation (2.3), i.e., that allows for values of $d \neq 2/3$ (but still requires that $d < 1$) is the *generalized* VBGF. For length, it can be written:

$$L_t = L_\infty (1 - e^{-KD \cdot (t - t_0)})^{1/D} \qquad ...(2.8)$$

with the parameter D, which is added to account for values of $d \neq 2/3$ and is defined by $D = 3(1\text{-}d)$.

For weight growth, the generalized VBGF is:

$$W_t = W_\infty \cdot (1 - e^{-KD \cdot (t - t_0)})^{b/D} \qquad ...(2.9)$$

where b is the exponent of the length-weight relationships for the individuals in question and $D = b(1\text{-}d)$.[60] Through the additional parameter D, the generalized VBGF allows for a better fit to data for most species, especially fish that can reach large sizes and whose values for d deviate strongly from 2/3.

As the generalized version of the VBGF has found fewer adherents than the special one, we will use the special (or standard) VBGF for the descriptive examples used in this book. The special VBGF allows for a reasonably good fit to length-at-age data in most cases and provides L_∞ estimates similar to maximum lengths (L_{max}) observed in most populations of fish and other WBE. Only in exceptional cases, for example, in bluefin tuna (*Thunnus thynnus*), which has a high scaling value, i.e., $d = 0.9$, does fitting the special VBGF lead to estimates of L_∞ that are much higher than the largest observed specimens (Pauly 2021a).[61] Using the special VBGF will allow comparisons with VBGF parameter estimates already in the literature, e.g., in FishBase for fishes and in SeaLifeBase for other WBE.

Mechanistic foundations I: What is 'catabolism' in Pütter's model?

We have now presented the mathematical foundation of the GOLT's basic growth model. The Pütter-von Bertalanffy growth model, further abbreviated as VBGF, is not unique to the GOLT but is widely used in fisheries science. This model is also shared by several other biological growth theories, including the *Ontogenetical Growth Model* of West *et al.* (1997), the various versions of the so-called *Dynamic Energy Budget* model as developed by Kooijman (2000), and many individual studies that do not explicitly adhere to a specific theoretical framework.

In many approaches that work with the Pütter-von Bertalanffy growth model, it is not clear if its mathematical foundation refers to real life-history parameters and offers a mechanistic account of the observed growth curves or if the model is just used in a merely descriptive way (as a phenomenological model, see, e.g., White and Marshall 2019). As discussed in Chapter 1, the GOLT adopts a different approach, emphasizing the importance of identifying a mechanism that can account for the observed patterns. As Kearney (2021) puts it: "*The Gill-Oxygen Limitation Theory (GOLT) of Pauly (1979, 2010) represents a more mechanistically explicit version of Pütter and von Bertalanffy's ideas about the balance between anabolism and catabolism,*" mainly by identifying the gills or other respiratory surfaces of water-breathers as the limiting factor for their oxygen supply and hence growth.

Mechanistic explications are indeed crucial for the GOLT, and, in our view, they should inform every form of scientific explanation that seeks to go beyond the description of statistical patterns.[62] The following three sections of this chapter provide a more precise account of the GOLT's mechanistic interpretations of the VBGF's key parameters. We will first define the negative or "*breakdown*" term of Pütter's equation and discuss how it differs from various definitions of 'catabolism' (which, incidentally, is a term Pütter never used).

As discussed earlier in this chapter, growth theories since the nineteenth century started to conceptualize growth as the result of a relationship between two opposing processes: on the one hand, consumed energy is invested in building new organic material; on the other hand, an inevitable consequence of life (and being alive) is that elements and cellular components become dysfunctional. The system's disintegration requires continuous maintenance and repair activities, and the larger an organism grows, the higher the energy cost of maintenance.

But what are these costs, and how do they scale with other energy expenditures in a living organism? Any mechanistic theory of growth will need to define the processes that require repair and maintenance, which limit the production of additional new material. Pütter (1920), whose growth equation explicitly modeled both uptake and expenditure of metabolic energy, decided to define this as broadly as possible and to include all processes that contributed to processes of disintegration and decay in the cells: "*A slow transformation of the building materials is continuously taking place in every living system and its ultimate result is the dysfunctionality of parts of the system. These transformations occur slowly in relation to the conversions of substances in the energy metabolism. Whether the conversion of building materials takes place in a visible form and turns a fully capable cell into a pile of rubble of disintegrating organic compounds, or whether they occur on a submicroscopic level is of lesser importance for the general theoretical consideration that is provided here – they all lead to the destruction of the smallest parts of the living system.*"

In this view, adult organisms that do not grow anymore can therefore be described as remaining in a "*dynamic equilibrium*." They still produce new organic mass, but all of this new material is needed to "*replace*" the elements that are constantly lost. "*The adult organism produces new cells on a constant basis: red and white blood cells, intestinal epithelia, epidermal cells, etc., and yet the current state remains constant because the newly produced cells only replace the continuously decaying elements*" (Pütter 1911, p. 127).

A similarly broad definition of the negative term of the growth equation was given by von Bertalanffy, whose "*Theoretical Biology*" (1951) defined breakdown metabolism as equivalent to Rubner's "*rate of wear and tear*" and argued that this rate was roughly proportional to the breakdown of protein and nitrogen structures – a theme we address further below. Like Pütter (who did not elaborate on this in detail), von Bertalanffy stated that these breakdown processes should be proportional to body mass: "*From a quantitative perspective, we have to primarily consider the loss of proteinic structures. The rate of wear and tear can therefore be measured as the metabolism of protein and nitrogen loss. Empirical physiological experience shows that the rate of wear and tear or the loss of organismic substance is – at least approximately – proportional to body mass, i.e., a constant fraction of body mass is lost per unit time*" (1951, p. 279).[63]

The term "*catabolism*," which was not used by Pütter and occurred only in the English translations of von Bertalanffy (1960), can be misleading in this context. Both authors referred to "*Abbau*" ("*breakdown*;" Pütter 1920; von Bertalanffy 1951), "*Zerfall*" ("*decay*;" Pütter 1920), "*Zerstörung*" ("*destruction*;" Pütter 1920), or "*Verlust körpereigener Bausubstanz*" ("*loss of organismic substance*;" von Bertalanffy 1951).[64] These definitions are incompatible with the notion of catabolism as the production of metabolic energy through the breakdown of larger molecules into smaller ones, provided these molecules are taken up as food and are not yet part of the organism itself.[65] The processes involved in catabolism, defined in this way, cannot be included in the negative term of Pütter's equation, especially those that consume oxygen and generate ATP by breaking down complex molecules into simpler ones (Pauly and Lam 2023; see below and Chapter 4). The crucial difference between these two concepts of 'breakdown' is that Pütter and von Bertalanffy (and the GOLT) define it as the breakdown of cellular material belonging to the organism ("*körpereigene Substanz*") rather than breaking up molecules that are taken up in the form of nutrition.

The GOLT agrees with Pütter's broad definition of "*breakdown*" and with von Bertalanffy's specification that such breakdown processes mainly involve protein and nitrogenous substances. Even though it would not strictly be necessary to know which substances are primarily affected by these processes, as long as they require repair and/or maintenance activities, this definition of "*breakdown*" must exclude all those processes that require oxygen (which is nevertheless required for repair or re-synthesis).

The most important factors that contribute to the negative term of Pütter's equation can be deduced from the repair costs. For the GOLT, and indeed for any interpretation of Pütter's equation, the factors of interest here are limited to those processes that subtract anything from the positive, left term of the equation, i.e., the buildup of new cellular material. The costliest part of such processes is the synthesis of native proteins, which can fold into the required tertiary and quaternary structures. Consequently, the same must be valid for the re-synthesis of vital structures: the costliest process is the repair (or, more frequently, the replacement) of proteins, especially those with quaternary structures, for example, hemoglobin or ribosomes. The details of these processes at the molecular level are only relevant to the GOLT with respect to the maintenance activities that must be subtracted, as the 'catabolic term,'[66] from the anabolic term.

These processes will be discussed in more detail in Chapter 4, which examines the impact of temperature on maintenance metabolism. For now, it is sufficient to define the relevant parameter that subtracts from the anabolic term of Pütter's equation; we can equate this term with all the maintenance activities that are necessary to repair and re-synthesize vital molecular structures, of which protein re-synthesis is the most important one. What follows is that the synthesis of new materials and the re-synthesis of degraded materials can be understood as similar processes, with the difference that one *adds* new organic mass and the other *maintains* what is already there. While re-synthesis must occur throughout the entire organism and increase in proportion to body mass, the synthesis of new structures increases in proportion to the surface. The following section will discuss which surface is the limiting factor in producing new organic structures and growth.

Mechanistic foundations II: What is the limiting surface?

Since all organisms are three-dimensional beings, the uptake of resources from their surroundings must occur through a surface. Metabolic processes depend on different metabolites; thus, they typically involve multiple surfaces. While nineteenth-century physiologists recognized the importance of surfaces for metabolic processes, the question of which surface acted as a limiting factor for resource uptake (and consequently growth) remained open. Following Rubner (1894), many researchers – implicitly or explicitly – assumed that intestinal surfaces might present a limitation for metabolic processes.[67] The GOLT assumes that the growth-limiting factor in water-breathing organisms is not the limited uptake capacity of food per unit of time but of oxygen. In Chapter 15 ("*Inferring causality in metabolic scaling*"), we will discuss the methodological problems of inferring causality in biology in more detail. For now, it is essential to provide a basis upon which the subsequent chapters of this book, as well as the taxon- or process-specific chapters, can be built.

Pütter's 1920 seminal paper did not explicitly indicate which specific surface limited the uptake of vital resources. However, the final pages of his "*Comparative Physiology*" (1911) indicate that one central "*physiological similarity*" in organisms that set a limit to growth and performance was the capacity to take up sufficient oxygen (p. 698).[68] Earlier in this book, he distinguished between two types of animals, one in which the uptake capacity of the respiratory surfaces was lower than that of the cells, and one in which cells consumed less oxygen than the respiratory surfaces could provide (p. 170). It was unclear what limited metabolism in the second type, but Pütter speculated that the resorption of nutrients might indeed be the bottleneck. In organisms of the first type, it could be shown that oxygen consumption scaled proportionally to respiratory surfaces; Pütter's examples (p. 162 to 165) were fishes, aquatic crustaceans, and amphibians in their larval stage.

The search for a causal factor behind the intensity of metabolic processes is interesting, given Pütter's relationship with his mentor, Verworn, who had developed an epistemological theory that postulated the unknowability of causes and effects in the natural world. This view, which he later called "*conditionism*," dismissed attempts at a causal explanation as the result of a human "*causality instinct*" ("*Kausalitätstrieb*;" Verworn 1901, p. 34) and instead called for an analysis of the conditions that enabled phenomena to occur (Stang 2023).

In this view, it would be impossible, in principle, to determine which surface limits metabolic processes or if metabolic rates are low because of the limiting role of either digestive or respiratory surfaces. Even though Pütter did not explicitly criticize Verworn, the final passages of his "*Comparative Physiology*" provide an argument for the possibility of inferring causes and effects. While it was undeniable that an organism's energetic exchanges depended on multiple surfaces, causality could be inferred if the "*weakest link*" in a chain of physiological processes could be identified. While there were animals for which the absorption of dissolved nutrients might set such a limit, for a wide range of phyla, the bottleneck was their capacity to absorb oxygen.

This feature allowed for broader generalizations for the physiology of all those types of organisms "*in which oxygen consumption per surface unit reaches a value that is determined by physical properties, and thus marks an absolute capacity limit*" (Pütter 1911, p. 698). Despite the absence of an explicit identification of the limiting surface in question in his other publications, it may have been clear to contemporaries who cited Pütter that his work on metabolism and growth referred to gills and oxygen uptake as a limiting factor for energy uptake and growth (see, e.g., Babák 1910).

As already mentioned, later authors involved with the Pütter-von Bertalanffy growth model were typically unaware of the original German texts, many of which do not exist in English, and the interpretation of the limiting surface in question adds another example of resulting misunderstandings. Some of the twentieth-century pioneers in modeling fish growth and assessing fish stocks have misinterpreted Pütter's and von Bertalanffy's growth function as a model that assumed food and gut surface as the limiting factors (see e.g., Beverton and Holt 1957; Ricker 1975), which was understanding, as von Bertalanffy himself was sometimes ambiguous about intestinal surfaces as a potentially limiting factor (see von Bertalanffy 1951, p. 284).[69] The idea that gut surface plays a fundamentally limiting role is, moreover, implausible for another reason than the surface proportionality of respiration detected by Pütter: a crucial difference between gills and intestinal surfaces is that gills are constantly in contact with oxygen through the medium in which breathing occurs (see also Chapter 3). Intestines, by contrast, are not constantly in contact with food and in fact, growth continues for quite some time after ingestion has taken place (Iles 1974; Parry 1983; Rosenfeld *et al.* 2015). Given that surface limitation is a factor that involves time, the continuity of the respective absorptive processes that are taking place can indicate whether a surface in question is likely to present "*the weakest link*" in a chain of physiological processes. While we can conclude that intestinal surfaces are unlikely to act as a limiting factor before gills or respiratory organs limit metabolic supply in water-breathing animals, we will return to this point in Chapter 3. While oxygen limitation could theoretically occur at any stage of the oxygen cascade, we will apply our method of identifying 'the weakest link' to determine whether limitation mainly occurs in the gills, the cardiovascular system, or at the cellular level.

Mechanistic foundations III: Differences between Pütter, von Bertalanffy, and the GOLT

As has become clear, the GOLT is deeply indebted to the work of both Pütter and von Bertalanffy, and Kearney's abovementioned characterization of this theory as a "*mechanistically explicit version of Pütter and von Bertalanffy's ideas*" is apt. However, it also departs from the work of these two pioneers in several important points. The crucial difference is, of course, the fact that the model presented here goes beyond the '2/3 rule'(in animals of "*growth type 1*"): as discussed in Chapter 1, the GOLT is not a merely constraints-based theory, but its explanatory foundation is based on both constraints *and* optimization (because values of $d > 2/3$ can be explained as adaptive optimizations). Pütter and von Bertalanffy paid little attention to adaptive explanations – partly due to an anti-Darwinian reflex that may be understood as a reaction to the simplistic adaptationist models prevailing in the first half of the twentieth century, and which they and many others considered to be *just-so stories.*[70]

Another critical difference between the GOLT and the Pütter-von Bertalanffy growth model, as interpreted by von Bertalanffy himself, is the mechanistic foundation on which the growth curves of air- and water-breathers are based. While Pütter (1920) developed his growth equation, he explicitly restricted it to modelling the growth of water-breathing ectotherms, i.e., our WBE. Von Bertalanffy, by contrast, applied the same model to mammals and fish alike, whom he believed to conform to what he defined as "*growth type 1*," which is limited by a surface growing isometrically (see above).

Although the special VBGF has been used to model the growth of birds (Brunner *et al.* 2021; Karpouzi and Pauly 2008)[71] and mammals (Brunner and Kühleitner 2020; Fernandes *et al.* 2019), a mechanistic underpinning for this equation in fitting bird or mammal growth data is lacking.[72] As discussed at the beginning of this chapter, (most) fish and other WBE are indeterminate growers, and their growth curves are asymptotic. Whereas mammals and birds typically do not grow anymore after maturation and first reproduction, fish growth continues, and indeed, the inflection point in the growth curves (in weight) of many species occurs well after first reproduction (Froese and Pauly 2023; Pauly and Liang 2022a). Even though the VBGF can provide an adequate phenomenological description of the growth curves of some homeotherms, the nature of their growth is fundamentally different.

An often-heard objection to growth models based on the VBGF is the so-called 'Reproductive Drain Hypothesis,' which posits that maturation and reproduction reduce and eventually end growth because the energy that would otherwise be available for further growth is now invested in the formation of gonadal tissues and eggs or sperm. This objection, which, incidentally, is refuted by every single goldfish in its bowl that stops growing even though it never reproduced, and by the fact mentioned above that fish growth in weight tends to *accelerate* after first maturity is reached, will be discussed in more detail in Chapter 9, but for now it suffices to conclude that a necessary condition for the GOLT to apply is that a given species exhibits asymptotic growth. This is due to the GOLT's mechanistic interpretation of the VBGF, which predicts that growth can continue as long as the anabolic term of the equation outweighs the catabolic term. The VBGF can model growth in some air-breathing vertebrates (amphibians, reptiles, birds, mammals), but only if the growth function is understood as a phenomenological, rather than a mechanistic model.

A final point in which the GOLT differs from von Bertalanffy's work is the growth of larvae and early juveniles. Since their values of d are usually > 1, their growth and metabolic rate cannot be limited by the surface area of their respiratory organs. Pütter only hinted at this point in his 1920 essay, stating that his growth model could only be applied to organisms whose geometrical and morphological proportions did not change markedly (p. 324). Despite von Bertalanffy's stress on questions of form and morphology (see, e.g., 1951, p. 332-350), he also neglected this issue. The growth of larvae and very young juveniles cannot be modeled using the VBGF because its prerequisite is a surface limitation.[73] The growth of fish larvae (Overnell and Batty 2000) and early juveniles (Guan and Cao 2007), which is exponential rather than asymptotic, provides another powerful argument for the hypothesis that gills and not intestinal surfaces are the limiting surface.

A model of the growth of fish and other WBE

The GOLT's conception of growth is presented as a graphic model **(Figure 2.1)**, based mainly on chapters 6 and 7 of von Bertalanffy (1951).

The fish feeds, and its food (represented by proteins) is assimilated, that is, broken down into amino acids, in which form it joins the 'amino-acid pool' in the fish's body. Part of the amino acids of this pool is 'burnt' (see Excretion I in **Figure 2.1**); the chemically bound energy thus obtained is used to synthesize native proteins, the building material being drawn from the amino-acid pool. Simultaneously with this synthesis process, the body's native proteins undergo continuous denaturation (see Chapter 4).

On the other hand, the rate of synthesis of body substances (hence, also the replacement of denatured proteins) is limited both by the rate of replenishment of the amino-acid pool (to which the degradation process itself partly contributes) and by the amount of oxygen available for the oxidation of the substances from the amino-acid pool. A high oxygen supply will synthesize a maximum amount of body substance from the amino acid pool. A low oxygen supply will allow for only a limited rate of synthesis, and a part of the amino-acid pool spills over and *"is excreted by the gills and kidney as incompletely oxidized nitrogenous compound"* – the latter point quoted from Webb (1978), who cited the works of Forster and Goldstein (1969); Savitz (1969, 1971); Olson and Fromm (1971) and Niimi and Beamish (1974). This phenomenon, whose existence was also confirmed by Kajimura *et al.* (2004), is called 'Excretion II' in **Figure 2.1**.

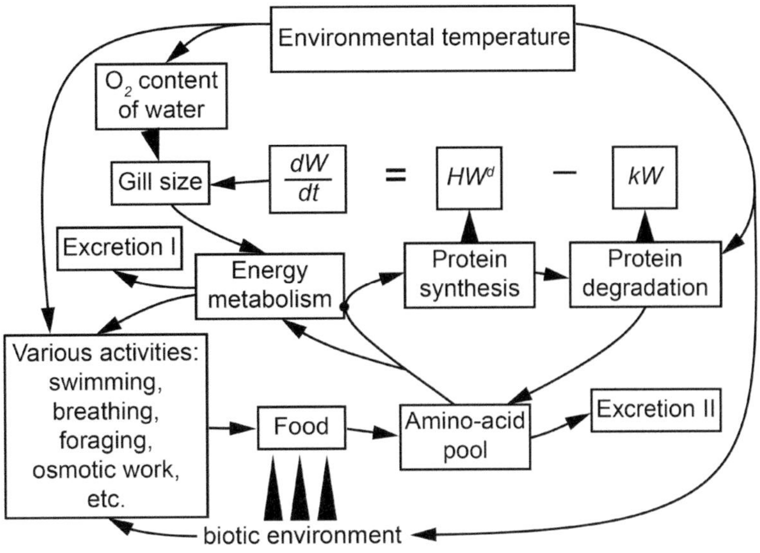

Figure 2.1. Model of the growth of fish and water-breathing invertebrates derived by Pauly (1979, 1981) from his reading of von Bertalanffy (1951) and other authors (see text). Note the emphasis on the role of gills and oxygen demand and supply for protein synthesis.

A portion of the available oxygen and protein is used to form gonad products, which, once the fish reach a certain size, are periodically released from the body (see Chapter 9). Also, the sum of synthesis minus breakdown of body substance, when positive, results in body growth, which, among other things, also increases gill surface area and, therefore, the total amount of oxygen that can enter the body per unit of time. Body weight tends to increase faster than gill surface area, causing relative gill surface area (= gill surface area/ body weight) and oxygen supply per unit weight to decline with increasing body size.

Oxygen supply per body weight steadily diminishes as body weight increases, resulting in large fish with a lower energy metabolism per unit weight and a lower synthesis of new material. However, the amount of body substance degraded per unit of time increases in a manner directly proportional to body weight, and growing fish gradually reach a point where the synthesis of body substances is sufficient to replace degraded substances. Webb (1978) confirms this last point by stating that losses of incompletely oxidized nitrogenous compounds via the gills increase with *size*, and are very low in actively growing, small fishes. The same concept can be expressed by calculating food conversion efficiency (food intake/ growth increment) for fish of different sizes. Such experiments consistently demonstrate a decrease in conversion efficiency with increasing size (Jones 1976; Kinne 1960; Gerking 1971; Menzel 1960). This point will be discussed in greater detail in Chapter 9.

Necessary criteria for the GOLT to apply

To conclude this chapter and link its implications to the following chapters on different taxa and phenomena, we summarize the main criteria that are required for the GOLT to apply:

1. Growth in late juvenile and adult water-breathing ectotherms is asymptotic;
2. Metabolic rate scales proportionally to gill surface area;[74]
3. Gill surface area scales with body mass with an exponent $d < 1$ in adults[75]. If this were not the case, the GOLT's proposed mechanism for asymptotic growth (see Statement 1) would not apply since oxygen would remain available to adults at the relative rate as in juveniles.[76]

The architectural design of the GOLT is illustrated in **Figure 2.1**, which sets its different components in relation to one another. The next chapter will relate the growth parameters resulting from the GOLT's interpretation of Pütter's equation to the growth and the geometry of gills. There, we will outline how limiting surfaces determine the growth patterns described here and then discuss how the different mechanisms of growth, reproduction, and maturation are impacted by the changing ratios of oxygen supply in the life histories of fishes and aquatic invertebrates.

3

Channeling the fire of life: fish gills and their geometry

Synopsis: This chapter examines the geometric structure of the gills of fish and other water-breathing ectotherms (WBE) and explores how their shape and growth influence oxygen uptake, metabolism, and, consequently, the growth of WBE. It discusses the relationship between gill surface area, body size, and oxygen supply, highlighting the constraints these factors impose on metabolism. By analyzing the functional implications of gill morphology, this chapter provides a foundation for understanding how respiratory limitations shape growth and energy allocation in aquatic water-breathers.

Respiration, combustion, and surface areas

Thinking about respiration is not limited to contemporary science; indeed, every human culture has found ways to conceptualize breath and the act of breathing. Even before a proper mechanistic understanding of gas exchange was established, humans worldwide developed knowledge about respiratory processes that enabled biological organisms to survive. In some cultures, breathing was associated with life itself and with the 'soul.'[77] In the ancient Mediterranean world, respiration was often understood as a form of cooling.[78] The most prominent proponent of this idea was Aristotle (384 – 322 BC), who likely adopted it from Plato (? – 348 BC). According to these philosophers, organisms constantly produce internal heat, which must be dissipated through respiratory organs. While Aristotle realized that animals died when denied access to fresh air, he ascribed this to the fact that the exhausted gas contained too much heat to cool down the body sufficiently. This cooling process was thought to apply to both lung- and gill-breathers, implying that the respiratory organs in question needed to have surfaces large enough to cool the entire body volume of the animal.[79] While not all Greek authors shared the assumption that respiration was a form of cooling – perhaps most notably Herophilus (335 – 280 BC), who argued that the lungs supplied the body with life-giving "*pneuma*" ("*πνεῦμα*") – they underlined the relationship between respiratory surfaces and the volume of the bodies they had to keep alive (von Staden 1984).

The understanding of respiration as a vital substance exchange rather than a cooling mechanism became more widespread in late Graeco-Roman antiquity and the Islamic Middle Ages. Authors from Galen (129 – 216) to Ibn al-Nafis (1213 – 1288) reflected this transition, describing pulmonary circulation in more detail and speculating on the distribution of vital substances through the body (West 2015). This was a crucial step toward the modern discovery of gas exchange and the identity of respiratory gases. Another step was the idea that breathing was related to a form of combustion. Seventeenth-century Europe marked a turning point in this respect, despite earlier anatomical discoveries on respiration as a part of 'burning' food (Wilson 1960). Vacuum experiments showed that neither respiration nor combustion could occur without air. These findings led to the insight that air has weight and density, too, and that *"we live submerged at the bottom of an ocean of the element air,"* as Evangelista Torricelli (1608 – 1647) famously put it (cited after West 2013a). Torricelli calculated that air was 400 times lighter than water, which can be considered a step in the right direction, even though he underestimated the density of water, which is, in fact, around 850 times denser than air. The existence of vacuums was still controversial in the seventeenth century – and famously denied by Descartes and others – but was then empirically demonstrated by Torricelli and Otto von Guericke (1602 – 1686). Von Guericke's famous vacuum demonstration at the Imperial Diet of Regensburg in 1654, for example, helped connect this emerging knowledge to questions about the role of gases in combustion and respiration (Schaffer and Shapin 1985).

Two English researchers who also conducted vacuum experiments, Robert Hooke (1635 – 1703) and Robert Boyle (1627 – 1691), noted that igniting a flame within a container free of air was impossible. As Boyle also observed, insects that were put into a vacuum would die off like a flame, and Boyle concluded that there must be *"some kind of resemblance betwixt fire and life"* (Boyle 1662). A similar experiment was performed by the English contemporary John Mayow (1641 – 1679), who observed mice in sealed glasses and noted that the lack of fresh air in the container caused their eventual demise. As he concluded, both respiration and combustion depended on the same substance to survive, which he called *"spiritus nitro-aereus"* (Severinghaus 2016).

Inspired by these findings, early modern ideas about respiration began to adopt a new metaphor, describing respiratory processes as a form of *combustion*. That breathing played an essential role in the *"fermentation"* of nutrition, and the organism's capacity to effectively transform food in a way that the body could utilize it was already articulated by Marcello Malpighi (1628 – 1694), albeit in a more intuitive form.[80] Connecting these ideas to the observation that access to air had similar effects on the survival of living organisms as it had on candles and flames paved the way for understanding energy metabolism as an equivalent of fire.

A century after Boyle, Hooke, Malpighi, and Mayow, chemists in England and France made further progress in identifying what we now know as oxygen. Joseph Priestley (1733 – 1804), a dissenting Protestant minister and polymath who became interested in natural phenomena such as electricity, optics, and respiration, designed an experimental setup that enabled him to understand gas exchange in different organisms. In an elaborate experiment involving mice and peppermint plants, he demonstrated that gas exchange in plants and animals showed opposing trends and that the peppermint cuttings produced the gas that enabled the mice to survive in a sealed container (Kutschera and Niklas 2013).[81] This gas – the equivalent of Mayow's *"spiritus nitro-aereus"* – was then called *"oxygène"* by the chemical pioneers Marie-Anne Paulze-Lavoisier (1758 – 1836) and Antoine Lavoisier (1743 – 1794). The Lavoisiers showed that the seventeenth-century identification of respiration as a form of combustion was more than just metaphorical: in an experiment on a live guinea pig, they measured how respiration also produced heat and demonstrated that oxygen was needed to *"burn"* food (Lavoisier and de la Place 1780; Fitting 2005; Koehler *et al.* 2011;

West 2013b). This breakthrough was key: without oxygen, the energy contained in ingested food cannot be accessed; thus, oxygen can be considered a "*resource*" (Kramer 1987).[82]

Important additional steps in the study of respiration in aquatic animals were undertaken by researchers around the turn of the eighteenth and nineteenth centuries, most notably by Lazzaro Spallanzani (1729 – 1799), Auguste Broussonet (1761 – 1807), Alexander von Humboldt (1769 – 1859), Jean-Michel Provençal (1781 – 1845), and Paul Erman (1764 – 1851), the last of whom was also the first known researcher to describe air-breathing in fish (Erman 1808). As these early physiological investigations demonstrated, aquatic respiration could be understood as analogous to breathing in terrestrial animals. Based on earlier work by Broussonet (1785), Spallanzani (1803) found that dead and live fish consumed oxygen and produced carbon dioxide, and concluded that both metabolism and the decay of dead flesh resembled each other regarding gas exchange. Humboldt and Provençal (1809) successfully quantified gas exchange and metabolic rates in fish by measuring the oxygen, carbon dioxide, and nitrogen concentrations before and after placing a fish in a glass container for a given time. While their work prepared the ground for modern respirometry on aquatic animals, their data were, in themselves, of little use as the authors did not indicate the body sizes of the fishes on which the experiments were conducted. Still, as all these new findings confirmed, gas exchange in aquatic and terrestrial animals was similar, and respiration depended on processes resembling the oxidation of organic matter.

If combustion was more than just a metaphor for organic metabolism, nineteenth-century biologists concluded that the physical conditions on which respiratory and metabolic processes depended should be taken seriously. All oxidative processes crucially rely on specific geometrical relationships. Like combustion, respiration depends on surface/volume ratios that are sufficient to allow for an effective exchange between oxygen and its reactive counterpart. Lighting a campfire is not workable with one big firewood log, but it is easy when chopped into kindling. Similarly, mechanical engineers experimenting with early combustion engines recognized that combustion largely depended on optimized surface/volume ratios. Similarly, the quest to develop carburetors that could be used to create a flammable gas mixture suitable for a car's engine depended on an effective method of diffusion, which essentially meant the maximization of surface in relation to volume: the smaller the drops of vaporized liquid fuel (typically realized through high-pressure injection), the faster and more efficient the explosion in the combustion chamber.

Respiration in living organisms depends on similar principles, and oxidative processes require sufficiently large surfaces to supply organs and organisms with energy. Just as the wood of a log will only catch fire if its surface is increased through fragmentation into small pieces, large organisms need to enlarge their surfaces in relation to their mass. In this context, it is not surprising that forms and shapes were a major theme in debates involving Johann Wolfgang Goethe (1749–1832) and Alexander von Humboldt, as well as among early Darwinists such as Ernst Haeckel (1834–1919) and Carl Semper (1832–1893) (Nyhart 1995).

The contributions of thinkers like D'Arcy Thompson, Pütter, and von Bertalanffy can best be understood in this context, and the centrality of morphology in nineteenth and early twentieth-century physiology should not be underestimated. Questions of form and shape continued to be influential in debates on physiology and growth, and they were at the very heart of what would later become known as mathematical biology. Morphological relationships were the core theme of Thompson's seminal and pioneering "*On Growth and*

Form" (1917). They became important again in a branch of biology that sought to go beyond simplistic explanatory accounts of adaptation that neglected the geometrical and physical structures in which adaptation took place, and that were themselves subject to natural selection. To cite J.B.S. Haldane again, if the shape of an organism or organ "*is unaltered its surface will be increased only a hundredfold, and ten times as much oxygen must enter per minute through each square millimeter of skin, ten times as much food through each square millimeter of intestine. When a limit is reached to their absorptive powers, their surface has to be increased by some special device. For example, a part of the skin may be drawn out into tufts to make gills or pushed in to make lungs, thus increasing the oxygen-absorbing surface in proportion to the animal's bulk*" (Haldane 1926).

The paradox of optimization

Specialized adaptive mechanisms that increase respiratory (and other) surfaces were crucial for all organisms to reach sizes beyond the unicellular level (Schulz and Jørgensen 2001). While fish larvae do not yet depend on gills since their surface-to-volume ratio is still such that cutaneous respiration is sufficient for oxygen uptake, this is only possible up to a length of 2-3 cm.[83] From that size onwards, the organism needs structures whose surfaces grow disproportionately to the increase in body mass, and skin-breathing is no longer sufficient.[84] There are still geometrical limits to the extent to which this increase is possible. Surfaces can drastically increase by forming pores, folds, filaments, and tufts, but the effectiveness of these shapes is not infinite.[85] Every surface that exhausts a fluid or gasiform medium from a specific substance requires an effective removal of the substances that are no longer needed. Gills can only use inhaled water once, and their function implies an inherent paradox: the more effective they are and the greater their increase in surface, the quicker their resources are exhausted. That means an increase in surface is only effective in width and height, but not in depth, since the water, once it is flushed through the gills, is largely devoid of oxygen (see **Figure 3.1** for a similar dimensional constraint in cars' radiators).

Fish gills allow for a highly efficient transfer of oxygen between water and blood. The blood flow runs in the opposite direction to the water that is pumped through the gills. This countercurrent exchange of oxygen enables the organism to efficiently extract the dissolved oxygen from the water, which, past the gill lamellae, often has an oxygen saturation of less than 10-15%. Paradoxically, it is this high efficiency of gas exchange that underlies the geometrical constraint of fish gills, and, in that respect, it would be apt to state that the construction of gills falls victim to its own success. The exhaust water can no longer be used once it leaves the lamellae, which also means that increasing their 'depth' (i.e., their size in the dimension parallel to the water flowing through gills) would not result in a higher oxygen uptake. Nothing that is located behind the gill lamellae has any effect on respiratory performance.

This mechanism can be explained by comparing it with the function of a radiator used to cool down a car's engine. The radiator's function is to exchange heat through a sufficiently large surface. Once the engine heats up above an optimal temperature, a thermostatic valve opens and cooling water is pumped through the radiator. The water is then cooled as it flows through the interior of small lamellae that are exposed to the airflow while driving (**Figure 3.1**), which may be enhanced by a fan that blows more air through the radiator.

Whereas gills pump oxygen-rich water across a highly maximized surface and almost entirely deplete it of oxygen, car radiators blow cool air through lamellae, which absorb

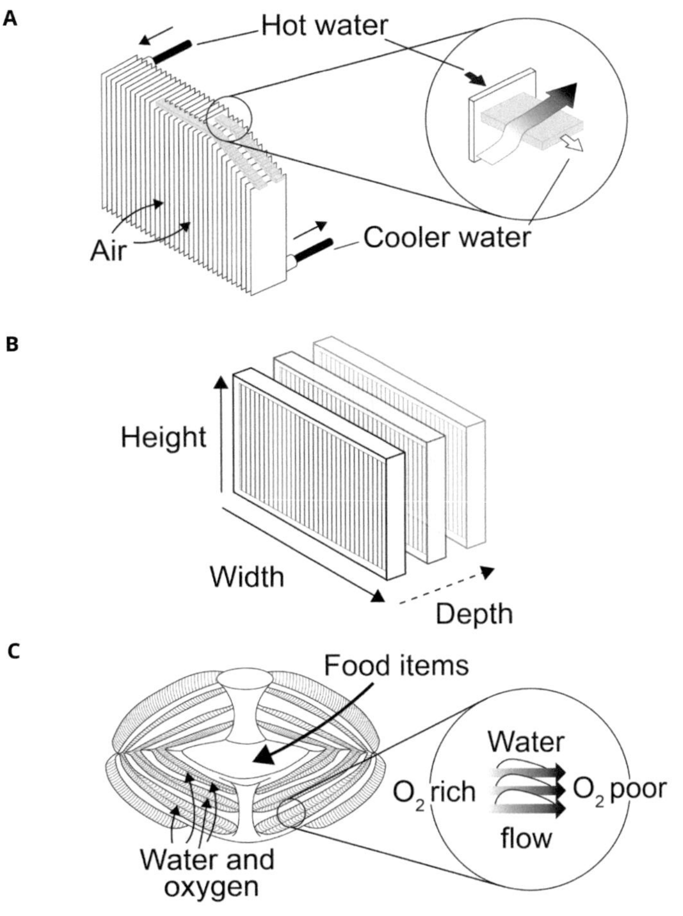

Figure 3.1. Illustrating the structural parallels between cooling systems in car engines and fish gills. **A:** Schematic representation of a car radiator, with one of its lamellae on the right. **B**: The radiator can grow in width and height but not in depth because air, once it has passed the first layer of lamellae, cannot pick up more heat from a second or third layer. **C:** Frontal view of the interior of a manta ray's mouth (*Mobula japanica*, adapted from figure 3.20A of Wegner 2015), illustrating the 2-dimensionality of breathing, i.e., the fact that once water has passed through the gills, it is largely depleted of oxygen, and it would be useless for it to pass through more gill structure.

the heat before it leaves. Like fish heads, which need sufficient space to pick up oxygen, car hoods have a geometrical problem in dissipating heat. If the radiator's efficiency is to be enhanced, it is impossible to use the 'depth' of the car hood since the air that would leave the first row of lamellae is already hot and cannot absorb more heat. The same is true for gills, where the highly efficient countercurrent exchange system in *Mobula japanica* (**Figure 3.1)** or *Nautilus* sp. (**Figure 3.2**) would not deliver more usable oxygen if it had more than 2 gills on each side or if their lamellae were longer.

Stacking gills behind each other would make as little difference as stacking radiators behind each other (**Figure 3.1B**). While gills must increase with positive allometry because they are surfaces that need to keep pace with a volume that requires oxygen, this increase can only occur on the surface that directly faces the water flow. Despite their allometric or

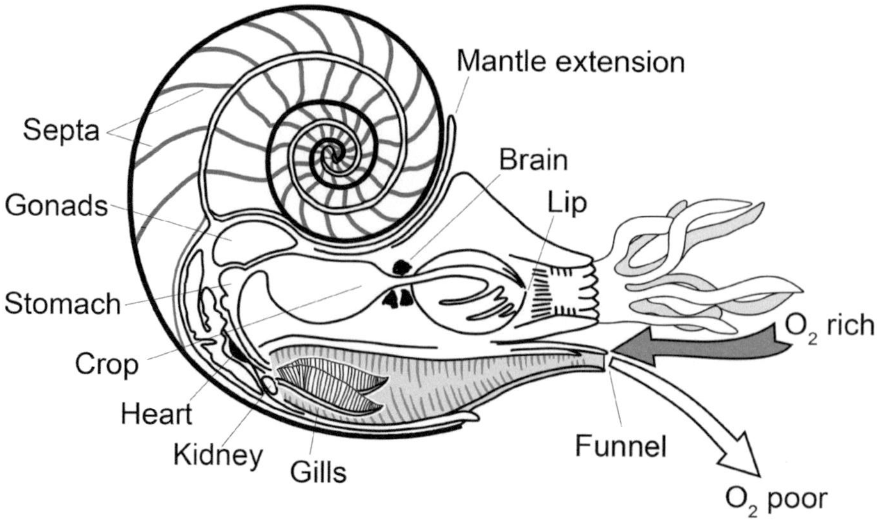

Figure 3.2. The geometrical constraint that reduces the effectiveness of longer gill lamellae can be perceived in the gill structures of *Nautilus* sp., which breathe by pumping water in and out (arrows) of the mantle cavity (grey) through its 'funnel' (Ward 1987, p. 13). This cavity is almost filled in width by the gills, but not in length. If the gills (or their lamellae) were to grow longer, the effectiveness of their distal parts would be drastically reduced, as the water reaching them would already be depleted of oxygen.

hyper-allometric growth, the oxygen supply that they provide reaches a point where it can no longer match the oxygen demand for further growth.

What can positive allometry effectively do?

As discussed in Chapter 1, the gills of most WBE defy the 2/3 rule proposed by Pütter and von Bertalanffy in that the scaling exponent (*d*) of fish gills can vary widely. Small fishes whose growth is initially fast but then stagnates often tend to show values of *d* around 0.6. These values are typically closer to 0.9 in large, active fish, such as tuna or the largest shark species. In adult fish, *d* never reaches 1. Values of *d* =1 or higher are only possible in young larvae with a minimal body mass and, therefore, a relatively large body surface area. Their soft skin allows for cutaneous respiration in larvae, and the large body surface can provide sufficient oxygen for exponential growth.[86] The scales that cover the bodies of most adult fish no longer allow for respiration through the skin.

The relatively high values of *d* in large, active species have sometimes led to the impression that these values (for example, *d* = 0.9) were sufficient to overcome oxygen limitation (e.g., in Lefevre *et al.* 2017). It is often overlooked how much difference the exact scaling exponents make (see Chapter 4). Even with such high values, the increase in respiratory surface area continues to decrease relative to body size. In the case of Atlantic bluefin tuna (*Thunnus thynnus*), where *d* = 0.9 (Muir and Hughes 1966), a 10,000-fold increase in body mass only results in a 4,000-fold increase in gill surface area, which means that the respiratory capacity in a 500 kg adult is 60% lower than in a 50 g juvenile. Intuition does not necessarily serve us well when appreciating these relationships, as illustrated by **Figure 3.3**.

Another example is provided by the cyprinid *Labeo catla*, whose adults have about ¼ of the gill surface area relative to the body size of juveniles (see **Figure 3.4**). While a fingerling

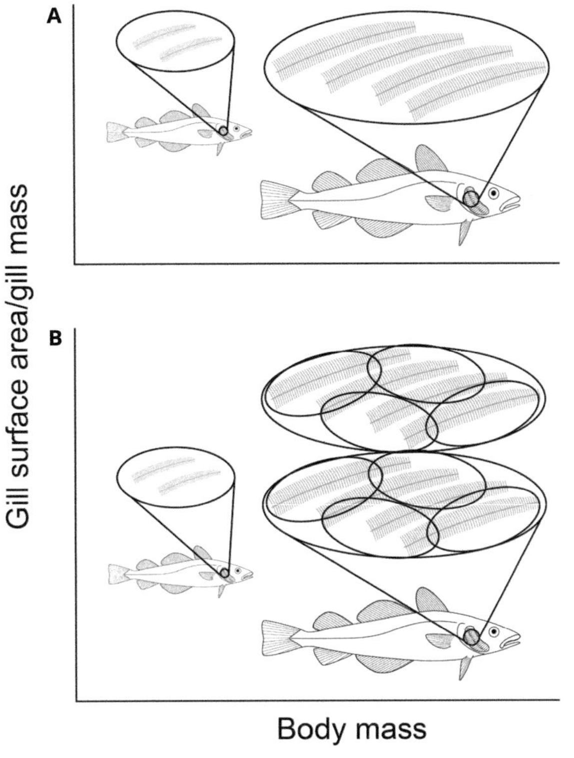

Figure 3.3. Two schematic illustrations of the perceived effect on its gill 'size' (i.e., surface area) of doubling the length of a fish. **A**: Lefevre *et al.* (2017, figure 1) used this schematic to illustrate (originally in color) how easily a larger fish – an Atlantic cod in this example – could grow larger gills as required when its length doubles; this graph suggests (via the number of gill symbols in the ellipses) an increase by 4. **B**: Given a doubling in length and the same overall shape, a large cod is 2x2x2 = 8 times heavier than a small cod. To maintain the same O_2 supply per unit weight as the small cod, the larger one would therefore need 8 times the gill surface area of the small one. While the drawing in panel **A** is not quantitative, it reflects how Lefevre *et al.*'s intuition prevents them from perceiving the problem (modified from Pauly 2018b).

of 17 g had a gill surface area of 8.3 cm^2 per gram of body weight, this was only 1.9 in an adult of 17 kg (Kunwar and Munshi 1988).[87]

Only extreme forms of surface hyper-allometry would effectively overcome the constraint posed by the dimensional tension between surfaces and volumes, but why are these values not realized in adult fish? As we have seen above, increasing the depth of the gill lamellae would not solve the problem, since the countercurrent exchange of oxygen between water and blood is so efficient that very little oxygen is left in the water that exits the space between the lamellae.

Some physiologists have attempted to explain this problem away by questioning the need to increase surfaces in relation to volume. As Lefevre *et al.* (2017) have argued, "*in our field, it is generally accepted that a species' oxygen demand determines the size of their respiratory surface area, not the other way around.*" Aside from the question of how such reasoning would be evaluated in other research fields,[88] it is noteworthy that not all kinds of organisms share the same ratios between energy demand and respiratory surface, as

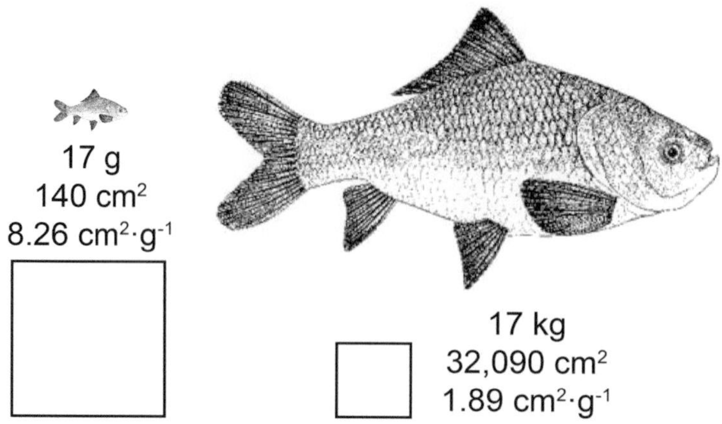

Figure 3.4. Changes in relative gill area and thus O_2 supply in the cyprinid *Labeo catla* during growth. Left: a fingerling weighing 17 g. Right: an adult weighing 17 kg, corresponding to a scaling factor d_G = 0.79. The available gill surface area per gram of body weight in adults is only 23% of that in the fingerlings (data from Kunwar and Munshi 1988).

has been shown since the early 20th century (see e.g., Pütter 1909). Explaining values of d < 1 with decreasing metabolic demand rather than geometrical limitation, as done by Lefevre *et al.* (2017), would require providing a reason why fish should allow their metabolism to decline if they could do otherwise,[89] or at least a precise identification of the energy savings that would allow for a reduced gill surface area. While adult fish of most species do indeed become quieter with increasing size and swim more efficiently (Bakun 2022), these differences are not substantial enough to explain the sharply declining relative supply of oxygen. The most important factor is decreasing energy allocation to growth in adult fish. This implies a different causality, suggesting that the relative decline in gill surface area per body mass limits growth rather than *vice versa* (we will return to this point in Chapter 16).

Optimal gill size is a result of tradeoffs and constraints

The highly optimized structure of gills and the efficient strategy of gas exchange between water and blood through countercurrent blood flow seem to defy the notion that evolutionary adaptation may be highly constrained. A surprisingly widespread view of optimization conceives of adaptive processes as capable of achieving perfection and, in the process, surpassing all the constraints that stand in the way. As White *et al.* (2022) argue, "*the invocation of constraints is unnecessary to explain the ontogenetic trajectories of metabolism and growth.*" Such an explanation neglects the fact that a set of parameters must define the spectrum in which processes and traits can be optimized. The example of gills perfectly illustrates why optimization- and constraint-based explanations should not be seen as opposites and why optimizations depend on constraints. Philosopher of science Arno Wouters has used the example of gills to outline his proposal for "*design explanation*" as a mode of scientific explanation that examines "*the constraints on what can be alive*" (Wouters 2007). In contrast to functional explanations that ask questions like "*What is the function of an organ?*", "*design explanations*" ask questions about the features on which processes rely as necessary conditions. As Wouters points out, above a given body size, fishes' lifestyle and respiratory medium make gill-like structures necessary because of the constraints that structure the possible adaptations that enable life in water.[90]

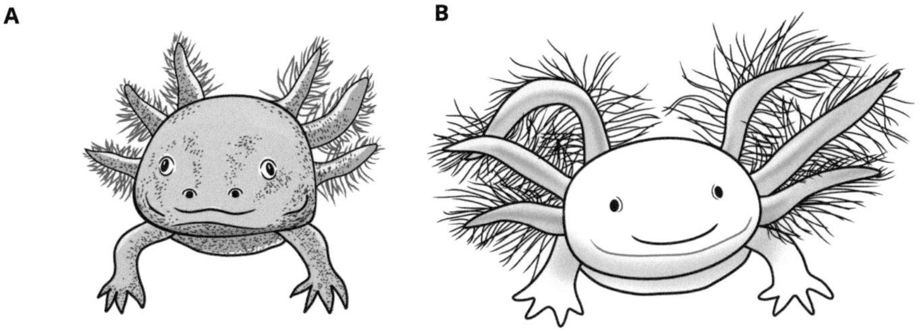

Figure 3.5. Two axolotls (*Ambystoma mexicanum*), adapted from various photos. **A**: Wild type, with short external gills. **B**: Domestic breed of the same species. Note the difference in gill size, which suggests that smaller gills are adaptive in the wild types because long filaments would be too vulnerable.

As we have seen in Chapter 1, it is impossible to define ultimate optimization without predefining the phase space of possible phenotypes (Maynard Smith 1978). Given that fish and water-breathing invertebrates must function in a world that poses different ecological, life-historical, and biomechanical challenges, we can evaluate the constraints on which possible (or actual) optimization can act. Suppose we evaluate these constraints sequentially. In that case, it becomes clear that adaptive features can act as constraints, and it is sometimes impossible to distinguish between adaptive and non-adaptive constraints.

As the analogy between the car's radiator and gills illustrates, the latter cannot expand in depth because a three-dimensional increase (thus including depth) would not allow for more oxygen absorption. To become more efficient, gills must expand in height and/or width. In fish, this is only possible within the limits of the *Bauplan* of the organism's anatomy and obvious hydrodynamic constraints.[91] In both bony fishes and elasmobranchs, the shape of the head limits the space available for gills stacked in rows above, below, and beside each other. The operculum is an adaptive constraint to further redesign fish gills in bony fishes (in invertebrates or water-breathing amphibians, different constraints act on gills' shape and maximal size).

Opercula allow more effective ventilation by guiding and constraining the water flow through the narrow buccal channel. They also have a protective function: gills are highly vulnerable and delicate structures that can easily be damaged when exposed to the organism's surroundings. Opercula are features that both protect and constrain gills' growth and have an essential mechanical function.

The lack of opercula does not mean that open external gills can grow unconstrained, as the case of water-breathing amphibians with external gills clearly illustrates. In the paedomorphic salamander known as the 'Mexican axolotl' (*Ambystoma mexicanum*), the gills are not protected by opercula and are attached externally to the head and not directly limited by geometry. However, axolotls are highly vulnerable to attacks by predators or mechanical damage that can occur during foraging or hiding under rocks. Severe damage to these filamentous structures would result in the asphyxiation of the animal.[92] A limitation of the size of these gills (and of body size) is one obvious trade-off, superbly illustrated by the diminutive shape of the gills in wild axolotl (**Figure 3.5A**) vs. the exuberant gill development in domesticated conspecifics (**Figure 3.5B**).

Another adaptive constraint to gill growth is vulnerability to parasites. The evolution of gills has also created a beautiful evolutionary niche: the inside of a fish operculum is well-oxygenated and an ideal site for small filter feeders that can thrive in the continuous water flow. The structures of the gill membranes must also be soft to allow for efficient gas exchange, thereby allowing parasites to bore or gnaw into and feed on the well-perfused gill tissue (Lo 1999). The vulnerability of fish gills to parasites often correlates with their size, which presents another disadvantage for developing large gill surface areas (see, e.g., Buchmann 1989; Grutter 1994; Lo 1999). This effect may be exacerbated in gill structures with larger empty spaces, which may make ventilation more effective but also offer more opportunities for, and possibly larger parasites, to infect the gills.

Finally, we should not forget that gills are multipurpose organs with many functions besides oxygen uptake and carbon dioxide excretion (Gonzalez and McDonald 1992; Evans *et al.* 2005). It has even been hypothesized that gas exchange was only a later and secondary function of gills in fishes (see Rombough 2007; Fu *et al.* 2010; Baker *et al.* 2015). Water-breathing animals depend on them for osmoregulation, ion and acid-base regulation, ammonia excretion, and other functions, some of which are probably not fully understood. The function of osmo- and iono-regulation was first described by August Krogh (1874 – 1949) in the 1930s (Krogh 1937, 1939). The multifunctionality of gills presents a constraint on their size and growth, as large gills would allow for more effective oxygen uptake but could also result in increased ion loss or too high excretion rates of other substances. Hence, the "*osmoregulatory compromise*" requires a whole set of trade-offs and imposes limits on gill size that depend on multiple factors, including the specific properties of the biochemical and ecological environment of the organisms, which determine the relationship between these different traits (see, e.g., Wood and Eom 2021).[93]

In an excellent summary of the factors that favor adaptive selection for smaller gill surface area, Nilsson *et al.* (2007) list the following points: "*Indeed, several problems are likely associated with a large lamellar surface area:*

1. *The 'osmorespiratory compromise,' which emphasizes that the gills are the main sites for water and ion fluxes, and that an increase in the respiratory surface area results in elevated fluxes that have to be counteracted by energetically expensive ion pumping* [...]. *Indeed, estimated energetic costs for ion-regulation/osmoregulation in fish varies between 10 and 50% of the energy budget of the animal.*
2. *The water may contain toxic substances, and a large lamellar surface area may facilitate the entrance of many of these into the fish. Naturally occurring substances such as ammonia, algal toxins and metal ions may have provided a selection pressure for small gills for millions of years, and since the industrial revolution, manmade pollutants may provide additional selection pressures for small gills.*
3. *Pathogens that enter the fish over the gills and parasites that attach to the gills should make it advantageous to have small gills.*
4. *In fish, the whole cardiac output has to pass through the gills, making hemorrhage from gill injury immediately life threatening* [...]. *Large gills are likely to be more fragile than small ones, so having large gills probably increases the risk of dying from gill injury.*
5. *Since the gills take up space in the oral cavity, large gills may impede the capacity for feeding.*"[94]

In short, given all these constraints, we can conclude that fish gills are already fully optimized in terms of their different functions, and their size, growth, and shape can only be understood as the result of a series of multiple trade-offs.[95]

Again: what is the limiting surface? An oxygen cascade within Russian dolls

Mechanisms related to surface limitation in metabolic processes have been proposed for both food ingestion and respiration. In Chapter 2, we presented and summarized several arguments against the hypothesis that surface-limited resource supply plays a role in processes related to food ingestion or nutrient absorption. Accepting that surface limitation applies to oxygen rather than nutrient supply still allows for multiple interpretations since the oxygen supply from the gills to individual cells involves multiple surfaces, all of which could play a limiting role (see **Figure 3.6**). It may seem that the GOLT's focus on gills is an arbitrary choice. While early theorists of metabolic scaling and its dependence on the limiting effect of surfaces were often hesitant to explicitly identify the surface in question, the currently available data allow for reasonable inferences on which surfaces in the oxygen cascade could play the most critical role.[96]

Recent studies on oxygen limitation in water animals have produced evidence that supports the GOLT but also suggests that different mechanisms could play an additional role. Verberk *et al.* (2022), Hermaniuk *et al.* (2021), and others have demonstrated that

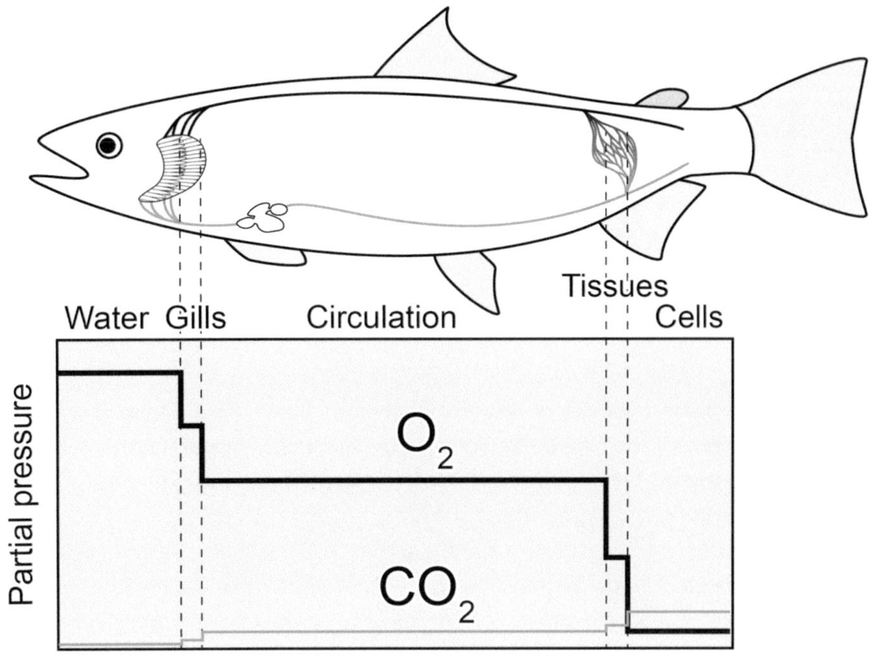

Figure 3.6. Schematic representation of the oxygen cascade in fishes (modified from Rummer and Brauner 2020). Each step of the O_2 cascade may be seen as representing one of the dolls in **Figure 3.7**. Note that the partial pressure inside the system may also be indicative of the respective 'bottlenecks' of transfer: if the cell surface were the main limiting factor, oxygen concentrations in the circulatory system should be expected to take values similar to those in the gill area.

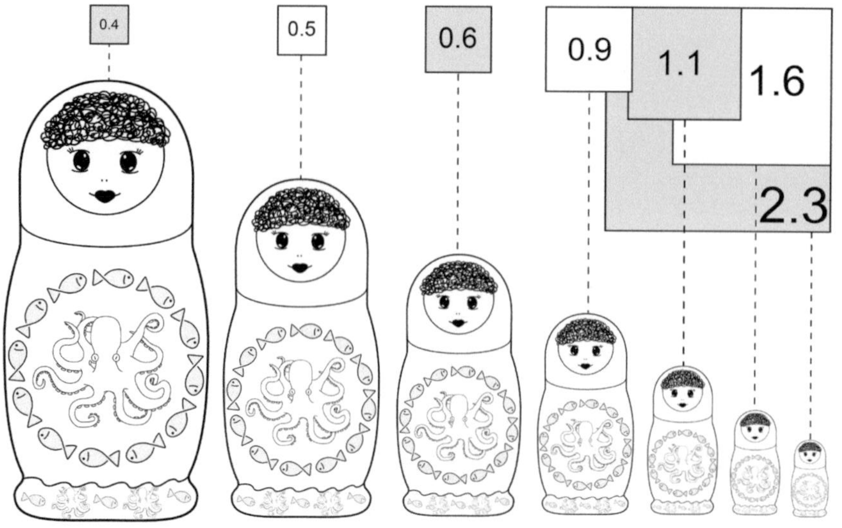

Figure 3.7. Seven 'Russian' dolls, where the six smaller dolls can be sequentially inserted into the next larger ones. Surface/volume ratio decreases with size, which means that the largest doll has the smallest surface area relative to its volume. Even if different degrees of surface allometry and morphological adjustments are involved, these relationships allow for causal inferences on surface limitation in physiological processes between different organs. Note that in organisms, the differences between the surface area of the different organs, cells, and organelles are much greater than those between the dolls.

hypoxia sensitivity is correlated with cell and genome size, suggesting that cellular surface-to-volume relationships provide an explanation. That would mean that surface-dependent oxygen limitation emerges at least partly at the cellular level. The idea that cell size plays a role in the capacity to transport and diffuse oxygen through the body is plausible and should not be dismissed offhand.[97] Two arguments for the prevalence of gills in this process stand out. The first is based on physiological and adaptive mechanisms, mainly the plasticity of the gill in relation to environmental stimuli.[98] The other argument is based on geometrical relationships, i.e., the relationship between whole organisms and their different organs and components is analogous to 'Russian dolls,' where different surfaces are placed over each other and, consequently, their respective surface-volume ratios decrease with each new layer (**Figure 3.7**). This makes it evident that there would be no point in evolving a highly efficient system for distributing oxygen to the smallest organelles because its supply is limited by the equivalent of the largest doll.

Oxygen entering the body through respiratory surfaces must be transported from the respiratory tissues throughout the entire body. Every membrane through which a specific amount of oxygen has to pass from the gills to the cells presents a limiting surface through which only a certain number of oxygen molecules can pass within a given amount of time. After oxygen enters the blood in the gill capillaries, it must be transferred from the blood plasma to the erythrocytes and then released again from the red blood cells to the plasma. Several additional membranes must be passed to enter the extracellular fluid and, finally, reach the cells and mitochondria. All these surfaces involve a tension between surface and volume, and each of these surfaces limits the eventual oxygen supply to the cells and mitochondria.

Despite the undeniable role of all the other limiting surfaces in the oxygen cascade, it follows from geometrical necessity that the first surface through which a substance enters the body has the greatest limiting impact. Suppose we imagine a body through which oxygen is transported as a multilayered Russian doll. In that case, we can see that the first barrier, the outer surface of the largest doll, is the largest, but it also envelops the largest volume, i.e., the entire body, with all the smaller dolls inside. Consequently, all the following surfaces are smaller and also cover smaller volumes. The point is that the surface/volume ratio increases with each step, reducing the limiting role of subsequent surfaces in the body. If gills are the first surface (the largest of the Russian dolls), then they present the greatest limiting factor, and the role of all the following surfaces (erythrocytes, body cells, etc.) is much smaller, if not negligible[99] (see **Figure 3.7**).

Ontogenetic changes in scaling and diffusion distance

One aspect of the literature on fish gills, and on the gills of other WBE, is that change in their scaling, while well-documented (see Glazier 2024 and references therein), is rarely considered. In debates about the growth of gill surface area relative to body weight (as discussed in Chapter 6), it is rarely explicitly mentioned that it is late juveniles and adult fish that are limited by their gills, and not larvae or early juveniles. **Figure 3.8**, which illustrates ontogenetic changes in scaling (OCS) in 5 fish species and in crustaceans, shows that the growth of their gills is very rapid at first, with $d_G < 1$ in juveniles, which becomes < 1 only in the terminal, adult phases (Pauly and Müller 2025).

We draw two conclusions from **Figure 3.8**: (*i*) that this pattern is generalizable, i.e., that OCS impacts on all WBE, and (*ii*) that it is the main reason why some authors report estimates of d_G or $d_Q \geq 1$, because they have included small (young) individuals in estimating a parameter meant to refer to adults (see also Chapter 6).

A related issue is the trend, admittedly not as well established as OCS, for the water-blood distance (*WBD*), also known as the 'diffusion distance' of gill lamellae, to increase as fish grow in size (Pauly and Müller 2025; **Figure 3.9**). An increase in *WDB*, i.e., of the thickness of the gill lamellae, probably to better resist the strong water flow across the gills of adult fish, will reduce the oxygen diffusion rate across these same gills (Pauly and Müller 2025; see also Equation (4.3)).

Jointly, the ontogenetic changes in scaling and in the increase in diffusion distance, which also occurs in Caribbean king crab and spiny lobster (Jarrett *et al.* 2025) suggest, if they can be generalized to fish and other WBE (and we don't know any reason why we shouldn't), that the gills of adult water-breathers limit their growth more than previously anticipated.

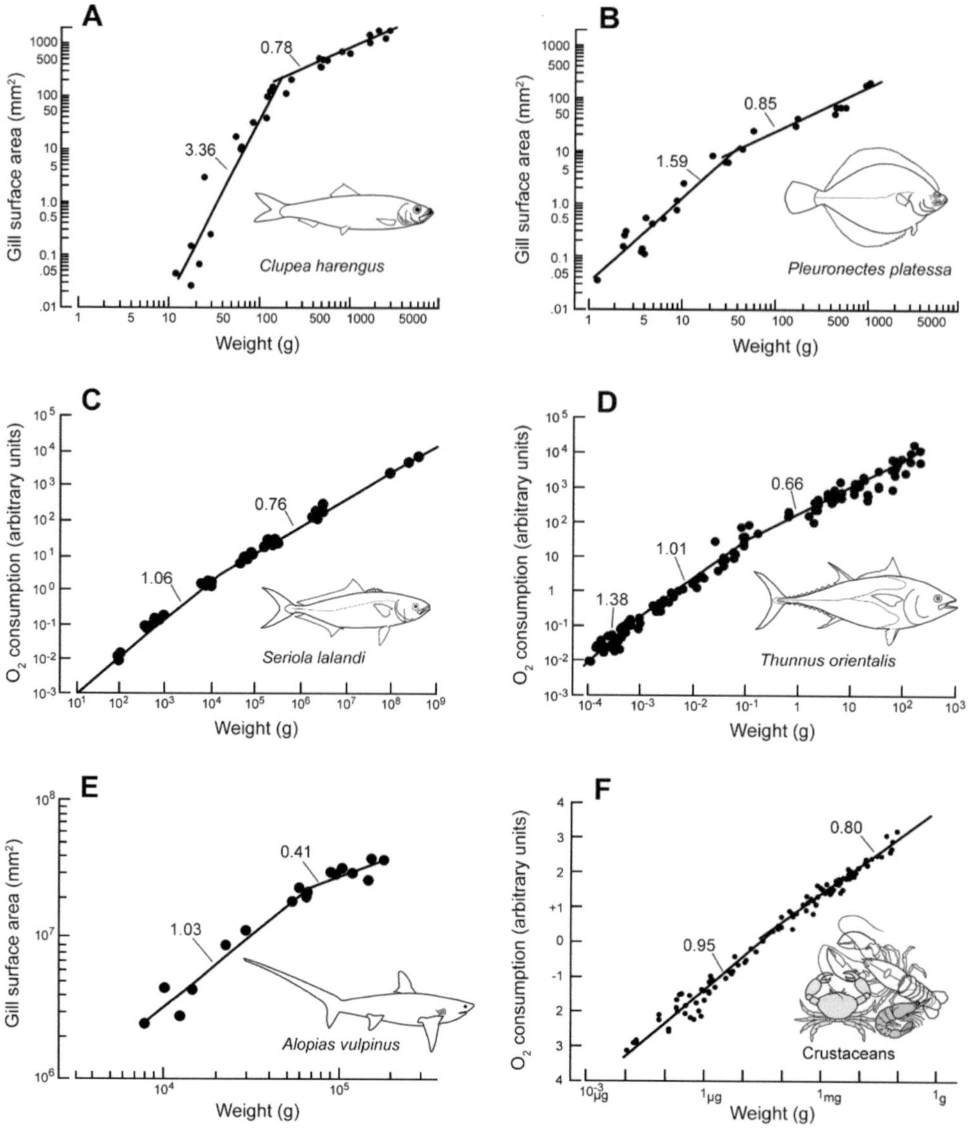

Figure 3.8. Illustrating ontogenetic changes in the scaling (OCS) of gill surface area or GSA (i.e., changes in d_G) and metabolic rate in fish and crustaceans (i.e., changes in d_Q). **A**: Development of gill surface area (*GSA*) in Atlantic herring, from De Silva (1974). **B**: Development of *GSA* in plaice from De Silva (1974). **C**: Increase in specific O_2 consumption in yellowtail amberjack, based on Moran and Wells (2007). **D**: Increase in specific O_2 consumption with weight in Pacific bluefin, based on Miyashita *et al.* (1999). **E**: Development of *GSA* in the thresher shark, from Pauly and Cheung (2017), based on Wegner (2015) and Wootton *et al.* (2015). **F**: Increase in the specific O_2 consumption of miscellaneous crustaceans, based on figure 3 in Zeuthen (1953).

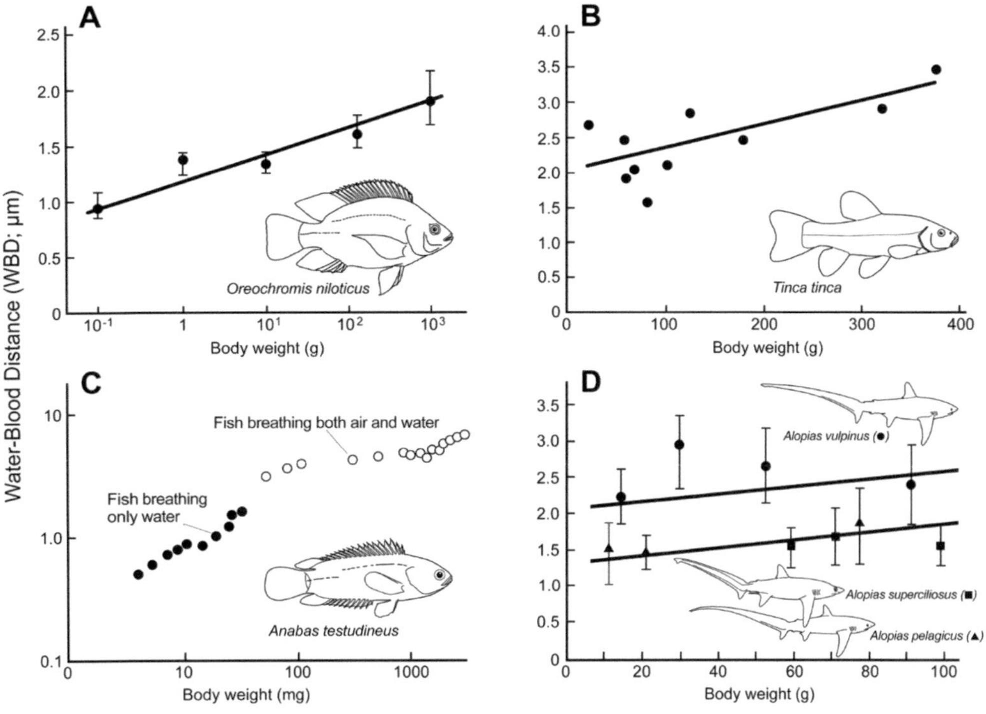

Figure 3.9. Increase of the water-blood (*WBD*, or 'diffusion distance) of the gills' lamellae in 6 fish species. **A**: Nile tilapia, based on Kisia and Hughes (1992); **B**: Tench, based on Hughes (1972); **C**: Climbing perch, based on Prasad and Prasad (1984), and **D:** 3 species of thresher sharks based on Wootton (2011).

Gill plasticity and remodeling as responses to hypoxia

A physiological mechanism often described as a response to hypoxia or increasing temperature in fish and aquatic invertebrates is the plasticity of gills that can be "*remodeled*" within a given period of time if oxygen demand increases (Nilson 2007; Sollid and Nilson 2006). This phenomenon has been described in several bony fishes of different orders, mainly in temperate freshwater species, which can *slowly* increase their gill surface area if oxygen demand increases *slowly*.[100] The range of species that can undergo such modifications is still unknown, but it mainly includes cyprinids and cichlids and, as well, amphibious killifish (Nilson 2007; Rutjes *et al.* 2009; Turko *et al.* 2012). In some species, for example, several African cichlids from Lake Victoria, the increase in gill surface area can be so substantial that even the cranial morphology and the shape of the skull need to be adjusted to enable the enlargement that becomes necessary in hypoxic or hot environments (Rutjes *et al.* 2009). Even in extreme cases like these, the increase in actual oxygen uptake capacity still lags far behind the increase in gill surface area.[101]

As we have seen above, the size of gills results from a series of necessary tradeoffs. That means that the range of what is geometrically and physiologically possible is narrowly constrained by multiple factors crucial for maintaining essential physiological functions. Even though other organs, and even single cells, also rely on such tradeoffs, the balancing act they must perform is less extreme for gills. Therefore, a certain degree of morphological gill plasticity is advantageous for regulating various functions, which range from iono- and osmoregulation to oxygen, carbon dioxide, and nitrogen exchange. The fact that at least some fish species can change their gill morphology in response to environmental conditions allows for inferences in the search for the surface with the most important limiting role in the oxygen cascade. At lower temperatures and higher oxygen concentrations, the surface area of gills can be reduced because a smaller surface area is advantageous. At higher temperatures, it can then be enlarged again. While the same may apply to certain parts of the cardiovascular system (e.g., veins and arteries), gills are the first most important organs in which such plastic responses occur – and some organs or structures do not seem to show any such plastic behaviors (van der Meer *et al.* 2005; Farhat *et al.* 2019).[102] If gills are the main site at which plastic morphological responses to higher oxygen demand occur, this suggests that they are the main bottleneck in the oxygen cascade occurring in the bodies of WBE. While every surface in the cascade can have a limiting role, a limitation is always an effect of the narrowest part, and the fact that individual plasticity first acts on gills rather than cells elsewhere in the body. Gill remodeling and gill plasticity can therefore provide important inferences about the causal mechanisms underlying oxygen limitation in water-breathing organisms.

It is important to note that the possibilities of individual morphological plasticity are not endless and that they are soon exhausted even in species that show extreme forms of gill remodeling and even forms of remodeling of the cranial structure to provide space for larger gills – see Rutjes (2006). An increase in the gill surface would pose severe challenges to other functions and could result in the unregulated loss or influx of ions. This may be a particular challenge in tropical marine fish species whose gill plasticity appears to be very limited. As Bowden *et al.* (2014) put it, *"marine species, in general, must also protect against a constant influx of ions from the surrounding hypertonic aquatic environment, which can be exacerbated by an increased gill surface area and result in higher energy costs to expel the unwanted ions."* In addition, many tropical fish species already live at the maximum of what is morphologically and physiologically possible, and further re-modification is therefore out of reach.

Why should the low heart rates in fish be a "mystery"?

In a review of the cardiorespiratory and cardiovascular physiology of fish, Farrell *et al.* (2009) provide an overview of the research on the impact of climate change and higher temperatures on the basic physiological functions of fish. While a number of their conclusions present a relatively coherent picture and show that increasing temperatures affect the aerobic scope of many species, several items puzzled these authors. First, as they concluded, arterial oxygen partial pressure was reduced at higher temperatures, but a mechanism was still lacking. Second, it was unclear why fish had lower heart rates than mammals: "*Why maximum heart rate is limited to a low rate in most fishes compared with similar-sized mammals, even when Q_{10} effects are considered, remains a mystery.*"

It may be worthwhile to address these three problems as related questions rather than in isolation and explore whether a solution might be found where they intersect and overlap. As we argue, an answer to one of these questions is an answer to *all*. The problem of limited heart rates in fish may seem secondary, but since it is the first "*mystery*" discussed in the review, it can be an entry point to our discussion. The two most striking differences between the physiology of mammals and fish are (*i*) a different medium from which each group extracts oxygen and (*ii*) endothermy vs. ectothermy. Endotherms have a disproportionately higher energy demand than ectotherms, and their vital body functions require more oxygen. At the same time, it is easier to breathe air than water. Also, (*iii*) the respiratory surfaces of mammals – lungs – are much larger than those of fish – gills – whose size is constrained by several factors (see above).

Higher heart rates have a real added value for mammals because they can effectively supply the body with more oxygen. The capacity of the heart to provide more oxygen to the body of a water-breathing organism is very different, even if this fact is obscured by the third assumption made by the authors, which is that sufficient levels of arterial oxygen concentrations at higher temperatures rule out the possibility of gill-oxygen limitation in fish. As Farrell *et al.* (2009) rightly argue, the high partial pressure of oxygen in the arteries demonstrates that temperature does not reduce the gills' capacity to take up oxygen. Indeed, what changes at higher temperatures is not primarily the diffusion capacity but rather the O_2 demand in the body. Therefore, measuring oxygen concentrations in the arteries tells very little about temperature-induced oxygen limitation in fish bodies. The arterial blood that leaves the gill may hold very high oxygen concentrations, but the increased demand creates a limitation.

As the GOLT postulates, gill-oxygen limitation results from the limited gill surface area in fish. This brings us to the authors' second (self-created) problem: low heart rates. Higher heart rates would only have an effect if the respiratory surfaces could enrich the circulating blood with more oxygen. If the surface area of the gills is the limiting factor for oxygen uptake, higher heart rates would only pump half-oxygenated blood through the body, further exacerbating hypoxic stress. Low heart rates are an efficient strategy to reduce energy demand under thermal stress while providing the body with as much oxygen as possible.[103]

The examples on which Farrell *et al.* (2009) base their central arguments are illustrative; for example, the data on Chinook salmon (*Oncorhynchus tshawytscha*), where, indeed, lower partial O_2 pressure was found in the arteries. Especially, "*in larger fish arterial PO_2 was reduced with increasing temperature (Clark* et al. *2008).*" This example perfectly illustrates our point: low heart rates ensure adequate oxygenation and distribution of oxygen in the arterial blood. Farrell *et al.* (2009) acknowledge that Chinook salmon exhibit cardiac arrhythmias at extreme temperatures, which may provide a plausible explanation for the lower partial oxygen pressure in their arteries. This phenomenon (which may also be found in other salmonids) illustrates what would happen if heart rates in fish resembled those of mammals, especially under thermal stress. Instead of being a "*mystery*," the interrelationships between limited gill surface area, oxygen limitation, and heart rate are predictable outcomes of the GOLT.

Conclusion: the GOLT and whole bodies

The basic geometrical observations presented in this chapter again show why it is crucial to consider entire bodies, rather than just isolated organs, cells, or even smaller units to understand organisms' responses to different temperatures or oxygen concentrations. This approach sets the GOLT apart from theories that may appear similar. However, we still need to emphasize that organisms are operational *wholes* and can only sustain themselves through the interactions of their various features and functions, as stressed by Longo and Montévil (2014). In our case, these interactions involve the relationship between gills and the whole organism, including all the components and organs that need to be supplied with oxygen.

4

Temperature and its biochemical impact on water-breathing animals

Synopsis: Since the pioneering work of August Pütter, outlined here, the well-established temperature-dependence of growth parameters and maximum sizes of fish and other water-breathing ectotherms (WBE) form the basis for various 'temperature-size rules' (TSR) for fish and other WBE. Numerous adaptationist interpretations of these rules exist, but their biochemical basis is largely ignored. One fundamental, but frequently overlooked component of the mechanism that leads to temperature-size rules is that proteins only 'work' if their native quaternary structure (or 'native folding') is maintained. However, proteins have half-lives that are U-shaped functions of temperature, which means that higher or lower than optimal temperatures increase their rates of spontaneous denaturation in aqueous solutions, i.e., within body cells. Proteins that lose their quaternary structures cease to function and need to be resynthesized. Protein denaturation (including 'cold denaturation' below 4 °C) explains why the metabolic rates of fish and other ectotherms increase with temperatures above 4 °C, at which hydrogen bonding in water is the strongest.

Why do animals grow larger at lower temperatures? Early answers

One of the crucial factors in the physiology, ecology and reproductive biology of all organisms is temperature. Attempts to identify patterns and rules that relate the growth and final size of animals to the respective temperatures of their environment have a long tradition in biology, ecology and biogeography. The last field, biogeography, was the first to address this topic; at least since Bergmann (1847), researchers started to look for correlations between temperature and size by either analyzing geographical distribution ranges or comparing individual growth curves.

The formulation of macroecological rules was initially based on the identification of underlying mechanisms, as seen in Bergmann's rule (Bergmann 1847) and Allen's rule (Allen 1877). The nineteenth and twentieth centuries saw a trend towards models that focused on generalized patterns deduced from statistically relevant numbers of observations. Simultaneously, older ecological rules were applied to groups of organisms in which the originally described mechanism could not occur. Perhaps the most prominent of all ecogeographical rules, Bergmann's rule, perfectly illustrates this trend: the trend to larger body sizes in colder climates was initially explained through thermoregulation, and it could only be applied to endotherms (see also Chapter 1). The mechanistic foundation of the rule eroded when various researchers tried to apply it to ectotherms, even though it was clear that the mechanism had to be different in animals that cannot keep their body temperatures at a constant level (see, e.g., Ray 1960; Mousseau 1997; Timofeev 2001; Partridge and Coyne 1997; Angilletta *et al.* 2004; Blanckenhorn and Demont 2004; Rollinson and Rowe 2018a, 2018b).

Generalizing any negative correlation between body size and temperature as a manifestation of Bergmann's rule is understandable, as this trend can be observed in both endotherms and ectotherms, including water-breathers. As previously mentioned in Chapter 1, thermoregulation has very different implications for ectotherms and cannot be invoked to explain the observed pattern.[104] An explanation for Bergmann-type size reductions in fish and other water-breathers at higher temperatures must consider different mechanisms. The pattern that ectotherms with indeterminate growth reach larger sizes at lower temperatures and also mature at larger sizes is now typically referred to as the 'temperature-size rule' (TSR; Atkinson 1994).[105] Even though the mechanisms behind this pattern remain disputed, this term is typically restricted to size differences resulting from ontogenetic growth plasticity rather than fixed adaptations to specific environments or lifestyles.

Early polar expeditions that focused on zoological surveys presented evidence of a phenomenon that is now often referred to as 'polar gigantism' in aquatic animals.[106] As researchers involved in such expeditions described in detail, at high latitudes, fish tend to grow to large sizes and small species to become rarer.[107] Even planktonic species grow to sizes that had no equivalent in tropical, subtropical or even temperate climates (see, e.g., Vosseler 1901; Rhumbler 1909).[108] Early twentieth-century zoologists speculated on the reasons for such forms of growth and offered different hypotheses: one possible explanation by Hesse (1913) was that late maturation would allow animals to grow to larger sizes before maturity, i.e., before further somatic growth would be slowed down by reproduction (see also Chapter 7 on the 'Reproductive Drain Hypothesis').[109]

Another hypothesis was the decreasing rate of "*wear and tear*" ("*Abnutzung*") at lower temperatures: if biological processes were slowed down by low temperature, this would allow organisms to live longer and then grow to a larger size. Physiologist Alexander Lipschütz (1883-1980) was skeptical about these hypotheses, but acknowledged the possibility that a temperature-induced slowdown of biological processes could be understood as pertaining to energy metabolism: if the processes of "*assimilation*" and "*dissimilation*" were both slowed down, this could mean that an organism would need more time to grow but also that the 'dissimilating' metabolic activities would only increase later in life and at larger size. This would then not only increase lifespan, but also final size (Lipschütz 1915).[110]

Pütter and the mechanistic origins of the 'temperature-size rule'

Pütter (1911, 1920) sought to present a more mechanistically defined account of the relationship between growth, size, and temperature in ectotherms. Even though his work was long neglected outside the German-speaking world, he can rightfully be considered the originator of what is now referred to as the 'temperature-size rule' (Kearney 2021; Müller and Koster 2024). Even before his seminal essay on "*Similarities of Growth*" (1920; see also

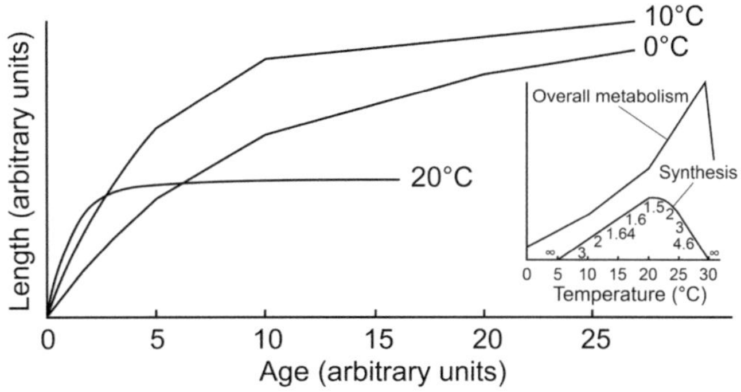

Figure 4.1. Predicted growth of a fish at 0, 10 and 20 °C (adapted from Pütter 1920). The insert, adapted from a figure in the same article, illustrates his findings about the relationship between temperature and both "*Betriebstoffwechsel*" (≈ energy metabolism) and "*Baustoffwechsel*" (≈ synthesis).

Chapter 2), Pütter started to theorize the impact of different temperatures on different metabolic activities. In his "*Comparative Physiology*" (1911), he presented the hypothesis that the catabolic term (of what later became 'Pütter's equation') had a different thermal sensitivity and a higher temperature coefficient (Q_{10}) than "*buildup metabolism,*" i.e., the anabolic term. Below a certain temperature, he argued, an organism could maintain its primary function, but not invest any energy in growth.[111] Only above a given temperature did growth occur, and steadily increased. The positive impact of temperature on growth then reached an inflection point above a thermal optimum beyond which growth decreased again (see insert in **Figure 4.1**).

In this early model, Pütter did not yet explicitly distinguish between the two different features that constitute the 'temperature-size rule': as he, and many other authors after him, noted, this pattern did not only become manifest in Bergmann-type size clines, but also in larger final sizes at lower temperatures. It also occurred in conjunction with another effect of temperature on growth, i.e., initial faster growth and earlier maturation as described in Pütter's aforementioned 1920 article.

It is often overlooked that Pütter, through his influential growth model (see Chapter 3), also presented the first mechanistic explanation for temperature-size relationships that could not be derived from thermoregulatory mechanisms (as in Bergmann's rule). Pütter was familiar with the results of the German 'Plankton-Expedition' of 1889, and he also noted that many invertebrate species and genera were larger in cooler waters. He confirmed this insight by laboratory observations on nematodes, which showed that body sizes in populations kept at 10°C were larger than at 25°C, but their early growth was slower (see **Figure 4.2** for a replication of these results).[112] This phenomenon, Pütter argued, could be modelled by his growth equation if the coefficients and exponents of the anabolic and catabolic terms could be estimated.

The observation that organisms grew faster at higher temperatures required an initial increase of the anabolic coefficient. Pütter assumed that the thermal sensitivities of synthesis and breakdown were different since the breakdown of cellular material was a chemical process, whereas its synthesis depended on the capacity of a living system to take up resources. Despite the acceleration in growth, the same mechanism also accounted for a decrease in maximum sizes. Therefore, faster growth could only be maintained for a

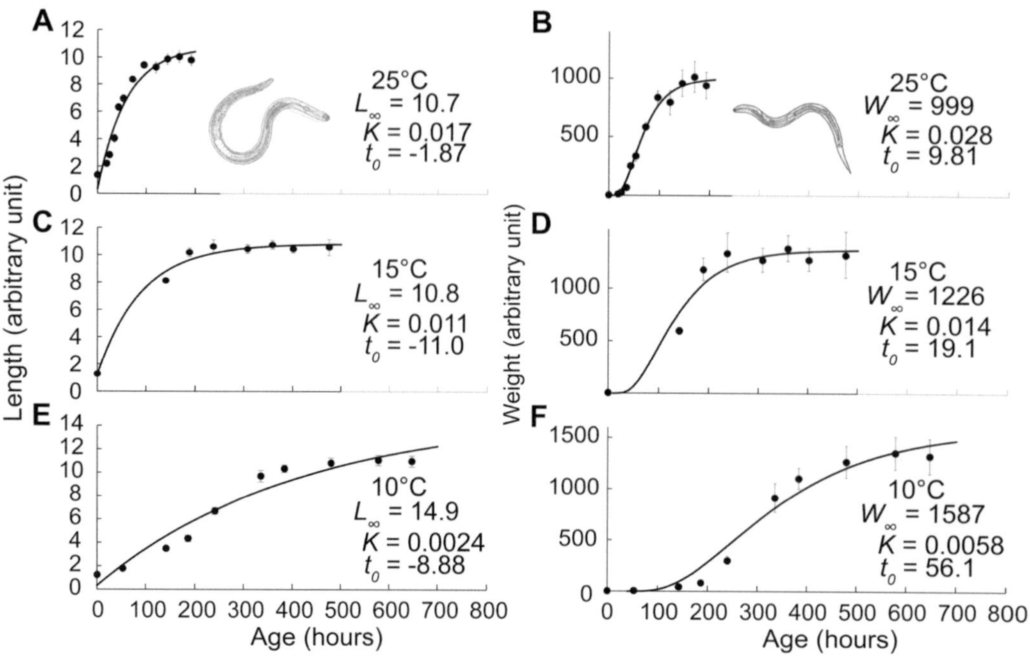

Figure 4.2. Growth in 'length' and 'weight' of the nematode *Caenorhabditis elegans* at different temperatures. Based on surface-at-age data in Van Voorhies (1995, figure 1A), in which the square root was taken to provide estimates of 'length,' with the 'weights' approximated by 'length' cubed[3]. The data points in A, C and E were fitted with the VBGF for length growth, and those in B, D and F with the VBGF for weight.

short period before anabolism was outpaced by the increase in repair processes that were required at elevated temperatures. Pütter reasoned that the descriptive power of his model depended on the correct determination of the coefficients of the anabolic and the catabolic terms (k and k' in Pütter or H and k in Beverton and Holt 1957; Pauly 1979, 2019; see also Chapter 2). To express the relationship between growth and temperature more specifically, Pütter teased apart the properties of the anabolic coefficient k (thus H in Beverton and Holt 1957) and defined its properties as follows:

$$k = ap / (p + q) \qquad ...(4.1)$$

Here, p stands for a diffusion coefficient, q represents maintenance metabolism and is thereby understood as a reaction constant (because Pütter assumed that maintenance metabolism increased exponentially with temperature, at least up to a limit; see insert in **Figure 4.1**) and a is an "*equilibrium concentration of a chemical reaction*" that describes the respective concentrations at which the involved reactants are in equilibrium. In contrast to p and q, which are extensive parameters, a must be understood as an intensive and surface-dependent variable. To illustrate the relationships between these three terms, Pütter used the model of a water barrel with an inflow and outflow to visualize the concentration of substances used in both growth and metabolism (**Figure 4.3**).

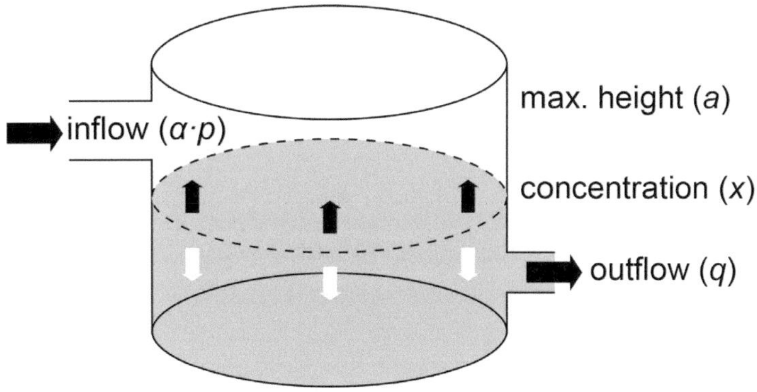

Figure 4.3. The water-barrel model used in Pütter's *"Similarities of Growth"* (1920).

The water level in the barrel is x, which should be equal to k. It represents the concentration of substances used in growth and metabolism and is directly governed by their in- and outflow. The inflow, which is proportional to the diffusion coefficient p, depends on the water height and the size of the barrel, a, as it can only occur between the top of the barrel and the water surface. The outflow, q, has a fixed value and is a measure of energy metabolism. The inflow of substances in an empty water barrel is larger than the concentration of substances used for energy metabolism, and the interaction between p and q leads to an increase in water level (see **Figure 4.3**).

As the barrel is filling up, the rate at which the water level rises will decrease and eventually decline to zero when the barrel is full. A full water barrel would describe the biological situation in which the organism would be able to attain the highest possible growth rate. If the inflow is reduced while metabolism continues, the water level will decrease again until the inflow and outflow are the same: the organism stops growing if this equilibrium is achieved.

Growth rates differ during the life-history of an organism and, as Pütter argued, they depend on ambient temperature in ectotherms. While p increases linearly with temperature, the increase in q should be exponential – Pütter estimated a value of $Q_{10} = 2$. The term a is then expected to be unaffected by temperature. Using a range of realistic values that p and q could possibly take, Pütter demonstrated that final sizes will be reduced in most scenarios (see growth curves in **Figure 4.1**).

The explanatory power of Pütter's mechanistic approach becomes clear when we compare his explanation of the relationship between temperature, size, and growth to postulates of ecogeographical rules whose validity relies on deductions from empirical observations. Such rules will inevitably be confronted with exceptions and irregularities; indeed, Pütter's mechanistic model already anticipated such apparent contradictions and was able to provide explanations for many of them. One example is that some organisms have specific growth requirements that cannot be met at low temperatures. Pütter (1920) refers to Rhumbler (1909), who had noticed that Foraminifera species, which require large amounts of calcium carbonate for their shells, are smaller in high northern latitudes because it is difficult to precipitate calcium carbonate at low temperatures (Clarke 1983).[113]

A more complex phenomenon, which seemed to contradict the trend toward larger body sizes at higher latitudes, arose from the way in which temperature-induced size distributions could be observed in wild populations. Due to the initially faster growth at higher temperatures, later growth reduction might often be difficult to observe when natural mortality is high.[114] This difficulty highlights the importance of developing a well-composed mechanistic model rather than relying solely on statistical data and observations (unless they correct for such underlying distorting factors).

Temperature and its impact on maintenance costs

While Pütter's mechanistic model explained the impact of temperature on growth and final size was a major step towards a better understanding of this phenomenon, and his argument that anabolism has a different Q_{10} than 'breakdown,' it can be simplified and still provide the same results. Even if the thermal sensitivities of the coefficients were the same, the concept of the limiting surface which underlies Pütter's reasoning, would still lead to the same outcomes: being a chemical process, breakdown could theoretically continue infinitely until the organism would collapse as a whole, but anabolism would still be limited by the surface that regulates uptake capacity. Hence, there will always be a point at which growth is no longer possible.

But how exactly does temperature increase the costs of maintenance, i.e., 'breakdown,' which needs constant repair? While Pütter did not explicitly define the different components of breakdown, but referred to them as the sum of all those activities that gradually turn *"a fully capable cell into a pile of rubble of disintegrating organic compounds"* (Pütter 1920). The definitions of maintenance costs in ectotherms still vary widely, and no general consensus seems to have been reached so far – with the unfortunate result that exact definitions are typically avoided. Metabolic maintenance costs in ectotherms arise from a wider number of biochemical processes, many of which remain poorly understood. One component of maintenance is mitochondrial proton leakage, which is believed to account for 20–30% of maintenance costs in endotherms and ectotherms and is highly sensitive to temperature (Pörtner 2001; Sokolova 2023). Furthermore, complex proteins that possess a quaternary structure, such as hemoglobin, ribosomes, DNA polymerase, or structural proteins like keratin or collagen, tend to lose this structure and need to be resynthesized. The energetic costs of re-synthesis may be proportional to the costs of the synthesis of new protein, which would represent growth. Thus, growth and maintenance are connected in our interpretation of Pütter's equation in the sense that synthesis creates new materials, while re-synthesis rebuilds what is constantly being lost.

A necessary constraint within this definition of maintenance is its limitation to repair activities applicable solely to processes that do not involve oxygen consumption. The spontaneous denaturation of proteins is such a process.

Pütter's equation and the spontaneous denaturation of proteins

When marine biologists, limnologists, or fishery biologists study the growth of WBE, they usually pay scant attention to the underlying biochemical processes. This is evident in recent publications where the choice of fish growth models (Gompertz, von Bertalanffy, logistic, polynomials, etc.) is a purely statistical exercise, in which, often aided by the criterion proposed by Akaike (1973), the growth curves that may be selected for the description of length-at-age data have no basis in biology (see **Figure 1.3**). Such approaches obfuscate, rather than elucidate, the relationships between growth and temperature.

This section aims to help overcome this state of affairs by describing the biochemical mechanism that gives rise to macroscopic phenomena described by various temperature-size rules purporting to describe and/or explain temperature-related size variations in WBE (e.g., Atkinson 1994).

Recall that Pütter (1920) conceptualized growth as the net result of two processes with opposite signs affecting the body weight (*W*; actually mass) of WBE, one adding to, and the other subtracting from it, which can be represented by

$$dW / dt = HW^d - kW^m \qquad \text{...(4.2)}$$

where *dW/dt* is the growth rate, and HW^d and kW^m are conventionally referred to as 'anabolic' and 'catabolic' terms, respectively.

The integration of Equation (4.2) with $d = 2/3$ and $m = 1$ yields the ubiquitous von Bertalanffy Growth Function (von Bertalanffy 1938), with growth ceasing when $HW^d = kW$. This implies that asymptotic (= final) length and weight will be reduced if temperature increases affect *k* more strongly than *H*, and because anything, irrespective of their 'starting point' (or intercept), growing in proportion to $W^{2/3}$ will eventually be matched by something growing in proportion to *W* itself. There is a huge number of studies of this specific temperature effect (see, e.g., Dimarchopoulou and Tsikliras 2022, Palomares *et al.* 2022, and references therein), including new evidence presented below. Before we delve further into the implications of Equation (4.2), we first recall some of the ideas presented in previous chapters about its exponents, *d* and *m*.

In animals, adenosine triphosphate (ATP) is required to synthesize the proteins from which body cells and tissues are made. ATP is derived from oxidative phosphorylation in the inner mitochondrial membrane, where the Krebs Cycle provides substrates, and where oxygen is required as the final electron acceptor (Nelson and Cox 2017). This requires an oxygen supply that must enter the body through a respiratory surface.

In biological organisms, respiratory surfaces (*S*) grow in proportion to the square of body length, i.e., $S \propto L^2$ in the case of isometry (i.e., the 'same,' or 'equal' dimension). The exponent can be < 2 (negative allometry) or > 2 (positive allometry), as elaborated in Chapters 1 and 3. The body weight (*W*, strictly 'mass') of WBE can increase isometrically with *L* (with $W \propto L^3$), or with negative (< 3), or positive allometry (> 3). Length-weight relationships of the form $W = a \cdot L^b$, with $b = 3$ or $b \approx 3$, are most frequent in fishes (Froese 2006), crustaceans (Pauly *et al.* 2022a), and other WBE (see SeaLifeBase).

Next, we must deal with how oxygen enters a WBE's body. Fick's law states that the rate of diffusion of oxygen through a respiratory surface can be quantified by

$$R = U \cdot S \cdot dP \cdot WBD^{-1} \qquad \text{...(4.3)}$$

where *R* is the oxygen uptake (i.e., supply to the body in $mL \cdot hour^{-1}$), *U* is Krogh's diffusion constant (i.e., the amount of oxygen (in mL) that can diffuse through membranes with an area of 1 mm^2 in one minute through a given type of material or tissue), *S* is the respiratory area (e.g., the sum of the lamellar area of gills, or *GSA*), *dP* the difference between the oxygen pressure on either side of the membrane in atm, and *WBD* is the water-blood distance, i.e., the thickness of the membrane in question (Fick 1855; De Jager and Dekker 1975; Pauly and Müller 2025).

If one assumes that, of the parameters in Equation (4.3), only S changes with the size of WBE, then oxygen uptake (R) can be considered directly proportional to S, i.e., $R \propto S$ (De Jager and Dekker 1974). Combining Fick's equation with the preceding considerations leads to the exponent of the anabolic term of Equation (4.1), HW^d, having a d-value of 2/3 when the respiration of a WBE is proportional to its length squared ($R \propto S \propto L^2$) and body weight to its length cubed ($W \propto L^3$).

Examples of WBE with isometric respiration are the guppies (*Poecilia reticulata*) that von Bertalanffy used to test his growth theory (Bertalanffy 1951) and subsequently, the brine shrimp *Artemia salina* (von Bertalanffy and Krywienczyk 1953). While such small WBE often have gills that grow near isometrically, or even lack gills and respire through their integument, where $S \propto W^{2/3}$ applies (as in chaetognaths; see Chapter 14), the gills of larger WBE usually grow with a positive allometry, i.e., with d-values commonly ranging between 0.7 and 0.9 (Hughes and Morgan 1973; De Jager and Dekker 1975; Pauly and Cheung 2017).

Important here is that the estimates of d pertain to late juvenile and adult WBE, which are always < 1, and not the d-values observed in fish larvae, which commonly exceed 1; see De Silva (1974) for larvae of Atlantic herring (*Clupea harengus*) and European plaice (*Pleuronectes platessa*), and in early juveniles. The growth of fish larvae, consequently, does not become limited by their oxygen supply as they gain weight (Bochdansky and Leggett 2001; Pauly 2019, 2021).

Given Fick's law, d can be estimated from respiratory or gill area studies (see De Jager and Dekker 1975; Pauly and Müller 2025). Studying only a small range of body sizes within a species of WBE, subjecting the studied WBE to various stresses, or not being aware of pitfalls in the study of gill areas (Hughes 1984), can easily produce erroneous estimates of $d \geq 1$.

In adult WBE, $d \geq 1$ would imply that the membranes through which oxygen is supposed to diffuse would fill up a 3-dimensional space. This would prevent water from flowing through the gills, which must have oxygen-rich water entering from one side, and oxygen-poor water leaving at the opposing side (Pauly 2021a). For fish larvae, the gills can expand rapidly within an initially 'empty' head, and $d \geq 1$ (see De Silva 1974), which then gradually transits to lower a value ($d \approx 1$) in early juveniles (Oikawa and Itazawa 1985; Pauly and Cheung 2017) before settling on the late-juvenile and adult values, i.e., $d < 1$.

The exceptions to $d < 1$ in adult WBE pertain to some fish, crabs, and other ectotherms that rely mainly on air-breathing, and which drown in well-oxygenated water despite having (small) gills (see Chapter 10). Here, we consider only water-breathing ectotherms, which comprise the majority of species of interest in aquatic biology and fisheries.

As already mentioned in Chapter 2, the catabolic term of Pütter's equation, kW^m, is where the real problems begin. This term has a multiplicity of names in German ("*Abbau, Abnutzung, Auflösung, Zerfall, Zerstörung*") in the writings of Pütter (1920) and von Bertalanffy (1951). In English, the terms 'degradation' or 'breakdown' are often used (e.g., in Jobling 1993), but without elaboration, or with elaborations that are seriously misleading.

For example, Ursin (1967, 1979) argued that perceiving catabolic processes as proportional to weight (i.e., $m = 1$) is "*too simple because it overlooks the fact that oxygen required by catabolic processes must enter the body through a surface*" (Ursin 1967, 1979). To overcome this self-generated problem, he reformulated Equation (4.2) with d as the exponent of the *catabolic* term, and invented a new exponent (n) for the *anabolic* term which did generate asymptotic growth curves in a variety of fishes, but only because n <

d.[115] This led nowhere, because conceiving the catabolic term as requiring oxygen resulted in numerous contradictions, notably that dead flesh would not rot.

Another definition of catabolism that may be appropriate for some uses, but which is not suitable in the context of Pütter's equation was implied by Hochachka (1969), who wrote that *"during initial phases of catabolism, large molecules are broken down to yield, apart from CO_2, and H_2O, a quite restricted group of small organic molecules, liberating about one-third of the available free energy in the process."* Here, the problem is that this definition of the catabolic term includes processes wherein the components of the food of a WBE (along with its denatured proteins) are used as substrate for the Krebs Cycle, at the end of which the ATP is generated that is used for the synthesis of new proteins, i.e., anabolism.

To avoid contradictory and overlapping definitions and the confusion they create, the catabolic term, at least in the context of Equation (4.1), must be conceived as 'removing' proteins from the stock of 'live' proteins through a process that does *not* require oxygen, or more precisely, that does not consume oxygen. Such a process exists, and although it is generally overlooked, it is ubiquitous in living organisms: the spontaneous denaturation of proteins, which occurs at all temperatures but increases at both high temperatures and very low temperatures.

The warm and cold denaturation of proteins

Disruption of the highly ordered quaternary structure or conformation of globular proteins, such as enzymes, from their natural or 'native' states is called ´denaturation.´ Denaturation occurs spontaneously in biological organisms, either due to the disruption of weak chemical links that stabilize the native state conformation or from preferential hydration of the denatured state. Higher-than-average denaturation rates are easily achieved upon heating (heat denaturation) via an increase of the kinetic energy of the random Brownian motion of proteins in aqueous solutions. Indeed, the link between temperature and denaturation is implied in the very fact that temperature reflects the average kinetic energy of the moving and colliding atoms of a substance. Denaturation also occurs at low temperatures (cold denaturation) due to hydrophobic interactions among protein nonpolar groups, which, somewhat surprisingly, favors their hydration in the denatured state (Privalov 1990).

Consequently, proteins in aqueous solutions, that is, in the bodies of WBE, have half-lives that are U-shaped functions of temperature, such that higher or lower than optimal temperatures increase their rates of spontaneous denaturation. Proteins that lose their quaternary structures as a result of a random collision with another molecule (which increases with temperature) or because the surrounding water is too cold cease to function, and in the overwhelming majority of cases, need to be resynthesized. Spontaneous protein denaturation explains why the metabolic rates of WBE increase with temperature, both above and below 4 °C (the temperature at which hydrogen bonding in pure water is the strongest and hydration of protein nonpolar groups is the weakest). In the aqueous solution that forms the bulk of the content of living cells, different proteins will have their stability optima at different temperatures, but their relationship to temperature will remain U-shaped.[116]

Once a protein has lost its relatively fragile quaternary structure, it becomes a *"random coil"* (Smith *et al.* 1996), i.e., an essentially useless jumble of amino acids (**Figure 4.4**). As such, denatured proteins become part of a WBE's amino-acid pool, where they join the amino acids originating from the food of that animal.[117] None of the oxygen supplied by

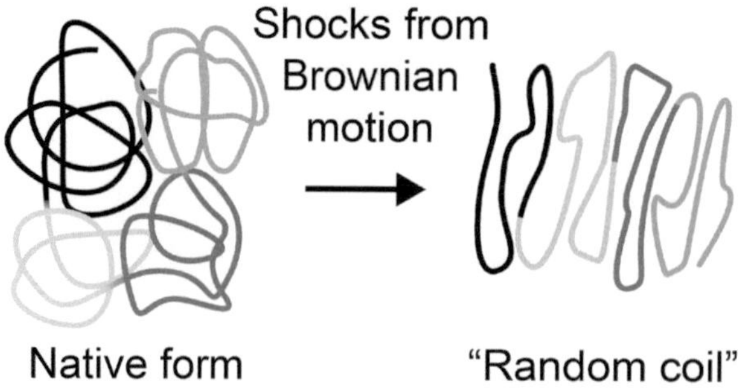

Figure 4.4. Schematic representation of how a globular protein with its intact quaternary structure is denatured into a "*random coil*" (Smith *et al.* 1996) of amino acid by the kinetic shocks of Brownian motion occurring at *all* temperatures, including 'benign' ones; see text.

a respiratory surface is required to remove a substantial part of the stock of 'working' proteins from the body of a WBE. These denatured proteins, however, must be immediately replaced if the organism is to maintain itself.

The biochemical basis of the mechanism of protein denaturation is summarized in the following:

(i) The quaternary structure or conformation of proteins is typically stabilized by weak chemical linkages, such as hydrogen and disulfide bonds, and hydrophobic interactions among nonpolar groups within the polypeptide chain (Privalov 1990). These weak chemical interactions are easily disrupted in aqueous solutions (i.e., within living bodies) by the random Brownian motion of water molecules surrounding the protein molecules, or the input of thermal energy that can overcome the strength of these bonds (~ 0.4 - 4 $kJ \cdot mol^{-1}$).[118] The breaking of many of the weak intramolecular linkages responsible for the highly ordered quaternary structure of a protein in its native state results in protein denaturation.

(ii) Protein denaturation typically proceeds to irreversible degradation and hence to biological inactivity. In some proteins, such as with electrostatic hinges, the reversible process of renaturation can occur in the presence of stabilizing ionic ligands, demonstrating the link between protein structure and function (Yan *et al.* 2018). Protein denaturation can be quantified by the time taken for a given quantity of proteins with well-formed quaternary structure to decrease by half, which is the definition of the aforementioned half-life of a protein. Protein half-lives decrease with increasing temperature, as increased Brownian motion disrupts the weak bonds which stabilize the quaternary structures. Half-lives can also decrease with decreasing temperature in the case of cold denaturation. Note also that different proteins have different half-lives and the same protein can have different half-lives in different environments (see, e.g., Kuhar 2009).

(iii) The quaternary structure of proteins is also stabilized by hydrophobic interactions, which consist of short-range attractive van der Waals interactions among protein nonpolar groups and long-range repulsive hydration of these nonpolar groups (Privalov 1990). Once disrupted, van der Waals interactions, being delocalized many-body interactions (Distasio *et al.* 2014), are not easily re-formed. As a result, once a protein's quaternary structure is lost, it is not easily repaired, i.e., it is thermodynamically possible, but kinetically improbable, as the stable structure requires achieving a conducive transition-state configuration of proteins. Once they have lost their quaternary structure, proteins usually need to be re-synthesized, which requires an input of energy (ATP) to overcome the (high) activation barrier for the protein synthesis reaction.

(iv) The stability of globular proteins is maximal at the temperature at which the degrees of entropy (disorder) of the native and denatured states are equal and the structure is stabilized only by the enthalpy (configurational energy) difference between these states (Privalov 1990). In aqueous solution, proteins are most stable at 4 °C, the temperature at which hydrogen bonding in water is the strongest and hydration of the protein polypeptide chain, which consists mainly of nonpolar groups, is the weakest. At temperatures above and below this point, protein stability decreases, i.e., protein denaturation occurs with both heating and cooling, yielding a distinct protein stability curve with respect to temperature. Spontaneous heat and cold denaturation reflect the thermodynamic balance between the configurational energy (or enthalpy), which is stabilized mainly by hydrophobic interactions, and the conformational entropy, or the tendency of the Universe to evolve toward disorder.

(v) Disruption of the native protein structure on heating, called heat denaturation, proceeds with heat absorption and, consequently, with increases in molecular enthalpy. Disruption of the native structure upon cooling, known as cold denaturation, proceeds with the release of heat and, hence, with enthalpy decreases (Privalov 1990). Heat denaturation requires only a small input of energy, on the scale of thermal fluctuations in Brownian motion, whereas cold denaturation releases a small amount of energy as heat. No additional external source of chemical energy, such as from ATP, is required, as the activation barrier to protein denaturation is low. Consequently, the folded and unfolded protein states often occur in a dynamic near-equilibrium state, with more denaturation than renaturation in growing organisms, and equilibrium in non-growing organisms.

(vi) Low temperatures (< 4 °C) also can cause loss of quaternary structure or "*cold denaturation*" (Privalov 1990). Cold denaturation is a general phenomenon caused by the specific and strongly temperature-dependent interaction of protein nonpolar groups with water. Hydration of these groups is favorable thermodynamically, i.e., the Gibbs energy of hydration is negative and increases in magnitude as the temperature decreases. As a result, the polypeptide chain, tightly packed in a compact native structure, unfolds at a sufficiently low temperature, exposing internal nonpolar groups to water.

The hydrophobic interactions stabilizing the quaternary structure of native proteins are an enthalpic balance between the temperature-dependent repulsive hydration of protein nonpolar groups and the temperature-independent attractive van der Waals interactions among these nonpolar groups. At low temperatures, the weakly attractive van der Waals interactions can be disrupted, leading to unfolding and denaturation. For example, Fraser *et al.* (2022), based on work on limpets and the notothenioid fish *Harpagifer antarcticus,* state that "*protein metabolism data for Antarctic invertebrates show low rates of protein synthesis and unusually high rates of protein degradation. Additionally, in Antarctic fish,*

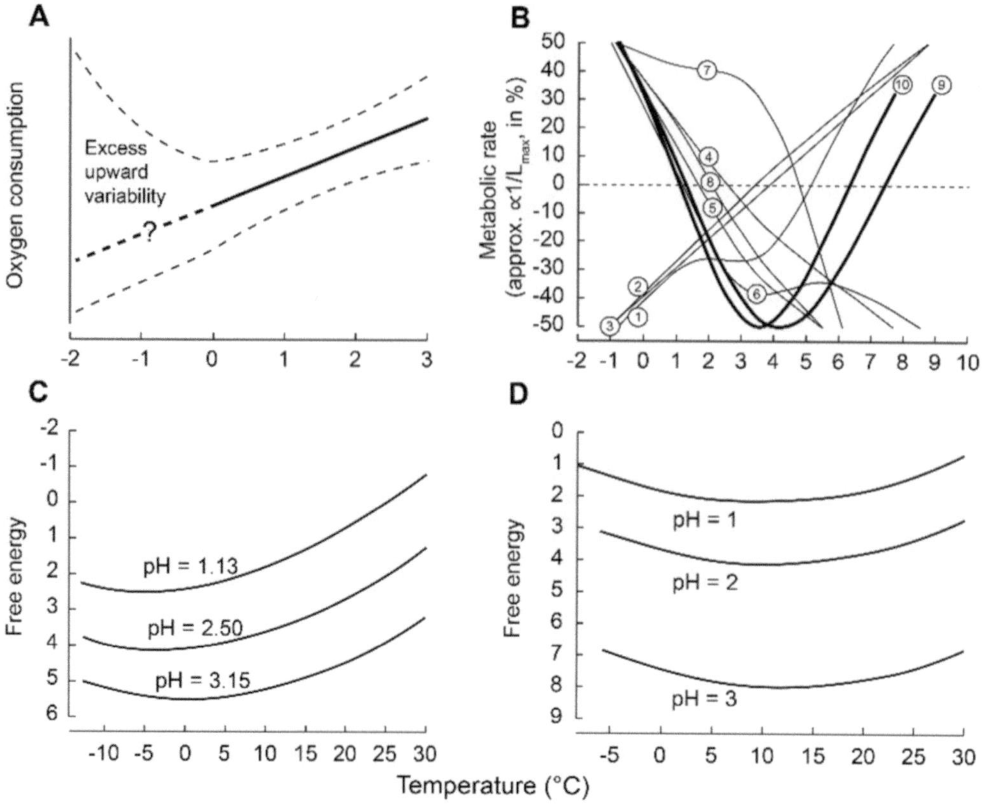

Figure 4.5. Relationships between oxygen consumption (and/or related processes) and temperature. **A**: Illustrating how Wohlschlag (1964) chose to present his data, with 'cold adaptation' interpreted as inducing *"variability."* **B**: Plots of non-linear regressions of maximum length vs. water temperature of the 10 fish species in Norwegian trawl surveys (adapted from figure 3 in Lavin *et al.* 2022), scale-inverted and rescaled as % deviation of their means, such as to make visible their relationships to plots of oxygen consumption vs. temperature. **C** and **D**: plots of the temperature-related stability (as a function of the Gibbs free energy) of two examples viz., ribonuclease, an enzyme (**C**) and chymotrypsinogen, an enzyme precursor (**D**). Note the U-shape of these plots, which were adapted from Brandts (1967).

increasing evidence suggests a lower frequency of successful folding of nascent proteins and reduced protein stability."

The 4 panels of **Figure 4.5** present evidence for the effects of protein denaturation on fish growth. **Figure 4.5A** illustrates that Wohlschlag (1964) interpreted his own data as increased *"variability"* at low temperature. This interpretation, for which he did not provide a mechanism, is probably one reason why his notion of *"cold adaptation"* became controversial. Another reason was the suggestion that his experimental set-up may have stressed the fish whose respiration he was measuring, and their elevated oxygen consumption was deemed an *"artifact"* (Holeton 1973).

Figure 4.5B presents new, if indirect, evidence for cold adaptation, i.e., elevated oxygen consumption, in 10 fish surveyed in cold Norwegian waters (Lavin *et al.* 2022). There, only three species exhibited a decrease in maximum observed length with increasing temperature, (as shown here as an inverse relationship) in line 1 (spotted wolffish,

Anarhichas minor)[119], 2 (cusk, *Brosme brosme*) and 3 (Norway redfish, *Sebastes viviparus*), compared with five species whose maximum length increased with temperature (here shown again as an inverse relationship), i.e., lines 4 (capelin, *Mallotus villosus*), 5 (Greenland halibut, *Reinhardtius hippoglossoides*), 6 (Golden redfish, *Sebastes norvegicus*), 7 (daubed channy, *Leptoclinus maculatus*) and 8 (polar cod, *Boreogadus saida*), while two species, bold lines 9 (Atlantic wolffish, *Anarhichas lupus*) and 10 (beaked redfish, *Sebastes mentella*) spanned the range from 1 to 10 °C with the very U-shaped relationships that best reflect the underlying biochemical mechanisms of protein denaturation.

Figures 4.5C and **4.5D**, finally, show two examples, among many that could be shown, of protein stability versus temperature (modified from Brandts 1967); **Figure 4.5C** refers to ribonuclease and **Figure 4.5D** to chymotrypsinogen. Note the U-shape of the protein stability curves (but also note the inverted scale).

At this point, we must clarify a generally misunderstood concept, whose relevance to the growth of WBE is generally missed. It is widely accepted that proteins in aqueous solutions undergo denaturation at very high temperatures (see, e.g., Nelson and Cox 2017). However, it is usually not realized that all proteins undergo spontaneous denaturation at *all* temperatures – even at their optimal temperature – and that this is why proteins have half-lives, ranging from a few hours to several days, even in homeotherms. This implies that while denaturation processes are minimized at a WBE's optimal temperature, a fraction of its stock of proteins, in any instant, still spontaneously loses the quaternary structures required to fulfill their specific roles, e.g., as an enzyme to catalyze a reaction, or, as in the case of hemoglobin, to transport oxygen.

Yet many theoretical explanations of the impact of temperature on ectotherm metabolism rely on categorical distinctions between biochemical mechanisms that are limited to physiological effects under 'stressful' or 'benign' conditions, as if these were two fundamentally separable categories (Lonthair *et al.* 2024a; Jutfelt *et al.* 2024; Angilletta and Dunham 2003; Angilletta *et al.* 2004). While it is clear that some biochemical effects of temperature on cellular processes will only become effective under extreme conditions, the GOLT points to the rich literature on protein half-lives that makes clear that these half-lives do not fundamentally, but often rather gradually, differ between benign and stressful condition (Bojkowska *et al.* 2011; Dice 1987; Zhou 2004). In fact, the very definition of certain thermal conditions as stressful depends on the reaction of cellular proteins at specific temperatures.

As shown above, one of the major drivers of maintenance costs in cells and organisms is the re-synthesis of denatured proteins, which have lost their quaternary structure. This is not a process limited to extreme conditions; it occurs constantly – the difference between 'benign' and 'stressful' thermal ranges is the acceleration of these denaturation processes under increasing temperatures. Rather than assuming protein denaturation as a factor contributing to increased maintenance costs under stressful and extreme conditions, the GOLT points to evidence for spontaneous denaturation occurring under *all* thermal conditions, but increases with temperature and, below 4 °C, with declining temperatures (see above).

Strictly speaking, the correlate of higher temperatures, i.e., an increase in Brownian motion, should not be understood as a *result* or a *consequenc*e of temperature. Rather, it is far more accurate to identify this increase in particle movement *as* a heat increase (Einstein 1905).[120]

The considerations above make sense from an evolutionary perspective, as organisms perform best when they are within a narrow optimal thermal range, representing the optimal energy trade-off between their rates of anabolism and their maintenance requirements. The underlying chemistry that supports this conclusion involves a similar trade-off between short and long-range forces of attraction and repulsion within and between biological macromolecules, as well as between them and their surrounding environment in the organism, which consists largely of water, other organic and inorganic molecules, and ions. The observed structure-function dualism of proteins (Yan *et al.* 2018) reflects the evolutionary adaptation of molecular processes within organisms to their biochemical environments, which include temperature, pH, ionic potential, solvents, etc.

The following points summarize what we believe is strong evidence for spontaneous protein denaturation being the essential contributor to the catabolic term ($-\ kW^m$) of Pütter's equation:

(1) Proteins undergo spontaneous heat and cold denaturation in aqueous solution, exhibiting stability curves as a function of temperature with a minimum at 4 °C. Denatured proteins in living organisms typically must be resynthesized, leading to elevated metabolic rates, increased oxygen demand, and smaller maximum sizes;
(2) In WBE, oxygen consumption is a concave function of temperature, with a minimum at 4 °C. If oxygen is limiting the metabolic rates of WBE and if spontaneous protein denaturation is responsible for the catabolism term of Pütter's equation, then their asymptotic growth (balance of anabolism and maintenance requirements) will reflect this temperature dependence;
(3) Oxygen demand in WBE increases with temperatures above 4 °C, reflecting elevated rates of protein re-synthesis in response to spontaneous protein denaturation. Above 4 °C, when temperatures increase, the maximum sizes of WBE decrease;
(4) Cold denaturation, which tends to occur below temperatures of 4 °C, also increases oxygen consumption and limits growth, owing to elevated metabolism required to resynthesize denatured proteins. Arctic and Antarctic fish occurring at temperatures below 4 °C are smaller than predicted by a projection of their size-temperature relationship to low temperatures (Pauly 1979, 1980a; Lavin *et al.* 2022);
(5) No other growth-negating process has been described in the literature that is consistent with Equation (4.1) and does not lead to contradictory conclusions about growth processes in WBE;
(6) Finally, sloughing skin and other mechanical processes also contribute to losses of body substances and weight, but these should generate 'replacement costs' that are likely minor compared to spontaneous denaturation.

Different proteins have different half-lives, but if it can be assumed that the protein composition of a young WBE is roughly similar to that of an older WBE of the same species, the abundance ratio of different proteins will remain similar. This implies that the same denaturation processes are occurring throughout the body and that the exponent of the negative term in Pütter's equation, $m = 1$, can be neglected. That is, maintenance requirements are proportional to the weight of WBE, in contrast to anabolism.

The effect of increasing protein denaturation provides a causal mechanism for temperature-size rules that state that temperatures higher than 4 °C lead to smaller

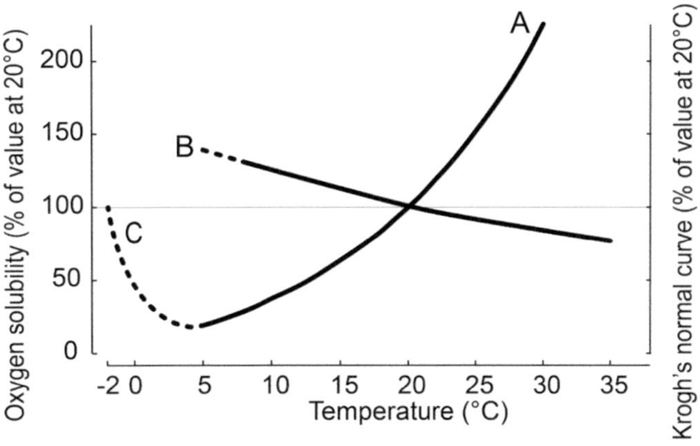

Figure 4.6. Above 4 °C, the oxygen demand by fish and other water-breathing ectotherms increases with temperature (line **A**) mainly due to increased protein denaturation, while the oxygen dissolved in the water declines (line **B**, with the difference between sea and freshwater being too small to show here). This reduces the oxygen supply to water-breathers and therefore their ability to re-synthesize denatured proteins. Below 4 °C, cold denaturation increases with decreasing temperature (dotted line **C**). Based on Ege and Krogh (1914) and Pauly (1979, 2023).

maximum sizes for WBE. More complex hypotheses to explain the smaller sizes of WBE at high temperatures are largely superfluous (Pauly 2021a), at least if Ockham's Razor still cuts.

The problems that fish and other WBE experience at extreme temperatures, especially now, given that oceans and freshwaters are warming and deoxygenating, are summarized in **Figure 4.6**. Not only does warmer water contain less oxygen (**Figure 4.6**, line **B**), but warmer water also increases the oxygen demand of fish and other WBE (**Figure 4.6**, line **A**).[121] On the other hand, at extremely low temperatures, there is a twist, derived by Pauly (1979; **Figure 4.6**, line **C**) from various growth and respiration datasets, including that of Wohlschlag (1964) based on nototheniid fishes. Wohlschlag's work was highly contested at the time (see, e.g., Holeton 1973), and therefore, the concept was left "*in profound hibernation*" (Pauly 2019). Nevertheless, signs of 'cold adaptation' were later detected in a set of natural mortality estimates (Pauly 1980a), which are also related to growth and temperature, and therefore to respiration and protein denaturation.

Now, given the contributions of Forster *et al.* (1987), Torres and Somero (1988), White *et al.* (2012), Sanfelicea and Temussia (2016), Fraser *et al.* (2022), Lavin *et al.* (2022) and Lin and Costello (2023)[122], we can assess why "*neither the careful experimental work of Holeton (1973, 1974) nor theoretical arguments (Clarke* [1980a], *1980, 1991, 1993) resulted in the demise of the concept of metabolic cold adaptation*" (Clarke and Johnson 1999). The reason for the resilience of the cold adaptation concept is that researchers continue to stumble upon its effects while conducting empirical studies of WBE, which primarily occur in waters below 4 °C. Thanks to Privalov's (1990) review, we now also have a mechanistic explanation that was previously lacking, and it is compatible with the GOLT.

5

Temperature and its impact on water-breathing animals

Synopsis: This chapter applies the theoretical insights from the previous chapter to real-world scenarios, exploring how temperature influences the physiology, ecology, and growth of fish. It examines the role of temperature in intraspecific and interspecific differences in growth rates, the widespread occurrence of 'Heincke's law' (the trend of larger fish occurring in deeper waters), and the impacts of heatwaves and hypoxic events on size-dependent survival. By connecting theory with empirical observations, this chapter provides a comprehensive understanding of how temperature shapes the size and age composition of fish populations in a changing climate.

Temperature, physiology, and ecology

Temperature shapes all the major traits of biological organisms, and of ectothermic animals in particular. In the previous chapter, we examined the biochemical foundations of the mechanisms by which temperature acts on metabolic processes. Identifying these processes allows for a mechanistic interpretation of the modified Pütter's equation that underlies our growth model, thereby 'translating' the biochemical level into the level of organisms' life histories and their ecological and environmental interactions. Bridging these different levels requires considering the diversity of life and its phenomenological manifestations in the real world.

In this chapter, we apply our explanatory model to physiological and ecological phenomena and connect these different levels of causation. However, a fundamental distinction has to be made before this general model can be used to interpret biological data regarding temperature differences between intra- and interspecific traits, thus *within* and *between* species. The impact of temperature on organisms with identical or similar genetic compositions is fundamentally different from that of species that have adapted to divergent thermal conditions and have evolved specific mechanisms to optimize their life-history traits in their respective environments. This chapter discusses the differential influences of temperature at intra- and interspecific levels, illustrating their consequences in different contexts and based on well-based phenomena.

First, we will examine why optimal growth temperatures decrease during ontogenetic growth trajectories and why water-breathing ectotherms grow better under colder conditions once they reach a larger size. Second, we explain the universally observed fact that bigger fish (within a given species) inhabit deeper water layers (and *vice versa*). Third, we will apply our model to interspecific and biogeographical patterns and discuss why the evolution of large body sizes is less likely under warmer conditions. The final sections of this chapter will examine the impact of temperature on aerobic scope and thermal tolerance. This topic is critical in the current climatic transition, where heatwaves are becoming increasingly likely to occur, and fish are often among the first victims (Cannell *et al.* 2019; Müller *et al.* 2023). We will conclude by discussing the widely observed phenomenon that fish follow their preferred temperature if they have the opportunity; for example, if their habitats warm up, they have to move to areas with the temperature these fish prefer. All these phenomena are already well-studied, and sufficient data are available; however, their description often lacks a more general theoretical foundation that allows for coherent mechanistic explanations. We argue that providing such a foundation and a better mechanistic understanding of the relation between temperature and ectotherm growth is possible. A more comprehensive and general explanatory model is needed in times of drastic climate change, and we believe the GOLT can and will play an essential role in this context.

Temperature and fish growth differences within species

The ichthyological literature is replete with reports of negative correlations between water temperature and optimal fish growth at greater sizes; as fish grow bigger, they seek lower temperatures. This pattern was confirmed by a meta-analysis based on 52 datasets on more than 2,000 species (Lindmark *et al.* 2022). The decrease of optimal growth temperature during ontogeny can be seen as one facet of the temperature-size rule, but it allows for interesting mechanistic inferences. While many authors working on this phenomenon are reluctant to relate their findings to more general underlying mechanisms, we argue that the modified Pütter's equation allows for a coherent interpretation of the decreasing optimal growth temperatures with increasing body size.[123]

Suppose we interpret the rightmost term of Pütter's equation as the sum of breakdown processes that require repair or re-synthesis of denatured protein. These processes are accelerated at higher temperatures. Since repair activities are necessary for the entire body (if we neglect ontogenetic allometries in the composition of body substances), we can assume that breakdown metabolism (kW) scales proportionally to body mass:

$$dW/dt = HW^{d} - kW \qquad \text{...(5.1)}$$

As previously discussed, anabolism scales, in adults, with an exponent lower than 1 if understood as a surface-limited process. On the other hand, biochemical breakdown processes – such as the denaturation of proteins, whose quaternary structure is directly impacted by temperature – will always cause maintenance costs (and thus repair activities) to be higher in warmer environments.

Because the anabolic term of the above equation (HW^{d}) is also temperature-sensitive, ectotherms will initially grow faster at higher temperatures. However, as increasing body mass also increases maintenance costs, this initial growth benefit is quickly undone—a dynamic that follows the temperature-size rule. But if habitats with lower temperatures are available during ontogeny, they would allow for a reduction in maintenance costs and maintain a positive balance in Equation (5.1), i.e., the fish would grow. The seemingly complex phenomenon of decreasing thermal growth optimum can be explained by the relationship of only a few parameters. The reason for the assumed complexity of the relationship between ontogenetic growth and temperature is the assumption – which we believe is invalid – that intraspecific metabolic scaling is not a result of decreasing energetic investment in growth but rather a reduction in maintenance costs (Lindmark *et al.* 2022; Marshall and White 2019). While a size-dependent decrease in maintenance costs is the foundation for differences in interspecific scaling (i.e., between species), the assumption of the same mechanism for ontogenetic scaling leads to erroneous results that only further complicate the relationship between available data on WBE of different life stages and metabolic requirements under different conditions.

Deeper water – bigger fish: Heincke's law is ubiquitous

The observation that larger fish are typically found in deeper and cooler water than their smaller conspecifics of the respective population has been noticed for centuries. It is sometimes referred to as 'Heincke's law' (**Figure 5.1**). German fishery biologist Friedrich Heincke

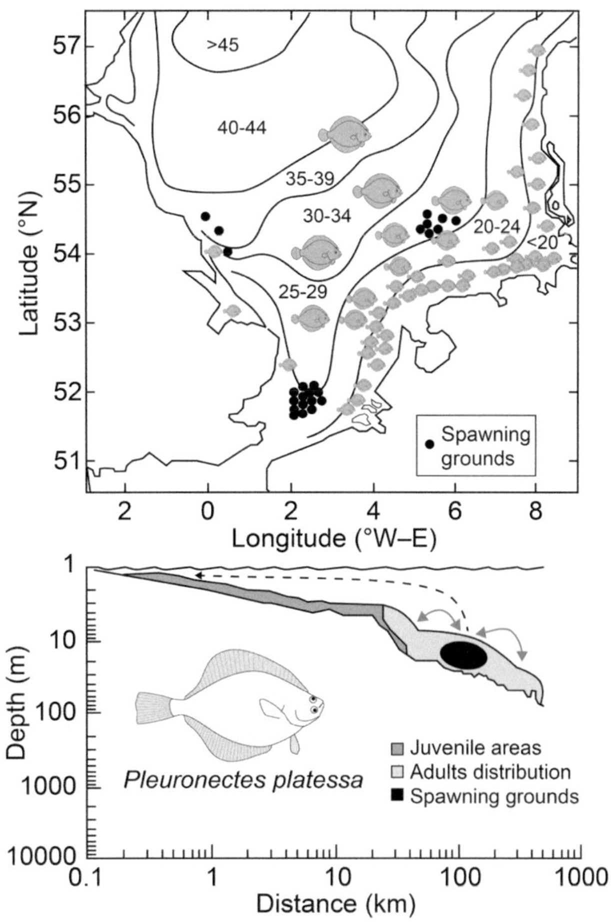

Figure 5.1. Illustrating Heincke's 'law,' i.e., the relationships between water depth and the size of fishes. **A:** Distribution of plaice (*Pleuronectes platessa*) in the North Sea, adapted from Garstang (1909), with mean sizes (TL; cm) given for each depth isobar. **B**: Typical depths transect (approx. 53° N, 80° E to 56° N, 3° E). Dark grey represents the juvenile distribution, light grey represents the adult range, and black indicates spawning strata (modified from Zeller and Pauly 2001).

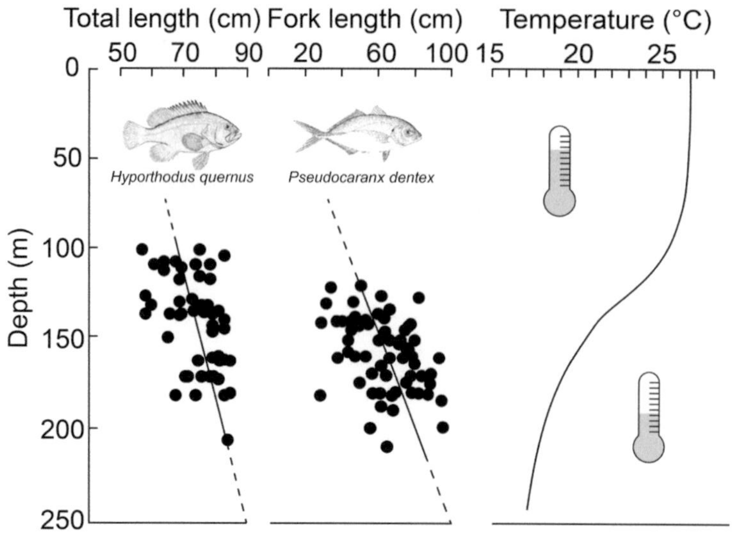

Figure 5.2. Increase of mean size in two species of reef fish as a function of depth and temperature at French Frigate Shoals, Hawaiian Islands; the dashed lines are extrapolations, which make the trends more visible (modified from Longhurst and Pauly 1987).

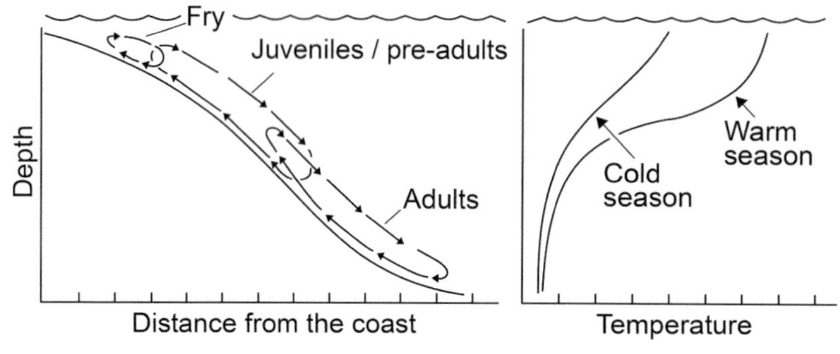

Figure 5.3. Seasonal offshore – inshore movements undertaken by non-migratory fishes to maintain themselves at roughly similar temperatures throughout the year. Note that larger fish tend to occur in deeper water than smaller ones, whatever the season (modified from Pauly 2019).

(1852–1929) described the size-dependent habitat use of fishes in the North Sea and found that the European plaice tended to appear in deeper zones of the North Sea with increasing body size (Heincke 1913; see also Pauly 2019).[124] (Also note, though, that larger plaice may occur in shallower water during the cold season, as generally suggested in **Figure 5.3**).

This pattern has been widely described and confirmed for many species (e.g., van Hal *et al.* 2016; Morita *et al.* 2010; Longhurst and Pauly 1987). **Figures 5.2 – 5.4** illustrate this pattern in tropical and temperate marine fish, including striped bass (*Morone saxatilis*) for the East Coast of the U.S.

The same pattern has also been widely documented for freshwater fishes and has sometimes been described as the *bigger fish–deeper* habitat pattern (Harvey and

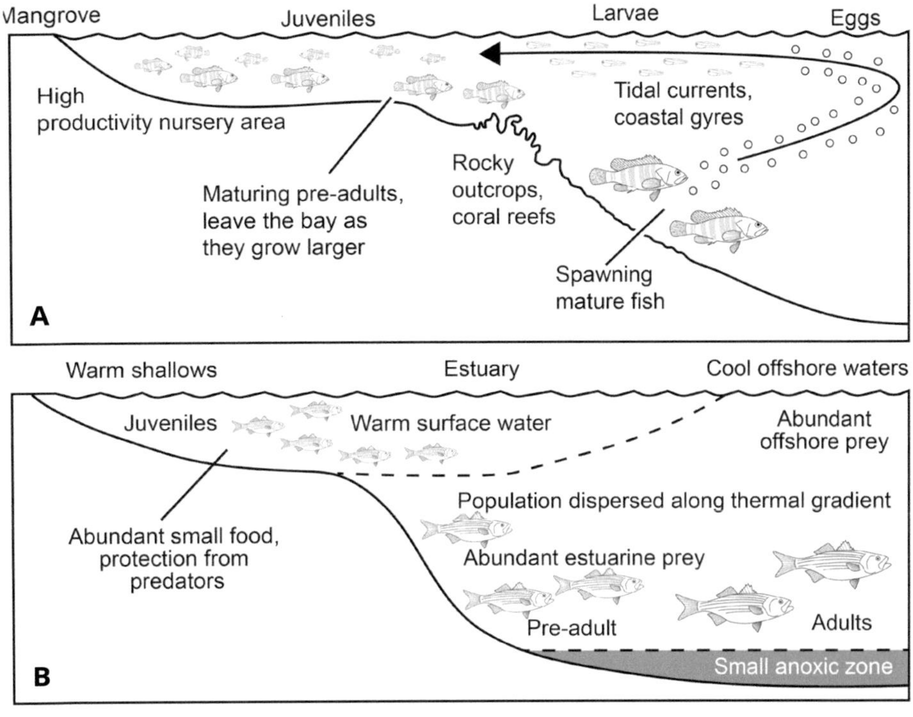

Figure 5.4. Depth distribution of different fish sizes. **A**: In San Miguel Bay, Eastern Luzon, Philippines (Pauly 1982b). **B**: along the U.S. East Coast, the habitat of the striped bass (*Morone saxatilis*); modified from Coutant (1986).

Stewart 1991; Maki-Petäys *et al.* 1997; Vehanen *et al.* 1999; Baran *et al.* 2005; Riley *et al.* 2006; Kaspersson *et al.* 2012; Arranz *et al.* 2015). Complementary to this observed regularity, others have postulated a *"smaller fish – shallower habitat"* pattern (Oliveira *et al.* 2022). In the context of our discussion of decreasing thermal growth optima during individual growth trajectories, the explanation for this phenomenon may seem straightforward, as deeper water strata are cooler and offer larger fish more opportunities to grow further. While similar mechanisms were already proposed in the literature of the early 1980s, mainly by Coutant (1985, 1986; Coutant and Carrol 1980), who primarily focused on varying oxygen concentrations in different water layers, later research on this topic explored a wide range of possible explanations – typically with inconclusive results. An influential hypothesis was the assumption that changing habitat use during ontogeny resulted from size-selective predator avoidance (see, e.g., Schlosser 1987; Harvey and Stewart 1991). While it is true that large fish species that depend on regular access to the water surface often develop particular adaptations to fend off terrestrial predators (e.g., in the air-breathing *Arapaima gigas*, whose skin is covered with extremely tough scales), size selectivity cannot explain why younger fish prefer the higher water strata – we should not forget that almost every coastal and freshwater ecosystem in the world accommodates birds that specialize on small fish as prey (see, e.g., Sheaves 2001).

Vehanen *et al.* (1999), who studied brown trout in an aquaculture set-up, suggested that the habitat of different size classes might be structured by intraspecific competition: *"This bigger fish – deeper habitat relationship is relatively common among stream fishes and has usually been interpreted as an antipredator tactic to avoid encounters with size-selective predators* [...]. *Another potential mechanism suggested to explain spatial segregation among differently sized brown trout is intercohort competition* [...]. *An obvious question is then: do small trout occupy low-velocity microhabitats along stream margins because they actively prefer these areas, or because they do not have access to more preferred habitats due to intercohort competition from larger trout?"*

While intraspecific competition could indeed be a factor that shapes size-structured habitat use in territorial and aggressive species, it is an unlikely candidate as a universal pattern for several reasons: first, not all fish species exhibit intraspecific aggression or territorial behavior. Second, territorial aggression is often directed against conspecifics of similar sizes, especially in the case of competition for shelter sites, which are only available to specific size classes (Harrington 1993; Shulman 1985). Most importantly, it is, at best, an incomplete explanation since it leaves open the question of why different size classes would compete for a particular depth layer.

For marine fishes, some authors have sought to explain the preference of larger fish for deeper waters, attributed to changing trophic relationships or size-selective exploitation by modern fisheries (see, e.g., Frank *et al.* 2018). Still, at least the latter point can be easily countered by the existence of this pattern in unexploited fish populations (see, e.g., Pauly 1980b).[125]

While trophic differentiation is another likely candidate to play a decisive role in size-structured habitat use – at least in freshwater habitats where many species are insectivores as juveniles – the factor that most clearly separates deep and shallow aquatic habitats – temperature – is often overlooked. The groundbreaking idea that temperature and oxygen concentrations, rather than food supply, shaped the size-dependent thermal niches was developed by Coutant and Carrol (1980) and Coutant (1985, 1986). It was explicitly formulated as "*a speculative hypothesis*" at the time (1985), and the author may not have initially realized the stringency of this argument.

As Coutant could show, niche specialization is not only a way to optimize growth conditions, but different size classes also critically depend on specific thermal niches. The affinity of larger fish to deeper water strata becomes a fatal liability once this specific niche suddenly becomes oxygen-depleted, and the fish can move between zones that are either too warm or contain insufficient oxygen (**Figure 5.5**).[126] Specific thermal preferences change with age and size. For fish at the lower end of their size range, this implies a decreased scope for growth. It is crucial to note that larger individuals, especially those close to their thermal maximum, are particularly vulnerable to lethal heat stress, a risk that we must consider to understand and protect fish populations. The impact of temperature on the size-specific aerobic scope will be further discussed below; clearly, Coutant's identification of oxygen- and temperature-specific niches was an essential step toward better understanding this phenomenon.

The well-established negative correlation between body mass and optimum growth temperatures may allow for predicting fish growth and a population's habitat. The higher the degree of a habitat's temperature stratification that provides for a fish population's thermal niche, the greater the scope for growth of its individuals throughout their life history. In other words, access to a broader thermal range increases the scope for growth.[127] Fishes specialized in hunting near the water surface may live in highly stratified thermal environments, but they will experience more constant temperatures as they never reach

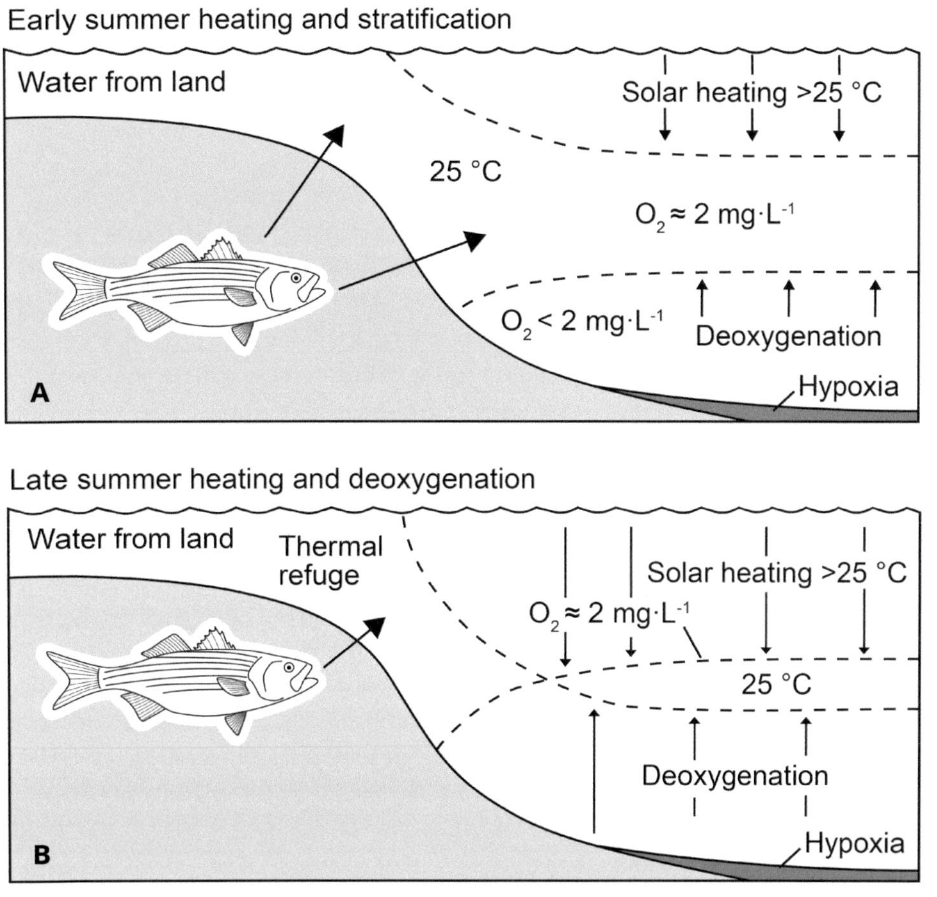

Figure 5.5. Schematic representation of the distribution of adult striped bass (*Morone saxatilis*) under two temperature and oxygen regimes along the US East Coast. **A**: Throughout the year, except late summer. **B**: In late summer, when the deoxygenation of their habitat forces them out of their size-specific thermal niche (modified from Coutant 1986).

the lower water regions. Another example is nocturnal species that hide under deep rocks during daylight and roam the water at night. They will experience a very different thermal environment compared to other fish in the same water body. In isolated habitats where the range of thermal niches (and seasonal oscillations) is minimal, such as shallow tropical pools, we should expect to see more species of similar sizes. Consequently, the lower the degree of temperature stratification of a fish population's habitat, the more limited the scope for growth in this specific habitat will be by the constant temperatures of their particular environment. They can grow fast as larvae and juveniles, but their growth will approach an asymptote earlier than in populations in more stratified habitats.

Temperature and fish growth differences between species

While temperature impacts the scope for growth *within* species, biogeographical patterns suggest similar effects on interspecific levels. Growth in temperate populations is typically slower than in tropical waters, but it eventually approaches a larger asymptote

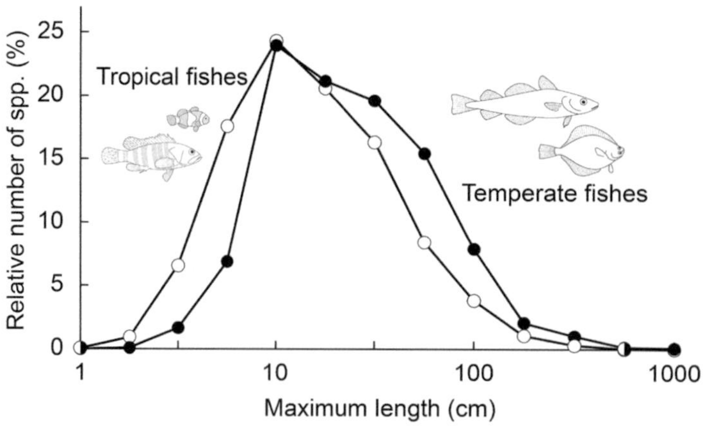

Figure 5.6. Relative numbers of species of tropical vs. temperate marine and freshwater bony fish species vs. their maximum length (in cm; log scale), showing that tropical fish are generally smaller than their temperate counterpart (data from FishBase, based on 20,062 tropical and 3,143 temperate species).

(Henderson 2005; Pauly 1998b). Maximum sizes in temperate bony fishes are consistently higher than those from tropical regions (**Figure 5.6**).

The body sizes of tropical freshwater fishes vary everywhere between specific habitats and ecozones. Examples of habitats with minimal temperature variation are tropical rainforests, whose diurnal temperature variation is limited by the isolating vegetation. Especially in smaller water bodies, the thermal niche of rainforest fishes is very narrow. Indonesia and the wider Malay Archipelago comprise 12,000 to 18,000 islands (depending on their definition) and a highly diverse freshwater fauna that differs strongly from island to island. Many freshwater fishes inhabit small and isolated creeks, peat swamps, and seasonal puddles, and some species can be found in barely submerged leaf litter during the dry season, a survival strategy that is possible mainly due to accessory breathing organs (Hui and Ng 2005). Fishes in such isolated habitats lack access to thermally stratified environments and experience high temperatures throughout their lives.[128] Unsurprisingly, the freshwater fish fauna of Indonesia consists of very small species: almost half of the freshwater taxa's maximum length is below 10 cm. The number of extremely small species is also remarkable, with more than 240 species having maximum lengths ranging from 0.8 to 5 cm (Hubert *et al.* 2014).

While the thermal range of tropical fishes is already lower than in temperate species, i.e., 2-5 °C as opposed to more than 10 °C in temperate regions (Pauly *et al.* 2019), the temperature range in isolated peat swamps and small tropical creeks is among the lowest of all aquatic habitats. Hypoxia-tolerant cyprinids and air-breathing catfishes, loaches, and gouramis are the only fish species that can inhabit such places and reach larger sizes (see Chapter 11). Complementing this observation, large tropical freshwater fish are typically found in the deepest (and coolest) waters they inhabit. While some of the largest freshwater species are tropical, they are either air-breathers or require access to very deep river pools, such as the giant Mekong catfish (*Pangasianodon gigas*) and the giant freshwater stingray (*Urogymnus polylepis*).[129] A comparative study on intraspecific fish sizes in the lower Mekong revealed that individual fish hiding in the deep-water pools of this river system were up to 78% larger than their conspecifics swimming closer to the surface (Baran *et al.* 2005).

While Pütter's equation (5.1) explains the impact of temperature on growth intra- and interspecifically, size and growth differences between species result from selection and evolutionary processes. As the asymmetry in size and growth between temperate and tropical species shows, the evolutionary pathways to larger body size are constrained by temperature, and tropical fishes often rely on specific adaptations to overcome these constraints. Some of these adaptations will be further discussed in Chapters 6 and 11, but reducing maintenance costs is necessary for further growth on both intra- and interspecific levels.

Being the wrong size during heatwaves and hypoxic events

Thermal niches that enable fish to grow larger and separate different size classes also shape other aspects of their physiology and often 'bind' individuals of a given size to a specific temperature and water depth. As Coutant (1985, 1986, 1990) argued, based on research on striped bass (*Morone saxatilis*), larger individuals critically depend on specific temperatures and oxygen saturations at the particular water depths that they inhabit. The rich evidence for this observation in the fisheries literature is also confirmed by a meta-review by Peralta-Maraver *et al.* (2021), who showed that heat tolerance in ectotherms scales intraspecifically with body size. As Coutant described, the dependence on specific thermal niches of different size classes could have fatal consequences if the fish were forced out of their preferred temperature zone due to deoxygenation events. Trying to escape to more well-oxygenated zones, they could suffer heat death in shallower water (Coutant 1985 and **Figure 5.5**).[130]

The prediction that larger fish will be more vulnerable to heat stress than smaller conspecifics is confirmed almost univocally in the fieldwork literature used in fisheries management. A review of a diverse body of fisheries literature, as well as physiological and ecological studies, revealed that 19 out of 20 instructional manuals produced by government agencies on temperature- and hypoxia-induced fish kills indicated the death of larger fish as a sign of hypoxia and/or thermal stress (Müller *et al.* 2023). Most of these manuals contained checklists with different recommendations. If mainly smaller fish were affected, the authors suggest a higher likelihood of chemical wastes or toxic algal blooms as the cause of the die-off. In contrast, if it was predominantly large fish that were affected, this pointed to oxygen depletion. As if to provide evidence supporting the GOLT, some field manuals explicitly referred to the dimensional tension between respiratory surfaces and body volume: "*Small fish usually survive because they have a greater gill size ratio than do larger fish. This enables them to extract enough oxygen from low dissolved oxygen water to survive*" (Missouri Department of Conservation 1994).

Others gave sensible reasons for why large gills would also be disadvantageous because of the higher uptake of substances at different surface/volume ratios (see also Chapter 3 of this book): "*The key difference is gill surface ratio to total body mass. Small fish, relatively more rapid in absorbing water solutes, withstand marginal oxygen depletion but succumb first to toxic materials*" (U.S. Environmental Protection Agency 1972).[131]

The literature suggests that some physiologists could benefit from such simple observations, confirmed repeatedly in the daily fieldwork experience of fisheries managers, fishers, and aquaculturists.[132] A 2008 review on this topic claimed that "*body size per se has little or no impact on the ability to take up oxygen during hypoxic conditions, primarily because the respiratory surface area matches metabolic rate over a wide size range*" (Nilsson and Östlund-Nilsson 2008). Surprisingly, these authors concluded that size should matter because larger individuals have a higher capacity for anaerobic ATP production, which allows them to survive longer during severe hypoxia. Under this scenario, given the high levels of lactate released due to anaerobic metabolism, larger individuals would have an advantage over smaller ones because they would be more resistant to the effects of these waste products.

An extensive body of literature confirms the specific observation of lactate production, as large individuals often rely on anaerobic metabolism (Somero and Childress 1980), while smaller conspecifics usually do not possess this capacity. Indeed, the need for larger fish to resort to glycolysis (or other forms of anaerobic metabolism) to maintain their vital functions is well documented. Goolish (1991), for example, noted that while smaller fish could perform energy-consuming sprints – up to their very physical limits – without using anaerobic resources, larger fish could not perform even moderate physical exercise without glycolysis.[133]

Nilsson and Östlund-Nilsson (2008) turned this observation on its head, concluding that larger fish were not oxygen-limited. While large (old) fish do indeed have a higher capacity for anaerobic metabolism than small (young) ones and are more tolerant of their waste products – their assertion that "*the ability to take up oxygen during hypoxia is independent of size*" goes against existing knowledge and everyday fieldwork experience in inland fisheries and aquaculture. Additionally, it contradicts most available studies on the size effects of hypoxia: as a recent review showed, studies on 26 out of 30 species for which size-specific responses to heat and hypoxic events were documented, suggested a decreasing tolerance to oxygen depletion with body size (Müller *et al.* 2023). For the 4 species that seemed to be the exception, it was unclear for how long the animals were exposed to the hypoxic environment. At the same time, body size also correlated positively with aquatic surface respiration in fish under respiratory stress, a response to aquatic hypoxia similar to air-breathing, but where only the oxygen-rich surface layer of the water is flushed through the gills.

Temperature and the elusive aerobic scope

The central tenet of the GOLT and its predictions on the growth and thermal tolerance of fishes and other water-breathing ectotherms is not that organisms grow until they lose their ability to perform essential maintenance functions, but that their potential or 'scope' for further growth gradually disappears. Since maintenance activities constantly consume energy that is unavailable for growth, this feature presents intrinsic protection against loss of aerobic scope if this parameter is understood as the difference between resting and maximum metabolic rates and the ability to perform routine activities (under optimal thermal conditions). The GOLT agrees with its critics, who have described this mechanism as "*aerobic scope protection*" (Jutfelt *et al.* 2021). We should not forget that routine activities in juveniles fundamentally differ from those of large adults who have developed different lifestyles, especially in sluggish species such as groupers in the sea or pikes in freshwater.

Body size strongly lowers aerobic scope under thermally stressful conditions. At higher temperatures, large (old) fish perform strenuous exercises anaerobically (which they can do only for a short time). At the same time, small (young) individuals can still rely entirely on their aerobic scope (Goolish 1991). The numerous studies on size dependence of anaerobic metabolism all reported a positive correlation with body size (Goolish 1989a, 1989b, 1991; 1995; Kieffer 1995, 2000, 2010; Kieffer *et al.* 1996; Kieffer and Tufts 1998; Ferguson *et al.* 1993, Gingerich and Suski 2012, Somero and Childress 1980; Clark *et al.* 2012 and many others).

If fish size did not influence the capacity to supply the body with oxygen, there would be no need to switch to anaerobic metabolism at larger sizes and/or higher temperatures.

Fisheries managers and anglers widely recognize the impact of temperature on aerobic scope. Awareness of this phenomenon underlies angling regulations in many regions of the world where temperate fish species can encounter temperatures beyond their thermal preferences, for example, in the Southwestern United States. Therefore, many fisheries districts and park services have installed so-called 'hoot owl regulations' that restrict daytime angling on hot days (e.g., Lamborn and Smith 2019; Meyer *et al.* 2023; Smith and Lamborn 2024). While fish can better withstand stressful temperatures at rest, their aerobic scope is severely limited at higher temperatures. The consequences of angling stress are greatly magnified, especially in larger fish, which show high lactate concentrations after being 'played' (Meka and McCormick 2005).[134]

While adult size reduction at higher temperatures may appear due to non-optimal adaptation to a specific thermal environment, reduced growth may also present an adaptive advantage. Final size reduction can also be regarded as a tradeoff that presents a safety margin for larger fish to protect themselves from lethal heat stress. This compromise also comes at a cost, for example, fewer and smaller eggs, since large-bodied fish, especially females, have the highest reproductive output (Hislop 1988; Chapters 7 and 9). However, heatwaves and hypoxic events illustrate the risk of being large (i.e., close to L_∞ in a given species or population) in warm environments, and this risk may outweigh other disadvantages. Such events drastically increased during rapid climate change, becoming a major threat to fish populations in the 21st century.[135] Staying small can, therefore, be seen as a safety margin that protects the aerobic scope of fishes but at the cost of reduced offspring. The evolutionary consequences of this relationship will be discussed in more detail in Chapter 15, but the "*physiological benefits of being small in a changing world*" (Clark *et al.* 2012; see also Daufresne *et al.* 2009) are already rewarded on the level of individual ontogenies.

Losing one's thermal niche or moving within the right temperature

Empirical observations and climate models indicate that mean global temperatures have been increasing over the last 100 years, as assessed in successive reports of the Intergovernmental Panel on Climate Change (or IPCC; see Pörtner *et al.* 2022), and biological responses to these changes have been observed in both terrestrial and marine biomes (Parmesan and Yohe 2003; Perry *et al.* 2005). These responses include changes in size and physiology, geographic range, and phenology at the population, species, community, and ecosystem levels (Cheung *et al.* 2008a, 2008b).

Bioclimatic envelope models have been widely used to predict the effects of climate change on the distribution ranges of terrestrial species (Pearson and Dawson 2003). The application of bioclimatic envelope models to marine ecosystems was initially limited by the scarcity of large-scale biogeographical and ecological data for most marine species. However, this was gradually overcome by the emergence of projects as *AquaMap* (Ready *et al.* 2010) for species distributions and the *Sea Around Us* for fisheries catches (www.seaaroundus.org), both of which utilize data from FishBase and SeaLifeBase, and more recently by *AquaMap* 2.0.[136] The maps provided by these initiatives can then be combined with model predictions of physical oceanographic data (such as temperature, salinity, ice cover) to project future distributions.

For example, a global bioclimatic envelope model was built which uses the temperature preferences of WBE and their affinities to various environmental and ecologically-relevant parameters of over a thousand species of marine fish and invertebrates to predict their climate-induced shift of distribution over a map composed of over 150,000 ice-free spatial cells of ½ degree latitude and longitude (Cheung *et al.* 2008a, 2008b, 2009). Specifically, this model integrates a species' bioclimatic envelope with its population and dispersal dynamics, thereby providing more realistic predictions than conventional bioclimate envelope models that only use environmental variables to predict changes in species distributions. Its first results were global maps of species immigration and extirpations and, by combining the two, of species turnover (Cheung *et al.* 2009). Unsurprisingly, these maps showed that species turnover was higher in polar regions, with extirpations being the dominant change in tropical areas. Moreover, the data in these maps also allowed for mapping the changes to 2050 in potential catches, which suggests possible 'winners' (cold temperate countries, e.g., Norway, Iceland) and losers (mainly tropical countries, e.g., Indonesia) in terms of fisheries catch potential (Cheung *et al.* 2010, 2013c).

The outputs of coupled atmosphere-ocean models suggest that, except possibly in the four major boundary currents, i.e., upwelling areas (Bakun 1990), future oceans will be more strongly stratified, while coastal eutrophication will increase (Rabalais *et al.* 2014). Hence, the future oceans will be warmer and contain less oxygen, while the frequency and intensity of hypoxic areas and events will increase. These effects are predicted to have increasingly severe impacts on fish resources (see contributions in Pauly and Dimarchopoulou 2022). One of these effects will be to amplify the trend toward reduction in the maximum size reached by many fish species, as caused by both fisheries and warming (Cheung *et al.* 2013b, Pauly 2019).

Documenting and understanding fish migrations and range shifts

The classic text by Harden-Jones (1968) on "*Fish Migration*" does not include the word 'temperature' in its index. Yet temperature is, arguably, the most important driver of the migrations of exclusively marine fish,[137] as illustrated in **Figure 5.7**.

The point is that, within a species, fish of given sizes usually have a narrow range of optimal temperatures. If these shift, e.g., latitudinally within the year as they do in Northwest Africa, then it is physiologically more efficient, at least for the adult of larger species (i.e., those with a smaller aerobic scope), to follow this temperature range than to stay put and endure temperatures outside of this range.[138] If, while doing so, these fish encounter successively good places to reproduce and feed, the migrating fish will grow and achieve high biomasses. Fisheries scientists will describe their 'migrations' as biological 'phases' with names such as 'overwintering,' 'spawning,' or 'feeding.' Fish avoid respiratory distress by moving *with* their preferred temperatures and doing their business along the way. Pauly and Keskin (2017) and Pauly (2019, p. 199) illustrate this with the migration of the mackerel *Scomber scombrus*, with the latter account covering other species, if more cursorily.

The range shifts (not 'migrations') that the bioclimatic envelope models simulate (notably that of Cheung *et al.* 2008a, 2008b, 2009) are also not based on fish deliberately choosing to 'migrate toward cooler temperatures,' as the phrase commonly goes. Rather, within their seasonally fluctuating distribution ranges (Lam *et al.* 2008), their (sub-)populations at higher and colder latitudes do better than those at the low-latitude, 'trailing' edge of their

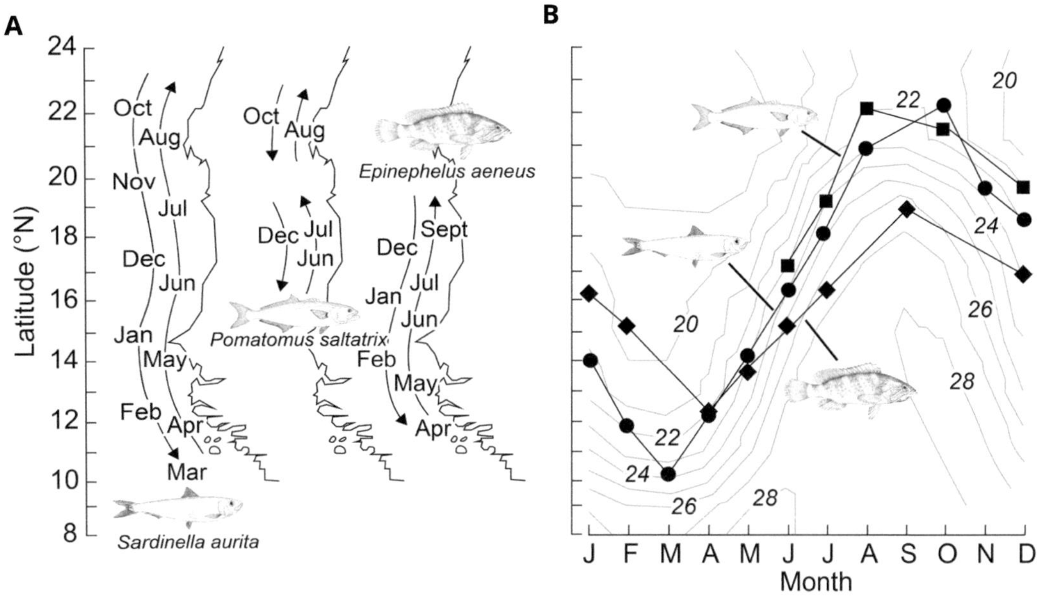

Figure 5.7. Seasonal latitudinal migrations of some Northwest African fishes (adapted from Pauly 1994c). **A**: Summary of information on the occurrence in space (latitude) and time (month) of 3 species, *Sardinella aurita*, *Pomatomus saltatrix,* and *Epinephelus aeneus* (from Boëly *et al.* 1978, Champagnat and Domain 1978, Boëly 1979, Barry-Gérard 1994). **B**: Same as in (A), but plotted against mean monthly temperature (data from ICOADS). The seasonal migrations result in the 3 species remaining in approximately the same temperature (and hence oxygen regime) throughout the year (modified from Pauly 2019).

ranges. Over generations, this leads to the appearance of poleward 'migrations,' although no directed migration took place. Instead, this phenomenon is entirely due to the fact that fish and other WBE cannot change their evolved temperature preferences in a few years or even decades, much loose talk about their 'adaptability' notwithstanding. The reason is that their temperature preference is the result of thousands of years of evolution fine-tuning their physiology, notably by coordinating and adjusting hundreds of enzymatic reactions, each involving hundreds of different proteins, of which each must perform best at the preferred temperature of the species or population in question (see Pauly and Lam 2023, and Chapter 4). One mutation or the other is unlikely to change the temperature preference of a WBE. Our intuition on the potential evolvability of 'warming-proof species,' which may partly be based on major handicaps in people being caused by a single gene's mutation, is misleading.

This persistence of biological traits, part of what Cury (1994) calls the 'obstinacy' of nature, can be used, on the other hand, for an index of the perception of warming by communities of fish and invertebrate WBE exploited by marine or freshwater fisheries (Cheung, *et al.* 2013c). This index, the Mean Temperature of the Catch (MTC), is defined by

$$MTC = \sum_i^n T_i \cdot C_{iyr} / \sum_i^n C_{iyr} \qquad ...(5.2)$$

where C_i is the catch of species i in a specific region in year yr, T_i is the temperature preference of species i, and n is the total number of species.

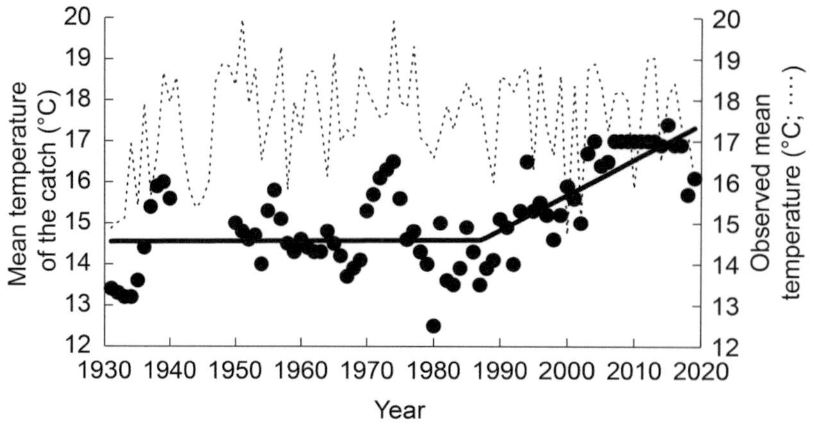

Figure 5.8. Segmented regression of the time series of MTC, illustrating that the two trend lines (the horizontal 1931 – 1986 line and the ascending 1987 – 2019 line) jointly explain over half of the variance in the data set, which documents the 'hockey-stick' concept in an exemplary manner (multiple R^2=0.53; adjusted R^2=0.51; modified from figure 2 in Kangur *et al.* 2021). The dotted line and the ordinate on the right refer to mean temperatures, whose long-term changes are less visible than those in the MTC.

Multiple local marine applications of the MTC now exist, which followed the nearly global coverage of Cheung *et al.* (2013c), notably Tsikliras and Stergiou (2014) from the Mediterranean Sea; Tsikliras *et al.* (2015) for Greek waters; Keskin and Pauly (2014, 2018) for Turkish waters; Khalfallah *et al.* (2023) for Egypt's Mediterranean waters; Alajmi *et al.* (2023) for the Persian Gulf; Liang *et al.* (2018) contrasting MTC trends in the South China, East China and Yellow Seas, and Dimarchopoulou *et al.* (2022) contrasting such trends in Japan, Indonesia and Eastern Australia. Another application to New Zealand allowed the detection of an MTC trend masking a 'fishing down' trend (Pauly *et al.* 1998), as both occurred simultaneously (Lavin *et al.* 2023). To date, only one application to a freshwater lake has been reported (Kangur *et al.* 2021). It pertains to a famous lake[139] shared between Estonia and Russia, and its MTC trend will be the last figure in this chapter, as it very well documents the 'hockey stick' pattern that global warming imposes on natural systems (**Figure 5.8**).[140]

Overall, while challenges will continue, the GOLT can explain the relationship between body size and temperatures that shape both the broad pattern of the global distribution of fish size, the differential effects of temperature-induce hypoxia on small (young) fish compared to large (old) conspecifics and the 'obstinacy' of fish regarding their temperature preferences, which itself allows for a better understanding of fish migration and range shifts, and the definition of an index enabling the detection of range shifts from fisheries catch data.

6

Objections to the theory and what to learn from them

Synopsis: Mechanistic explanations of the impact of climate change on fish growth are currently under debate. However, critical assessments of even the most prominent theories are not always based on accurate interpretations of their underlying mechanistic models. This chapter addresses some of the major misunderstandings that still prevent the Gill-Oxygen Limitation Theory from being examined based on its actual structuring elements and assumptions. As we argue, recent critiques of the GOLT are based on implausible interpretations of respirometry data that fail to distinguish between maintenance costs and the overhead costs associated with growth. We emphasize the fact that fasting young and growing fish for short periods of time is not sufficient to suppress energy (i.e., oxygen) allocation to growth. In the process of dealing with these issues, several cases of misunderstanding and one case of fabricated 'counter-evidence' are discussed. We recommend that testing the theory should focus on broad reviews or meta-analyses, e.g., on datasets of gill surface area and on the relationship of these data to growth performance under different temperature regimes.

Apparent problems with the GOLT

As outlined in the introduction to this book (Chapter 1), the origins of the GOLT can be traced back to the doctoral dissertation of Pauly (1979),[141] and its basic principles were formulated by the early 1980s (Pauly 1981, 1984). It took nearly 3 decades for the theory to receive more than perfunctory attention, which occurred in the context of debates about the impact of climate change on aquatic life and ecosystems. While key elements of the GOLT were independently tested and confirmed starting in the 2000s (e.g., Kolding *et al.* 2008; Tran-Duy *et al.* 2012; Diaz Pauli *et al.* 2017; Meyer and Schill 2021; Kuparinen *et al.* 2022; Lin and Costello 2023; Gunwardane *et al.* 2024) and further developed (e.g., Cheung *et al.* 2013b; Pauly 2019, 2021a), it became the topic of a heated debate between some physiologists on one side and ecologists and fisheries scientists on the other. The main point of contention in this debate was the mechanistic foundation of observed temperature-induced size reductions in fish and other aquatic ectotherms (Lefevre *et al.* 2017; Pauly and Cheung 2017, 2018; Pauly 2021a; Scheuffele *et al.* 2021; McKenzie 2023).

The debate surrounding the GOLT does not so much revolve around the observed phenomena or trends but rather the question of potential explanations. While some physiologists until recently stated that *"based on their respiratory capacities, we would not predict a change in the size of fishes in a warmer world"* (Lefevre *et al.* 2017), these size decreases are now universally observed and confirmed. While disputes around the GOLT continue, the limiting role of oxygen has become more apparent and is acknowledged more widely (e.g., Atkinson *et al.* 2022; Woods *et al.* 2022; Deutsch *et al.* 2024) and confirmed in many empirical studies (e.g., Tunnicliffe *et al.* 2020; Verberk *et al.* 2016 and 2021, as well as the studies cited therein). None of the studies that investigated the impact of temperature and oxygen on fish growth have presented their results in a broader and more generalizing theoretical framework, with the notable exception of the theory of oxygen- and capacity-limited thermal tolerance (OCLTT) as developed by Pörtner (2010), which comes to many conclusions similar to the GOLT.[142] Besides these two generalizing theoretical approaches, no alternative, widely applicable theoretical framework has been proposed, nor have other mechanisms been identified to explain why fish within a wild population (or in captivity with food *ad libitum*) remain smaller at higher temperatures.

While the GOLT's quest for generality implies that the possibility to discuss all specific details relating to the growth and metabolism of water-breathing ectotherms across phyla and taxa is necessarily limited, the debate around the theory's applicability is a welcome opportunity to examine more specific cases and questions, some of which would not have to come to mind without the critics of the GOLT. While every theory with the ambition to explain a broader range of empirical phenomena through a limited number of (ideally universal) mechanisms must necessarily encounter phenomenological challenges, these challenges often allow for new and unexpected insights.[143] While the GOLT faces several explanatory challenges, most of the criticisms raised so far have been easily addressed and produce new insights into the theory. In this chapter, we will discuss these counterarguments – to defend our viewpoint but also because these apparent challenges open perspectives that can enrich its explanatory potential.[144]

It is for this reason that here, we present one of the heuristics of the GOLT – the so-called P-diagram (Kolding *et al.* 2008) – in the context of the criticism that it has recently faced (Scheuffele *et al.* 2021; McKenzie 2023; Jutfelt *et al.* 2024). This criticism allows us to get to a hopefully more precise explanation of one of the critical elements of the theory. We will first discuss the definitions of some key parameters and features on which the GOLT's critics base their arguments, especially their interpretation of metabolic measurements in respirometry experiments. Then, we will explain the heuristic value of the P-diagram, followed by a discussion of some more specific arguments against the theory, notably the growth of fish under hyperoxic conditions, the growth of gills at different life stages, the argument that *"gills are folded surfaces, not spheres"* (Lefevre *et al.* 2017)[145], and the existence of large fish species in tropical waters, which has been used as an argument against the broad validity of the GOLT.

So far, we believe that none of these arguments pose an existential challenge to the GOLT and can be addressed straightforwardly. Still, we welcome further discussions of the theory and explorations of potential limits and obstacles to its explanatory scope. Since theory is not the same as dogma (and good theory is, *by definition*, not dogmatic), every scientific theory must open itself to counterarguments and take them seriously. This chapter attempts to do so, and as we will show below, it is enriched by the multiple questions posed by the GOLT's critics, many of which we would not have thought about ourselves. One central problem remains as a challenge: the question of ultimate causation. We will address this issue in Chapter 16 in more detail and discuss how the GOLT explains allometric metabolic scaling *in* and *between* species. For now, we use the opportunity to elaborate on some crucial features of the GOLT by addressing the major critiques of our model.

Fish growth, respiration, and temperature

Deriving general physiological models and drawing mechanistic principles rely on data from a wide range of studies that must be integrated and harmonized. In physiological studies, definitions of parameters and traits are of crucial importance, for allowing meaningful interpretations of data. In studies on the scaling of metabolic rates, definitions are often less clear-cut than reviews of the state of the art in the discipline suggest. This is particularly important if data are used to test more general hypotheses initially produced in different contexts and for other purposes, as is the case for the definitions of resting, standard, and active metabolic rates. Despite the instructional and widely cited review of this issue by Chabot *et al.* (2016), the problem of defining parameters remains unsolved. As this chapter shows, it often leads to serious misunderstandings.

The authors of some recent studies have argued that oxygen limitation cannot be expected to impact growth if either resting or active metabolic rates scale with a similar exponent as gill surface area (Lefevre *et al.* 2017; Scheuffele *et al.* 2021; Lonthair *et al.* 2024a, 2024b). As these authors argue, the fact that gill surface area scales isometrically with oxygen consumption implies that the capacity to meet oxygen demand is independent of body size. Hence, they assert that changes in the relationships between respiratory surfaces and body mass during ontogeny cannot explain the decline in growth rates at larger size or lower final sizes at higher temperatures.

This idea underlies many critiques of mechanistic growth models, which explain the asymptotic form of growth curves in fish due to the different scaling exponents of gill surface area and body weight (Lefevre *et al.* 2021). While Lefevre *et al.* (2017) initially argued against the impact of surface-volume relationships on oxygen uptake capacity *per se*, later studies revisited this argument and examined the relationship between the slopes of metabolic rate and gill surface area (Scheuffele *et al.* 2021; Lonthair *et al.* 2024a). As these authors argued, it is not body size itself that should be regarded as the relevant parameter in this context but rather metabolic rate, which also scales (with negative allometry) with body mass, as the extensive literature on metabolic scaling consistently shows (see, e.g., Schmidt-Nielsen 1984). As long as metabolic rate and gill surface area scale with the same exponents, these authors suggest that oxygen limitation can be excluded as a factor that could impact growth.[146]

For this reasoning to hold, it is necessary to exclude metabolic costs of growth from the respiratory measurements that underlie the model in question. The exact definitions of 'standard' and 'active' metabolic rates are crucial for the conclusions that can be drawn from the existing data: the standard metabolic rate (*SMR*) must then be defined in such a way that it does not account for the oxygen uptake that is required to support further growth.

In sharp contrast to this reasoning, the similarity between the slopes of oxygen uptake and gill surface area is a constitutive element of the GOLT (see, e.g., Pauly 2019, p. 36-37), based on Fick's law of Diffusion (Fick 1855) and the review of De Jager and Dekkers (1975) on the gill surface area and respiration of fishes (see also Pauly and Müller 2025). The notion on which the proposed 'tests' of the GOLT by Scheuffele *et al.* (2021), Lonthair *et al.* (2024a, 2024b), and its echoes in Jutfelt *et al.* (2024) are based is a parameter which these authors believe to be derived from the GOLT itself: the parameter b_S, which is defined as the difference between the scaling exponents of standard metabolic rate (b_{SMR}) and that of gill surface area (b_{GSA}). Thus:

$$b_S = b_{GSA} - b_{SMR} \quad \text{...(6.1)}$$

As both Scheuffele *et al.* (2021) and Lonthair *et al.* (2024a) assert, for the GOLT to apply, b_S should be ≤ 0. Therefore, the authors conclude that values of $b_S > 0$ would refute this theory, as the allometric slopes of the GSA would cause it to increase faster than *SMR* with increasing body size. The GOLT always assumed b_{GSA} and b_{SMR} to have the same or similar values; this is explicitly stated in Pauly (2019, p. 36-37), where b_{GSA} and b_{SMR} are referred to as 'd_G' and 'd_Q,' respectively, and are used as substitutes for one another throughout his book.

Values of $b_S < 0$ would be in contradiction of Fick's law, while values that exceed $b_S > 0$ would not make evolutionary sense (why would an organism invest energy in producing excess gill tissue, it does not need and which is vulnerable to parasites that can easily access the bloodstream through the extremely thin epithelia that separate the blood from the surrounding water?). The GOLT considers substantial deviations from $b_S = 0$ to be caused by methodological issues of no theoretical importance. Systematically occurring b_S values higher or lower than 0 would be a refutation of the GOLT; random deviations from 0 do not (see Pauly and Müller 2025).

The reasoning of Lefevre *et al.* (2021), Scheuffele *et al.* (2021), and Lonthair *et al.* (2024a) assumes that the GOLT, and its assumption that $b_{GSA} < 1$ (i.e., $d_G < 1$), implies that fish growth continues until vital body functions would be close to collapse, i.e., that fish are supposed to grow their way into asphyxia.[147] What $b_{GSA} \approx b_{SMR} < 1$ implies, in reality, is that growing fish have less and less surplus oxygen to allocate to growth until they reach the maximum size they can reach in a given environment (and temperature). They can do everything they normally do (forage, digest their food, escape predators, etc.) *except grow*. In other words, what the GOLT claims is *not* that old fish lose their aerobic scope and collapse because they cannot maintain their normal activities, but rather that they do not have any more surplus oxygen to devote to further growth.

As outlined in Chapter 2, the foundation of the GOLT is the modified version of Pütter's equation (1920),

$$dW/dt = HW^d - kW \quad \text{...(6.2)}$$

where the increase of breakdown processes (kW) with increasing body mass outpowers the anabolic processes (HW^d), which enables the production of new organismic materials as long as the slope d is < 1. This change in energy allocation from growth to increasing 'maintenance' (not reproduction, as we demonstrate in Chapter 9) is unrelated to the assumption behind Equation (6.1), which assumes that there will be a critical shortage of available oxygen for even basic maintenance functions during a fish's growth trajectory. From this perspective, Equation (6.1) is biologically meaningless and will only generate values that can be predicted to be identical across species, i.e., values ≈ 0. To understand the assumptions that underlie the reasoning of Lefevre *et al.* (2017; 2021), Scheuffele *et al.* (2021), Lonthair *et al.* (2024a), and Jutfelt *et al.* (2024), we will first examine the data used in these studies and the conclusions that can be drawn from them. We then propose a more appropriate interpretation of these data by introducing the so-called P-diagram as a heuristic that illustrates the relationship between growth and maintenance costs more intuitively.

Do fish stop allocating oxygen to growth when in a respirometer?

To test the conclusions of Lefevre *et al.* (2017; 2021) and Scheuffele *et al.* (2021) and their argument that oxygen limitation cannot impact fish growth as long as values of b_S are ≥ 0, it is worthwhile to take a closer look at the 54 studies cited to make this claim (**Table 6.1**). For the argument of these authors to hold, it would be necessary to exclude additional costs

Table 6.1. Data used to support the argument of Lefevre (2016; # 1-37) and Scheuffele *et al.* (2021; # 38 - 54). P = parameter; F = fasting period: h = hours; d = day(s). O2M = minimum O_2 consumption rate; O2C = O_2 consumption rate; Re = resting O_2 consumption; Rou = routine metabolic rate; and N = not stated.

#	P	F	Source	#	P	F	Source
1	*SMR*	36h	Brett (1964)	28	*SMR*	44-48h	Norin *et al.* (2014)
2	*SMR*	24h	Bruneaux *et al.* (2014)	29	Rou	N	Nowell *et al.* (2015)
3	RMR	48h	Chen *et al.* (2015)	30	*SMR*	3d	Pang *et al.* (2010)
4	*SMR*	24h	Claireaux *et al.* (2006)	31	RMR	2d	Pang *et al.* (2013)
5	*SMR*	24h	Claireaux *et al.* (2000)	32	In	N	Rosa and Seibel (2008)
6	O2M	N	Clark *et al.* (2011)	33	Res	N	Rummer *et al.* (2014)
7	O2C	2d	Clark *et al.* (2005)	34	RMR	N	Seth *et al.* (2013)
8	*SMR*	2d	Dickson and Kramer (1971)	35	*SMR*	N[c)]	Tirsgaard *et al.* (2015)
9	RMR	12-24h	Donelson (2015)	268	*SMR*	48h	Tu *et al.* (2012)
10	RMR	N	Donelson and Munday (2012)	37	N	96h	Vagner *et al.* (2015)
11	*SMR*	48h	Duthie (1982)	38	*SMR*	N	Al-Kadhomiy (1985)
12	*SMR*	N[a)]	Dwyer and Kramer (1975)	39	*SMR*	24h	Beauregard *et al.* (2013)
13	N	N	Eliason *et al.* (2011)	40	*SMR*	24h	Blasco *et al.* (2022)
14	*SMR*	10h	Ferreira *et al.* (2014)	42	RMR	24h	Bowden *et al.* (2018)
15	*SMR*	N	Fry (1947)	42	N	N	Clark (2013)
16	Res	24h	Gardiner *et al.* (2010)	43	RMR	45-72h	Clark *et al.* (2010)
17	*SMR*	40h	Gräns *et al.* (2014)	44	*SMR*	20h	Cutts *et al.* (2002)
18	Rou	N	Healy and Schulte (2012)	45	BMR	N	Degani *et al.* (1989)
19	*SMR*	24h	Khan *et al.* (2014a)	46-	*SMR*[)]	48h	Duthie (1982)
20	SMR	48h	Khan *et al.* (2014b)	47	*SMR*	24h	Herrmann & Enders ('00)
21	Rou	N	Lee *et al.* (2003)	48	RMR	??[d)]	Hughes (1978)
22	Res	N	Lefevre *et al.* (2015)	49	RMR	36h	Huang *et al.* (2013)
23	SMR	24h	Lefrançois and Claireaux (2003)	50	O2C	24h	Ken-Ichi (1992)
24	RMR	21d[b)]	Lowe and Davison (2006)	51	RMR	24+48h	Li *et al.* (2016)
25	SMR	4d	Mallekh and Lagardère (2002)	52	RMR	48h	Li *et al.* (2018)
26	SMR	24h	Marras *et al.* (2015)	53	O2C	24h	Miyashita (1999)
27	RMR	24h	Nilsson *et al.* (2009)	54	*SMR*	24 h	Seppänen *et al.* (2010)

a) # 12: No statement about fasting, but maintenance ration was provided prior to respirometry, even though the period is not stated; b) # 24: three weeks because specific dynamic action (SDA) in Antarctic fish can last 5-15 days post-feeding; c) #35: Not stated, but a maintenance ration was provided for 5 – 10 days; d) # 48: b) Fish were held *"generally without feeding"* (unspecified).

of growth from the measurements of metabolic rates. Including overhead costs would not allow interpretation of these data as standard metabolic rate or *SMR*, i.e., the parameter the authors believe works within their analysis.

As a comparative analysis of the definitions and methodologies of the 54 studies cited in Lefevre (2016), Lefevre *et al.* (2017), and Scheuffele *et al.* (2021) shows, the descriptions of the experimental set-ups are too divergent for consistent and realistic inferences of *SMR* in growing fish – if this parameter is understood as metabolic rate *excluding* activities, digestion, and especially growth (see **Table 6.1**). Additionally, most of these studies were not designed to distinguish between the metabolic activities of juvenile and adult animals, nor to identify the resulting differences in energy allocation. As numerous studies have shown, juvenile and subadult animals do not suddenly switch their metabolism away from growth to maintenance when confined in a respirometer. Rather, respirometry on smaller and younger individuals always incurs energetic overhead costs of growth (Parry 1983; Rosenfeld *et al.* 2015; Chabot *et al.* 2016).

This illustrates the difficulties of drawing inferences on the relationship between oxygen uptake capacity and growth from data interpreted as *SMR*, which are then expected not to include growth expenditures, as these differ greatly between juveniles (i.e., growing animals) and adult individuals (with low growth rates). In the context of the issue raised by the GOLT, an evaluation of the declining oxygen consumption rates in larger animals is only possible if reduced growth costs are taken into account.

As Rosenfeld *et al.* (2015) reminded us, fish physiologists follow different procedures and definitions than researchers who work on endotherms: "*because of the strong influence of growth on SMR, physiologists studying maintenance metabolism in birds and mammals generally work on adults; in contrast, fish physiologists commonly measure SMR on actively growing juveniles in laboratory experiments.*"

To obtain realistic values, fish must be fasted sufficiently before being put in a respirometer (Chabot *et al.* 2016). But how long must they be food-restricted to exclude the metabolic contribution to growth that would otherwise interfere with values understood as genuine *SMR*? Despite a growing body of literature and the instructive review of Chabot *et al.* (2016), the answer is far from straightforward, and some authors argue that *SMR* can only be reliably measured in non-growing, i.e., large adults (see, e.g., McNab 1997). Similarly, more recent studies define the term *SMR* as "*the metabolic rate of an adult, inactive, unstressed, postprandial ectotherm,*" i.e., for animals where no growth expenditures are expected to be included in measured metabolic rates (e.g., White and Marshall 2023a).

While many studies rely on 24-hour fasting periods before placing their fish in respirometers, this may not be sufficient to exclude the overhead costs associated with growth. As Rosenfeld *et al.* (2015) demonstrate, the measured values of *SMR* depend heavily on the rations consumed in the days and weeks before fasting. In their study on juvenile salmonids, animals fed heavily before a 35-hour fasting period showed overhead costs of growth of at least 67%. Therefore, Rosenfeld *et al.* (2015) recommended putting the fish on maintenance ratios for extended periods before respirometry, as even in fish fed only moderate rations for 2-3 weeks before the experiment, overhead growth expenditure fluctuated between 19% and 33%. These experiments show that juvenile fish do not abruptly cease allocating oxygen to growth when placed in a respirometer after fasting for only 24 or 48 hours.

If we examine the studies upon which criticism of the GOLT is based following the recommendations of Rosenfeld *et al.* (2015) and Chabot *et al.* (2016), we see that almost none of the research setups that underlie the studies in Lefevre (2016) and Scheuffele *et al.* (2021) conform to the recommended standards. In the 37 studies cited in Lefevre (2016), many authors did not indicate fasting periods; some only fasted the fish for 10 hours (**Table 6.1**). The only longer fasting period was maintained in a study on Antarctic fish, where the metabolic effects of specific dynamic action, or SDA, can last for 2 weeks or longer. In the more recent study by Scheuffele *et al.* (2021), the average fasting period was 36.6 hours, and some datasets did not indicate whether the fish were fasted.

Only one study (Tirsgaard *et al.* 2015) indicated how pre-fasting maintenance rations were calculated. As the experiments of Rosenfeld *et al.* (2015) show, information like this is necessary to infer if *SMR* data are realistic, i.e., represent metabolic rates that exclude growth expenditures. Only one study followed the suggestion of Chabot *et al.* (2016) to report feeding conditions in the weeks preceding respirometry and pre-respirometry fasting (Dwyer and Kramer 1975). Another study reported that the fish were on maintenance rations 5-10 days before the experiment (Tirsgaard *et al.* 2015). Overall, this suggests that the data cited by Scheuffele *et al.* (2021), Lefevre *et al.* (2017), and Lefevre (2016) are not adequate for addressing the question raised by these authors.

The time between food consumption and subsequent protein synthesis varies among species. While postprandial standard dynamic action (SDA) in most fishes occurs and also decreases rather quickly (polar species seem to be the exception here), animals from other phyla show SDA-related higher metabolic rates even weeks after feeding. According to Vahl (1984), postprandial SDA (which may at least partially reflect the costs of growth) could still be measured after 42 days in the common starfish *Asterias rubens*. Based on a similar experiment, also involving echinoderms, Broertjes *et al.* (1982) inferred that the impact of a single meal on tissue growth could still be reflected in metabolic rate measurements after 40-50 days. These extreme values may be unique to echinoderms and fail to reflect broader patterns in invertebrate WBE. Still, they demonstrate that standard procedures must take into consideration the metabolic patterns in different phyla, classes, families, and species.

The P-diagram, metabolic rates, and the cost of growth

As outlined in Chapters 1 and 2, the primary growth model that underlies this book is built on the assumption that the dimensional tension between gill surface area (*GSA*) and body mass (or simply 'weight') in fishes results in a decreasing capacity to take up oxygen in large adults compared to growing juveniles. A consequence of this changing relationship between respiratory surfaces and the three-dimensional bodies that require oxygen is a steady decline in growth rates. Even though gills are filamentous organs whose surfaces can increase hyper-allometrically, the increase in their respiratory surface is not proportional to the mass increase of a growing body. This limitation occurs even though oxygen demand may decrease with body size due to the hydrodynamic advantage of large size (Bakun 2019, 2022), the reduced number of potential predators (Hone and Benton 2005), and the greater ease of capturing larger, energy-dense prey (see Chapter 10).

The relationship between *GSA* and body weight at different sizes can be presented in the so-called P-diagram (Kolding *et al.* 2008; **Figure 6.1A**). In this diagram, the difference between the oxygen demand of maintenance metabolism – defined as the metabolism

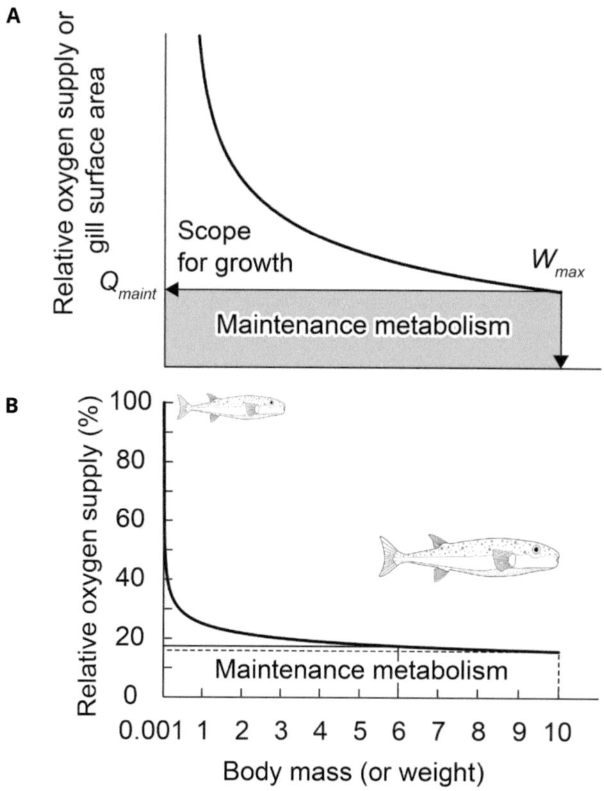

Figure 6.1. Illustrating the relationships between the metabolic rate(s) of fish and other WBE and their body weight. **A**: Original version of a P-diagram, documenting the placement and definition of its key elements, maintenance metabolism (Q_{maint}), and the scope for growth (Pauly 1984; Kolding *et al.* 2008). Note that this representation is schematic, i.e., not to scale. **B**: A more realistic P-diagram, based on $GSA \propto W^{0.8}$, and showing how the silver-cheeked toadfish *Lagocephalus sceleratus*, a Lessepsian species, could increase in maximum weight from 6 to 10 kg if it manages to reduce its maintenance metabolic rate by only 10%, as indicated by the dotted line (Ulman *et al.* 2022).

of a large (old) and hence non-growing animal – and the supply that results from the relationship between gill surface area and body weight is then visualized as 'scope for growth' (Pauly 2019).

Another P-diagram is illustrated as an insert in **Figure 6.2**, which documents the growth of gill surface area in common carp (*Cyprinus carpio*). It illustrates that, due to the rapid increase of the gill surface area in larvae and early juveniles (where $d > 1$), growth of carp – and other fish and WBE – is not oxygen-limited. From a weight (W) of about 1 g, gill surface area (S) grows according to $S \propto W^d$ with $d \approx 0.8$, a pattern that appears to last through adulthood.

Aside from explaining the reasons behind the cessation of growth at the maximum or asymptotic weight (W_∞ or W_{max}), the schematic versions of P-diagrams play a pivotal role in understanding why fish and other WBE tend to be smaller when subjected to higher temperatures (**Figures 6.1 – 6.3**)[148]. Also, in Chapter 7, we will further demonstrate the practical utility of P-diagrams in explaining the maturation and spawning patterns of fish and other WBE.

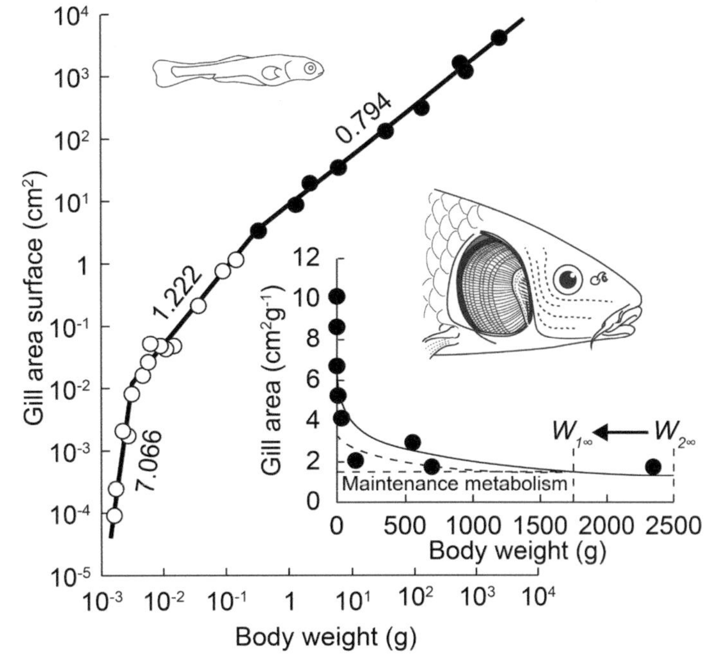

Figure 6.2. Growth of gill surface area in common carp (*Cyprinus carpio*), whose head is shown with the operculum removed. Note that its growth relative to body weight initially proceeds with $d > 1$, which then stabilizes at $d < 1$, here ~ 0.8. Also note that this implies (see insert) that at some values of weight, the oxygen supplied by the gills, while still sufficient to maintain an individual's activities (moving around, feeding, etc.), will be insufficient to support further growth, which will stop at $W_{\infty 2}$, or $W_{\infty 1}$ if, e.g., an increase in temperature elevates it metabolic rate (straight dotted line). Modified from figure 2 in Pauly and Cheung (2017) and figure 2 in Oikawa and Itazawa (1985). See also **Figure 3.8** for more cases of changing values of d_G with increasing size.

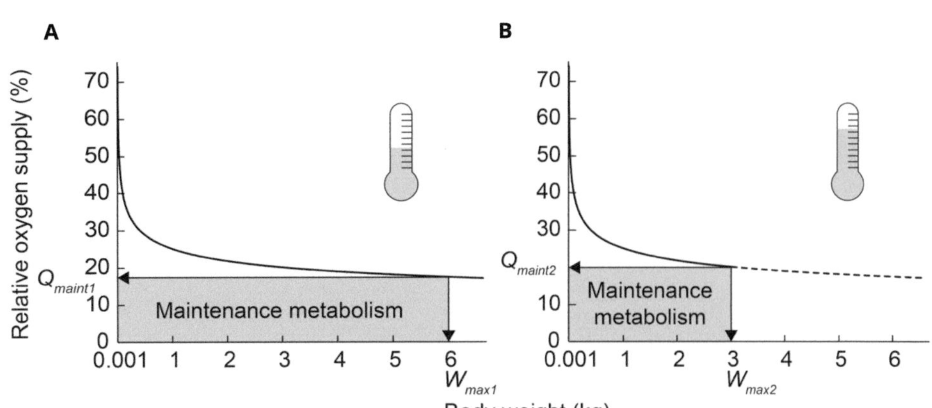

Figure 6.3. Using P-diagrams to illustrate the effect of an increase in ambient temperature on the terminal size (W_∞ or W_{max}) of fish or other WBE. **A:** Lower ambient temperature, implying low O_2 requirements (Q_{maint1}) and resulting in high W_∞ or W_{max}. **B:** Higher ambient temperature, implying (above 4-5 °C) higher O_2 requirements (Q_{maint2}), resulting in lower W_∞ or W_{max}.

Nevertheless, the concept of the P-diagram has been criticized by McKenzie (2023), who claimed that it leaves maintenance metabolism undefined – although a clear definition was provided in the book he discussed in his review.[149] Instead of using maintenance metabolism as a parameter to infer the remaining scope for growth (as visualized in the P-diagram), he suggested looking at 'aerobic metabolic scope' as the relevant parameter, i.e., the difference between standard and maximum metabolic rate. This is not the same as the scope for growth as described in **Figure 6.1**. Indeed, the aerobic scope does not allow for any inferences on growth. Still, according to McKenzie (2023), Scheuffele *et al.* (2021), and Lefevre *et al.* (2017), the fact that *SMR* scales with roughly the same exponents as gill surface area refutes the GOLT because the role of body mass in relative oxygen supply is, in their view, negligible.

The fact that metabolic rate and gill surface area have the same exponent is not only explicit in Pauly and Cheung (2017, table 2) and in many of the articles they cite, but also in *Gasping Fish* (Pauly 2019, p. 157) in which Rombough and Moroz (1990) are quoted, viz: "*Mass exponents for total surface area (d 0.85) and metabolic rate (*d = *0.8-0.9) were not significantly different for the larger fish.* [...] *Studies of scaling in juvenile and adult fish have shown that gill surface area and aerobic capacity expand at roughly the same rate as fish grow (Muir and Hughes 1969; Holeton 1976; Hughes 1979, 1984; Schmidt-Nielsen 1984). This is usually interpreted as indicating that aerobic capacity is effectively limited by gill surface area. Recent ablation studies tend to support this hypothesis (Duthie and Hughes 1987).*"

None of these authors comes even close to suggesting that measured oxygen uptake scales more steeply with body mass than gill surface area, and none of the GOLT-related publications assert this, either. Instead, the GOLT, based on earlier authors, such as De Jager and Dekkers (1974), relies on the scaling exponents of metabolic rates and a gill surface area being similar or identical, which is itself due to Fick's law (see Fick 1855).

The GOLT proposes that juveniles must invest a substantial amount of their energy (i.e., food + oxygen) into growth. Growth expenditures are minimal or absent in older adults, and the similarity of the scaling exponents of gill surface area and metabolic rates in juveniles and adults can only be understood if this is considered.

Even if *aerobic scope* would remain the same over a wide range of body sizes (see **Figure 6.4**), it would still not correspond to the *scope for growth* (**Figure 6.1A**). In other words, aerobic scope remains self-similar over a wide range of body sizes, which is only possible because the costs of growth decrease with size, and the energy that would otherwise be invested in producing new tissues is increasingly used for their maintenance.[150]

The issues addressed above strongly suggest that it is debatable whether *SMR* in juveniles and growing fish would solve the problems associated with measuring the overhead costs of growth. Indeed, a comparison between Rosenfeld *et al.* (2015) and the data cited in **Table 6.1** shows that this assumption is not justified. An earlier paper by McKenzie *et al.* (2003) indicates that respirometry conducted after 24 hours of fasting does not necessarily reflect genuine *SMR*. In their experiment, tilapia hybrids (of *Oreochromis* spp.) treated with a growth hormone showed significantly higher metabolic rates than the control group, an observation that the authors ascribed to ... the overhead costs of growth.

Maintenance metabolism, *as defined in the GOLT* (**Figure 6.1A**) and used for various inferences (e.g., **Figures 6.1B, 6.2,** and **6.3**), is a theoretical variable that can only be defined retroactively. In contrast to *SMR*, it cannot be directly measured in growing fish at a given time because a *growing* organism will never be in a state where all its metabolic energy

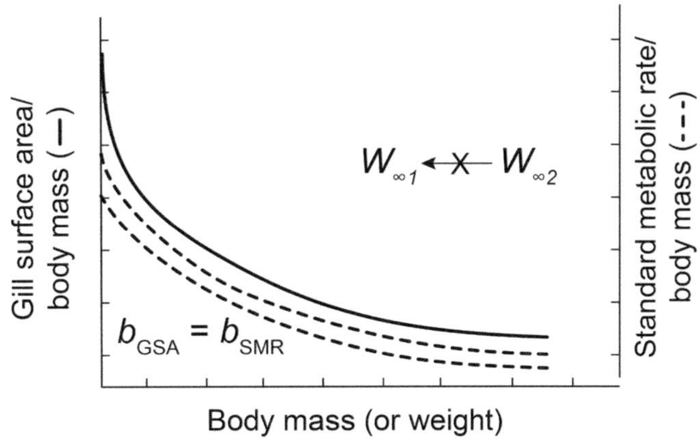

Figure 6.4. Alternative to the P-diagram proposed by Scheuffele *et al.* (2021), with the lower dotted line representing the standard metabolic rate (*SMR*) at a low ambient temperature and the upper dotted line representing *SMR* at a higher temperature, and "$W_{\infty 1}$ ←X -- $W_{\infty 2}$" is supposed to mean that increased temperatures do not reduce asymptotic size. The original caption reads as follows: *"Representation of the scaling relationship of SMR* [...], *where* b_{SMR} *has the same value as* b_{GSA} ($b_{GSA} = b_{SMR}$) *despite temperature modifying the intercept of the SMR regressions. The resulting 'scope for growth' thus never becomes 0, and the corresponding ratio S* [...] *stays constant throughout the entire body mass range. In this latter scenario, maximum body mass declines cannot be explained by GOL."*

will be invested *exclusively* in maintenance activities. We suggest that **Figure 6.4**, which summarizes an alternative to the logic of the P-diagram, is profoundly mistaken.

The primary reason for this is that *SMR*, as defined by proponents of the notion illustrated in **Figure 6.4,** includes the cost of growth (of which more is discussed below). Also, two corollaries of **Figure 6.4** are worth highlighting mentioning here: (*i*) its authors suggest that *"in the latter scenario, maximum body mass declines cannot be explained by GOL,"* i.e., temperature does not affect the maximum sizes reached by fish and other WBE and (*ii*) there can be no size so large that it would preclude further growth. Both of these corollaries are wrong.

Notably, the second of these corollaries seems to imply that fish would grow forever if not constrained by some external factor(s) that is (are) independent of their physiology. One such factor is food availability, explicitly mentioned, e.g., by Lefevre *et al.* (2021), who assert that *"many studies focus on food limitation as the causative factor for declining fish sizes in the field."* These studies did not demonstrate why food availability declines rapidly, for each fish species, in each of the world's ecosystems, at precisely the sizes that mimic well-established temperature-size relationships, i.e., patterns caused by oxygen stress.

Back in the real world, the suggestion that maintenance should be seen as a primarily theoretical parameter avoids the problems that occur when values are inferred from studies that only measure oxygen uptake at rest and then interpret this as *SMR* (even if the original publications did not use this term) – without acknowledging that these data should also involve overhead costs of growth. Using published respirometry results from experiments on juvenile fish to infer standard metabolic rates (tacitly understood as excluding growth) requires context and a closer examination of the methodology behind the respective experimental setup.

This raises the question of how the problem of convincing parameterization could be solved if the data on '*SMR*' presented in various publications do not eliminate the growth component of metabolism, as **Table 6.1** suggests. One possibility would be to develop a model that incorporates estimated growth expenditure and then explicitly uses these to correct the metabolism estimates for the cost of growth in juveniles, perhaps using approaches analogous to those of Rosenfeld *et al.* (2015) or Garcia-Gómez (2023). Pending the emergence of such a model, inferences based on simply *assuming* that a few hours of fasting eliminate the cost of growth will not suffice.

Fish growth under hyperoxia

Another critique of the GOLT was recently published, which was supposedly tested by fish (*Galaxia maculatus*) that were exposed to absurdly high (150%) levels of hyperoxia (Skeeles and Clark 2023), the idea being that if fish grown in hyperoxia do not grow faster than in normoxia, this should demonstrate that oxygen is not a limiting factor to fish growth.[151] This experimental setup is a surprising choice, given that it has been known for nearly a century that under hyperoxia conditions, oxygen can act as a stressor, and in higher concentrations even as a toxin (see Lane 2002, and the review by McArley *et al.* 2021).[152]

Much more realistic results can be produced in experiments where mild hypoxia is used to quantify the effect of oxygen on growth and reproduction (see, e.g., Kolding *et al.* 2008, or Diaz Pauli *et al.* 2017) – and they confirmed the GOLT's predictions.[153] Other experiments that would effectively test the GOLT could focus on the food conversion efficiency of fish (i.e., growth increment/food intake), which ought to decline with size (as illustrated, e.g., in Tran-Duy and van Dam 2012; see also Chapter 9), a feature that the GOLT assigns to a declining oxygen available for growth. This would be useful as the GOLT is incomplete, as it does not adequately address food and feeding-related issues (Pauly 2021a).

It is unclear what these authors expected from experiments with hyperoxia. As Skeeles *et al.* (2023) explain, they assume that hypoxia exposure would be an imperfect test for the limiting role of oxygen for growth. As they argue, "*depriving fish of normal oxygen levels may trigger a range of physiological processes (e.g., gill remodelling) that divert resources away from somatic growth, leading to a flawed understanding of the role of oxygen in the [temperature-size rule] under normoxic conditions*" (Skeeles *et al.* 2023). While gill remodeling as a response to hypoxia is indeed observed in many freshwater species (and it illustrates that the gills are a bottleneck in the oxygen cascade), this argument not only lacks support from empirical data, but its underlying assumptions are inconsistent. Suppose an organism is lacking a crucial resource like oxygen. In that case, it is inappropriate to consider only the processes involved in rescuing the organism from the stressful situation, but not the stressful situation itself. As experimental studies on the impact of increasing temperatures on fish have shown, stressful conditions activate mechanisms that reduce metabolic costs rather than incurring even higher energy expenditures (Wootton *et al.* 2022). This argument is not based on current physiological knowledge, illustrating that ectotherms can utilize plastic responses to reduce metabolic costs. These reduction strategies do indeed come at the expense of growth – which is what our theory essentially relies on.

Allometric scaling of gill surface area and research integrity

It is a core prediction of the GOLT that gill surface area (GSA) does not increase isometrically with body weight in fish beyond the larval and early juvenile stages. In other words, in

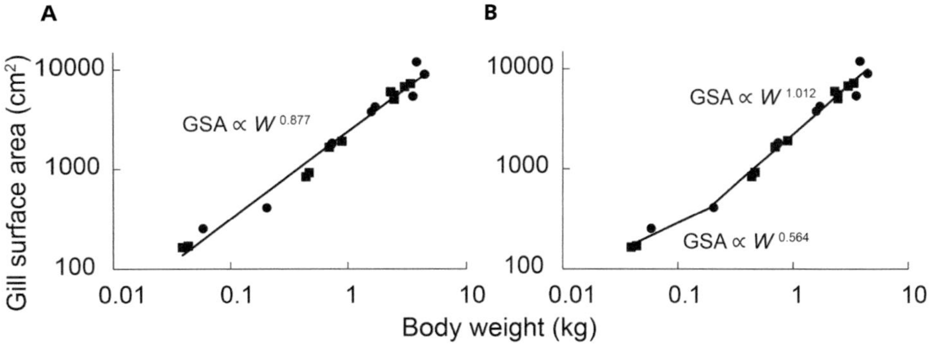

Figure 6.5. Scaling of gill surface area in relation to body weight in the horn shark *Heterodontus francisci*, based on data by Prinzing *et al.* (2023); ♀ = dots, ♂ = squares. **A**: Regression using all 19 fish studied, with the slope significantly lower than unity, as expected from the GOLT. **B**: Data fitted with a segmented regression, whose upper segment has a slope of 1.012, suggesting the gills of this shark grow such that they keep up with body weight, but whose lower segment suggests that the gills of small (young) horn shark grow very slowly – a feature for which there is no evidence in the literature.

adults, d must be < 1, a testable hypothesis. A recent paper on this issue again demonstrates that the positive results of such tests can be twisted such that they appear to be negative. As Prinzing *et al.* (2023) argue, based on their analyses of 19 individual horn shark *Heterodontus francisci*, GSA can be shown to increase proportionally to body weight, which, they then conclude, invalidates the GOLT.

As can be seen in **Figure 6.5A**, over the whole size range studied, the scaling exponent stayed solidly within the range predicted by the GOLT, with $d = 0.877$. To arrive at a value of $d \geq 1$, the regression had to be broken down into two segments. By using a segmented regression, with a separate low slope ('d' = 0.564) for the smallest three individuals, Prinzing *et al.* (2023) obtained a convenient slope of 'd' = 1.012 with the remaining 16 fish (**Figure 6.5B**), which they then used as an argument for vituperating against the GOLT.

Changes of slopes of *GSA* at different sizes are no anomaly and occur in many species – especially at early life stages (Pauly and Müller 2025; see Chapter 3). The usual pattern in the species for which comparative data are available shows a trend that points to the very opposite of what Prinzing *et al.* (2023) reported in their segmented regression that highlighted three individuals with seemingly lower values of d. In fish, and specifically shark species for which changing slopes of *GSA* scaling have been reported, e.g., in the thresher shark *Alopias vulpinus* (Pauly and Cheung 2017; see also **Figure 3.8E** in the section on '*Ontogenetic changes in scaling of gill surface area and diffusion distance*' in Chapter 3), it is at the high end of the fish size range where the d parameter takes lower values, when the gills can be assumed to be so large as to have difficulties increasing their lamellar area further.[154] This pattern, known as 'ontogenetic change of scale (OCS)', is well-documented (Glazier 2024 and references therein; see also **Figure 6.2**). One can assume that a segmented regression, which produced a standard error for 'd' (± 0.113) much higher than the honest solution (± 0.065), was used only because it generated what looked like an argument against the GOLT.[155]

Gills are folded surfaces, not spheres

The assumptions behind the idea that fish gills can grow with scaling exponents equal or even larger than unity (i.e., $d \geq 1$) are related to one of the more idiosyncratic points used as an argument against the GOLT. As Lefevre *et al.* (2017; 2020) repeatedly argued, surface-volume relationships do not apply to the relationship between gills and the body they must supply with oxygen. According to these authors, "*gills are folded surfaces, not spheres,*" which is true. They argue that this geometrical fact would invalidate the assumption of declining surface-volume relationships in growing bodies and that "*the scaling of surface area to volume is not constrained by spherical geometry.*" Even though this inapt statement is based on a misunderstanding of basic geometry, it reveals an interesting kind of confusion. As outlined in the introduction to this book and Chapter 3, gills typically have higher scaling exponents than expected in the early 20th-century literature on this topic (i.e., $d > 2/3$ in species that can reach lengths exceeding 5-10 cm). The argument that "*gills are not spheres*" does not pertain to the problem in question: if gills grew at the same rate as spheres, their slope relative to body size would not be $\approx 2/3$, but significantly lower (depending on the species and size range, it could be as low as <0.2). This argument would only apply to fish growth if gills had to support themselves and their own oxygen consumption. Since they have to supply an entire organism with oxygen, hyper-allometric slopes are needed to reach even exponents of $\approx 2/3$. This is nicely illustrated by Scott (2005), also cited by Lefevre *et al.* (2017), who use his study to argue that d can take values of 1.0. As Scott shows in the case of the bivalve *Solemya velum*, the scaling exponents of the gill surface relative to body mass are not 1.0, but 0.85. Instead, it is the scaling exponent of gill surface area relative to gill mass (excluding the rest of the organism's weight) that takes a value of 1.07 (which obviously has very different implications). Lefevre *et al.* (2017) state that "*in morphometric studies where both total lamellae area and gill mass have been measured, a linear scaling relationship (scaling exponent of 1.0) has been found in fishes (Gehrke 1987) as well as bivalves (Scott, 2005). Consequently, there is no geometric constraint that prevents an increase in body size (mass or volume) from being accompanied by a corresponding increase in gill mass and hence respiratory surface area.*" As outlined above, the slopes ascribed to Scott (2005) refer to the slope of gill surface area relative to gill mass. Even if the volume of gills grows with a slope of 1.2 relative to gill mass, the exponent of their surface area relative to body mass is still well < 1 (as illustrated for other bivalves in **Table 13.1**).[156] As for the value of $d = 1.04$ in spangled perch, *Leiopotherapon unicolor* reported by Gehrke (1987), it was based on early juveniles, below 12% of the maximum adult weight reached by this species (see FishBase), implying that they were in transition between the very high d values occurring in larvae and early juveniles and values < 1 occurring in adults, as implied by OCS, and illustrated here in **Figure 6.2**.

The improbable existence of large tropical fish

The GOLT implies that larger fish are generally constrained in their further growth by the oxygen uptake capacity of their gills, which limits the flow of oxygen to the body's interior. This is particularly true for large fish in the tropics, where the high temperatures increase oxygen demand. As we discuss in Chapter 11 on air-breathing fish, the constraints to sufficient oxygen delivery from the gills have resulted in several evolutionary developments that enabled tropical fish to take up oxygen from the air. Many of the large fish species in tropical freshwater ecosystems, such as those in the lower Mekong River and the Amazonian floodplains, are indeed air-breathers — for example, the Amazonian osteoglossid *Arapaima gigas*, whose anabolism is constrained by neither the size of its gills nor the oxygen content of the water it inhabits, and which shows highly atypical growth patterns (see Chapter 11).

Not all tropical freshwater 'giants' breathe air, and a notable example is the giant freshwater stingray (*Urogymnus polylepis*), which has recently been revealed to be the largest known freshwater fish extant (Hogan and Lovgren 2023). Even though there are reasonable explanations for why the aquatic megafauna of the Mekong is so extraordinary – and the giant fish species typically occur in the vicinity of deep holes where they can cool themselves (see Chapter 5) – critics of the GOLT have used the existence of these exceptions as arguments against the theory. As outlined elsewhere (Pauly 2019; 2021a), large tropical fish have developed strategies to maintain their body temperatures at levels that enable them to reduce their metabolic maintenance efficiently. The apparent problem of large tropical fishes reveals a more fundamental point of the GOLT, and the counterexample invoked allows us to address it even more concisely.

A fundamental implication of the theory is that inter- and intraspecific metabolic scaling depends on two different mechanisms. The decrease in the metabolic rates in growing fish of the same species differs from the higher metabolic rates in a guppy compared to a whale shark.[157] As Pauly (2020) stated, "*the GOLT does not claim to explain why minnow and anchovies are small while carp and tuna grow to much larger sizes. We have the theory of evolution for that. What the GOLT explains is why a minnow and a large tuna have gills which supply them, as they grow, with a declining oxygen per unit weight, and that this reduced oxygen supply reduces their growth rate until finally, they stop growing.*"

While this may seem trivial, it is, in fact, of crucial importance for our mechanistic model. Decreasing oxygen consumption rates *within* species are caused by slower growth rates at larger body sizes, and *not* decreasing maintenance costs, as assumed by Lefevre *et al.* (2021) and White and Marshall (2023a). This is not the case in the differences *between* species. As we show in Chapters 1 and 15, the GOLT is a theory that connects optimization- and constraint-based explanations. Since metabolic scaling between species cannot be the consequence of ontogenetic growth trajectories, it must be explained as the result of evolutionary selection. Since large animals cannot consume and metabolize the same amounts of energy as smaller ones, the evolution of their sizes depends on selection for low maintenance costs. In short, an elephant with the metabolic rate of a mouse would be a physical, chemical, and physiological impossibility. If it could find enough food to fuel its extreme demand, it would immediately overheat and collapse (and then probably explode). A large fish with high maintenance costs would quickly suffocate; however, its relatively lower metabolic rates allow it to exist.

What follows from this is that evolutionary roads to large body sizes are more unlikely to occur and more difficult to navigate in tropical waters since they rely on more refined mechanisms that allow for the reduction of metabolic rates, e.g., by performing regular cooling dives,[158] and/or by relying on different enzymes in metabolic processes. The evolution of large water-breathing ectotherms is in itself not in contradiction to the GOLT's underlying model: while the theory provides a good explanation for the strikingly close correlation between latitudes and average fish sizes, the existence of whale sharks or large sunfish is not a real challenge. In contrast, their relative rareness, along with their unique adaptations to life at low latitude neatly illustrates the GOLT's predictive value.

Objections to the GOLT and the future of mechanistic models

As the recent debate around mechanistic physiological models illustrates, the question of how published data on metabolic rates should be integrated into one's perspective depends on one's coherent interpretation. As shown here, the currently available evidence fails to support the idea that oxygen uptake capacity is independent of body size (see **Table 6.1**). Even though the costs of growth cannot be directly measured during respirometry, they must be assumed to impact respirometry data in growing animals. The difficulty in isolating overhead costs of growth from measured metabolic rates does not imply that the GOLT is untestable (as has subtly been suggested by Wootton *et al.* 2022). What it shows, rather, is that empirical data only allow for biologically meaningful interpretations if they are analyzed within the framework of coherent scientific models.[159]

What colleagues interested in testing the GOLT may therefore consider is confronting its prediction of patterns with the result of meta-analyses, comprehensive reviews, and/or global analyses while using single-species studies to identify the limits of its applicability. For example, the study of *Galaxia maculatus* mentioned above, which identified hyperoxia as a limit to the positive role of oxygen in the growth of fish, would have been helpful if it had been framed as testing when and where ambient oxygen, for fish, ceases to be an asset and may turn into a liability. As the authors of this study do not mention, the studied species is also an oxyconformer, which is rare among bony fishes, and its exceptional lack of oxyregulatory capacity has been intensively studied (see, e.g., Urbina *et al.* 2012; Urbina and Glover 2013, and the many other studies of these authors on inanga).

The question of how climate warming impacts the physiology and, eventually, the population structure and ecosystems of fishes and other WBE may be the most pressing current issue in all the different fields in which this group of animals is studied, and the need for valid models is urgent. Taking this question seriously requires methodological stringency, conceptual transparency, and, as the regrettable case illustrated by **Figure 6.5B**, integrity.

7

Why bony fishes mature and spawn when they do

Synopsis: Among fishery biologists and ichthyologists, maturation and spawning of fish are viewed as processes that use 'energy' that would otherwise be applied to somatic growth, which is supposed to explain why post-maturity growth in length tends to decline. This widespread conceptualization may be called the 'Reproductive Drain Hypothesis' (RDH). When growth is correctly viewed as involving body mass, and is expressed in weight, post-maturity turns out (in iteroparous bony fish whose maximum length exceeds 10 cm) to accelerate after first maturity, despite its energy cost. This, and other common observations, flatly contradict the RDH, and it is time to withdraw this hypothesis. As a contribution towards this task, we propose a reconceptualization of fish spawning that is consistent with current knowledge of fish biology. This revision serves as the basis of a model that quantifies changes in the phenology of spawning due to ocean and freshwater warming. It also allows for tentative predictions of the timing of spawning within a diurnal oxygen cycle. As a contribution to this task, we propose an alternative reconceptualization of fish spawning that is consistent with current knowledge of fish biology. This reconceptualization serves as the basis for a model that quantifies changes in the phenology of spawning due to ocean and freshwater warming and also enables tentative predictions of spawning timing within a diurnal oxygen cycle.

An old myth

Among fishery biologists and ichthyologists, fish maturation and spawning are viewed as processes that use 'energy' that would otherwise be applied to somatic growth, which is supposed to explain why post-maturity length growth tends to decline (**Figure 7.1**).

This widespread (mis-)conception (see **Figure 7.2**), which may be called the 'Reproductive Drain Hypothesis,' can be traced back to the early twentieth century (see, e.g., Lea 1913; van Oosten 1923; Hubbs 1926; Jones 1976; Lagler *et al.* 1977; Sebens 1987; Day and Taylor 1997; Charnov 2008; Quince *et al.* 2008). While its validity has long been criticized (see, e.g., Iles 1974), it still plays a central role in optimization-based theoretical models on metabolic scaling (see, e.g., White *et al.* 2022 and the replies to their piece). The reason for the persistence or re-emergence of this idea may be that fish growth slows down *in length* around the time of first spawning. When growth is correctly viewed as involving

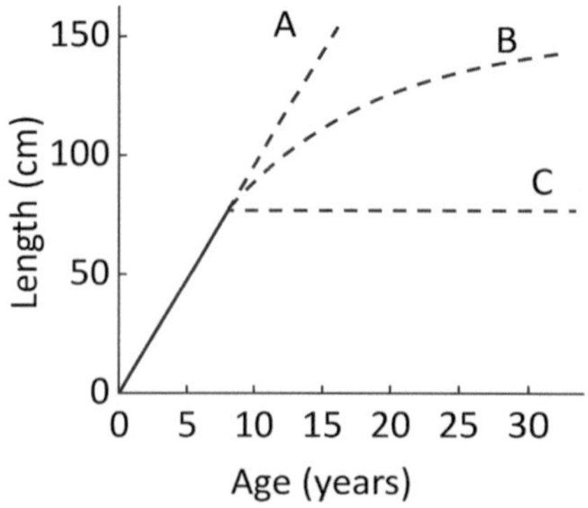

Figure 7.1. Representation of the 'Reproductive Drain Hypothesis,' i.e., the notion that reaching the size at first maturity causes previously 'linear' growth (line A) to decline due to 'energy' previously used for somatic growth being transferred to the elaboration of gonads, with line B implying a small, and line C a strong transfer of 'energy' (modified from figure 2 in Lester *et al.* 2004).

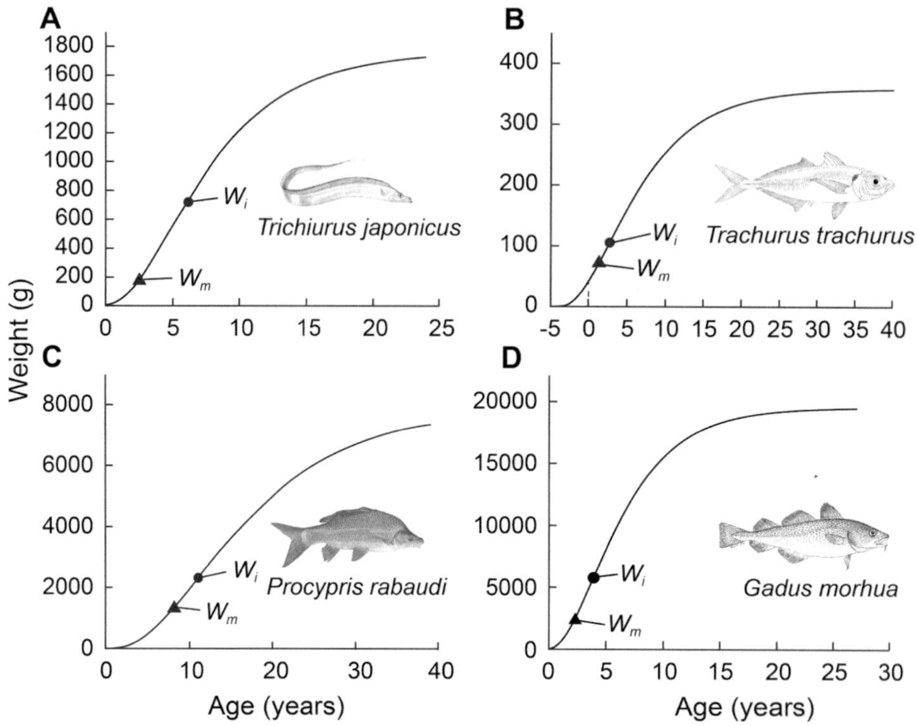

Figure 7.2. Illustrating the fact that when growth in weight is considered, first maturity, in fish reaching more than about 10 cm maximum length, the weight at first maturity (W_m) is reached at a size where growth is accelerating, i.e., well below the weight at which maximum growth rate is attained (at *Wi*); **A**: Japanese cutlassfish (*Trichiurus japonicus*), based on data in Shi *et al.* (2020); **B**: Atlantic horse mackerel (*Trachurus trachurus*), based on Froese and Pauly (2023); **C**: Rock carp *Procypris rabaudi*), based on data in Wang *et al.* (2015); **D**: Atlantic cod (*Gadus morhua*), based on data in Le Franc (1970).

body mass and is expressed *in weight*, post-maturity *accelerates* after first maturity, despite the energy costs associated with maturation and subsequent reproduction (**Figure 7.2**).[160]

Another line of evidence against the 'Reproductive Drain Hypothesis' can be drawn from the literature on triploid fish, which are unable to reproduce and therefore do not invest in reproduction. As early optimistic studies postulated, triploids should grow faster because they can allocate a much larger portion of their energy to growth (see, e.g., Galbreath *et al.* 1994). As aquaculturists hoped, this would yield much better economic results. As it soon emerged, however, triploidy itself had little impact on somatic growth, and in some cases, it even negatively impacted growth (Maxime 2008; Trippel *et al.* 2014; Farrell *et al.* 2025). Triploidy remains attractive in commercial aquaculture because it is possible to produce more female fish, which grow better and bigger, and males are usually quieter, as they do not invest energy in competition for females (Galbreath *et al.* 1994). Interestingly, triploids sometimes show better growth results as larvae and juveniles, before any investment in reproductive activity occurs (Bowden *et al.* 2018). Evidence for faster or more efficient triploid growth in the absence of reproduction is lacking (Maxime 2008; Farrell *et al.* 2025). These observations flatly contradict the 'Reproductive Drain Hypothesis'[161], and as we argue here, the time has come to retire it.[162]

In the process of the reconceptualization of fish spawning, this chapter provides an answer to the related question as to why longer-lived fish (and other WBE, by extension) 'choose' to reproduce for the first time at a given year (and/or spawning season). **Table 7.1** shows that this question is not trivial because longer-lived fish may experience, over years (and sometimes decades), multiple 'spawning seasons' without spawning and without having perceived the environmental cues or 'stimuli' that supposedly trigger the hormonal cascade leading to the elaboration of gonadal products.

The hormonal cascade that leads to maturation and spawning in fish, which is typically initiated by environmental stimuli that trigger reproduction (see, e.g., Bhattacharya 1992, 1999; Pankhurst 2016; Taranger *et al.* 2010), is not self-starting. This was also noted by Thorpe (1986, 1990), who suggested that an element was missing that would, when present, cause juvenile individuals' internal readiness to experience these environmental stimuli in the same manner that adult fish do. Thorpe (1990) suggested that this element is the growth

Table 7.1. Age at first maturity (t_m; years) and longevity (T_{max}; years) in 6 species of long-lived fishes, i.e., species whose pre-adults experience multiple 'spawning seasons' without spawning (from Pauly 2021b)

Species	Location	t_m	T_{max}	Reference
Epinephelus marginatus	Southern Mediterranean	12	50	Tsikliras and Stergiou (2015)
Acipenser fulvescens	Nottaway River, Canada	20	154	Anderson (1954); Magnin (1966)
Huso dauricus	Amur River Estuary, Russia	21	80	Koshelev and Ruban (2012); Sytova *et al.* (2004)
Acipenser transmontanus[a)]	Fraser River, B.C., Canada	11-14	104	Semakula and Larkin (1968); Rien and Biemesderfer (1994)
Albatrossia pectoralis	North Pacific	23	56	Tuponogov *et al.* (2008)
Hoplostethus atlanticus	New Zealand	34	93	Doonan (1994); Andrews *et al.* (2009)

a) The estimates for *A. transmontanus* are confirmed by Doroshov *et al.* (1997), who also mention that "*males reach puberty at a younger age*" and that "*cultured females reach puberty at a considerably younger age compared to wild fish.*"

performance of fish, i.e., *"exactly how the fish monitors its performance is unknown, but I have suggested elsewhere (Thorpe 1986) that it is physiologically aware of its growth rate through its rate of accumulation of surplus energy, and through hormone kinetics associated with storage of that energy."*

We demonstrate below that a specific relative size, roughly proportional to Thorpe's *"rate of accumulation of surplus energy,"* must be reached by pre-adult fish for environmental cues or stimuli to be sought and/or perceived, triggering the hormonal cascade that leads to maturation and spawning[163]. In the second part of this chapter, we then present a within-year, temperature-driven mechanism that impacts this oxygen-related trigger and show how it is affected by climate change, leading to changes in phenology. Finally, in the last section of this chapter, we propose the hypothesis that the apparent preponderance of spawning at dusk may also be explained by an oxygen, or rather hypoxia-related mechanism.

The crux of the problem

Figure 7.1 illustrates the conventional conceptualization of fish spawning. It shows that it is perceived as a costly process wherein 'energy' is transferred from somatic to gonad growth, thus abruptly impacting somatic growth. Hence, somatic growth is slowed or even stopped depending on the extent of the 'energy' transfer.

As already suggested, the first and perhaps most important reason for the survival of this conceptualization – the 'Reproductive Drain Hypothesis' – is that its outward plausibility rests on the representation of growth as proceeding *in length*. Length, while it can *represent* growth, is, however, a process involving mass, better represented by weight; 'energy' certainly does not have the dimension of length, or length per time. Once somatic growth is – correctly – viewed in terms of weight per time (**Figure 7.2**), the visual argument (in **Figure 7.1**) for the 'Reproductive Drain Hypothesis' is refuted. Except for species whose maximum length doesn't exceed 10 cm (and we shall deal with these species further below), fish tend to reach first maturity at sizes *below* that at which they experience their highest growth rate (*dW/dt*).

Figure 7.1 doesn't support the notion that 'energy' being transferred from somatic growth to the elaboration of gonads is the *cause* for growth in length declining after first maturity. Rather, it illustrates a case of the '*post hoc, ergo propter hoc*' fallacy, which considers that an event (1) is the cause of an event (2) simply because event (1) occurred before event (2).

Earlier authors, notably Iles (1974), noted that the usual narrative does not make sense. He wrote about the age and size at first maturity of fish that *"despite the fact that at some time during this stage of the life history large quantities of protein, ultimately derived from the same food sources that sustain body growth, will be newly required for gonad development, there is no indication that the growth pattern is disrupted or disturbed. Indeed, under 'normal' conditions it appears that it is singularly unaffected by the new physiological and metabolic demands which the fish is called upon to meet. The fact that, for most species of fish, unlike those of mammals and birds, growth continues after the attainment of the adult stage is one of the most characteristic features of the growth of fishes."*

The *"programme"* that Iles (1974) proposed as an alternative did not explain, however, how fish manage to spawn at the 'right' size, i.e., at the size that is a predictable fraction of the maximum size they can reach in a given environment (Pauly 2019).

The special and generalized von Bertalanffy growth function

We start again with Pütter's equation, i.e., with the growth rate (*dW/dt*) of fish and other WBE conceived as a differential equation

$$dW/dt = HW^d - kW \qquad ...(7.1)$$

where W is the body weight (i.e., mass), H is the rate of synthesis of proteins and other body substances, $0 < d > 1$ is the exponent of a relationship of the form $S = a \cdot W^d$, which limits protein synthesis, and k the rate of breakdown, whose main component is protein denaturation (see Chapter 4).

When $d = 2/3$, corresponding to $S = a \cdot L^2$ and $W = a \cdot L^3$, the integration of Equation (7.1) into a growth curve is the von Bertalanffy Growth Function (VBGF), which for growth in length has the form

$$L_t = L_\infty \cdot (1\text{-}e^{-K(t-t_0)}) \qquad ...(7.2)$$

where L_t is the mean length at age t of the animals in question, L_∞ is their asymptotic size, i.e., the mean size attained after an infinitely long time, K is a growth coefficient (here in year^{-1}; with $k = 3K$), and t_0 is a parameter adjusting for the fact that the VBGF usually fails to describe the growth of larval and early juvenile fish and other WBE.

For growth in weight, this becomes

$$W_t = W_\infty \cdot (1\text{-}e^{-K(t-t_0)})^3 \qquad ...(7.3)$$

where W_∞ is the weight corresponding to L_∞, all other parameters being defined as above, and the exponent ($b = 3$) is justified by the fact that it is the most common exponent of the length-weight relationship in fish (Froese 2006; Hay *et al.* 2020; see also FishBase). Equation (7.2) has no inflexion point (*dL/dt* declines linearly with length), but Equation (7.3) has an inflexion point (i.e., *dW/dt* is at a maximum) at $W_i = 0.296 \cdot W_\infty$.

In Chapter 4, we interpreted Pütter's equation in terms of the oxygen required for the synthesis of native protein, i.e., the first term on the right side of Equation (7.1), and to replace denatured proteins, which drives the second term on the right side of that equation. In this interpretation, ***d*** refers to the scaling exponents of gill surface area or other respiratory organs, i.e., 2-D structures whose growth cannot keep up with a 3-D body that requires oxygen.

As outlined in Chapter 2, when $d \neq ⅔$, the VBGF is 'generalized' and becomes

$$L_t = L_\infty \cdot (1\text{-}e^{-K \cdot D(t-t_0)})^{1/D} \qquad ...(7.4)$$

for length, with $D = 3(1\text{-}d)$, and

$$W_t = W_\infty \cdot (1\text{-}e^{-K \cdot D(t-t_0)})^{b/D} \qquad ...(7.5)$$

for weight, where $D = b(1 - d)$ and b is the exponent of a length-weight relationship of the form $W = a \cdot L^b$, with b generally 3 or near 3 (see above).

Equation (7.4) has an inflection point when $D \neq 1$, at age $ti = t_0 - (\ln(D)/K \cdot D)$, and length $L_i = L_\infty \cdot (1-e^{-(\ln(D))})^{1/D}$, while Equation (7.5) has an inflection point at age $t_i = t_0 - (\ln(D/b)/K \cdot D)$ and weight $W_i = W_\infty \cdot (1-(D/b))^{b/D}$, which implies

$$W_i / W_\infty = (1-(D/b))^{b/D} \quad ...(7.6)$$

In practice, the difference between the standard VBGF (Equations 7.2 and 7.3) and the generalized VGBF can be neglected when fitting a set of age-at-length, tagging-recapture, or length-frequency data with a growth curve. This applies especially to fishes whose gill surface area scales in a manner not differing much from the 2/3 value assumed in the standard VBGF (i.e., up to $d \approx 0.8$). This is the case for small fish, e.g., coral fishes, and small pelagic fishes, such as herring, sardine, or anchovies, where $d \approx 0.75$, and medium-sized species, e.g., carp or cod, where $d \approx 0.8$ (De Jager and Dekker 1975; see also **Figure 7.3**).

Only with a higher value of *d* does the fit of the standard VBGF become problematic, e.g., in the case of Atlantic bluefin tuna (*Thunnus thynnus*), where $d = 0.9$ (Muir and Hughes 1969), notably because the estimates of asymptotic length (L_∞) that are generated by fitting the equation to reliable age-at-length data tend to be much larger than the maximum length (L_{max}) in a given population, which is not the case when the generalized VBGF is used (Pauly 2021a).

As most applications of the VBGF involve species in which *d* ranges from 0.7 to 0.8 (see FishBase), which is still relatively close to 2/3, the fact that the standard VBGF, in such cases, differs slightly from a physiologically correct equation is usually ignored. This is also what is done here.

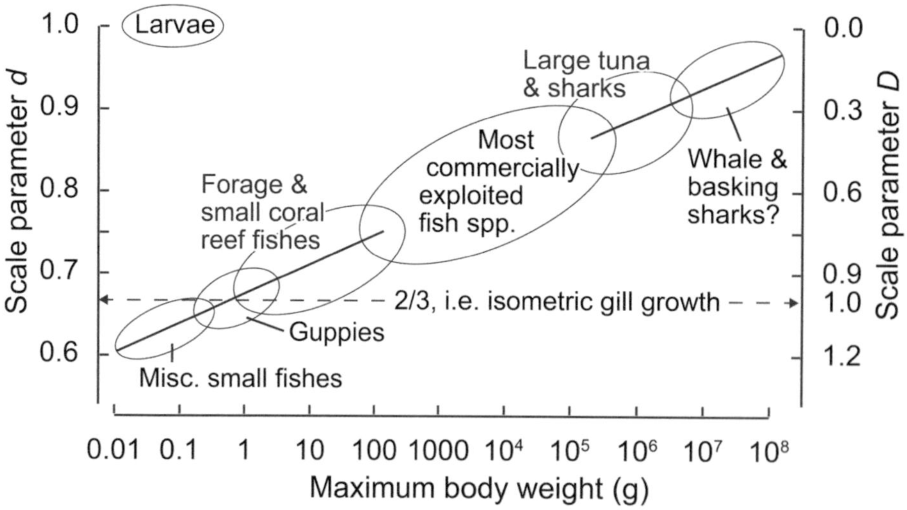

Figure 7.3. Showing that in fish, the parameters *d* and *D* are related to the maximum weight reached by various taxa (modified from figure 2 in Pauly 1981); the straight line is based on the relations $d = 0.674 + 0.0357 \cdot \log(W_{max})$, with live weights in g based on 27 estimates of *d* from studies of gill surface or respiration in bony fishes, all documented in table 2 of Pauly (1981).

Spawning and the 'reproductive load' concept

It is well-established that the mean length at first maturity (L_m) of the individual in a given population of iteroparous bony fishes is a predictable fraction of their asymptotic length (L_∞) or maximum length (L_{max}), which generally ranges from 0.4 to 0.6 in fish that reach larger sizes and from 0.6 to 0.8 in fish that remain small (Beverton and Holt 1959; Pauly 2021a; Prince *et al.* 2015).[164] This ratio does not deserve the name 'Beverton and Holt *invariant*' that Charnov (1993) gave it. Instead, here we use the term 'Reproductive Load' for the L_m/L_∞ (or the similar L_m/L_{max}) ratio used by Cushing (1981), which resonates with the 'Reproductive Drain Hypothesis' mentioned above.

That reproductive loads are not 'invariant' but vary systematically among fishes of widely different sizes was demonstrated by Froese and Binohlan (2000), who, based on data from 265 fish species from 88 Families and 27 Orders in FishBase, derived the relationships:

$$\log(L_m) = 0.898 \cdot \log(L_\infty) - 0.0782 \qquad ...(7.7)$$

where L_m and L_∞ are lengths in cm.

This model suggests that fish with an asymptotic (or maximum) length of approximately 10 cm will have a length at maturity (L_m) of 6-7 cm, while fish with L_∞ (or L_{max}) values of approximately 100 and 1000 cm will have L_m values of near 50 and 400 cm, respectively.

First reproduction precedes growth acceleration

Because the gill surface area of late juvenile and adult fish, which scales with $d < 1$ and supplies oxygen to their growing bodies, cannot keep up with their weight, and hence, with their oxygen demand, relative oxygen supply declines with body weight. This results in internal hypoxia, which manifests as an increase in the role of glycolytic enzymes. As individual fish grow, these enzymes gradually replace oxidative enzymes (Somero and Childress 1980; Burness *et al.* 1999; Norton *et al.* 2000; Davies and Moyes 2007).[165]

This decline in relative oxygen supply reaches its limit when $HW^d - kW = 0$, i.e., when the oxygen supply only meets the requirements for maintenance (Q_{maint}), i.e., normal life, including various activities, including reproduction, but with nothing left for further somatic growth. The second part of this chapter and Chapter 9 deal with the seasonal trade-off allowing the elaboration and release of fish reproductive products without much impact on their multiyear growth.

Several physiologists have perceived our definition of 'maintenance' as implying that fish, when they have attained maximum size (L_{max} or W_{max} or approached their asymptotic sizes, L_∞ or W_∞), would have to become completely passive, e.g., could not move to search for food or escape potential predators, nor reproduce (Seibel and Deutsch 2020; Scheuffele *et al.* 2021).[166] Our definition of maintenance encompasses all the activities that fish usually have in the wild or in captivity; the only thing they cannot do is grow (see Chapter 6).

Since $d < 1$, their relative oxygen consumption at first maturation (Q_m) will be higher than when they have attained their maximum size, and their oxygen supply will be sufficient only for their maintenance (Q_{maint}). Thus, $Q_m > Q_{maint}$. Also, it can be shown algebraically that, given length-weight relationships of the form $W = a \cdot L^b$, and gill surface area – body weight relationship of the form $S = a \cdot W^d$, the ratio $L_{max}{}^D/L_m{}^D$ is equivalent to the ratio Q_m/Q_{maint} (Pauly 1984).

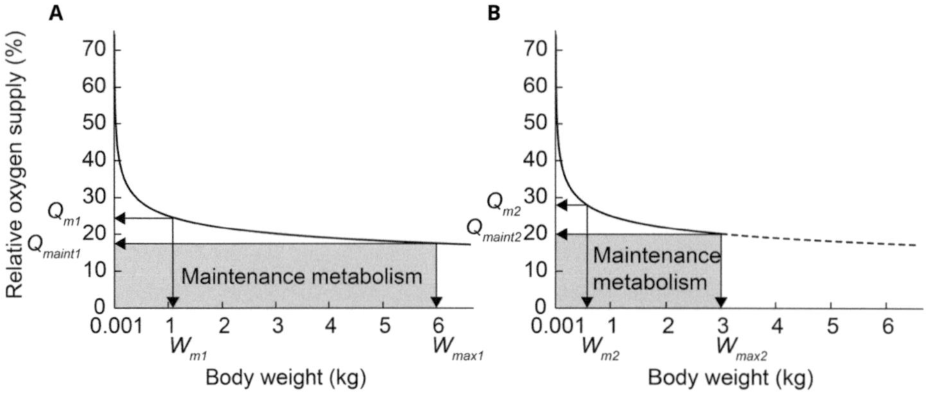

Figure 7.4. A P-diagram representing schematically the major metabolic features of two populations of the same fish species (or other WBE) in two different habitats, one **(A)** with cooler water, leading to a high value of W_{max1} and a relatively low value of the corresponding Q_{maint1} (which must be lower than Q_{m1}), the other **(B)** with warmer water, leading to a lower of W_{max2} and a relatively high value of the corresponding Q_{maint2} (which must be lower than Q_{m2}). The question is now: is the Q_{m1}/Q_{maint1} ratio the same as Q_{m2}/Q_{maint2} ratio? Or, more generally, is Q_m/Q_{maint} 'constant' in fish and other WBE?

The question now becomes: does a Q_m/Q_{maint} ratio estimated for a population whose fish reach a given L_{max} (and hence the corresponding Q_{maint} value) remain the same when the population is exposed to a different temperature regime – one that results in lower L_{max} values and, consequently, higher Q_{maint}? (See also **Figure 7.4**).

The questions in **Figure 7.4** were answered by the study of Pauly (1984), which covered 34 species with 56 pairs of L_{max}^D and L_m^D values,[167] included species ranging in L_{max} from 19.2 to 250 cm, and produced an estimate of the mean $L_{max}^D/L_m^D = 1.36$ with a 95% confidence interval later estimated as 1.22 – 1.53 using the method of Fieller (1940; see **Table 7.2**). The implications of that study were largely ignored (except by Imsland 1999), and its replication by Meyer and Schill (2021) came as a complete surprise. These authors were concerned about the warming-induced reduction of L_{max} and L_m in freshwater salmonids of Idaho (Meyer *et al.* 2023), which support lucrative recreational fisheries, when they came across the 1984 study. They replicated it with 4 species in 51 populations, and surprisingly, they also obtained a mean L_{max}^D/L_m^D ratio of 1.35, i.e., virtually the same as Pauly (1984) estimated for unrelated marine fish.

This inspired several similar studies on other groups of iteroparous bony fishes (**Table 7.2**), confirming that sexual maturation is initiated in these fish when the spawning threshold L_{max}^D/L_m^D ~1.35 is reached.

These findings (and Table 7.2) allow for interesting generalizations, of which

$$L_m = L_{max}/1.35^{1/D} \qquad ...(7.8)$$

may be the most useful.

Other generalizations are possible; for these, we use the label $A = L_{max}^D/L_m^D = 1.35$ for compactness, but only in this chapter.[168] Given the definition of A and length-weight relationships, $A = (W_{max}^{1/b}/W_m^{1/b})^D$, whose inverse ($A^{-1}$, i.e., L_m^D/L_{max}^D) is $A^{-1} = (W_m^{1/b}/W_\infty^{1/b})^D$. Thus, $A^{-1} = (W_m/W_\infty)^{D/b}$ or

Table 7.2. Studies establishing the existence and similarity of the numerical value of a spawning threshold (i.e., L_{max}^D/L_m^D) in various groups of bony fishes. The estimates (± standard errors) were calculated using Microsoft Excel for the slope of a linear regression with the intercept set at zero.

# Bony fish group	L_{max} or L_∞ range (cm)	# of species	# of 'cases'	Estimate (± s.e.)	Reference
1- Miscellaneous (mainly) marine species[a)]	1.9 -250	34	56	1.36 (± 0.06)	Pauly (1984)
2- Freshwater salmonids in Idaho, US.	11.0 -48.1	4	51	1.35 (± 0.04)	Meyer and Schill (2021)
3- Cichlid fish from 3 continents	4.2 - 55	7	41	1.35 – 1.40	Amarasinghe and Pauly (2021)
4- Chinese freshwater and marine fish	5.3-102	117	241	1.40 (± 0.021)	Chen *et al.* (2022)
5- Freshwater and marine fish in or around Turkey	2.5 - 219	57	120	1.44 (± 0.035)	Keskin and Pauly (2023)
6- 22 of 25 sturgeon species[b)]	116-323	22	22	1.35 (± 0.03)	Chu and Pauly (2024a)
7- Indian Ocean pelagics	26.2 - 311	18	18	1.36 (± 0.04)	Gunwardane *et al.* (2024)
8- Indo-Pacific and Atlantic coral reef fish	1.8 - 176	131	207	1.35 (± 0.02)	Chu and Pauly (2024b)
9- 'All' bony fishes[c)]	1.8 - 311	194	379	1.37 (± 0.02)	See **Figure 8.1A.**

1: a) A table including the raw data used for this study was included in the Supplementary Materials of Pauly (2021a); #6: b) This excludes the Syr-Darya and Chinese sturgeons, both extinct, and beluga, a manifest outlier; # 9: c) Combining the studies # 1, 3, 4, 6 and 8.

$$W_m = W_\infty \cdot (1/A)^{b/D} \quad ...(7.9)$$

By combining equation (7.6) and (7.9), we obtain

$$W_i / W_m = (1\text{-}(D/b))^{b/D}/A^{-b/D} \quad ...(7.10)$$

When $W_i > W_m$, we also have:

$$(1\text{-}(D/b))^{b/D} > A^{-b/D} \quad ...(7.11)$$

which implies (1-(D/b)) > A. Now, given the definition of D = b·(1-d), we have

$$d > A^{-1} \text{ implies } W_i > W_m \quad ...(7.12a)$$
$$d \approx A^{-1} \text{ implies } W_i \approx W_m \quad ...(7.12b)$$
$$d < A^{-1} \text{ implies } W_i < W_m \quad ...(7.12c)$$

This all means that, for d < 1/1.35 (i.e., smaller than the value of d = 0.74, as generally occurs in small (reaching less than about 10 cm) and short-lived fishes; **Figure 7.3**), spawning occurs after their growth rate (dW/dt) has started to decline. This is not the case in larger iteroparous, longer-lived fish where d > 0.74. This explains why the first maturity of, e.g., cod (*Gadus morhua*) and similarly large (and well-studied) species occurs well before these fish have experienced their fastest growth, which, we should remember, occurs at W_i (**Figure 7.2**). One cannot but conclude that the 'Reproductive Drain Hypothesis' is based on an invalid assumption and must be retired.

Reconceptualizing maturation and spawning in fish

The case made above is that a heuristic (*sensu* Budaev *et al.* 2019) determines the overall readiness of fish to mature and spawn as a function of their metabolic rate (Q_m) relative to their maintenance metabolism (Q_{maint}). Specifically, this heuristic, in longer-lived iteroparous fish, readies them to perceive seasonal (i.e., within-year) environmental maturation and spawning stimuli only when the threshold A is reached (see **Table 7.1** and the above equations for the definition and properties of A) and not earlier (Pauly 2021b). A framework for understanding the concept of a 'spawning season' is still required. For simplicity, we will assume a single spawning season annually, in spring (Pauly and Liang 2022b).[169]

As will be shown, the processes taking place can be described in a way that various aspects of fish spawning, which had remained unexplained or treated as anomalies, can be straightforwardly accounted for, i.e., without *ad hoc* hypotheses. Still, a second heuristic is required, the 'cusp catastrophe.'

The cusp catastrophe, or 'cusp,' is one of the 7 topological entities which, as shown by Thom (1975), are sufficient to describe the transitions among a maximum of 4 'control factors' *qualitatively* and to distinguish areas where the transitions would be smooth from areas where it would be sudden (thus the term 'catastrophe,' the cause of many misunderstandings). With its two control factors, the cusp can easily represent sudden transitions between, for example, 2 biologically relevant variables (Woodcock and Davis 1978). The growth curve in **Figure 7.5** illustrates our use of the cusp (see insert) and the two control factors that adapt it to maturation and spawning events as functions of body

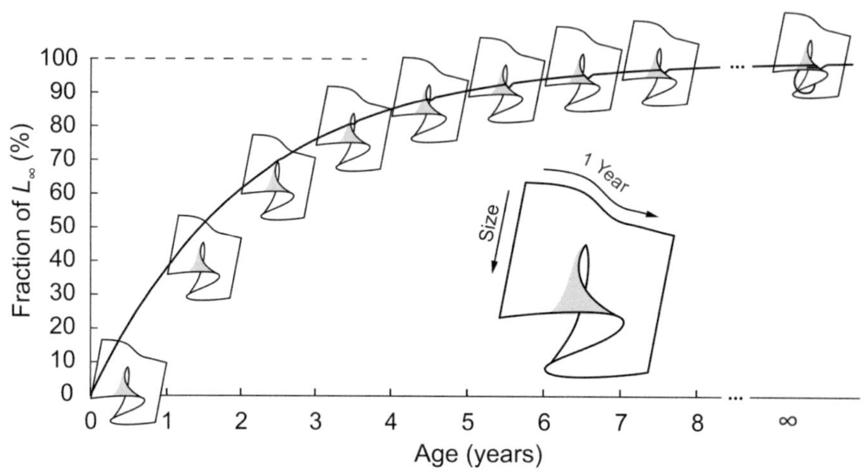

Figure 7.5. A representation of bony fish growth as a succession of cusps, with the youngest fish being too small to enter the region (shaded grey) where spawning occurs. This region is approached in pre-adults, who may undergo an (abortive) maturation but fail to spawn. The following year, they will again reach a size associated with Q_m/Q_{maint} ~ 1.35, and maturation and spawning will proceed (as illustrated by the growth curve overlapping with the grey zones representing both the spawning seasons and the seasonal incentives to spawn). In subsequent years, size increases are associated with a spawning season that starts earlier and ends later until maximum size is approached, when spawning often occurs repeatedly and the body weight is reset at a lower value, allowing growth to be re-initiated.

size. Our cusps have body size and time (here, 1 year, given one spring spawning season per year) as control factors.

Young fish (ages 1 and 2 in **Figure 7.5**) are too small for their L_{max}^D/L_m^D ratio to have fallen to 1.35: they avoid the cusp region where maturation and spawning occur. At 3 years, the largest fish of a cohort will mature and spawn, the smallest will not, and the fish of intermediate size may undergo an abortive maturation, i.e., elaborate gonadal products which are resorbed and not shed (Iles 1974; review in Rideout *et al.* 2005).

This 'skipped spawning,' performed "*more often by young and small fish*" and often when food is scarce (Jørgensen *et al.* 2006), is therefore neatly explained without requiring elaborate models of trade-offs between reproductive output, growth, and survival, and which individual fish, in any case, could not use as an individual heuristic to 'decide' whether to spawn or not.

In this example, the subsequent growth in year 4 and beyond pushes the fish deeper into the shaded area in **Figure 7.5**, implying that the maturation of larger (older) fish should start earlier than in smaller individuals and end later in the season. In contrast, smaller (younger) fish mature (and spawn) only during the peak of the season. Numerous authors confirm this, for example, Rijnsdorp (1989) for plaice *Pleuronectes platessa* in the North Sea, and Trippel *et al.* (1997), who summarized their review of this phenomenon by stating that *"the importance of female size to recruitment success is reinforced by the observation that large females commonly start spawning earlier in the season* [...], *continue for longer, and produce larger eggs with higher viability than smaller females."*

The cusp often implies that once a phase transition has occurred, the system may display 'hysteresis,' wherein the behavior in question (here: spawning) loops repeatedly (at the highest age in **Figure 7.5**, above the ∞ symbol). This is precisely the behavior of large, old fish, which may spawn repeatedly during a spawning season, while small adults usually spawn only once or even skip spawning. This behavior can be interpreted as follows: once a large fish has spawned, i.e., lost some of the tissues it had to supply with oxygen, it has a higher gill area/body weight than before spawning. Hence, when it returns to feeding, its food conversion efficiency will be high. This leads to increased body weight and renewed lowering of the gill area/body weight ratio, which can cause (at least in large individuals) rapid re-maturation and repeated spawning (Trippel *et al.* 1997; see also Chapter 9, where this mechanism is elaborated in more detail). This cycle can repeat within a spawning season until, e.g., the temperatures drop and respiratory stress declines, at which point the reproductive season ends.

In addition to providing a graphic metaphor for spawning, the cusp links three phenomena that, to date, had not been tied to a common explanatory framework, i.e., skipped spawning, size-dependent reproductive seasons, and spawning hysteresis. **Figure 7.5** thus represents an integrated view of the life of a long-lived fish as a succession of cusps, each 'entered' at another size (corresponding to successive ages), implying a different set of responses by the fish in question.

Maturation and spawning are obviously more complicated than presented in this account. Notably, these processes involve the release of numerous hormones in response to environmental stimuli (Pankhurst 2016). As documented in **Table 7.1**, long-lived fish species fail to respond to such seasonal stimuli during many years (even decades) of pre-adulthood during which they could perceive these stimuli.

The 'hormonal cascade' leading to maturation and spawning is not self-starting. For maturation and spawning to be initiated, an internal state is linked to fish size (**Figure 7.5**), i.e., to their relative oxygen supply. The mechanistic simplification presented here aligns with the observed phenomena, regardless of the underlying biochemical complexity of the hormonal cascade involved in the processes of maturation and reproduction.[170]

Finally, the reconceptualization presented here implies that, rather than being analogous to humans and other uniparous mammals, the life-threatening and often debilitating event of giving birth is, in fish, a seasonally *liberating* event. It frees females from a quivering mass of eggs that must be supplied with scarce oxygen and whose removal enables them to grow again.

Spring spawning and global warming

Global change, or global warming, has modified much of the Earth's flora and fauna's reproductive phenology and is expected to continue for decades and centuries. These modifications take various forms, including earlier flowering in some trees and earlier bird migrations (see, e.g., Ahas and Aasa 2006; Cohen *et al.* 2018), many of which are due to sinusoid seasonal temperature oscillations with a higher annual mean (**Figure 7.6**).

Over the past few decades, the temperatures of marine ecosystems have increased (Belkin 2009), as has been observed in freshwater bodies (Carpenter 1992; Kangur *et al.* 2021). Increasing temperatures, on the other hand, will alter the physiology of fish, particularly their reproductive processes, as is expected in ectotherms (Alix *et al.* 2020; Neidetcher *et al.* 2014).

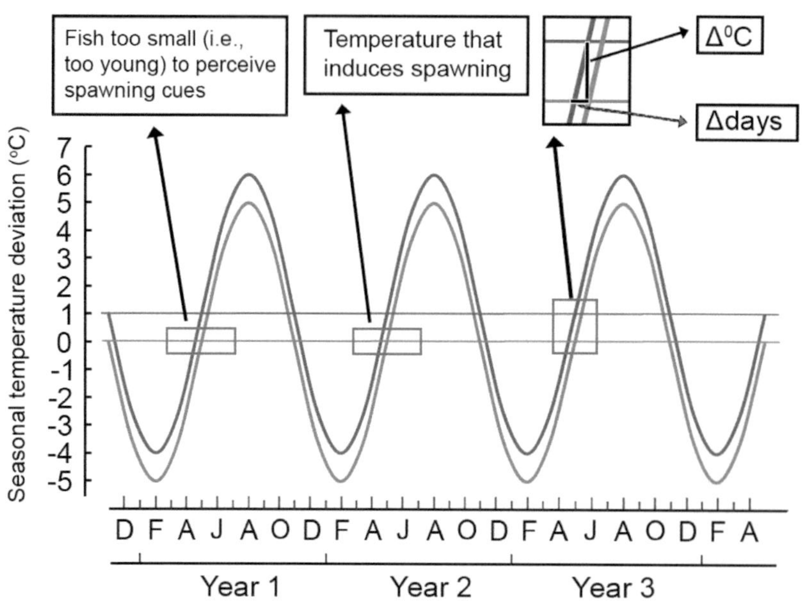

Figure 7.6. Schematic representations of the effect of global warming on the phenology of spring spawning fish via two sine curves. Here, the lower sine curve represents the baseline climate, with a temperature threshold that acts as an evolved trigger for maturation and spawning. In contrast, the upper curve represents a warmer climate, leading to the threshold temperature being reached earlier (see insert on the upper right).

The fact that ocean and freshwater warming is modifying the phenology of maturation and spawning in fishes, i.e., the timing of their reaching puberty, has been widely reported. A meta-analysis of observed phenological shifts suggested that seasonal events of marine species advanced by an average of 4.4 days per decade during the late 20th Century (Poloczanska *et al.*, 2013). Additionally, larvae of 17 of the 43 fish species in the California Current ecosystem occurred earlier in recent decades relative to the last 58 years (Asch 2015). Zooplankton, one of these fish's main food items, did not shift their phenology synchronously with most fishes (Asch 2015).[171] Such a mismatch often disrupts the fine-tuned adaptation that usually results in fish larvae being hatched at the time when potential prey is abundant, which leads to high larval mortality due to starvation or increased predation on slower-growing larvae (Cushing 1990; Asch *et al.* 2019).

Here, we present a simple method, based on Pauly and Liang (2022b), for quantifying the effect of global warming on the onset of maturation and spawning in the spring-spawning freshwater and marine fish of temperate latitudes (roughly 24 °N - 67 °N; 24 °S - 67 °S), for which ample documentation exists. This method consists of a simple model that predicts how spawning in marine and freshwater fish is modified by temperature changes. Its use could stem the proliferation of adaptationist *ad hoc* hypotheses that are currently being proposed to explain why, given the warming of their surrounding water, spring-spawning fish now spawn earlier.

As proposed here, fish can monitor their growth performance, which is strongly dependent on their oxygen supply, by monitoring the ratio of their metabolic rate (Q) relative to their maintenance metabolism (Q_{maint}). Given the fact that the respiratory area of fish gills, as a 2-D surface, cannot keep up with the growth of the 3-D bodies that they supply with oxygen (see Chapter 3), this ratio must decline and, as individual fish grow, reach a value (Q_m) at which maturation is triggered. The ratio Q_m/Q_{maint}, i.e., the spawning threshold for bony fishes, was estimated at ~ 1.35 (see above).

When fish are small (young), their $Q_m/Q_{maint} \gg 1.35$, and no maturation and spawning occur, even during the season that adult conspecifics perceive as 'spawning season.' As the small fish grow, they will reach a size at which Q_m/Q_{maint} declines to 1.35, which triggers maturation and spawning. If this size coincides with a period of increasing temperature, as occurs in spring spawners, the increase will further reduce the Q_m/Q_{maint} ratio (because the catabolic term of the underlying Pütter equation will increase). If the rise in temperature occurs earlier, they will also mature and spawn earlier.

Figure 7.6 defines the problem for which a solution is provided based on two sine curves. These sine curves represent typical seasonal oscillations of sea surface temperatures (SST) or temperature oscillations in (northern) temperate freshwater bodies, where mean monthly temperature tends to oscillate with an amplitude (Amp) of approximately 10 °C (see, e.g., figure 5 in Genner *et al.* 2010). These curves require a shift of 6 months in the southern hemisphere.

The lower sine curve represents the oscillation of the 'baseline' climate, which usually exhibits minimum temperatures in February in the northern hemisphere. The upper sine curve represents a warmer climate, here with a temperature increase of 1 °C. The ascending parts of the upper curve imply differences in the time at which various threshold temperatures are reached. This difference has its annual minimum (Δd_{min}, with the subscript also referring to the 'minus' sign of Δd, and 'd' here referring to days) when the slope of the sine curve is highest. However, Δd does not change much within a range of 3-4 months.

The derivation of an equation allowing for the computation of Δd_{min} is facilitated by shifting the time axis by 6 months. It intercepts the ordinate in August (**Figure 7.7A**), with the curves now turned into cosine curves. The equation of the lower curve in that figure is then:

$$Y_1 = (Amp/2)\cdot\cos(X) \qquad \text{...(7.13)}$$

with *Amp*/2 being half of the summer-winter amplitude.

When $Y_1 = (Amp/2)\cdot\cos(X) = 0$ (g_1 and g_2 points in **Figure 7.7A**), then $\cos(X) = 0$, and $X = \arccos(0) = \pi/2$. The equation of the upper curve is:

$$Y_2 = (Amp/2)\cdot\cos(X) + B \qquad \text{...(7.14)}$$

with *Amp* being the oscillation amplitude, as defined above, and *B* being the temperature difference between the two cosine curves. We then have $Y_2 = (Amp/2)\cdot\cos(X) + B = 0$ (r_1 and r_2 points in **Figure 7.7A**), then $\cos(X) = -2B/Amp$, and $X = \arccos(-2B/Amp)$.

The distance from r_1 to g_1 and from r_2 to g_2 on the X-axis (**Figure 7.7A**) is $[\arccos(-2B/Amp) - (\pi/2)]$. When the frequency of the sine curves is one year (365.24 days), each day corresponds to $2\pi/365.24$. Therefore, in areas where the spring spawning fish experience strong temperature changes between summer and winter, and the water temperature has become warmer on the average (*B*), spawning should occur earlier, by a time predicted by $\Delta d_{min} = [\arccos(-2B/Amp) - (\pi/2)]/(2\pi/365.24)$, or simplified

$$\Delta d_{min} = [\arccos(-2B/Amp) - (\pi/2)]/0.0172 \qquad \text{...(7.15)}$$

which allows Δd_{min} to be calculated for spring spawners, given *Amp* and *B*, or the ratio *Amp*/*B*. **Figure 7.7B** illustrates the range of values that can be taken, given different amplitudes of seasonal growth oscillations (*Amp*) and warming-induced increases of mean annual temperature (*B*).

Table 7.3 shows that the empirical estimates of $\Delta d/\Delta$°C, which can be computed from the numbers in various explicit statements in the literature (mean 12.7 ± 1.6 days/Δ°C) correspond, on the average, to the theoretical estimates of Δd_{min} derived from Equation (7.15), i.e., 11.7 days if the amplitude of the seasonal temperature is 10 °C and the overall temperature increase is 1 °C, as occurred 1 to 2 decades ago when most studies documented in that table were performed.

The sinusoidal model presented here, and Equation (7.15) derived from it, are simplifications of complex phenomena. Still, we are not helpless in the face of this complexity. We know that seasonal temperature oscillations, driven by the Earth's obliquity and highly regular movements around the Sun, generate and maintain the sinusoidal oscillations of many natural phenomena, including seasonal changes in the temperature of water bodies. We also know that baseline temperatures have increased throughout much of the world. Combining these two facts with the knowledge that increasing temperature triggers maturation and spawning in spring-spawning fish leads to a robust model for predicting the extent to which spawning will be shifted back in time, as long as the warming under consideration does not exceed 2 °C when the amplitude of the seasonal oscillation is < 10 °C.[172]

The model may be used to assess whether an observed change in the phenology of spawning is small or large compared to a prediction from Equation (7.15b). Another potential

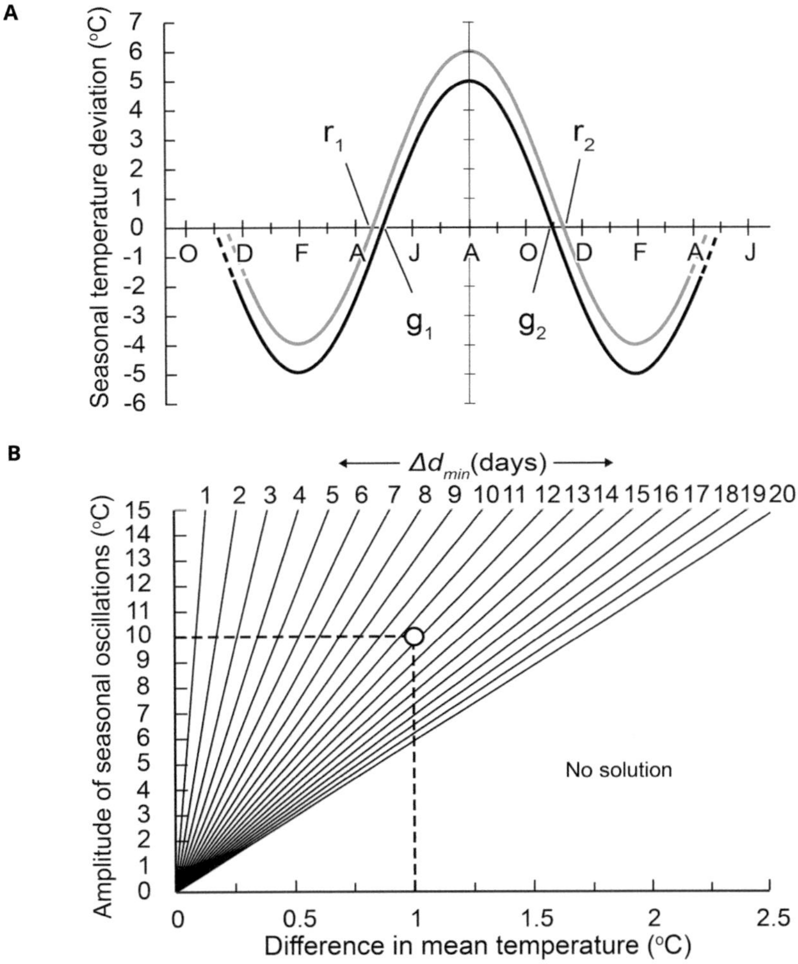

Figure 7.7. How the shift toward earlier temperature-induced spawning is estimated and can be evaluated. **A**: Using trigonometry, with the variables g_1 and g_2 referring to the lower curve and r_1 and r_2 to the upper curve; see also Equations (7.13) – (7.15) and text; **B**: The nomogram illustrates that the solutions of Equation (7.15) are robust, but only within a limited warming range, and becomes useless when the mean annual temperature increase exceeds 2 °C. The dotted lines and the open circle refer to the example in panel **A**, which describes a case where Δd_{min} = 11.7 days.

use of the model would be to assess whether it is justified to evoke the match-mismatch hypothesis (Cushing 1990) or complex or *ad hoc* hypotheses often presented to explain the temporal shift in spawning in terms of adaptation to temporal shifts in the emergence of prey species. Note that this model doesn't account for the delayed maturation of spawning of fall spawning temperate fish, for which ample evidence also exists (Koenigbauer *et al.* 2025) and for which the GOLT currently has no simple, mechanistic explanation.

Diel spawning periodicity and hypoxia

If, as stated above, getting rid of oxygen-consuming eggs is experienced as a *liberating* event by female fish,[173] which begs the question of similar consequences for the shorter timeframes that shape the reproduction of fish and other WBE. Many marine and freshwater species

Table 7.3. Instances of fish maturing and spawning earlier in the 'spring' reproductive season due to higher water temperature. *Δ°C* refers to the difference in temperature between the years, and *Δdays* refers to the time difference in maturation and/or spawning. The standard error and standard deviation (s.d.) in the penultimate row are underestimates, as only the midranges of the *Δd/Δ°C* values were used.

Species	Location (years)	*Δ°C*	*Δdays*	*Δd / Δ°C*	Source
Abramis brama	Tjeukelmeer Lake, Netherlands, 1961 vs. 2006	1.94	20	10.3	Mooij *et al.* (2008)
Acipenser schrenckii	A pond in China, 2001	3 vs. 6	30/50	8.3-10.0	Zhang *et al.* (2012)
Gadus chalcogrammus	Gulf of Alaska (1979 vs. 2015)	--	--	5.0	Roger and Dougherty (2019)
Gadus morhua	North and Irish Sea, 1985 vs. 2015	1.1-1.4	--	8.1-27.4	McQueen and Marshall (2017)
Schizopygopsis selincuoensis	Selinco Lake, Tibetan Plateau, China, 1970 vs. 2000s	~1	11.7	11.7	Tao *et al.* (2008)
Rutilus rutilus	Meuse River, below a nuclear power plant, 1977	3	21	7	Mattheeuws *et al.* (1981)
Rutilus rutilus	Lake Geneva, 1983 vs. 2001	~1	14	14	Gillet and Quétin (2006)
Scomber scombrus	North Sea, 1968 vs. 2008	--	--	7.4-11.4	Jansen and Gislason (2013)
Solea solea	North and Irish Seas, 1970 vs. 2010	1.72	40	23.3	Fincham (2013)
Tachysurus fulvidraco	Aquaculture ponds, China, 2013	5	25-30	23.3	You (2015)
13 marine species	English Channel, spring and summer spawners, 1975 vs. 1987	1.0	17	17	Genner *et al.* (2010)
Multiple spp. (larvae)	South. California, 1949 vs. 2000	1.3	--	17.7	Asch (2015)
10 freshwater spp.	Estonian waters, 1951 vs. 1998	2.65	10-30	3.8-11.3	Ahas and Aasa (2006)
All previous spp.	Marine and freshwater fish in temperate regions	--	--	12.7±1.6)	Mean of above estimates; s.d. = 6.2 days
Spring spawners	Marine and freshwater fish in temperate regions	--	--	11.7	Predicted for *Amp* = 10 °C and *B* = 1 °C

spawn at dusk, meaning that females release their eggs at a time when oxygen availability is at its lowest (Xu and Xu 2015). This is particularly the case in freshwater habitats with high densities of plants and algae, which oxygenate the water in the daytime but require oxygen at night when photosynthesis is impossible. It is at dusk that the competing oxygen demand of plants and algae becomes more acute (Correa-González *et al.* 2014).

The widespread pattern of dusk-spawning in fishes still needs to be fully explained, and traditional answers typically involve predator avoidance or light stimuli (Lobel and Neudecker 1985; Samoilys 1997). While these factors may play a role, it is remarkable that spawning – a highly energy-consuming activity – takes place when fish experience the most severe oxygen limitation of the entire diurnal cycle.[174] Suppose the experience of a changing ratio of oxygen demand and supply indeed induces spawning. In that case, a similar mechanism may be at play at the exact moment of the diel cycle at which spawning occurs.

This hypothesis still needs to be tested, but experiments on small cyprinids such as the zebrafish (*Danio rerio*), a model species well-known as a dusk-spawner, could be easily performed. A setup in which zebrafish groups are exposed to different concentrations of dissolved oxygen in the morning hours (preferably with different light regimes) could test the hypothesis that spawning represents a process that *liberates* fish from tissues that are maintained at high energy cost.

8

The growth and reproduction of Chondrichthyes

Synopsis: The size at first maturity of most Chondrichthyes (sharks, rays and chimaeras), when expressed as a fraction of their maximum size, is larger than in bony fishes and most marine invertebrates. Here, based on the fact that gill surface area, in adults, always scales with $d_G < 1$, which causes relative oxygen supply to decline with size, and body fluids and tissues to become hypercapnic and acidic, we test the hypothesis, presented in the previous chapter, that it is lowered pH inside a fish body which triggers the hormonal cascade leading to maturation. This test is made possible by the urea that all sharks and chimeras use for osmoregulation, which is a base and therefore tends to mute a signal caused by the low pH resulting from hypercapnia. Also, it turns out that in rays, decreasing urea retention correlates with the fraction of size at first maturity relative to maximum size, with freshwater rays approaching the values occurring in bony fishes. These results support the existence of a critical maturation size threshold, while also suggesting a role for urea that was previously undocumented.

Why should Chondrichthyes differ from most other water-breathers?

Given that the GOLT appears to apply to taxa as evolutionarily remote from each other as bony fishes (Chapter 7) and crustaceans (Chapter 12), and particularly that they also seem to rely on a critical value of the $L_{max}{}^{D}/L_{m}{}^{D}$ ratio as a heuristic to trigger their first reproduction, expanding the GOLT to cover the class of Chondrichthyes, or cartilaginous fishes, was unavoidable. The task of testing this specific aspect of the GOLT was entrusted to a brilliant graduate student, and her MSc thesis (Warren 2023) was subsequently developed into an article (Warren and Pauly 2024), whose pre-submission version formed the basis of this chapter.

As detailed in several previous chapters of this book, the Pütter growth equation provides the starting point for our analyses of growth and reproduction in water-breathing ectotherms. This model also underlies the considerations presented in this chapter. In addition, we also demonstrate that the argument presented in Chapter 7, regarding the need for a size-related trigger to set the perception of environmental stimuli, which themselves initiate the hormonal cascade leading to maturation and eventual reproduction, also pertains to Chondrichthyes and their various sub-classes. As it turns out, Chondrichthyes illustrate the pattern discussed in Chapter 7 in a surprising way, offering additional support for the proposed mechanisms that underlie the correlation between body size, oxygen supply, and maturation. As mentioned in Chapter 7, the GOLT's explanation of this pattern is based on the fact that Pauly (1984) estimated that $L_{max}{}^{D}/L_{m}{}^{D} \approx 1.36$ based on 56 populations in 34 species of bony fishes, ranging in L_{max} from a few cm in guppies to over 2 m in tuna, and on the fact that the number of subsequent studies, involving hundreds of bony fish and invertebrate species confirmed not only the existence of a 'spawning threshold' ($L_{max}{}^{D}/L_{m}{}^{D}$) but its specific value, estimated back in 1984 (see Chapters 7 and 12).

The ubiquity of this ratio suggests that it evolved early in the evolution of WBE as a handy heuristic (Budaev *et al.* 2019; Morbey and Pauly 2022), enabling them to mature when the relative surface area of their gills can still supply both their somatic growth and the energetically costly production of gonadal products. This inference is supported by experimental evidence showing that external hypoxia causes maturation at sizes lower than in normoxia (Kolding *et al.* 2008). Therefore, if maturation is triggered by a critical oxygen supply and its corresponding low pH threshold, it may be hypothesized that any WBE possessing the ability to buffer against low pH may be able to delay maturation to a larger fraction of their maximum size (i.e., $L_{max}{}^{D}/L_{m}{}^{D} < 1.36$).

Despite their ancient origins, only 5% of all cartilaginous fish species are known to be euryhaline, typically entering freshwater seasonally or during juvenile stages (Constance *et al.* 2023). Another 3-4% of all Chondrichthyes live solely in freshwater and cannot tolerate strong fluctuations in salinity (Ballantyne and Fraser 2012). This contrasts with bony fish, of which about 40% occur only in freshwater (Ballantyne 2016), with numerous neritic species easily entering freshwater bodies, especially in the tropics (Longhurst and Pauly 1987).

A reason for the rarity of freshwater cartilaginous species may be found in their osmoregulatory mechanisms. In marine environments, aquatic organisms must be able to tolerate an environment that is hyperosmotic to their body fluids (Greenwell *et al.* 2003). Marine bony fish 'osmoregulate' through their seawater intake and active excretion of excess salt ions (Ballantyne and Fraser 2012). In contrast, cartilaginous species prevent excess water loss by retaining high urea levels in their tissues, matching the osmolarity of their surrounding environment, a strategy also known as 'osmoconforming' (Ballantyne and Fraser 2012).

While critical for osmoregulation, urea is also known to aid in acid-base regulation in Chondrichthyes and requires carbon dioxide (CO_2) for its synthesis, given its chemical composition, i.e., $2NH_3 + CO_2 + CO(NH_2)_2 + H_2O$ (Ballantyne and Fraser 2012; Withers 1998). Since marine Chondrichthyes need to produce urea to maintain an appropriate osmotic environment within their tissues (Ballantyne and Robinson 2010), it is reasonable to assume that they possess the ability to remove CO_2 at a greater rate compared to bony fish, which can only do so via diffusion across their gills.

This adaptation presents Chondrichthyes with a challenge in environments of reduced osmolality, as high urea concentrations create a hypoosmotic internal environment prone to a rapid influx of water (Ballantyne and Fraser 2012). Despite this, there have been incursions of marine Chondrichthyes into freshwater environments, the first of which is thought to have occurred 25 million years ago (Ballantyne and Fraser 2012) and involved mainly rays. The main adaptation of rays to reduced salinity is a substantial reduction in urea retention. While marine stenohaline species of rays retain upwards of 400 mM urea, euryhaline species have reduced their concentrations to 80-150 mM, and freshwater stenohaline species have lost the ability to retain urea, with tissues containing only negligible amounts, < 1 mM (Ballantyne 2016).

With obligate freshwater, saltwater, and euryhaline species possessing varying urea concentrations, the Chondrichthyes may be an ideal group of aquatic organisms to test how the L_{max}^D/L_m^D ratio may vary. If a given spawning threshold (L_{max}^D/L_m^D) makes WBE respond to external stimuli, therefore inducing maturation, it can be expected that ureosmotic species will mature at a greater fraction of their maximum size than non-ureosmotic species. Specifically, it is hypothesized that L_{max}^D/L_m^D will differ depending on salinity, with reduced urea retention (i.e., reduced ability to remove CO_2) being correlated with an increase in mean L_{max}^D/L_m^D ratios.

Moreover, there is an additional oxygen-related factor to consider when dealing with the reproduction of Chondrichthyans: as recent surveys have shown, the vast majority of sharks and rays are viviparous (Brooks *et al.* 2024). Given that large embryos that develop in the mother's uterus substantially add to the oxygen demand of the maternal organism, this raises the question of how large sharks can cover the metabolic costs of viviparity. Recent research suggests that some live-bearing sharks developed strategies to cover the additional oxygen demand by 'flushing' the uterine fluid with freshly oxygenated seawater.

The pups in these species can take up oxygen through their gills while still in the womb, thereby relieving the mother from supplying them with high volumes of additional oxygen (Tomita *et al.* 2016). It is still unclear how many species are capable of uterine flushing, and even in confirmed cases, this resource is not endless; pregnant females still provide their pups with oxygen that passes through their gills (Tomita *et al.* 2012, 2017). This suggests an additional oxygen strain on viviparous females, absent in males or oviparous females. It can be hypothesized that the additional oxygen demands associated with viviparity will increase L_{max}^D/L_m^D, independent of phylogeny or habitat. To test these hypotheses, life-history parameters of all known cartilaginous species with suitable data (L_m and L_{max} pairs) have been compiled through a literature review, and their L_{max}^D/L_m^D ratios have been examined, while also accounting for other life-history parameters.

Finally, this chapter addresses a more fundamental methodological point about integrating data into models designed to elucidate interspecific patterns. It is often stated that, in interspecies comparisons, data points cannot be treated as independent values due to the shared evolutionary history among species (Revell 2010). Indeed, some authors suggested that analyses without phylogenic considerations are both "*flawed*" and "*biased*" (Ricklefs and Starck 1996), a topic addressed further below. After discussing the life-history invariances of growth and reproduction in bony fishes and chondrichthyans, the result is contextualized with a brief discussion of the 'living fossil' *Latimeria chalumnae*, or the Coelacanth. This species, which shares features with both ray-finned fish and elasmobranchs, illustrates how different phylogenies lead to comparable patterns and reflect the universality of the fundamental constraints that underlie water-breathing and its specific functional consequences for aquatic lifestyles in aquatic vertebrates.

Creating a suitable database[175]

To detect life-history invariances in Chondrichthyes, information was collected for all cartilaginous species for which at least one L_{max} and L_m data pair could be obtained from reliable sources such as scientific papers and textbooks, as well as 'grey' literature, including theses, technical reports, and museum collection databases.[176]

Length measurements for all species were recorded in centimeters (cm) and weights in grams (g); details on the taxa-specific measurements are provided in Warren (2013) and Warren and Pauly (2024), but they are not relevant here, as for all three groups of Chondrichthyes covered here, the length type (TL, FL, etc.) has no influence on the results because it is the L_{max}^{D} to L_{m}^{D} ratios which were of interest, not the absolute values of L_{max} or L_m. Warren (2023) should also be consulted for other assumptions made while assembling the required estimates of L_{max}^{D} and L_{m}^{D} for females and males Chondrichthyes, as well as their reproductive traits and habitats. This also applied to their length-weight relationships, which were used, jointly with an L_{max} estimate for each species, to derive W_{max}, from which d values were derived using

$$d = 0.674+0.0357\cdot\log(W_{max}) \qquad ...(8.1)$$

an equation (already encountered in **Figure 7.3**) based on a robust regression method of 27 estimates of maximum body weight (in g) in various bony fish taxa and estimates of d from gill area and/or respiratory studies. Equation (8.1) also provides reasonable estimates of d in sharks (see d values in table 1 of Bigman *et al.* 2018).[177]

The estimates of d thus obtained were converted to values of D using the formula $D = 3(1\text{-}d)$, which allowed for calculating the L_{max}^{D}/L_{m}^{D} ratio for each elasmobranch species. The average ratios between L_{max}^{D} and L_{m}^{D} for various groups of elasmobranchs were then estimated by regressing L_{max}^{D} against L_{m}^{D} with the intercept forced through zero using Microsoft Excel, which also produced a 95% confidence interval (C.I.) for each of the ratio estimates.[178]

Finally, to determine whether shared evolutionary history introduces bias into regression models, the steps mentioned previously were repeated while also accounting for the effect of phylogeny. Here, only one L_{max}^{D} and L_{m}^{D} data pair per species was used in the analysis, while the impact of phylogenetic affinities was accounted for by first obtaining a complete phylogeny from VertLife (www.vertlife.org); the few species that did not have a resolved phylogenetic position on this tree were removed from the analysis. The phylogenetic signal (Pagel's lambda) was then determined using the *fishtree* package in R v.4.0.2 and integrated into each regression model.

Of the 10,000 possible phylogenetic trees available on VertLife, 10 were randomly selected for analysis, with the above steps being repeated for each tree. This was done to generate an average slope, ensuring that tree selection did not affect the results. A comparison of regression models, both with and without phylogenetic considerations, was then conducted to ensure that the results of this study were not affected by random effects due to faulty individual measurements (Chen *et al.* 2022).

A wild hypothesis and surprising test results

In total, 1622 L_m and L_{max} data pairs in 918 species of Chondrichthyes were collected, representing 414 species of sharks (vs. 561 shark species in FishBase), 470 species of rays (vs. 717), and 34 species of chimaeras (vs. 55); given these high numbers a bias due to a small sample size is unlikely.

Forty-nine shark species lacked a resolved phylogenetic position on the trees obtained from VertLife and were therefore excluded from the analysis (see Warren 2023, Table S4). The remaining species (n = 438) were analyzed in separate regressions, both accounting and not accounting for phylogeny. Combining all the L_{max}^{D} and L_{m}^{D} data pairs (i.e., without considering phylogeny) yielded a slope $L_{max}^D = 1.13 \cdot L_m^D$ (±0.01). Accounting for phylogeny resulted in a significantly lower average slope, i.e., $L_{max}^D = 0.98 \cdot L_m^D$ (±0.06), or $L_{max}^D \approx L_m^D$.

This is an absurd result; while cartilaginous species are known to mature at a larger fraction of their maximum sizes than bony fishes, with $L_m \approx 0.75 \cdot L_{max}$ (Cortés 2000; Ebert *et al.* 2006), this fraction is not $L_m \approx L_{max}$, as phylogenetic analyses would suggest. Other authors have noted that phylogenetic methods can result in *"poor statistical performance"* (Revell 2010). Weak correlations and wide confidence intervals have also been reported using these methods (Rohle 2006), and it is suggested that the inclusion of a phylogenetic signal is unnecessary in (*i*) phylogenetically diverse datasets and (*ii*) whose raw data exhibits strong correlations (Ricklefs and Starck 1996). Given that the dataset used in this research meets both of these criteria and the biologically nonsensical slope produced by the inclusion of phylogeny, its results are not considered further. Only results analyses performed without consideration of phylogeny are discussed below.

The finding that the values of L_{max}^D and L_m^D were correlated ($p < 0.001$) across each of the major groups of Chondrichthyes was prominent. That an average L_{max}^D/L_m^D ratio of 1.13 (± 0.01) was obtained for non-viviparous marine species (n = 994; 821 species; **Figure 8.1B**) was not obvious, as it was significantly lower than the mean ratio $L_{max}^D/L_m^D = 1.37$ (± 0.04), obtained by combining the L_{max}^D vs. L_m^D data pairs in Amarasinghe and Pauly (2021), Chen *et al.* (2022), Chu and Pauly (2022), and Pauly (1984) **(Figure 8.1A)**.

This result confirms that marine non-viviparous Chondrichthyes, i.e., the majority of cartilaginous fishes, experience their first maturity at a significantly larger fraction of their maximum size than bony fish (Chapter 7). This cannot be explained away by stating that 'sharks are different' or some other facile (non-)explanation, as some authors have suggested, because crustaceans and other invertebrate WBE (as shown in Chapters 12 and 14) have L_{max}^D/L_m^D ratios similar to those of bony fishes, although they have an evolutionary history very distinct from that shared by bony and cartilaginous fishes.

In viviparous marine species, of which 349 L_{max}^D and L_m^D data pairs were collected, representing 262 species of sharks and 79 species of marine rays, the values of L_{max}^D and L_m^D were also correlated ($p < 0.001$) across each group. More importantly, they yielded an average L_{max}^D/L_m^D ratio of 1.16 (± 0.01) for sharks (**Figure 8.2B**) and 1.24 (± 0.03) for rays **(Figure 8.2C)**, the latter differing significantly from non-viviparous marine Chondrichthyes

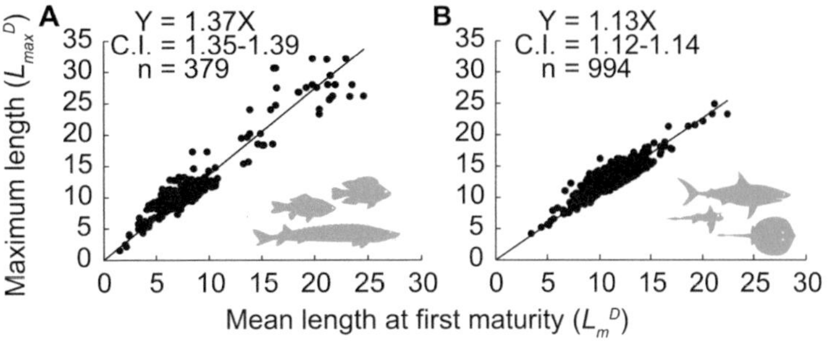

Figure 8.1. Plots of L_{max}^D vs. L_m^D and 95% C.I. in two clades of fishes. **A**: Bony fishes, combining data in the studies of Amarasinghe and Pauly (2021), Chen *et al.* (2022), Chu and Pauly (2022), and Pauly (1984). **B**: Non-viviparous marine cartilaginous fishes (sharks, rays, and chimaeras). Note that these regressions have significantly different slopes.

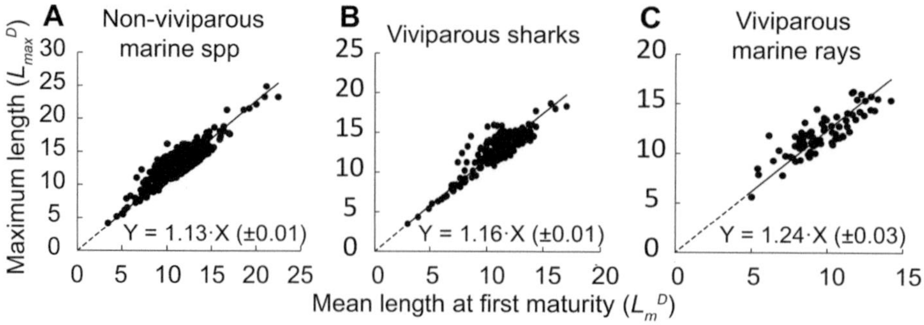

Figure 8.2. Plots of L_{max}^D vs. L_m^D and 95% C.I. for different groups of cartilaginous fishes. **A**: Non-viviparous marine species (sharks, rays, and chimaeras). **B**: Viviparous sharks. **C**: Viviparous marine rays. Note that these regressions have significantly different slopes.

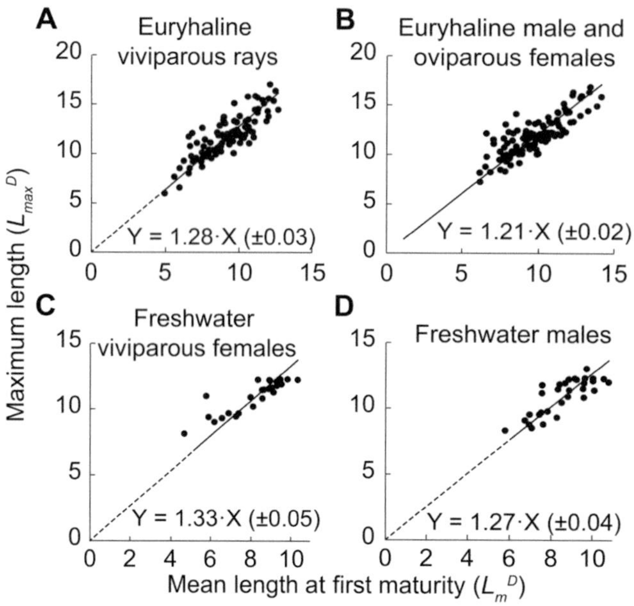

Figure 8.3. Plots of L_{max}^D vs. L_m^D and 95% C.I. for rays in different habitats. **A**: Euryhaline viviparous species. **B**: Euryhaline males and oviparous females. **C**: Freshwater viviparous species. **D:** Males of freshwater species.

(Figure 8.2A) and suggesting that viviparity imposes an oxygen burden on female sharks, resulting in a reduced ability to delay maturation. This is probably what led to the ability of female viviparous sharks to flush their uterus with oxygen-rich seawater, which their embryonic pups can breathe. Uterine flushing does not occur in viviparous rays, which otherwise could have explained their higher L_{max}^D/L_m^D ratio.

Finally, for non-marine (i.e., euryhaline and freshwater) species, 279 L_{max}^D and L_m^D data pairs were assembled, representing 124 species of euryhaline and 29 species of freshwater rays; the two variables were correlated ($p < 0.001$) across both euryhaline and freshwater species (viviparous and non-viviparous). In contrast to oviparous marine rays, which have a mean $L_{max}^D/L_m^D = 1.13$ (**Figure 8.1B**), euryhaline and freshwater rays have high L_{max}^D/L_m^D ratios, whether they are viviparous (**Figure 8.3A**) or oviparous (**Figure 8.3B**), and independently of sex (**Figure 8.3 B-D**). Most importantly, it is in freshwater that the means L_{max}^D/L_m^D reaches its highest values for Chondrichthyes, i.e., 1.33 ($\pm$ 0.05), which is statistically the same as that in bony fishes and multiple invertebrate WBE.

Comforting implications

Similar to bony fish, the results of this research demonstrate that L_{max}^D/L_m^D is also conserved across marine non-viviparous cartilaginous species ($L_{max}^D/L_m^D \approx 1.13$), which implies that L_m can be predicted from L_{max} using:

$$L_m = L_{max}/1.14^{1/D} \quad ...(8.2)$$

The 1.13 ratio is lower than that in bony fish and various invertebrate species (for which $L_{max}^D/L_m^D \approx 1.36$), confirming that marine Chondrichthyes delay maturation to a greater fraction of their maximum size than other WBE. This is supported by the fact that most cartilaginous species tend to mature at around 60-90% of their maximum length, typically averaging around 75% (Cortés 2000; Ebert *et al.* 2006), a ratio that is higher than typically observed in iteroparous bony fishes (see Beverton 1963; Prince *et al.* 2015). Similarly, it is well recognized that Chondrichthyes are late to mature, a life-history characteristic that has been well investigated due to the high susceptibility to fishing pressure it causes (Dutilloy and Dunn 2020). Looking specifically at L_m/L_{max}, Tsikliras and Stergiou (2015) also noted a significantly smaller ratio for bony fish (0.59) compared to sharks (0.70) and chimaeras (0.77) in the Mediterranean Sea.

The CO_2 required for urea production likely enables ureosmotic species to remove CO_2 from their bodies faster than those who can only do so via diffusion across the gills. Consequently, ureosmotic species maintain a relatively high pH in the bodies' tissues and fluids despite their relative oxygen supply declining to a level that would trigger maturation in bony fishes. They delay the critical maturation and spawning threshold to a greater fraction of their maximum size. Non-viviparous euryhaline rays also seem to confer some benefit from intermediate urea levels, removing CO_2 at a greater rate than freshwater rays and bony fishes (**Figures 8.1** and **8.3**).

The reduction of urea concentration seems to be reflected in the L_{max}^D/L_m^D ratio (≈ 1.21) in these species, which, although lower than in freshwater rays (≈ 1.27), is higher than observed in marine rays and other Chondrichthyes, where $L_{max}^D/L_m^D \approx 1.13$ (**Figures 8.1** and **8.2**). These results emphasize the role of (lack of) oxygen in triggering maturation and the existence of a critical hypercapnic (elevated CO_2 in the blood) consequent increase in acidity, which affects maturation and spawning threshold. Additionally, while it has long been established that Chondrichthyes are late to mature, these results provide a mechanistic explanation for this phenomenon.

A pattern confirmed by the Coelacanth

Having discussed the two largest groups of vertebrates commonly called 'fish,' the differences and parallels raise the question of how the observed patterns pertain to oddities such as the Coelacanth (*Latimeria chalumnae*), which is often described as a 'living fossil.'

Although it is a lobe-finned fish, *L. chalumnae* shares many features with cartilaginous fishes, including similarities in reproduction (Smith *et al.* 1975) and osmoregulation. Thus, *L. chalumnae* is also ureosmotic (Griffith 1991), maintaining over 350 mM of urea within their tissues, a level similar to that of marine cartilaginous species (Pickford and Grant 1967). If the timing of maturation can be delayed due to the physiological effects of urea, it would be expected that *L. chalumnae* would also display high values of L_m/L_{max} and low values of their $L_{max}{}^D/L_m{}^D$ ratio. Available measurements allow for estimates of L_m/L_{max} corresponding to 0.74 for males (Froese and Palomares 2000; Mahé *et al.* 2021) and 0.75 for females (Froese and Palomares 2000), similar to sharks (0.70) and chimaeras (0.77) and significantly higher than bony fishes (see above) with an estimate of $D = 0.44$ corresponding to $d = 0.85$, as estimated from a maximum weight of 98,000 g and Equation (8.1), while an estimate of $L_{max}{}^D/ L_m{}^D = 1.13$ is obtained for *L. chalumnae.*

The value of $L_{max}{}^D/ L_m{}^D = 1.13$ is similar to that of non-viviparous marine rays, sharks, and chimaeras and significantly lower than bony fishes as well as euryhaline (≈ 1.21) and freshwater rays (≈ 1.27). The low $L_{max}{}^D/L_m{}^D$ ratio values in *L. chalumnae* and marine Chondrichthyes suggest a shared triggering mechanism for maturation and a shared ability to delay maturation compared to other WBE. The commonality suggested here is the presence of tissues suffused with urea and the resulting ability to offset the critical hypercapnic maturation and spawning threshold through consistent CO_2 removal.

Further supporting the existence of a critical maturation and spawning threshold are viviparous females, whose increased oxygen demands associated with this form of reproduction results in an increase in $L_{max}{}^D/L_m{}^D$ (i.e., $L_{max}{}^D/L_m{}^D >$ males and oviparous females), independent of phylogeny or habitat (**Figures 8.1** and **8.2**). Despite this, the significant differences between viviparous sharks and marine rays (**Figure 8.1B** and **C**), with the higher $L_{max}{}^D/L_m{}^D$ displayed by marine rays, suggest a higher maternal (oxygen) investment than in viviparous sharks.

We suggest that these results add to a better understanding of the oxygen uptake modes in pups. In-utero studies are often challenging, and oxygen uptake in viviparous species appears to be based on a minimal number of species (Carter and Soma 2020; Tomita *et al.* 2016). In sharks, it has been suggested that oxygen diffusion across the uterus appears sufficient during the early stages of development. The increasing oxygen requirements and diffusion rate limits suggest this is inadequate for late-stage embryos (Tomita *et al.* 2016). During this developmental stage, shark pups are thought to receive only 15-30% of their oxygen requirements from their mother, with the remaining supply provided by periodic flushes of the uterus with seawater, as documented in *Squalus acanthias* and *Orectolobus ornatus* (Tomita *et al.* 2016). Similar inferences have been made in *Somniosus microcephalus* (Carter and Soma 2020). While uterine flushing has not been observed in this cryptic species, it is suspected that oxygen diffusion across its uterus is insufficient to support the oxygen demands of mid-to-late-stage embryos or even early-stage embryos in larger litter sizes (Carter and Soma 2020).

In contrast, uterine flushing has not been documented in ray species. While it cannot be stated with certainty that it does not occur, current knowledge suggests that rays do not employ this strategy or provide an appreciable oxygen support to their pups. It seems likely that the development of uterine trophonemata, highly vascularized structures involved in gas exchange and nutrient supply (Hamlett and Hysell 1998) and unique to viviparous stingrays, allows females to support their pups with abundant oxygen throughout the entire gestational period (de Sousa Rangel *et al.* 2020). While modifications of the uterine lining in sharks do suggest a facilitation of gas exchange, direct maternal-pup contact is absent or minimal compared to rays in which the entire uterus acts as a respiratory exchange surface via dense trophonemata which has even been observed to enter the mouth, gill slits, and spiracles of ray pups in utero (Hamlett and Hysell 1998).

While sharks and rays are viviparous, their maternal investment and metabolic (oxygen) requirements are not comparable. Nevertheless, the increased L_{max}^D/L_m^D ratios associated with viviparity and the fact that these ratios generally increase with decreasing urea retention, which is identical in males and oviparous females (**Figures 8.1** and **8.2**), implicate both oxygen and urea in the timing of maturation.

While the main conclusion of this chapter is that urea production enables various species to delay the critical hypercapnic (and/or acidic) maturation and spawning threshold, this conclusion was not experimentally tested. Therefore, empirical investigations into the role of urea and other aspects of chondrichthyan physiology in relation to internal hypoxia and maturation would be of great interest to the GOLT, especially if they are extended to coelacanths and other ureosmotic species, including invertebrates.

Additionally, the rising CO_2 levels in global seas and freshwater bodies, along with their acidification, may affect the functioning of urea as an osmoregulator, as suggested for other compounds with similar properties (see Giareta *et al.* 2023). Major consequences could include changes to previously known urea functions, such as acid-base regulation and osmoregulation, as well as, as suggested in this chapter, size at first maturation. Such changes may be especially important for ecologically and economically important species of elasmobranchs, of which 37% are currently being threatened with extinction (www.iucnredlist.org; see also Constance *et al.* 2023).

Nevertheless, this research has demonstrated the applicability of the GOLT via a detour that elucidates the masking role of urea, which, in a warming and deoxygenating world, will be required to explain changes in the performance of fish and other WBE.

9

Post-spawning acceleration and its implications

Synopsis: The conventional view of spawning in iteroparous bony fishes, i.e., the 'Reproductive Drain Hypothesis,' is based on the observation that somatic growth (in length) slows down noticeably at approximately the time fish attain maturity, and hence the assumption is made that investment in gonadal development slows down growth. In contrast, when this is translated as growth in weight, the weight at first maturity (or puberty) is usually smaller than the weight at which growth rate is highest, i.e., weight growth accelerates after first maturity. We solve this conundrum, with some emphasis on female cod (*Gadus morhua*) by proposing that the substantial loss of body mass experienced by fish as a result of spawning is quickly compensated for by increased somatic growth after the spawning period, notably because of the increase in food conversion efficiency resulting from a sudden loss of body weight, which unavoidably leads to a large increase in relative oxygen supply via the gills.

Tackling 'compensatory growth'

In the previous two chapters, and especially in Chapter 7, we presented several arguments against the conventional view of the relationship between fish growth and reproduction, i.e., that fish grow rapidly until they begin to mature and spawn, with reproduction being the cause for the reduction in growth (see Chapter 7). As we previously demonstrated, this view is untenable, and the assumption that growth slows down due to energetic investment into reproduction produces unrealistic results when related to real-life data. The relationship between growth, maturation, and reproduction is the opposite of the conventional view; Iles (1974) has summarized this concisely: *"far from growth depending on maturity, on the contrary, maturity depends on growth."*

This raises the question of how the fish that spawn produce gonadal products, i.e., egg and sperm, in large quantities without their apparent effects on their overall growth curves, as described by the VBGF. Here, based on Pauly *et al.* (2023), we present evidence that the substantial loss of body mass experienced by fish, especially females, as a result of spawning is compensated by increased somatic growth after the spawning period, providing a simple explanation for this phenomenon.[179] We propose an alternative hypothesis for the decline of growth with increasing body size and show that fast growth after spawning is one of its predictions.

A revision of the idea that reproduction reduces or stops overall growth is also urgently needed because the 'Reproductive Drain Hypothesis,' continues to inform current debates about the impact of climate change on fishes and other aquatic animals. The global trend to smaller body sizes at higher temperatures is often explained by a redistribution of energy to earlier reproduction (see, e.g., Wootton *et al.* 2022; Audzijonyte *et al.* 2019). This perspective on the interaction between growth and reproduction at increased temperatures is often based on resource allocation models (Kozłowski *et al.* 2004) or on interpretations based on the *Dynamic Energy Budget Theory* or DEB (Moreira *et al.* 2022; Lavaud *et al.* 2021). According to these approaches, adult growth continuously competes with reproduction, and the allocation of energy to both processes is a key factor in life-histories.

The gonadosomatic index of temperate fish can be very high

The work on the contribution from which the bulk of this chapter is adapted (Pauly *et al.* 2023) began with a search for reports published on the gonadosomatic index (*GSI*) of fish, i.e., the weight of ovaries (or testes) relative to female (or males) body weight in iteroparous gonochoric bony fishes, which explicitly covered both the periods before and after spawning events. This literature search focused on temperate latitudes and yielded **Table 9.1**.

Table 9.1 was complemented with original data on relative gonad weights of the 6 main commercial species of fish from the Western Baltic Kiel Bight, Germany (**Figure 9.1**), from a monitoring survey conducted in collaboration with commercial fishers in Kiel Bight, Germany (Froese *et al.* 2020).

The *GSI* data in **Table 9.1** and **Figure 9.1** were complemented with a literature review on the immediate post-spawning growth of 'spring spawning' fish, i.e., a phenomenon typical for most fish species at temperate latitudes.

As shown above, the gonadosomatic index of females of temperate species, such as herring (*Clupea harengus*), plaice (*Pleuronectes platessa*), and cod (*Gadus morhua*) can reach extreme values, sometimes exceeding 30% of the total body weight. After spawning, the females lose most of this extra volume, typically followed by an intensive feeding period (Trippel *et al.* 2014; Shul'man 1974). The extreme values of *GSI* in temperate fishes suggest that, for a while, they allocate their resources to gonadal rather than somatic material. After spawning, the converse applies, and the body weight reduction that spawning implies causes a massive increase in food conversion efficiency, which, combined with intensive post-spawning feeding, leads to a growth spurt that can last until the end of the respective

Table 9.1. Weight of ripe gonads in % of the body weight of (mainly) female iteroparous fish from temperate seas and freshwaters, in descending order of (midrange) % *GSI* values; the + symbol indicates the existence of some higher values.

Species	*GSI* (%)	Source
Scophthalmus maximus	30+	Pauly *et al.* (2023)
Platichthys flesus	27-30+	Pauly *et al.* (2023)
Clupea harengus	25-30	Pauly *et al.* (2023); also '20+' in Iles (1974)
Pleuronectes platessa	25-30	Pauly *et al.* (2023)
Coregonus lavaretus	25-27	Rösch (2000)
Gadus morhua	20-30	Trippel *et al.* (2014) and Pauly *et al.* (2023)
Limanda limanda	20-27	Pauly *et al.* (2023)
Perca fluviatilis	16-28	Le Cren (1951) and Heibo (2005)
Acipenser fulvescens	8-12	McKinley (1998)

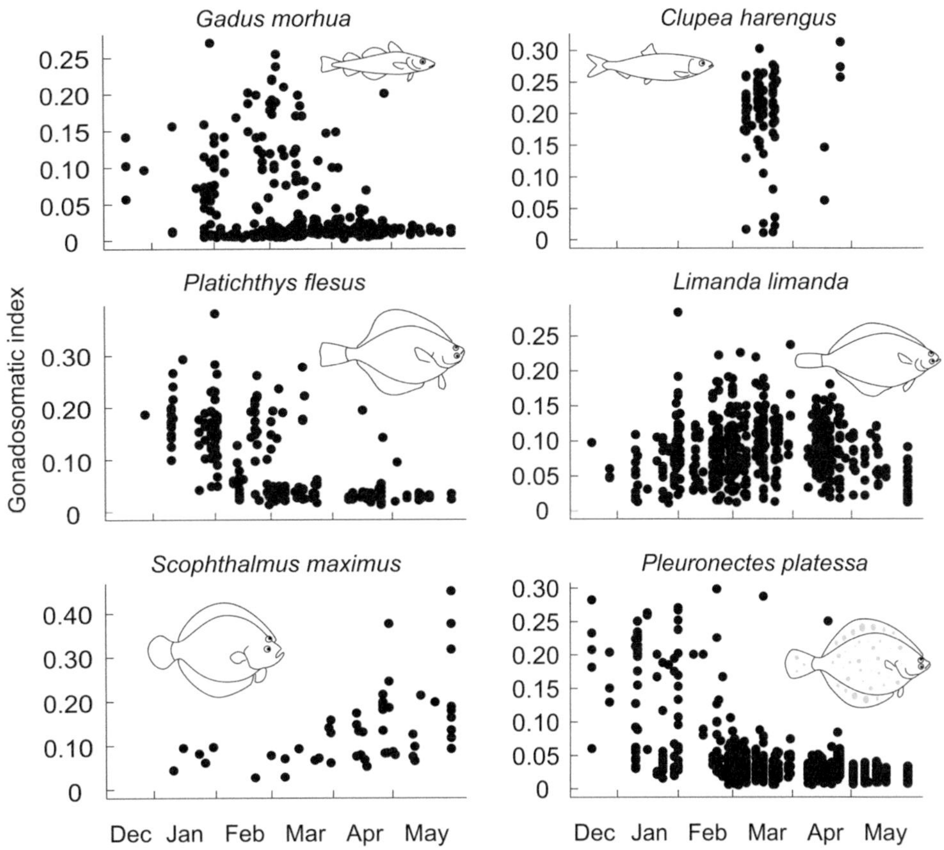

Figure 9.1. Gonadosomatic index (ovary weight relative to whole body weight) for females of the main commercial fish species in Kiel Bight (Germany) in 2020, 2021, and 2022. Note that all species spawn in spring, with flounder (*Platichthys flesus*) and plaice (*Pleuronectes platessa*) starting in January and turbot (*Scophthalmus maximus*) starting in April and/or May, and that peak ovary weight is near or above 30% of body weight in all 6 species (see Froese *et al.* 2020 for a description of the project that generated the data used to produce this figure).

season. However, to understand the dynamics of growth and reproduction, taking this change in food conversion efficiency seriously is crucial, as will be discussed further in the section below.

Food conversion efficiency – an often-misinterpreted concept

The *GSI* data cited in **Table 9.1** and their relationship to subsequent post-spawning events can be better understood when analyzed using our interpretation of Pütter's growth model (see preceding chapters). We recall that this model, jointly with $d < 1$, implies that as fish grow, less oxygen per unit body weight is available to turn the ingested food into metabolically usable 'energy' (e.g., ATP), which causes large fish to either reduce their relative food intake per unit of time and/or causes them to excrete food-derived nitrogenous compounds via their gills (see Webb 1978 and Chapter 2).

Numerous experiments have demonstrated that in addition to this reduction of relative food consumption and/or excretion of unassimilated compounds, the food conversion efficiency (*FCE*) of fish declines with increasing body weight (see Gerking 1952, 1971). This decline starts at very high *FCE* values, e.g., 0.93 in rainbow trout larvae 'feeding' on their yolk sack (From and Rasmussen 1984) to $FCE = 0$ at W_{max} (or W_∞), when a fish that has reached the maximum size possible in its environment continues to eat, but does not grow anymore. This *FCE* decline was conventionally described by the Equation

$$FCE = a\ W^{\beta_1} \qquad ...(9.1)$$

where a and β_1 are estimated empirically from a linear regression of several log(growth increments/food consumption) vs. log(*W*).[180] Note that *FCE* implies that the growth increments and the food ingested should be expressed in comparable units.[181]

Pauly (1986) proposed an alternative relationship between *FCE* = growth increment/food ingested, which can be expressed as

$$FCE = 1\text{-}\ (W/W_\infty)^{\beta_2} \qquad ...(9.2)$$

where *FCE* is predicted for a given weight (*W*) from W_∞ (or W_{max}) and β_2, the latter being estimated as the slope (with the sign changed) of -log(1-(growth increments/food consumption)) *vs.* log(*W*). W_∞ can then either be estimated as the intercept of the regression with the abscissa or entered externally to obtain a compatible estimate of β_2. Note that either choice implies that *FCE* at W_∞ should be zero, not an assumption but a fact, by definition (Equation (9.2).

Note that Equation (9.2) avoids the prediction of $FCE > 1$ at very small sizes and $FCE > 0$ when $W>W_\infty$, both of which can be predicted from (9.1), but are inconsistent with the definitions of *FCE* and W_∞.[182] Also, Equation (9.2), which provides a better fit to empirical data than the model it replaces (Pauly 1986), is fully compatible with the special and generalized VBGF (Silvert and Pauly 1987).

To estimate β_2 for Atlantic cod (*Gadus morhua*) and to predict *FCE* as a function of weight in this species for a specific estimate of W_∞, a dataset from Table II in Edwards *et al.*

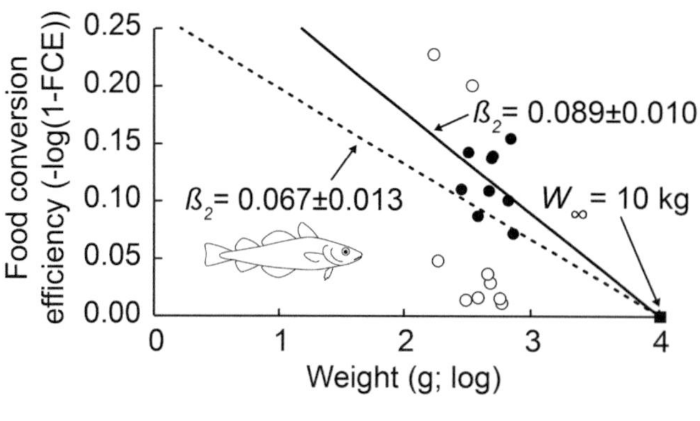

Figure 9.2. Food conversion efficiency (*FCE* = growth increment/food ingested) of Atlantic cod *Gadus morhua* (from table II in Edwards *et al.* 1972; open circles, referring to very high and very low FCE values were considered outliers) fitted with the equation $FCE = 1\text{-}(W/W_\infty)^{\beta_2}$, with $\beta_2 = 0.089$, and W_∞ = 10,000 g (see also **Table 9.2** and **Figure 9.3**).

Table 9.2. Estimates of the parameter β_2 relating food conversion efficiency to the weight of fish (see Equation 9.2), with the asymptotic weights (W_∞) related to these estimates.

Species	β_2	W_∞ (g)	Remarks and Source(s)
Sarotherodon melanotheron	0.036	215	Pauly *et al.* (1988), based on field data
Holacanthus bermudensis	0.040	800	Pauly (1986), based on Menzel (1958)
Serrasalmus nattereri	0.067	1000	Pauly (1994a) based on various data sets
Limanda limanda (♀)	0.073	756	Pauly (1986), based on Pandian (1970)
Channa striata	0.077	1290	Pauly (1986), based on Pandian (1967)
Limanda limanda (♂)	0.089	149	Pauly (1986), based on Pandian (1970)
Gadus morhua	0.089	10,000[a)]	See **Figure 9.2**
Epinephelus guttatus	0.136	1148	Pauly (1986), based on Menzel (1960)

a) Assuming an asymptotic length of 100 cm for *Gadus morhua* and a length-weight relationship with a = 0.01 and b = 3, i.e., likely values for Atlantic cod (see FishBase); see text and **Figure 9.2**.

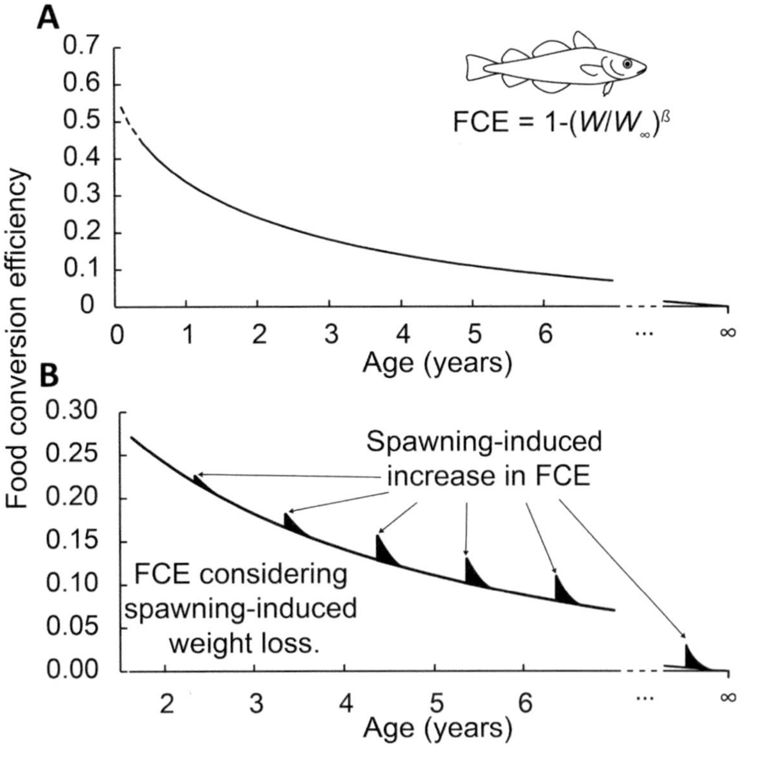

Figure 9.3. Plots of food conversion efficiency (*FCE* = growth increment/food consumed) vs. age for a fish with the parameters W_∞ = 10,000 g, K = 0.2 year^{-1}, t_0 = - 0.1 year, and β_2 = 0.089. **A**: Without accounting for spawning. **B**: Accounting for a spawning-induced loss of 10% of body weight at first spawning (i.e., at 2 years of age), 20% at 3 years, and 30% at 4 years and thereafter, which generates *FCE* increases of 3% at 2 years, 9.4% at 3, 21% at 4, 28% at 5, and 36% at 6, while the percent increase at $t = \infty$ could be defined only indirectly. Also, the schematic accounts for a 3-month post-spawning period weight-recovery period, after which *FCE* is again at the 'predicted' level.

(1972) was fitted with Equation (9.2). As **Figure 9.2** shows, this yielded β_2 = 0.089, and as **Table 9.2** illustrates, this value fits well with earlier estimates of this parameter.[183]

The results of the *FCE* plot in **Figure 9.2**, combined with Equation (9.2), leads to **Figure 9.3**, which illustrates in **A** how the *FCE* of Atlantic cod declines with age, and hence with size, and in **B**, how its annual spawning events, by suddenly releasing gonadal material, makes the fish up to 20-30% lighter, which, because *FCE* is weight-dependent, results in a sudden increase of *FCE*.[184] This, combined with *"ravenous"* feeding (Trippel *et al.* 2014) or *"hyperphagia"* (Shul'man 1974) immediately after spawning, leads to faster growth than would have occurred if spawning had not occurred.[185]

Another example is presented in **Figure 9.4**, where the shape of an Atlantic salmon immediately post-spawning is contrasted with its shape 51 days later. **Figure 9.4A** illustrates the loss of weight and the resulting change in body shape (*c.f.* = 0.8) following spawning in an (iteroparous) Atlantic salmon (*Salmo salar*) in Newfoundland (see Reddin *et al.* 2011). **Figure 9.4B** illustrates the rapid regain of weight and body shape (*c.f.* = 1.0) resulting from less than 2 months of post-spawning feeding with an increased *FCE*.

A conundrum resolved

The suggestion in Chapter 7 that "spawning in fish is a seasonally liberating event," which frees females from a quivering mass of eggs that must be supplied with scarce oxygen, and enables them to grow again was here turned into a testable hypothesis. Because of the sudden reduction of the live weight (or mass) of fish caused by spawning, which can be as high as 30% and beyond (**Table 9.1**), their oxygen supply per body weight will increase because their gill surface area is not affected by spawning, which increases their food conversion efficiency (**Figure 9.3B**).

If the fundamental feature of the GOLT is correct in its assertion that the ratio between gill surface area (and hence, oxygen supply) and the fish's body demand for oxygen impacts their growth, then post-spawning fish ought to experience a phase of accelerated somatic growth immediately after spawning. Such an accelerated growth phase is well documented in the literature, although the cause(s) for this distinct phase is (are) usually not identified. Of course, this phenomenon may often be masked by several factors, including rising water temperatures (and reduced oxygen) that frequently occur after springtime spawning and/or increased activity of post-spawning fish, including nest guarding for the same species.[186]

The post-spawning accelerated somatic growth period is due to a greater allocation of oxygen being devoted to growth. The total weight of fish bodies – which require oxygen for maintenance – is reduced while their gill surface area is not. This necessarily increases the *FCE*, as schematically shown in **Figure 9.3**, which then converts increased feeding into a much-increased growth until the entire pre-spawning somatic growth loss is compensated for.

Accelerated growth in fishes and other ectotherms is well-documented, notably by von Bertalanffy (1934, 1951). Endocrinological studies have sometimes described this phenomenon as "*compensatory growth*," but often remained ambiguous about the energetic prerequisites for the increase in somatic mass, or only discussed food supply as the decisive factor that enables further growth (e.g., Won and Borski 2013; Caldwell *et al.* 2013; Jobling *et al.* 1994). Ali *et al.* (2003) highlight the role of hyperphagia after a period of reduced growth, but they do not address changes in food conversion efficiency. For more food to be converted into growth, an increase in oxygen supply is required, which occurs due to the

Figure 9.4. Illustrating one aspect of fish spawning that is usually neglected. **A**: Representation of an Atlantic salmon (*Salmo salar*) adapted from a photo taken immediately after spawning of a lean 'kelt' of 63.8 cm and weighing 2.06 kg. **B**: The same fish 51 days after it spawned, with a length of 66.7 cm and a weight of 2.25 kg. The gill surface area of the kelt (in **A**), which was not reduced by spawning, had a low body mass to supply with oxygen; hence, the kelt's food conversion efficiency was high (based on photos by Kristin Bøe, supplied by Ian Fleming, Memorial University, St. John's, New Foundland).

abrupt loss of live weight caused by spawning (during which the head retains its shape, and, one can safely assume, the same gill surface area; see, e.g., **Figure 9.4**).

As mentioned above, this chapter does not address the seasonal growth of fish, whose oscillation can usually be related to temperature changes (see Chapters 12 and 13). Also, **Figure 9.3B** is incomplete, as it does not account for the effect on growth of the increase in temperature that follows upon spring spawning, nor for the dynamics of the fat accumulation.[187] The point of this chapter is to show that the simple fact of a loss of live weight at spawning is sufficient to explain post-spawning growth acceleration. This also implies that *ad hoc* hypotheses to explain why fish feed intensely after, but not before, spawning may be superfluous. One such hypothesis (Trippel *et al.* 2014) is that "*body size cavity restriction associated with hydrated ovaries may also play a role in limiting ingestion during the gravid state (Weeks 1996).*"

Michalsen *et al.* (2008) commented as follows on a study by Hoar *et al.* (1983), which also included this hypothesis: "*They suggested that feeding activity could be affected by less space being available in the body cavity for food, or possibly associated with a change in hormone levels during spawning, which might reduce an individual's appetite. As the feeding of males in our study was suppressed during spawning to an even greater degree than that of females, the thesis of lack of space does not seem convincing.*"

The suggestion of possible "*changes in hormone levels*" is yet another example of an *ad hoc* hypothesis, which wouldn't be helpful even if it tested positively because it would be impossible to know, after establishing that such changes occurred, that this was the cause for the decline in appetite and not an intermediate step. Rather, it is sufficient to note that pre-spawning, gravid females are so challenged by having to supply oxygen to their ovaries that they have little oxygen to direct to the assimilation of ingested food, and they stop feeding in the months before spawning. Then, post-spawning, they can catch up, and as their lower body mass is well supplied with oxygen, and for 2-3 months, their somatic growth is faster than would be predicted without spawning having taken place.[188]

The argument presented above dealt overwhelmingly with fish species occurring at relatively high latitudes, exposed to strong seasonal fluctuations in temperature and resource availability, resulting in well-defined and relatively short spawning periods. In lower latitudes, where such fluctuations are generally less pronounced, fish spawning tends to occur over more extended periods, often related to monsoon winds (Longhurst and Pauly 1987; Navaluna and Pauly 1988). This may be the reason why fishes from lower latitudes, even though they are likely to also rely on the *FCE*-driven mechanism presented above, tend to have a lower *GSI* than fish of higher latitudes, e.g., 4-6 % in bluefin tuna *Thunnus thynnus* (Heinisch *et al.* 2008), 3-4 % in skipjack *Katsuwonus pelamis*, or even 1.3-1.5% in yellowfin tuna *Thunnus albacares* (Stéquert *et al.* 2001).[189]

The recent debate surrounding reproduction and growth in a warming world appears to have overlooked the implications of seasonal fluctuations in fish growth. As a group of biologists and ecologists state in a recent paper, "*one key problem with the growth efficiency approaches that rely on a von Bertalanffy function* [...] *is that they ignore the single evolutionary goal of every organism – reproduction*" (Audzijonyte *et al.* 2019). This statement doesn't account for the fact that the asynchrony between seasonal changes in body weight and relative gill surface area generates periods during which food growth efficiency is very high and which, combined with heavy feeding, can quickly compensate for the cost of producing gonad material.

While they do this implicitly, the von Bertalanffy growth equations, when not modified to include seasonal changes, cannot explicitly account for interactions between growth, feeding, and reproduction. Seasonalized versions of the VBGF for both length and weight growth exist (reviewed in Pauly 2019); they should allow for re-expressing the VBGF as a model of growth explicitly accounting for reproduction and responding to a frequent critique of the standard VBGF (and indirectly of the GOLT).

10

Gigantism in fish

Synopsis: In this chapter, we distinguish between two forms of gigantism in fish: 'Brobdingnagian' and 'Goliathan.' The former refers to populations in which all individuals are larger than is typical for the taxon; the latter applies to exceptional individuals that attain unusually large sizes within otherwise normally sized populations. While Brobdingnagian gigantism is largely explained by various ecological and evolutionary rules, Goliathan gigantism is not. A mechanistic hypothesis is proposed to explain Goliathan gigantism which reduces oxygen requirements in individual fish via moving to cooler temperatures and/or the acquisition of larger, more energy-dense prey. These shifts enable some individuals to grow larger and, in the process, sometimes generate bimodal size distributions that may represent gradual forms between Goliathan and Brobdingnagian gigantism. This mechanism, which relies on the way their gill surface area grows, is more likely to operate in fish that can get big in the first place than in fish that remain small. Finally, we suggest that a Brobdingnagian iteroparous giant (and extinct) salmon, *Oncorhynchus rastrosus,* may be the ancestor of the recent semelparous Pacific salmon species via the paedomorphosis of their offspring.

Two types of gigantism

As a theory of growth, one of the GOLT's central themes is the question of body size and the constraints that set a limit to this trait. Extreme and extraordinary sizes are, therefore, a topic that allows for a more thorough exploration of these physical and physiological constraints and inferences on the mechanisms behind their evolution. In this chapter, mainly based on Pauly *et al.* (2024a), we will discuss this trait and relate it to the central mechanisms proposed by the GOLT.[190] Both the ichthyological literature and fisheries reports are full of mentions of individual specimens that reach enormous sizes, and their occurrence in populations of otherwise much smaller fish is a continuous source of fascination for anglers as well as public media and their audiences. Unfortunately, these observations do not appear to have been used to formulate a generalized hypothesis that would explain gigantism in fishes. This is possibly due to the confusion created by the fact that gigantism has two meanings, not differentiated in the literature.

One meaning is that *all* individuals of a single population within a species (Herczeg *et al.* 2009), or species within a genus or higher taxon (Chapelle and Peck 1999) became very large in certain habitats, which may be referred to as 'Brobdingnagian gigantism' (Swift 1726). The other meaning is that single individuals within (or a small fraction of) a population become much larger than most other adult individuals in the same habitat (see, e.g., Le Cren 1992; Ulman *et al.* 2022), which we may call 'Goliathan gigantism' after the Bible story (King James Bible, 1 Samuel 17). These two forms of gigantism have different causes and consequences. Although they are not always clearly separable, we will attempt to distinguish the central mechanisms that are responsible for these phenomena. We will first discuss Brobdingnagian and Goliathan gigantism as distinct forms of gigantism and explore their wider implications for the evolution of large body sizes in fishes and the ecological and physiological conditions on which they rely. While the GOLT's implications for evolutionary and adaptationist modes of explanation will be addressed in more detail in Chapter 15, the dynamics between evolution and morphological plasticity are also a central theme in this chapter, which will illustrate the main implications of the GOLT for the evolution of body sizes in water-breathing ectotherms.

Brobdingnagian gigantism

Examples of Brobdingnagian gigantism are very large stickleback (*Gasterosteus aculeatus*) living in ponds devoid of predators (Herczeg *et al.* 2009), or instances of 'polar gigantism' (Chapelle and Peck 1999). In the former case, the absence of predators allowed the evolution of placid fish, which, similar to domesticated (= farmed) fish, can devote their scarce oxygen to acquiring food and growing, as opposed to incessantly stressing about potential predators (Vincent 1960; Walters 2000; Pauly 2019). In the latter case, some fish and other WBE may reach large sizes, either because low temperatures generally reduce oxygen demand (Krogh 1941; Chapter 4), as seen in the giant isopods (*Bathynomus* spp.) or the amphipod *Alicella gigantea*, which live in the deep sea. Another example is provided by the sea spiders (Pycnogonida) of Antarctica, whose large sizes may (also) have been useful in countering the negative effect of the viscosity of very cold water on their respiration (Verberk and Atkinson 2013).

As Vermeij (2016) has reminded us, *"the single most important trait that affects the body size of organisms is metabolic rate."* Whereas air-breathing animals can grow to gigantic sizes by maximizing food consumption — because their metabolic rate can keep up with the oxygen demand of assimilating that food — gigantism in water-breathers follows the opposite pattern. Aquatic giants are typically 'gentle giants': animals with low metabolic demand, sluggish habits, slow growth, and long lifespans.[191]

Given the well-established evidence for the 'gigantization' of entire populations within one species, or single species within a higher taxon, Brobdingnagian gigantism is not discussed further here. This form of gigantism is actively studied and largely explained by the ecological and evolutionary adaptations of WBE populations, species, or higher taxa (see e.g., Shimada *et al.* 2021; Pimiento *et al.* 2019; Vermeij 2016; Chapelle and Peck 1999; van Roy *et al.* 2015).

These adaptations also provide an answer to the question as to why "*there* [are] *no really big bony fishes*" (Freedman and Noakes 2002), which is that this question is wrong as posed, as there *are*, or more precisely, there *were* extremely big bony fish, in the form of the giant Jurassic *Leedsichthys problematicus*.[192] This fish, which may have reached a length of 16.5 m (Liston *et al.* 2013), implying a maximum weight of about 20 tonnes, was planktivorous, similar to that of the largest extant fish, the whale shark *Rhincodon typus* (Pauly 2002a).[193]

Goliathan gigantism and 'parasitic castration'

One hypothesis to explain Goliathan gigantism is implied by the phenomenon of 'parasitic castration.' If one assumes that somatic growth and reproduction are antagonistic, i.e., that adult fish stop growing because they put all their 'energy' into the elaboration of gonads (Hubbs 1926; Lagler *et al.* 1977; Charnov 2008; Quince *et al.* 2008),[194] then castrated individuals may reach sizes larger than non-parasitized, fertile individuals. Heins *et al.* (2010) formulated this as follows: "*As a consequence of castration, host energy provides for growth of greater parasite volume and may promote host gigantism* (*Ebert et al. 2004; Bonds, 2006; Hall et al. 2007*)."

As Merilä and Eloranta (2017) demonstrated in a population of parasitized sticklebacks (*Pungitius pungitius*) with giant individuals among them, there was no correlation between body size and parasitic infestation: "*As noted by Hall et al. (2007), diversion of energy away from host reproduction by parasites such as* P. pungitius *may result in gigantism. However, in our data, there were no differences in mean sizes of parasitized versus unparasitised individuals, suggesting that parasitic infections did not lead to improved growth of fish host. Furthermore, none of the largest (>100 mm) individuals in our data were parasitized* [...]." Instead, the authors found a strong correlation between cannibalistic behavior and the giant size of individual sticklebacks.

Parasitic castration is a phenomenon well-documented in invertebrates, such as marine gastropods (Rothschild and Rothschild 1939), but it is not mentioned for fish in the early review by Boudoin (1975). Since that review, the roach *Rutilus rutilus* and three-spine stickleback *Gasterosteus aculeatus* appear to be the only two fish species in which the effects of parasitic castration on growth have been studied in detail. The results of these studies are confusing, notably because the weights of the parasite(s) and those of the fish host(s) were not always assessed separately.

When this rather basic error was corrected, the supposedly growth-enhancing effect of parasitic castration disappeared. Barber and Svensson (2003), working with *G. aculeatus* parasitized by the cestode *Schistocephalus solidus*, noted that "*when including the plerocercoid, infected females gained weight more quickly than controls, but when plerocercoid weight was removed, this trend was reversed.*"

Barber (2005), working with the same host and parasite species, found that "*the closest predictor of parasite mass was body size-corrected host growth rate, indicating that the fastest growing fish developed the largest parasites. Faster growing hosts therefore apparently provide ideal environments for growing parasites.*" Pioneering work on hitherto unknown biological phenomena is often prone to explanations that confuse cause and effect. In this case, rather than a heavy parasite enhancing the growth of its host, fish that are fit and grow well can be assumed to also tolerate a heavier parasite load.

Moreover, castration is not the only effect of parasitic castrators; they also reduce the evasive behaviors of their hosts vis-à-vis predators, which often are the parasites' final hosts. This reduced activity level, combined with hyperphagia (Barber and Svensson 2003), would result, given that darting about uses a large fraction of the oxygen required by fish (Koch and Wieser 1983), in larger body sizes of parasitized fish without their 'energy' having to be shunted from reproduction to growth. For all these reasons, and because it appears to occur – if at all – in only small fish species (Lafferty and Kuris 2008), parasitic castration will likely remain peripheral to a general explanation of Goliathan gigantism.

A metabolic interpretation of Goliathan gigantism

To examine the issues of Goliathan gigantism, we rely, as in previous chapters, on Pütter's definition of growth:

$$dW/dt = HW^d - kW \quad \text{...(10.1)}$$

where dW/dt is the growth rate, HW^d is the rate of protein synthesis, requiring oxygen that diffuses across the gills, and kW is the rate at which proteins are being spontaneously denatured, a process not requiring an oxygen supply (see Chapter 4).

Growth ceases when $HW^d = kW$, which must occur when W increases and $d < 1$; however, the fish can continue living, given that they can maintain their metabolic rate and other life-supporting processes. The low weight-specific oxygen consumption (Q) at this maintenance level is Q_{maint}.

This interpretation of Equation (10.1) focuses on gill surface area as a limiting factor for the oxygen supply to fish, rather than on 'internal' processes such as hematocrit, circulating blood volume, heart size, or heart stroke rate, because oxygen in the water entering a fish's mouth first passes over the gill surface area before being transported to the cells where it is used. If the gill surface area becomes limiting as the fish grows (and it does), it does not matter whether the internal oxygen transfer mechanisms are highly efficient or not; the first barrier is still a limiting factor (see Chapter 3).

Given Fick's law, relative gill surface area and relative oxygen consumption are directly proportional. Accordingly, the ratio of weight-specific oxygen consumption rate (Q) from a smaller length (L_1) or weight (W_1) to a larger length (L_2) or weight (W) is given by 1- $(Q_{L2}/Q_{L1})^3$ for length and 1- (Q_{W2}/Q_{W1}) for weight. This leads to:

$$\text{Percent reduction } O_2 \text{ consumption} = 100\cdot(1-(L_{S1}/L_{S2})^{3(1-d)}) \quad \text{...(10.2a)}$$

for length, and

$$\text{Percent reduction } O_2 \text{ consumption} = 100\cdot(1-(W_{S1}/W_{S2})^{(1-d)}) \quad \text{...(10.2b)}$$

for weight, if the length-weight relationship of the fish in question can be assumed to have an exponent $b = 3$, as is common in fish (Froese 2006). These two equations provide the framework for our considerations of the Goliathan gigantism.

Figure 10.1, which is based on Equation (10.2a) for growth in length (**Figure 10.1A**) and Equation (10.2b) for growth in weight (**Figure 10.1B**) establishes that fish species that grow to large sizes (and have high values of *d*) can increase their size at a relative cost that is much lower than for small species (with low values of *d*). The question then becomes: how are extraordinary individual growth performances realized?

In the case of Nile perch (**Figure 10.2A**), we know that they switch, at sizes of about 50 cm and beyond, from feeding on zooplankton, a diffuse, low-energy prey, to fish, an energy-dense prey (Lauzanne 1976; Soriano *et al.* 1992). This may be sufficient to reduce the oxygen they require for their maintenance (including chasing prey) by 10%, which allows them to 'hop' on a second growth curve, and reach a weight that is more than 3 times the asymptotic weight of their first growth curve.

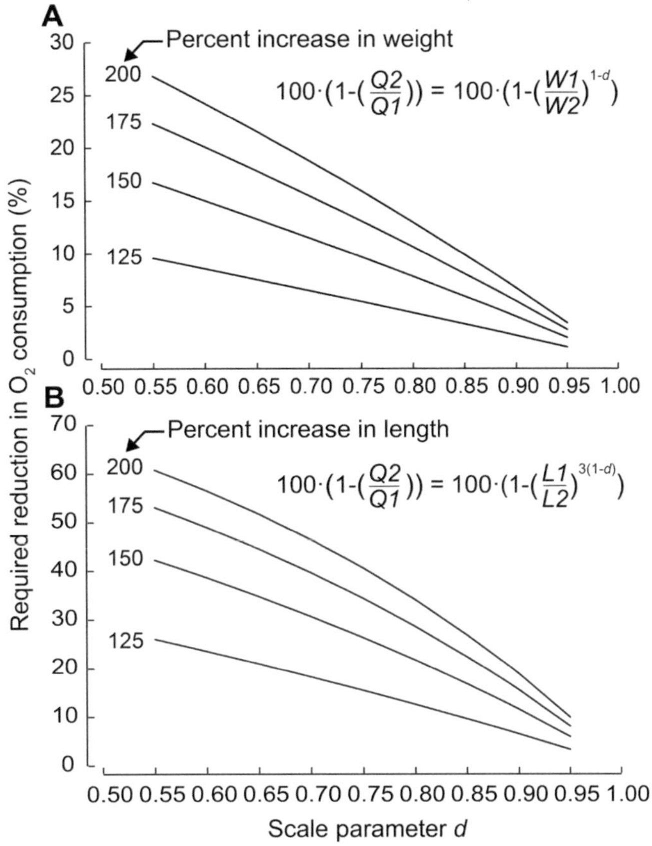

Figure 10.1. Illustrating the relationship between the scale parameter *d* (of Equation 10.1) and the % reduction of oxygen demand (*Q*) that fish must achieve to grow from a maximum size (length L_1 or weight W_1) to a larger maximum size (L_2 or W_2). **A**: % reduction of metabolic rate required for a 125, 150, 175, and 200 % increase in maximum weight. **B**: The same as required for increases in length. Note that the latter reductions are much higher because a doubling in length implies an eightfold (2x2x2) increase in weight, and similarly for other length increases. Also, note that for both length and weight, the reduction of oxygen demand required for a size increase is smallest when *d* is high.

Similarly, yellowfin tuna (**Figure 10.2B**), which feed on zooplankton when small (Maldeniya 1996) and on nekton when larger (Sabatié *et al.* 2003), switch from diffuse, low-energy prey to more energy-dense prey, as do white sturgeon (**Figure 10.2C**). The latter feed on zooplankton and invertebrate zoobenthos when small (Radtke 1966), but switch to fish when they become larger, i.e., "*herring, anchovies and Eulachon during the annual spring smelt run*" (Pietsch and Orr 2010), which is equivalent to the transition of food type in Nile perch and yellowfin tuna.

A similar phenomenon has been described as "*cannibalistic polypheny*" in fish and amphibians, where a few individuals start feeding on their conspecifics and grow to larger sizes (Claessen *et al.* 2000). It is well documented that the "*exceptionally big perch*" in Lake Windermere, England (**Figure 10.2D**), and on other lakes in Europe are cannibalistic, i.e., feed on other perch (Le Cren 1992; Claessen *et al.* 2000), which provide an energy-dense food, more so than the planktonic and benthic invertebrates that smaller perch feed on (Allen 1935).

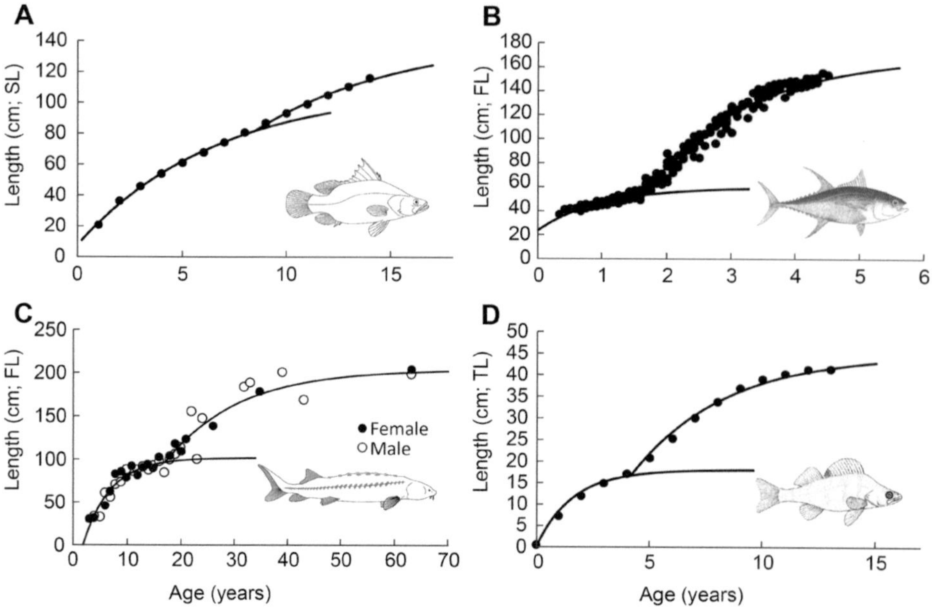

Figure 10.2. Biphasic growth curves in 4 fish species. **A**: Nile perch *Lates niloticus* in southern Lake Chad (from Soriano *et al.* 1992 and Loubens 1974); **B**: Yellowfin tuna *Thunnus albacares* in the Eastern tropical Atlantic (Gascuel *et al.* 1992); **C**: White sturgeon *Acipenser transmontanus* in the Fraser River, British Columbia (Semakula and Larkin 1968), and **D**: Perch *Perca fluviatilis* in Lake Windermere, England (Le Cren 1992).

Table 10.1. Parameter estimates for the two growth phases of the 4 fishes in **Figure 10.2.**

Parameter (units)	Perch (TL)	White sturgeon (FL)	Nile perch (SL)	Yellowfin tuna (FL)
First-phase growth				
L_∞ (cm)	18.1	102	110	60
K (year^{-1})	0.62	0.499	0.153	1.026
W_∞ (kg)	0.055	7.85	21.7	3.49
Second-phase growth				
L_∞ (cm)	46.3	204	172	170
K (year^{-1})	0.242	0.188	0.39	0.628
W_∞ (kg)	1.45	69.8	76.7	73.0
Other parameters				
a (of $W=a{\cdot}L^b$)[a)]	0.00233	0.0039	0.0404	0.0224
b (of $W=a{\cdot}L^b$)[a)]	3.478	3.14	2.81	2.92
Scaling parameter d	0.82[b)]	0.85[b)]	0.85[c)]	0.90[d)]
Ratio of W_∞ estimates	1 : 26	1: 8.9	1: 3.5	1: 21
O_2 demand reduction (%)[e)]	32	28	13	16

Note in 'Other parameters': a) from FishBase; b) from Equation (10.3); c) from Soriano *et al.* (1992); d) from Muir and Hughes (1969); e) as required for the transition from first to second of W_∞ (in % of the first estimate).

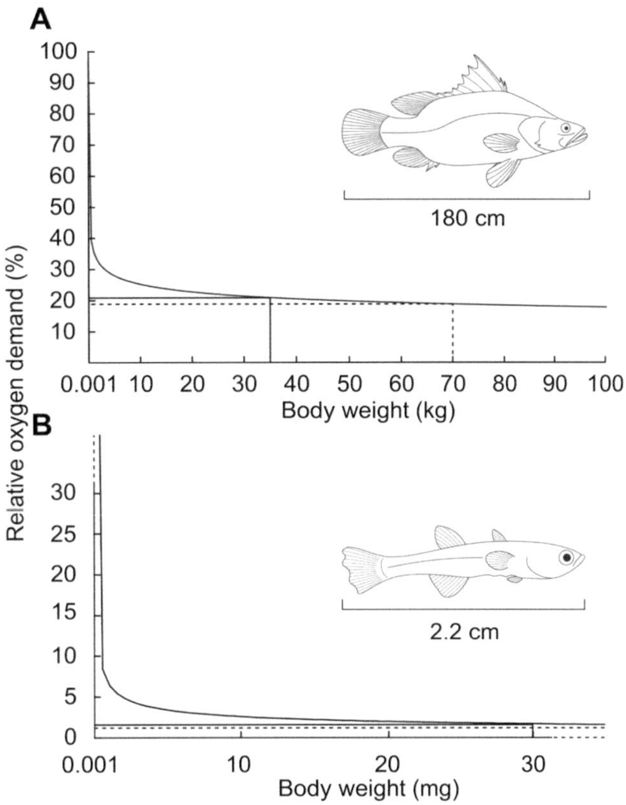

Figure 10.3. Relative oxygen demand of two fishes at weights W_1 and $W_2 = 2 \cdot W_1$, but with different values of the scale parameter *d*. **A**: Nile perch, with W_1 and W_2 approximately corresponding to **Figure 3A** and **Table 10.1**, and requiring, for *d* = 0.85 a 10% reduction of its oxygen demand to be able to double its weight. **B**: The Philippine goby or *sinarapan* (*Mistichthys luzonensis*), which, with *d* = 0.60 (Pauly 1982a), would require a decline of its oxygen demand of 24 % to double is maximum weight of 30 mg (see insert, with a reduced ordinate scale).

Overall, **Table 10.1** suggests that the reduction of metabolic rate required for the transition from the first to the second growth curve in the 4 cases of **Figure 10.2** is likely, given the energy density of prey fish compared with diffuse prey such as zooplanktonic and zoobenthic invertebrates.

On the other hand, it is hard to conceive of small fishes such as *Mistichthys luzonensis* as being able to find, among the tiny prey that they can feed on, one that would reduce their oxygen requirement by 24%, as required to double their weight (**Figure 10.3B**), given their *d* value of 0.6 and the constraints expressed by **Figure 10.1A**.

Another common solution for fish to reduce their oxygen requirements is moving to a different, cooler part of their ecosystem, a phenomenon also discussed in Chapter 5. This explains the depth distributions of many fish species, which usually move to cooler deeper waters as they grow (Heincke 1913; Longhurst and Pauly 1987; Pauly 2019; Ulman *et al.* 2022), and is aligned with the result of a recent review, which showed that optimum growth temperatures universally decline with increasing body size in fish (Lindmark *et al.* 2022; **Figure 10.3A**). Another case is very large fishes (including rays) of the Mekong River, which occur mainly around and in deep pools (Viravong *et al.* 2006; Campbell *et al.* 2020), where the water is much cooler than near the surface. Here again, it is difficult to imagine that tiny fishes would be able to migrate to cooler waters to reduce their metabolic rate similarly.

The points above argue against Goliathan giants among smaller fish species, but cases exist. As the data in Merilä and Eloranta (2017) show, stickleback gigantism may not only involve Brobdingnagian forms (see above) but also Goliathan forms, which are typically a result of cannibalistic behavior in only a few individuals within a population. Another example is a report of a male Madeiran sardinella *Sardinella maderensis* measuring 37.3 cm (FL) in a species supposedly reaching up to 32.1 cm (FL; Whitehead 1985; converted from 30.0 cm (SL).[195]

We did not discuss the seasonal variability of environmental parameters, notably temperature, as the seasonal growth oscillations that it induces (Pauly 2019) do not appear to be related to the maximum sizes reached by fish species. This is different for very small metazoans such as pteropods, whose maximum sizes appear to depend on the temperature during their first months of life (Wang *et al.* 2017; Chapter 14). Very small fish, such as *sinarapan* (*Mistichthys luzonensis*) in freshwater (Te Winkel 1945) and dwarfgobies (*Eviota* spp.) on coral reefs, inhabit tropical environments where seasonal temperature fluctuations are minimal.

Goliathan gigantism may lead to Brobdingnagian gigantism

To our knowledge, the distinction between Brobdingnagian and Goliathan gigantisms has not been made previously, at least not in ichthyology, and the evidence for their separate existence is tentative. It could well be that the two-phase growth in three of the four cases in **Figure 10.2** is, in reality, a case of Brobdingnagian gigantism, with only the perch (**Figure 10.2D**) being a clear case of Goliathan gigantism. Indeed, it could well be that the successful reproduction of one or a few Goliathan giants in a specific habitat may result in a Brobdingnagian population, if the necessary traits are stabilized by natural selection.

This may explain the existence of the various giants described in salmonids, for example, Arctic charr (*Salvelinus alpinus*) and brown trout *(Salmo trutta*), and it may be common in even more fish taxa (see Claessen *et al.* 2000, 2002; L'Abée-Lund 1992; Frandsen *et al.* 1989 and references therein). The transition from Goliathan individuals into Brobdingnagian population may be likely to occur in species with a wide variety of polyphenic or bimodal types, for example in salmonids where individuals or parts of a given population switch to piscivory and cannibalism (Grey 2001; Claessen *et al.* 2000; Hughes *et al.* 2019; Griffiths 1994, table 9; L'Abée-Lund 1992). So-called 'ferox trout,' for example, earlier described as a separate species because of their huge head and jaw, are now seen as giant morphs of *Salmo trutta* that feed on conspecifics, which themselves may have been piscivorous (Grey *et al.* 2002; Mangel and Abrams 2001). Gigantism in species with such polyphenic and bimodal growth and sizes suggests that the distinction between Goliathan and Brobdingnagian gigantism might be gradual, but the implications of the relationship in **Figure 10.2** appear inescapable nonetheless.

Incorporating oxygen and reduced metabolic costs into the discussion of fish gigantism may also help solve a riddle addressed by studies whose findings struggled to explain bimodal size distributions in populations that would qualify as gradual forms between Goliathan and Brobdingnagian gigantism. Claessen *et al.* (2002) point out that the existence of only a few large "*cannibalistic giants*" ('Goliaths' in our terminology) can be explained by invoking a food-based model. Brobdingnagian cases, i.e., populations in which a large group of individuals reaches the prey shift that allows them to grow further, require different explanations. To solve this puzzle, the authors hypothesized the existence of an intermediate food source that could smooth the transition between the two growth stages.

In our model, this assumption is not strictly necessary, as the reduction in metabolic costs associated with foraging should occur as soon as the first larger prey is available. If this is the case, the second growth stage becomes available for more individuals even without an intermediate food source. While the new prey may be scarcer due to the presence of a larger number of competitors, it allows the preying individuals to reduce their activity levels, compensating for the relative scarcity of the new food resource. This assumption would need to be tested further, but it may help explain how bimodal size distributions within stable populations can emerge.

The GOLT, 'Cope's rule,' and the evolution of body sizes

'Cope's rule,' the observation that species tend to increase in body size over evolutionary time, is often invoked to describe an evolutionary trend that drives the members of various animal genera and families, both terrestrial and aquatic, toward larger body sizes.[196] The mechanisms behind this pattern have long been debated, and its validity has been challenged as a statistical or even a "*psychological*" artifact (Gould 1997; Jablonski 1996). For many fish groups, evidence for this pattern is mixed (Knouft and Page 2003). Regardless of the validity of this 'rule,' it is clear that its postulate can only be maintained if the boundary conditions of its applicability are well-defined, since most organisms with larger body sizes have smaller ancestors even within shorter phylogenetic lineages.[197] But even if Cope's rule is not universal, the selective advantage of larger body size can be detected in many taxa, and the possible mechanisms that drive species and genera to larger sizes are numerous.[198] Such advantages may include an increase in predation success and a greater range of available food resources, more effective defense mechanisms against predators and competitors, increased longevity, and several other traits (see, e.g., Hone and Benton 2005). While all of these advantages are indeed likely evolutionary drivers of larger body size, size-based selection still requires a certain degree of phenotypical plasticity that produces the development of a range of sizes from which the largest can then be selected. The existence of individual 'Goliaths' in a population of smaller-sized fish could have adaptive consequences: if these larger individuals carry inheritable traits that facilitate earlier prey shifts – and therefore reduced maintenance costs at an earlier age or size – this may constitute a scenario in which the occurrence of individual Goliaths can result in the formation of Brobdingnagian populations.

This proposed mechanism for size increase in water-breathing ectotherms aligns with the view proposed by Kingsolver and Pfenning (2004), who identified selection on an individual level as the main driver of evolution towards larger body sizes. It may also illustrate an evolutionary mechanism described as "*plasticity-led evolution*" (Levis and Pfennig 2020; West-Eberhard 2003). This implies that phenotypes that are shaped by developmental plasticity or the occupation of individual (or ontogenetic) niches, such as access to earlier prey shifts, can be translated into phylogenetic and evolutionary trends.

As Kingsolver and Pfennig (2004) argued, evolutionary drivers of larger body size are likely countered by selection for decreased development time. Individuals that grow and mature early would then have more reproductive success than their slower-growing competitors and suppress the emergence of Brobdingnagian populations. As discussed in Chapters 4 and 5 and the rich literature on temperature-size effects in aquatic ectotherms, higher temperatures correlate with faster growth and development rates but also with lower final sizes. This would imply that lower temperatures should favor the transition from Goliathan to Brobdingnagian gigantism. Should the climate become warmer, slow-growing individuals that would have developed into giants would be outcompeted by conspecifics that develop and mature earlier. How this pattern translates into wider biogeographical patterns will need to be examined further. Still, the dynamics between development and growth rate, on the one hand, and 'giant' growth on the other, may serve as a promising hypothesis for further research on this topic.

Our proposed mechanism for Cope's rule and the factors that act as checks and constraints on the evolution of larger body sizes may have even wider evolutionary implications. As shown by researchers who studied increasing body sizes over evolutionary time in mammals, a surprisingly consistent pattern emerges in many families and genera; while Cope's rule can indeed be confirmed in many (mammalian) taxa, body size turned out to correlate positively with extinction risk (Cardillo *et al.* 2005; Clauset and Erwin 2008). Despite the short-term advantages of larger body size, larger mammal species face a higher risk of extinction than their smaller relatives. The mechanism behind this pattern is assumed to be related to metabolic demand and trophic specialization: a giant homeotherm is dependent on great amounts of food to survive – if the available resources are dwindling, its earlier advantage over smaller competitors is reversed and it is now the smaller individuals that are less affected by resource scarcity. Another factor that makes larger species more vulnerable to extinction is their relatively lower population density, which tends to be inversely correlated to body size (Cardillo *et al.* 2005). Losses in numbers can therefore lead to the extinction of species consisting of larger individuals more easily than those with smaller ones.

A similar mechanism may occur in fishes: as outlined in Chapter 5, larger body sizes can pose risks for many fishes, especially in a warming climate. While food availability may not normally be the limiting factor, large-bodied water-breathing ectotherms are more susceptible to oxygen limitation than smaller ones (see also Müller *et al.* 2023). As biogeographical surveys indicate, fish size is negatively correlated with low oxygen levels and temperature, and smaller species are more numerous in warmer climates (Adhikari 2015; Blanchet 2010). While this pattern clearly emerges from gradual adaptation to specific thermal environments, paleozoological research suggests the existence of a similar mechanism as in the above-mentioned mammals, since fish species and genera with larger body sizes are more prone to extinction than smaller ones (Sallan and Galimberti 2005; Friedman *et al.* 2010; Shimada *et al.* 2023). For adult fish, the primary driving factors are more typically temperature and oxygen availability, rather than food limitation (Shimada *et al.* 2023; Nätscher *et al.* 2021)[199]. As recent surveys have shown, the long-term patterns in the evolution of body sizes in WBE reflect climatic conditions and oxygen saturation levels (Troyer *et al.* 2022; Vermeij 2016; Verberk and Atkinson 2013; Ward 2006).

In the light of our proposed physiological mechanism, individual 'Goliathanism' and its risks can be explained coherently: if a reduction of maintenance costs and a high value of *d* (see **Figure 10.1B**) enable organisms to grow larger because more energy (i.e., oxygen) is available for growth, this opportunity for size increase disappears when temperatures – and maintenance costs – increase. In this case – and most cases of climate warming – being large can be a liability rather than a selective advantage. Under warm conditions, the roles are reversed: other organisms can enjoy the *"physiological benefits of being small in a changing world"* (Clark *et al.* 2012) and outcompete the giants whose body sizes earlier allowed them to explore new food sources, to grow older, and to reproduce at higher rates. We will return to the evolutionary implications of the GOLT in more detail in Chapter 15, but as this chapter has shown, connecting physiological adaptations to inferences on their evolutionary success can elucidate complex phenomena on a mechanistic level that would otherwise appear as mere statistical patterns.

Brobdingnagian gigantism and the transition from itero- to semelparity

Semelparity, i.e., death right after first spawning, is not very widespread in fish. Most temperate species that reach sizes > 10 cm are iteroparous, meaning they spawn repeatedly, typically at annual intervals (Breder and Rosen 1966). One group within which semelparity is the rule are the 8 species of Pacific salmon in the genus *Oncorhynchus,* whereas the 8, generally smaller 'trouts' in this genus are iteroparous (see FishBase). While the semelparity of the Pacific salmon species is well established and commented upon in multiple scientific articles, there is a dearth of accounts on how they may have acquired this trait, if assuming, as suggested, e.g., by Crespi and Teo (2002), that it is a derived trait, with iteroparity being the original condition.

The extinct planktivorous spike-toothed salmon *Oncorhynchus rastrosus* (Cavender and Miller, 1972) was a Brobdingnagian giant which reached a maximum length estimated at 240 cm and a weight of 200 kg (Stearley and Smith 2016; see also Cleason *et al.* 2024 and Sankey *et al.* 2016), and it probably was iteroparous (Crespi and Teo 2002). *O. rastrosus* lived from the mid-Miocene (i.e., 13.0 – 14.8 million years ago; Crête-Lafrenière *et al.* 2012) to the early Pleistocene when the cold climate would have been propitious to the emergence of giant fish species, as suggested in the previous section for various fish (see also **Figure 10.3A**). *O. rastrosus,* reported from fossil beds spanning the entire North Pacific from California (Sankey *et al.* 2016) to Japan (Takakuwa 2021), is therefore a possible or likely ancestor of the Recent species of Pacific salmon.

Given **Figure 7.1**, *O. rastrosus* may have had a gill surface area and hence an oxygen consumption which scaled with their body weight with $d \approx 0.85$, which is compatible with the estimates of d for Recent *Oncorhynchus* species, which range from $d_Q = 0.80$ (Meyer and Schill 2021) to $d_G = 0.90$ in *O. mykiss* (Morgan 1971; De Jager and Dekkers 1975).

If it is assumed that the ancestral Pacific salmon used the same triggering threshold as recent bony fishes for 'choosing' the size (and the age) at which to first mature and spawn, it follows that a value of $d = 0.85$ (and $D = 0.45$) would lead, via Equation (7.8) to an estimate of $L_m \approx 100$ cm in *O. rastrosus*. This result would remain largely unchanged if the uncertainty around the estimate of d and L_{max} in *O. rastrosus* were considered. The estimates of 100 cm and its corresponding weight ($W_m \approx 15$ kg) for *O. rastrosus* are roughly equal to or higher than all but one of the reported maximum sizes of Recent *Oncorhynchus* species (see FishBase).

This suggests that Recent *Oncorhynchus* species may have been derived from paedomorphic forms of *O. rastrosus*; the genus *Oncorhynchus* is, within the Salmonidae, isolated enough for the trend toward semelparity to have remained unique (Crête-Lafrenière *et al.* 2012; Zhivotovsky 2015).

This account is here proposed as a tentative hypothesis, and we cannot elaborate on the details of how the putative early paedomorphic species descended from *O. rastrosus* may have evolved. They did manage to maintain their ancestors' diadromous habits while gradually adapting themselves to the lower, central, or upper sections of the rivers in which they spawn and die, all the while enjoying, as Clark *et al.* (2012) noted, the "*physiological benefits of being small in a changing world.*"

11

Air-breathing fish: between two worlds

"Now suppose that Nature wishes to adapt a fish, which breathes by gills, to life in the air; she does not create an organ specially for this purpose, but utilises the moist gill-chamber (e.g., in Anabas scandens*), modifying it in certain ways so that the fish can take advantage of the oxygen it contains. But this gill-chamber lung is at best a makeshift, and when she comes to the more definitely terrestrial Amphibia nature gives up the attempt to use the gill-chamber as a lung, and creates a new organ, the true vertebrate lung, specially adapted for breathing air."*
E.S. Russell, *Form and Function* (1921, p. 199).

Synopsis: This chapter elaborates on the major implications for fish of breathing air as opposed to breathing water. It covers anatomical differences that can vary between the air- breathing groups, e.g., lungfishes and catfishes. Emphasis is given to the growth performance of a few large air-breathing fishes, notably the Arapaima (*Arapaima gigas*) of the Amazon, whose growth curve cannot be fitted with the asymptotic von Bertalanffy growth function (VBGF), and to the Mekong giant catfish (*Pangasianodon gigas*) whose growth curve appears to be asymptotic, and roughly comparable with that of bluefin tuna. Such high growth performance can only be achieved because *P. gigas* breathes air. Other aspects of air-breathing fishes are discussed, notably that small species, such as *Betta* spp., can handle confinement in small water volumes much better than water-breathing fish. These observations limit the scope of the Gill-Oxygen Limitation Theory which applies only to water-breathing fish and invertebrates.

The growth of air- and water-breathing fish

Not all aquatic ectotherms rely entirely on water as their source of oxygen – several thousand species of fish and a large number of crustaceans are capable of breathing air, a medium that contains not only far more oxygen, but also makes gas exchange incomparably easier, as we have seen in Chapter 1. The GOLT claims to be able to generalize and theorize life-history traits across all phyla of water-breathing ectotherms. How does the theory then relate to animals that rely on bi-modal respiration? As should have become apparent from the previous chapters, oxygen limitation is a critical factor in the life-histories of water-breathing organisms. Still, if the constraints of aquatic respiration represent the ultimate limitation to growth and energy expenditure, we should expect radically different growth patterns between water-breathing and air-breathing fish.

These differences are indeed hard to miss: many air-breathing species belong to the fastest-growing fishes in the world, some with growth rates that can be compared to those of endotherms (Bayley 2020). *Arapaima gigas*, for example, can gain up to 80 cm in length within only one year, and their growth is strikingly different from the common pattern found in other fishes (Queiroz 2000; Arantes *et al.* 2010). Another huge freshwater fish, the giant Mekong catfish (*Pangasianodon gigas*), can gain as much as 150 to 200 kg in only six years (Rainboth 1996). We will discuss the growth of these two species in more detail further below.

Such growth rates are rare in exclusively water-breathing freshwater species, even though freshwater can contain up to 25-30% more oxygen than saltwater under ideal conditions (Debelius *et al.* 2009). The fast growth rates of large freshwater species with the capacity of aerial respiration seem to confirm the prediction of the GOLT that access to more oxygen should result in higher growth rates. In the following, aerial respiration is discussed with reference only to fish; however, as the recent literature on arthropods suggests, many of the generalizations we present should be easily transferable to air-breathing crabs, e.g., to the coconut crab *Birgus latro* and other crustaceans with amphibious lifestyles (Drew *et al.* 2010; Giomi *et al.* 2014).

Oxygen and growth in freshwater and marine fish species

Several suggestions have been offered to explain the disparity in growth and metabolism between marine and freshwater species (Winberg 1960). Given the higher solubility of oxygen in freshwater, marine fishes' comparatively higher growth rates seem to contradict theories that link growth to gill surface area and oxygen availability. Another counterintuitive observation in this context is the unequal distribution of accessory breathing organs between freshwater and marine species. Air-breathing is an adaptation that is extremely rare in fishes that are exclusively marine and non-amphibious, i.e., that do not occasionally enter hypoxic freshwater (as do the tarpons, i.e., *Megalops atlanticus* and *M. cyprinoides*) or hop on mudflats, as do mudskippers (Graham *et al.* 1978; Turay *et al.* 2006). Except for some coral reef fishes (Nilsson *et al.* 2007b), non-amphibious marine fishes only use air-breathing as juveniles in fresh- or brackish water, for example, in tarpons.

The lower oxygen solubility in saltwater does not seem to matter much, as marine waters tend to be better oxygenated than most freshwater fish habitats. There are at least three factors that limit the oxygenation of freshwater bodies: (*i*) tropical rivers, lakes, and stagnant pools are often warmer than marine waters of the same latitudes; (*ii*) freshwater habitats, except fast-flowing hill streams and some lakes, are smaller and tend to be more stagnant than the open seas, which limits gas exchange between water and atmospheric air. Whereas organic decay processes in pools and smaller rivers can consume all available dissolved oxygen, larger water bodies are less affected by aerobic rotting processes because of the relatively smaller bottom surface and amounts of organic materials relative to the water volume and (*iii*), freshwater habitats are generally more acidic than marine water bodies which, again, limits oxygen saturation and produces higher CO_2 levels.

The average pH of oceans is currently 8.1 (8.2 before the industrial age), whereas such high levels are comparatively rare in freshwater. Many air-breathing fishes inhabit neutral and acidic water bodies, and some are adapted to niches where extreme acidity is favorable (with levels down to pH 3 and sometimes lower), such as the small gouramis of the genera

Sphaerichthys and *Parosphromenus*. Other air-breathing fishes also tolerate such low pH levels, even though they do not critically depend on them (for example, *Channa* spp. or *Osphronemus* spp.). Air-breathers are often among the rare fish species that inhabit acidic habitats, sometimes in the company of very small cyprinids.

While many air-breathing fishes show indeed impressive growth performances and large final sizes, not all species with the capacity of aerial respiration show this trend – some are only a few centimeters long and belong to the smallest known vertebrates, such as the smallest gourami species, which remain smaller than 2 cm (Finke and Hallmann 2013). Another assumption about air-breathing bony fishes that is challenged by more detailed observations is that air-breathing impacts the aerobic scope of fishes – positively or negatively. While earlier research suggested that air-breathing fishes tended to be more "*sluggish*" than others (Johansen 1970), this is a perspective that is no longer accepted and is certainly not valid for migratory species like the pangasiid catfishes of the great rivers of South and Southeast Asia (Bayley *et al.* 2020). Despite these revisions, air-breathing does not necessarily add energetic scope for growth or activity. In many cases, it requires several physiological mechanisms to regulate its effects on the organism's overall energetic balance.[200]

In this chapter, we will discuss the opportunities and limitations posed by bimodal respiration in aquatic animals, as well as the conditions under which air-breathing can become a mechanism to overcome the constraints of oxygen limitation in water-breathing ectotherms. As we demonstrate, aerial respiration is an adaptation that enables air-breathing fish to colonize specific niches that would otherwise be unavailable and is a mechanism that can enhance both their aerobic scope and scope for growth (see Chapters 2 and 6). This adaptation cannot 'solve' the problem of oxygen limitation entirely, and the physiological modifications that involve the combination of two different modes of breathing pose their own energetic challenges. Therefore, the central tenets of the GOLT still apply to fishes and crustaceans that evolved bimodal respiration. On the other hand, the adaptation of air-breathing also illustrates several features of the GOLT *ex negativo*, i.e., by showing to what extent some biological traits of air-breathers are indeed exceptional.

Air-breathers defy the rules

As an adaptation to low-oxygen environments or water bodies that can heat up quickly during the day, air-breathing fish can occupy habitats inaccessible to obligatory water-breathing fish. More than 99% of all air-breathing fish species are distributed in tropical regions, and only the North American gars (Lepisosteidae), the northern snakehead (*Channa argus*), the Alaska blackfish (*Dallia pectoralis*), and some blennies (Blenniidae) inhabit cold temperate or even Arctic regions. A similar pattern emerges in the distribution of air-breathing crustaceans, which has inspired the hypothesis that the evolution of bimodal respiration is an adaptation that allowed tropical arthropods to tolerate heat stress more effectively (Giomi *et al.* 2014).[201]

The differences in oxygen demand between temperate and tropical species are substantial, despite the necessary adaptive strategies of tropical species to reduce their maintenance costs. As Brett and Groves (1979) estimated, tropical species in the weight range of 10 and 100 g require between 60% and 70% more oxygen for maintenance than temperate species of the same size (see also Kramer 1983).

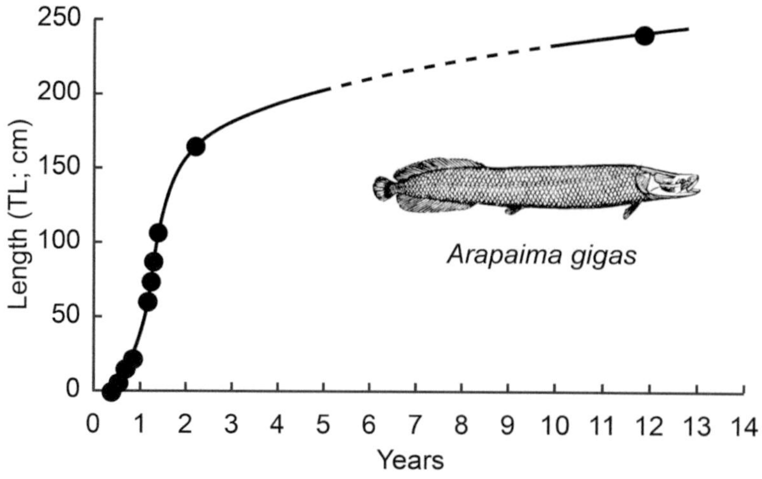

Figure 11.1. Length-at-age data on the air-breathing fish *Arapaima gigas* stocked in a lake of the Peruvian Amazon (modified from Wosnitza 1984). The (eye-fitted) growth curve is incompatible with all forms of the von Bertalanffy growth function. The head of *A. gigas* is small, limiting the space available for gills, in contrast to that of large marine fishes capable of reaching large sizes.

As comparisons between large numbers of genera and species have shown, the growth of water-breathers from tropical and temperate regions differs in at least two major ways. First, tropical fishes grow faster due to the lack of seasonal influences and often mature much earlier than species in temperate regions. According to Henderson (2005), the average values of K in the VBGF that describe their growth range between 0.8 and 1.6 year^{-1}. Second, despite their initially faster growth, water-breathing tropical fish typically reach lower maximum sizes (Henderson 2005). Especially tropical forest streams, such as Amazon tributaries or strongly acidic peat swamps in Southeast Asia, tend to have species of small sizes, which are less likely to occur in cold temperate or Arctic regions (Britz and Conway 2009). Such ecosystems contain not only the world's smallest fish species but also the greatest diversity of small fish species on the planet (Henderson and Walker 1986; Thornton 2018). Such biotopes do not occur in temperate regions, where fish usually grow more slowly but where small species are comparatively rare.

Air-breathing fishes are a notable exception to this pattern: not only do they include the biggest freshwater species of South America, tropical Asia, and southern Africa (*Arapaima gigas*, *Pangasianodon gigas*, *Heterobranchus longifilis*, and *Clarias gariepinus*), they are also among the few aquatic ectotherms that can grow very fast and *yet* reach large sizes (Wosnitza 1985; Queiroz 2000; Arantes *et al.* 2010). Pauly (2019), after several attempts, concluded that the growth of *Arapaima gigas*, as documented in **Figure 11.1**, cannot adequately be fitted with the VBGF.

Only marine species such as the Atlantic bluefin tuna (*Thunnus thynnus*) show similar growth rates and maximum sizes (see below). The fact that air-breathing freshwater fishes can still maintain growth rates like those of the fastest-growing marine species under conditions where hypoxia is frequent, for example, in the habitats of pangasiid catfishes and snakeheads, highlights the role of their accessory breathing organs.

Their resilience to hypoxia and high temperatures is one of the primary reasons for the recent increase in air-breathing fish species in aquaculture (Bayley *et al.* 2020). Air-breathers such as snakeheads, knifefish, or arapaima often exhibit the highest growth rates near the maximum of their thermal range. One study on *Pangasianodon hypophthalmus* even shows a six-fold increase in growth rates when exposed to temperatures close to the tolerated maximum (Bayley *et al.* 2020). As a recent study has highlighted once again, access to air increased the thermal tolerance of air-breathing fishes, in this case, two Amazonian catfish species (de Lima *et al.* 2025)

In contrast to unimodal water-breathers, species with accessory breathing organs can often be cultured in poorly aerated containers and under high stocking densities. Studies on big obligate air-breathing species such as *Arapaima gigas*, *Pangasius bocourti*, and *Channa striata* show that high stocking densities in aquaculture have far fewer effects on growth than in unimodal water-breathers, and sometimes a negligible effect on total length (Jiwyam 2011; Aminur Rahman *et al.* 2012).[202] Even though the setup of some of these studies included open cages, in which high stocking densities did not necessarily lead to drastically lower oxygen levels, the results were comparable and suggest that the effect of poor oxygenation is less important.[203] Aminur Rahman *et al.* (2012) report undiminished growth rates in *Channa striata* at oxygen levels between 5.28 and 5.66 mg/L.

In the following, we focus on three specific issues related to the features of air-breathing fish that differ from those of water-breathing fish in a manner relevant to the GOLT, i.e., ones that supports its major tenets and highlights some of its corollaries, sometimes surprisingly. These issues are why air-breathing fish can still grow well in small and poorly aerated containers, and whether the growth of large, air-breathing freshwater fishes can approximate that of the fastest-growing marine fishes. We will answer the latter question by comparing the growth of the Mekong giant catfish (*Pangasianodon gigas*) to that of Atlantic bluefin tuna (*Thunnus thynnus*) and then close with a discussion of the factors by which most air-breathing fish are still constrained in their oxygen uptake capacity, regardless of their ability to breathe air.

Air-breathing fish and their growth in small containers

A point of debate among aquarium hobbyists and aquaculturists is the problem of growth in small containers. This question has also caught the attention of animal welfare organizations, and many European countries and local city administrations have recently banned goldfish bowls, citing stress and lack of swimming space. This raises the question of which container size would provide sufficient space for goldfish (*Carassius auratus*), a species that can reach sizes up to 35 – 40 cm, with Murdy *et al.* (1997) even reporting a maximum size of 48 cm. Even though smaller aquarium fish, due to the absence of predators and other causes of natural mortality, often get bigger in captivity than in the wild, it is clear that goldfish, like many other larger aquarium species, cannot reach their maximum sizes in a bowl.

A traditional argument to defend goldfish bowls or smaller aquaria is that fish somehow 'adjust' their size to the volume of the container, even though no coherent mechanism has ever been presented that would provide an explanation. Perhaps a better argument is that fish cannot get old in small bowls or aquaria. The average maximum age of goldfish in bowls or aquaria does not even approach half of the 41 years reported in FishBase. Jointly, these answers may solve at least part of the problem: (1) The 'adjustment' argument has some support in the growth-limiting impact of nitrogen levels (as ammonia, nitrite, or nitrate) that are typical for overstocked tanks; (2) large fish in small containers are often exposed to stressful situations for extended periods, which might explain their earlier demise.

Air-breathing fish in small containers or heavily stocked grow-out tanks do not appear to suffer from confinement similarly. As several aquaculture surveys have shown, the growth of large air-breathers such as *Pangasius bocourti, Channa striata,* or *Arapaima gigas* does not seem to be affected in heavily overstocked tanks or small water volumes (Jiwyam 2011; Paredes-López *et al.* 2021; Aminur Rahman *et al.* 2012; Cavero *et al.* 2003; De Oliveira *et al.* 2020).[204]

Observations from the ornamental fish trade further support the critical role of water oxygenation and oxygen uptake for fish growth. One of the most popular aquarium species, the air-breathing Siamese fighting fish (*Betta cf. splendens*), is commercially produced under circumstances that seem to contradict all the above hypotheses, including the idea of container size as a constraint on growth. In Thailand, Cambodia, and Indonesia, professional Betta breeders typically use hundreds of small rum or whisky bottles as grow-out containers for their fish. This method of 'jarring' is also used by hobbyists in China, the United States, and Europe, even though organizations such as PETA call for it to end.[205]

The fact that this method of growing fish is commercially viable is interesting in itself: if we assume that the container size limits fish growth, it would make little sense to 'jar' juvenile fish in bottles with a volume between 0.5 and 0.7 liters and even lower net water volumes. The 'jarred' fish are transferred into these bottles at sizes between 15 and 25 mm and remain there until they are big enough for sale. Container volume cannot be the limiting factor, at least not for these air-breathing fish. The reason for using such bottles is that fighting fish start to exhibit aggression at around 20 mm, and growing out in groups might lead to fights that damage their tails and fins, for which they are valued in the aquarium trade. Furthermore, being 'quiet' in a bottle drastically reduces the organism's energetic costs – more of its food can be used for growth. The fact that small container sizes benefit the growth of air-breathing fish suggests that oxygen is the limiting factor for the growth of water-breathing fish in small containers. While air-breathing fish seem unaffected by lower water volumes, water-breathing fish would quickly reduce the dissolved oxygen level and perhaps survive for a while, but without growing. What might look like an 'adjustment' to smaller water volumes is, in fact, oxygen limitation. The fact that traditional fish bowls, due to their shape, have a strongly reduced water surface for replenishing their dissolved oxygen also contributes to the low availability of this crucial element in such containers.

Comparing two giants

The differences in growth between exclusively water-breathing fishes and air-breathing species can best be illustrated in large tropical air-breathers, such as the Mekong giant catfish (*Pangasianodon gigas*), which is not only among the fastest-growing but also among the largest freshwater fishes. Graham (1997, p. 256) suggested that "*the growth rate of* Pangasianodon gigas *is said to rival that of pelagic species such as tuna, a rare phenomenon for a freshwater species.*" As we will show below, this comparison is apt, but only because *Pangasianodon gigas* is an air-breathing species.

The Mekong giant catfish reaches a length of 300 cm (Baird *et al.* 1999) and a weight of 350 kg (Kottelat 2001), data that are similar to the maximum sizes reported for the three bluefin tuna species and for yellowfin tuna (i.e., *Thunnus* spp., see FishBase). The notion that *P. gigas* reaches these large sizes as rapidly as these tuna species is a strong claim because the latter evolved gills with very narrow interlamellar spaces and huge large surface area (Muir and Hughes 1969), which provides them with a large amount of oxygen required to sustain their elevated metabolism; Mekong giant catfish do not have such extraordinarily large gills.

In contrast to tuna, whose growth has been extensively studied by numerous authors (see FishBase), the growth of Mekong giant catfish has not been well studied. Its growth in captivity has been reported upon (Lorenzen *et al.* 2006), but no studies document its growth in the wild. Roberts and Vidthayanon (1991, pp. 119 – 120) mention specular growth rates in the early juvenile stages, but their information is based solely on personal communications. According to their account, the growth in length of these fish accelerates during their second year, which is atypical for both marine and freshwater fishes and has so far only been demonstrated in *Arapaima gigas* (Wosnitza 1984; Pauly 2019).

In addition to these more anecdotal accounts, Lorenzen *et al.* (2006) presented two rough estimates of the parameters of the von Bertalanffy Growth Function (VBGF) for *P. gigas,* from which a single growth curve for the remaining few specimens in the wild may be inferred. The special VBGF for length and weight is used here, along with parameter Ø' (Pauly 1998a), defined by the equation

$$Ø' = \log(K) + 2 \cdot \log(L_\infty) \qquad ...(11.1)$$

which has a normal distribution when applied to numerous populations of the same species, e.g., in skipjack tuna, *Katsuwonus pelamis* (see figure 9.4 in Longhurst and Pauly 1987). When comparisons between species with different body shapes are made, it is better to use Ø, which is defined by

$$Ø = \log(K) + 2/3 \cdot \log(W_\infty) \qquad ...(11.2)$$

The first estimation of growth parameters in Mekong giant catfish is based on Lorenzen *et al.* (2006), who wrote that *P. gigas "are widely stocked into 'semi-natural' reservoirs in Thailand, where they appear to survive and grow well but are not known to mature or spawn. Stocking and recapture data were analysed for Sirikit reservoir in Thailand* [and growth parameters] *were estimated as* $L_\infty = 210$ *cm and* $K = 0.2$ *year*$^{-1}$. [...]. *The stocked fish thus appear to grow at a higher rate but to a lower asymptotic size than the wild fish in the Mekong."*

Lorenzen *et al.* (2006) provide no textual details beyond the statement that *"no data were collected for the first 7.5 years after release,"* and they did not specify the method they used to estimate growth parameters from the growth increment data they casually mention. The graph they presented suggests that the stocked and released fish, whose length increments were used to calculate growth parameters, were relatively large and did not exhibit much growth. This should result in a somewhat uncertain estimate of the parameter *K* because of the more rapid growth of small (young) fish, which usually stabilizes the estimation of this parameter.

For their second estimate, Lorenzen *et al.* (2006) used the maximum size (they) recorded for *P. gigas*, i.e., 290 cm, as an estimate of L_∞.[206] To estimate *K*, they plotted the values of L_∞ and *K* in Pauly (1980a) as ln(*K*) vs. ln(L_∞), from which they derived a regression equation. Solving this equation for L_∞ = 290 cm (TL) yielded $K = 0.08$ year^{-1}. They thought these values were reasonable and used them to assess the (dire) state of their population in the Mekong basin.

Applying Equation (11.2) to the first and the second estimate of growth parameters of *P. gigas* yields estimates of Ø' = 3.94 and 3.83, respectively, with an average Ø' = 3.88. Combined with an estimate of L_∞ = 300 cm, the mean Ø' yields an estimate of K = 0.085 year^{-1}. An estimate of t_0 is provided, in the absence of any other applicable method, by the empirical equation of Pauly (2019), which suggests that

$$\log(-t_0) = -0.3922 - 0.2752 \cdot \log(L_\infty) - 1.038 \cdot \log(K) \quad \text{...(11.5)}$$

which here yields t_0 = -1.12 years.

These growth parameters for *P. gigas* in the wild, i.e., L_∞ = 300 cm (TL), K = 0.085 year$^{-1,}$ and t_0 = -1.12 years, are very tentative. Still, they are based on broader considerations than the previous estimates and may be more accurate. Lorenzen *et al.* (2006) presented a length-weight relationship for *P. gigas*, i.e., $W = 0.04 \cdot L^{2.8}$, where weight is in g and length in cm. Based on the previous considerations, the VBGF for the growth of *P. gigas* is

$$W_t = 345{,}000 \cdot (1-e^{-0.085 \cdot (t+1.12)})^{2.8} \quad \text{...(11.6)}$$

which leads, via Equation (11.4), to an estimate of Ø = 2.62, which is close to estimates of Ø for bluefin tuna (*Thunnus thynnus*; black dots in **Figure 11.2**), and within the ellipsoids that could be drawn around the Scombridae family (grey dots in the same figure).

Figure 11.2 suggests that the Mekong giant catfish *Pangasianodon gigas* (black star) grows almost as fast as Atlantic bluefin tuna (black dots), commonly and justifiably seen as the fastest-growing tuna. The Mekong giant catfish is indeed a very fast-growing fish. A total length of 300 cm in *P. gigas* roughly corresponds to a fork length of 266 cm, smaller than the asymptotic fork lengths estimated by various researchers for Atlantic bluefin tuna. Using the mean value of Ø' = 4.00 in Atlantic tuna, which refers to LF, an estimate of K can be

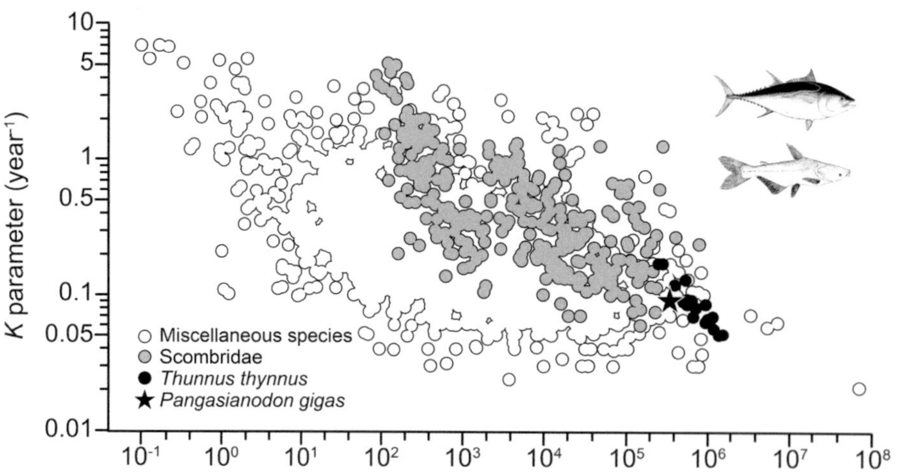

Figure 11.2. Auximetric plot of the fish with weight growth parameters in FishBase, with each dot representing a K-W_∞ pair, i.e., the mean growth curve of the individuals of a fish population. The non-Scombridae are represented by white dots ('Miscellaneous species'), the fast-growing Scombridae (mackerels, tuna, etc.) by grey dots, and Atlantic bluefin tuna by black dots. The star represents the tentative growth curve for *P. gigas* (W_∞ = 345 kg and K = 0.085 year^{-1}), which is suggested here to grow almost as fast as bluefin tuna.

derived for a hypothetical bluefin with an asymptotic length of only 266 cm LF. This estimate of K would be 0.14 year^{-1}; inserted into Equation (11.5), these estimates yield $t_0 \approx -0.2$ years.

Figure 11.3A compares the growth curves in length for these two species, indicating that their growth rates are not very different. When comparing growth in weight between these two species (**Figure 11.3B**), what appears to be a small difference turns out to be higher, with young tuna growing faster. This fast growth is made possible by huge gills, which supply the oxygen required to sustain a metabolic rate almost equivalent to that of a similar-sized mammal. Even by breathing air, *P. gigas* does not match the growth of Atlantic bluefin tuna, although it does grow very fast.

Lefevre *et al.* (2013), presumably because their first author fervently believes that fish always meet the oxygen requirements required for further growth through their gills, failed to understand why they devoted energy to breathing air. They write, *"though air-breathing is usually considered a beneficial behaviour, we show that surfacing indeed incurs an energetic cost. Though air-breathing is important to swimming in hypoxia (McKenzie et al. 2012), it remains unclear what mechanisms drive the air-breathing behaviour in* Pangasianodon hypophthalmus *in normoxia."* This self-created conundrum is quickly resolved: air breathing is not "*a beneficial behavior*" in *Pangasianodon* species, but a *necessary* behavior, because large individuals of this genus cannot meet their oxygen requirements by relying

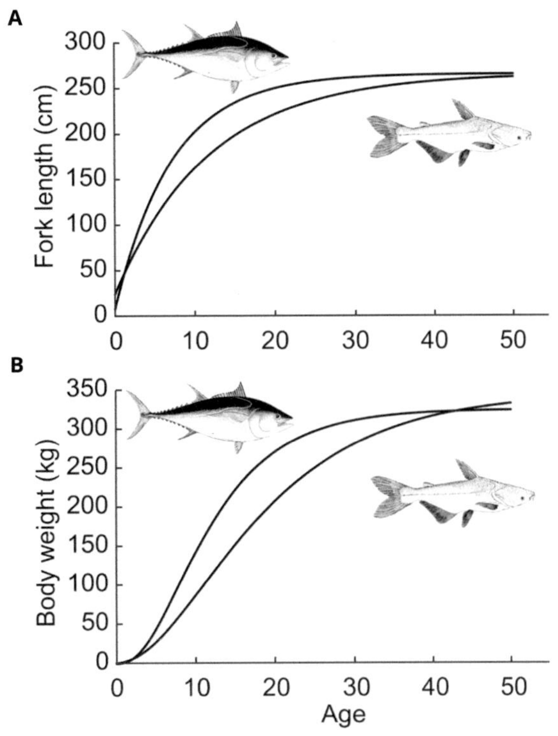

Figure 11.3. Growth curves in length (**A**) and weight (**B**) of (hypothetical) bluefin tuna and Mekong giant catfish, both with an assumed asymptotic length of 266 cm (FL; see text).

only on their gills at higher temperatures, even in normoxic water (see Mitamura *et al.* 2009, figure 4).

As a migratory species, *P. gigas* experiences a relatively wide range of temperatures compared to other tropical freshwater fishes: it spawns in the cool regions of Chiang Rai and Chiang Kong in Northern Thailand (Eva *et al.* 2016; Ngamsiri *et al.* 2007; Pholprasith and Tavarutmaneegul 1997; Pholprasit 1994) and moves downstream into the lower delta region. The temperatures at its most northern habitats range from 18 to 26 °C between January and August (Zhang *et al.* 2007), whereas the water in the delta is typically 10 °C warmer. Some authors (Ngamsiri *et al.* 2007; Hogan *et al.* 2001) assume a historical distribution of *P. gigas* in the Chinese province of Yunnan, where water temperatures can be as low as 14-16 °C in the cold season and rarely exceed 22 °C in the summer (Wang *et al.* 2014; Zhang *et al.* 2007). Hogan *et al.* (2001) and Hogan (2004) hypothesize that *P. gigas* might still be present in southern China and that its decline in Yunnan could be attributed to deforestation and the construction of dams in this region. Even in the warmer areas of the delta, *P. gigas* may not be exposed to the upper extremes of its thermal tolerance range.

Like other large species of the Mekong Delta, *P. gigas* is often reported near, or even inside, deep river pools (Eva *et al.* 2016). Some of these holes are as deep as 90 m (Gupta and Liew 2007), but it is unclear to what depths the Mekong giant catfish can dive. Even in stagnant and warm water bodies, such as the Mae Peum reservoir in Thailand, depths of only 6 to 12 m provide temperature differences of up to 6 °C between the water surface and the bottom (Mitamura *et al.* 2009). In this reservoir, *P. gigas* frequently surfaces in the hot season but less so in January and February, when the deeper water columns are well-oxygenated (Mitamura *et al.* 2009, figure 6).

As for Graham's claim of 1997, which appears credible when growth is expressed in length (**Figure 11.3A**), it is less convincing when growth in weight is considered (**Figure 11.3B**), at least when bluefin tuna is used as representative of 'tuna.' Still, the fact that a fish that inhabits warm and often deoxygenated freshwater bodies can reach growth rates that equal those of large marine fishes is remarkable.

Is air-breathing a way out of oxygen limitation?

The rapid growth of many air-breathing fishes and their ability to thrive in hypoxic environments suggest that the central tenets of the GOLT do not apply to species with their anatomical and physiological adaptations. The growth of most air-breathing can still be adequately fitted by the VBGF, which raises the question of why the evolution of air-breathing organs did not universally result in different growth patterns, e.g., as those exhibited by *Arapaima gigas* (**Figure 11.1**). Early studies on air-breathing in fishes often assumed that bimodal respiration was an evolutionarily unfinished business, or "*at best a makeshift*," as E.S. Russell wrote in his book on *Form and Function* (1921). Early descriptions of the air-breathing organs of fishes typically stress the rudimentary nature of these physiological adaptations.[207] While this view had to be revised in the light of newer research, it foreshadowed a later finding that would become important for an adequate understanding of the physiology and biomechanics of air-breathing fishes.

Given the distinct physical properties of air and water and their fundamentally different impacts on respiration, the combination of both modes of respiration in the same organism creates a different challenge. Especially under conditions of aquatic hypoxia, when oxygen cannot be taken up from the water in sufficient amounts but is readily available through

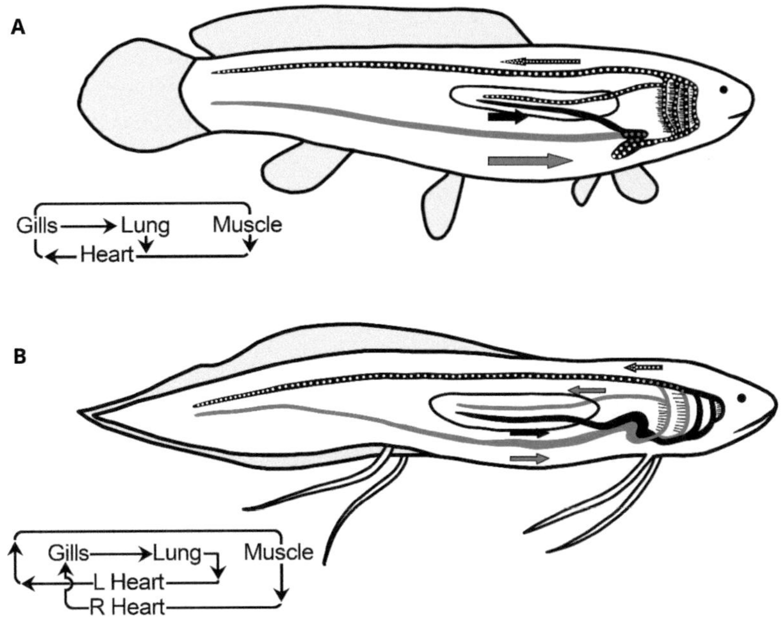

Figure 11.4. Schematic representation of the circulatory system of two air-breathing fishes, *Amia calva*, a facultative air-breather (**A**), and *Protopterus annectens,* an obligatory air-breather (**B**); modified from Farmer (1999). In **A**, after being oxygenated in the air-breathing organ, the water flows through the heart and back to the gills, where branchial oxygen loss can occur. In **B**, the oxygenated blood can be bypassed from the gills through a shunt such that it first arrives at the tissues and is absorbed before flowing back to the gills.

aerial respiration, a mismatch may occur between the two modes of breathing, leading to an adverse effect. Since the gills are both connected to the cardiovascular system and to an environment that is often hypoxic, oxygen can be lost through the gills. This risk of oxygen excretion through gills may explain the often extremely small gills (and thus small head) in large species with highly efficient and well-developed air-breathing organs, such as *Arapaima gigas*.

To avoid such branchial oxygen loss, many air-breathing fishes rely on a mechanism that allows them to temporarily disconnect their gills from the circulatory system. The body plans of the different air-breathing families and genera are not all organized alike. In many facultative air-breathers, for example, *Amia calva* (see **Figure 11.4A**), the cardiovascular system is permanently connected to the gills, so the blood that leaves the heart first passes through the gills. Its flow is then divided between the air-breathing organ and the tissues that need to be provided with oxygen (which is different from what occurs in obligatory air-breathers such as *Protopterus annectens*; see **Figure 11.4B**).

The cardiovascular system in *Amia calva* (**Figure 11.4A**) implies that either the heart mixes oxygenated and deoxygenated blood – which then reduces the capacity of the gills to take up more oxygen due to the mechanism of countercurrent exchange (see Chapter 3) – or that most of their tissues never benefit from the oxygen supply that is generated in the air-breathing organ. Earlier research interpreted this body plan as a maladaptation and contributed to the verdict that air-breathers were generally "*sluggish*" due to their inefficient respiratory architecture (see, e.g., Johanson 1970). Newer research proposes a different functional mechanism and interprets the function of the air-breathing organ in such a body as additional oxygenation to the heart to protect cardiovascular performance (Farmer 1999). Fishes with this *Bauplan* remain oxygen-limited in the same way as unimodal water-breathers.[208]

Other families of air-breathing fishes show very different body plans and can use their air-breathing ability to oxygenate their entire body efficiently. To do so, they must minimize the risk of branchial oxygen loss, which can be realized through a shunt that opens or shuts and bypasses that allow the blood to circumvent some of their gill arches to keep the oxygen from the air-breathing organ within the cardiovascular system (Bassi 2010; Aaskov *et al.* 2023). Another strategy is to minimize the gill surface area entirely, which appears to be typical for large species with highly developed air-breathing organs, such as *Arapaima gigas* (Aaskov 2022; Pelster 2020; Brauner 2004). Gills remain necessary in such cases because the osmoregulatory compromise limits the minimum and maximum gill surface area possible (see Chapter 3). Fishes that cannot effectively regulate branchial oxygen loss through shunts or other mechanisms can be considered oxygen-limited in the same way as unimodal water-breathers, even if the scaling exponents of their gill surface area (*d*) can reach values >1, as reported for mudskippers by Palzenberger and Pohla (1992). Such values are only possible because the gills are very small.[209] Like most amphibious fish, mudskippers lack a regulatory mechanism to bypass oxygenated blood from their gills (see Graham 1997, p. 120).

Why did not all fishes evolve lungs?

While air-breathing can present a highly effective adaptation to some fish families, the previous section has shown that it has its limitations, and the additional resources it provides are not endless. In addition to these limitations, aerial respiration also has its costs. Water animals that depend on breathing air from the water surface must invest energy into surfacing, which is costly and dangerous, especially for species that inhabit benthic or profundal zones, such as catfishes or snakeheads (Kramer 1983). The energetic costs of locomotion are exceptionally high for bottom-dwelling species such as loaches and catfishes and greatly exceed those of fishes in the limnetic zone. In species that inhabit the immediate surface of the water column, such as insect-feeding killifishes or gouramis, locomotion costs can be negligible. Still, the constant risk of predation from above requires continuous vigilance, which implies stress. For smaller benthic fishes, locomotion through the various water zones forms a great risk, and they often rely on synchronous surfacing with a bigger shoal (Kramer and Graham 1976). When denied access to the surface or in stressful circumstances without the safety of the shoal, *Corydoras* catfish reduced their food intake to reduce ingestion costs (Kramer and Braun 1983; Persaud *et al.* 2006).

The costs of aerial respiration are further illustrated by the fact that many air-breathing fishes have strong diving reflexes in the presence of (assumed) predators, and many have nocturnal lifestyles (Smith and Kramer 1986). Another correlate of air-breathing in fish is nocturnal or crepuscular activity, which strongly reduces the risk of predation from above (Graham 1997; Farmer 1999). The exception to these patterns may be very large air-breathers for whom these risks are much lower. Big surface-dwelling species such as *Arapaima gigas* only run a considerable risk of predation as juveniles. Like many fishes, they engage in synchronous air-breathing in their early life-stages, a strategy that is no longer necessary after adulthood is reached. As adults, they possess extremely hard scales that protect them efficiently from predation and represent *"one of the toughest flexible biological materials"* (Yang *et al.* 2019).[210] Being dependent on constant access to air, the selective pressure to develop such protective armor must have played an important role.

These species, such as *Arapaima gigas,* exhibit the greatest contrast to unimodal water-breathers, for whom aquatic oxygen limitation sets the major constraint to metabolic performance and growth. While other air-breathers can only partly free themselves from this constraint, the tradeoff with other risks and offsets remains, and shows why air-breathing is seldom the way out of oxygen limitation.

12

Growth, respiration, and reproduction of water-breathing arthropods

Synopsis: Gill surface area (S) and respiration (R) in adult water-breathing crustaceans scale with their body weight (W) such that $S \propto R \propto W^d$, with d ranging mostly between 0.6 and 0.9, but always <1, as in other water-breathing ectotherms. The growth of adult crustaceans, therefore, approaches an asymptote, whether or not seasonal growth oscillations are explicitly considered in the model used to describe that growth. On the other hand, the variation in asymptotic size (L_∞ or W_∞) among crustaceans is primarily determined by water temperature, which impacts the oxygen requirements of WBE. Through multiple examples, this and related aspects of the Gill-Oxygen Limitation Theory (GOLT), first developed for fishes, are shown to also apply to the growth of a wide range of crustaceans and other arthropods, including fossil forms. The GOLT also explains certain aspects of crustacean reproduction, such as the relationship between size at first maturity and maximum size, and, possibly, the feature that female crustaceans hold their eggs outside of their bodies instead of internally.

Extending the GOLT to arthropods

As presented in the earlier chapter, the Gill-Oxygen Limitation Theory (or GOLT) was initially developed to explain regularities in the growth, respiration, and reproduction of bony fishes (Pauly 1979, 1984) and predicts these patterns in a variety of fishes, from fast-charging tuna to shy seadragons (Pauly 2021a; see Chapters 1 – 6). Here, based mainly on Pauly *et al.* 2022a), we examine the GOLT's efficacy in predicting growth and reproduction patterns first in a major group of arthropods, the crustaceans, then in other groups, including two that have long been extinct.

We do not deal with the growth of insect larvae, some of which, unlike the metamorphosed adults, can possess gills that enable them to breathe water (Lee and Matthews 2021).[211] The reason is that larvae, whether of fish or invertebrates, are generally small enough for the oxygen that is supplied by their respiratory organs (primordial fin in fish, dedicated gills, or even parts of their integument) to process any amount of the food they can capture and ingest. Their growth is usually food-limited rather than O_2-limited, and thus exponential (though chironomid larvae may be an exception to this[212]).

Crustaceans[213] and horseshoe crabs are often perceived as different enough from fish — principally because of their moulting – to require different models to describe their growth and other life-history patterns (Breen 1994; Zeng and Wan 2000; Foo 2019, 2020), but we believe that this is largely unwarranted, as will be shown below. Key features of each phenomenon examined here (growth, respiration, and reproduction) are best interpreted in a generalized context, as opposed to taxonomic specificity. The generality of the GOLT applied to crustaceans is demonstrated here, along with a few examples from various other groups of water-breathing arthropods.[214]

The growth of crustaceans

As elaborated in earlier chapters, particularly the chapter on the fundamentals of organismic growth, Pütter's equation, which here also serves as the basis for analyzing the growth of arthropods, implies asymptotic growth because $d < 1$ in adults. Asymptotic growth, which is a necessary condition for the GOLT to apply to water-breathing arthropods is, however, not a sufficient one.

In fish, there is a strong tendency in species that remain small (e.g., in gobies and guppies) to have gills with d ranging from 0.6 to 0.7 (Pauly 1981, 1982a), whereas species capable of reaching large sizes (and long lifespans) have higher values (i.e., $d = 0.85 - 0.90$). This is similar in crustaceans, where the tiny brine shrimp *Artemia salina* has gills growing approximately according to the 'surface rule,' i.e., $S \propto W^{2/3}$ (**Figure 12.1**), whereas larger crustaceans (but not their larvae, where $d \geq 1$) can be shown to have values of slightly higher than fish of similar sizes (**Figure 12.2**; see also Hughes 1983).[215]

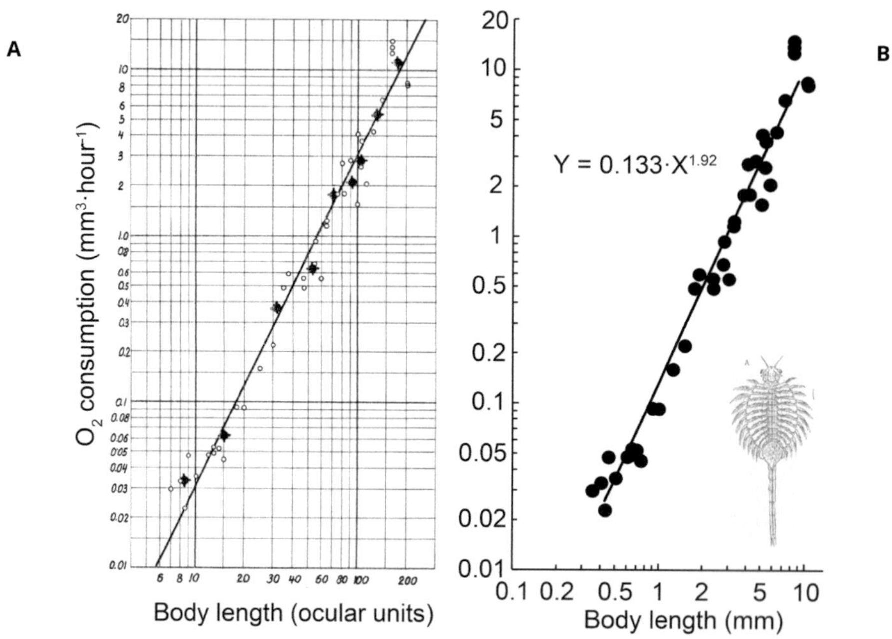

Figure 12.1. Oxygen consumption of the brine shrimp *Artemia salina* as a function of their length. **A:** Reproduction of the graph published by von Bertalanffy and Krywienczyk (1953), showing how they 'guesstimated' from the black dots (i.e., the means of the white circles) a slope of 2 for the log(O_2 consumption) vs. log(length) relationship. **B:** Plots of the actual data in **A** (small white circles, here big black dots), showing that the slope was 1.92, corresponding to $d = 0.64$.

The relationship between oxygen consumption and body weight is well documented for many crustaceans, e.g., Bridges and Brand (1980) for decapod crustaceans. This confirms the relationship between *d* and crustacean size.[216]

With this, the stage is set to claim that – even if their moulting has the effect of growth proceeding in steps – crustaceans, in general, grow asymptotically, with their maximum (L_{max} and W_{max}) or asymptotic size (L_∞ and W_∞) often corresponding to a terminal moult when such occurs (Vogt 2012). Asymptotic growth is best represented by the von Bertalanffy Growth Function (VBGF; see Chapter 2). For illustration, **Figure 12.3** documents asymptotic growth in populations of four very different crustacean species.

Growth tends to vary seasonally, which can be accommodated by the seasonally oscillating version of the VBGF. One of these, proposed by Somers (1988) and improved by Garcia-Berthou *et al.* (2012), is

$$L_t = L_\infty\{1 - \exp\text{-}[K(t- t_0)+S(t)\text{-}S(t_0)]\} \quad \text{...(12.1)}$$

where

$$S(t) = (CK/2\pi)\cdot\sin(2\pi(t - t_s)) \quad \text{...(12.2a)}$$

and

$$S(t_0) = (CK/2\pi)\cdot\sin(2\pi(t_0 - t_s) \quad \text{...(12.2b)}$$

with L_∞, K, and t_0 are defined as for the standard VBGF.

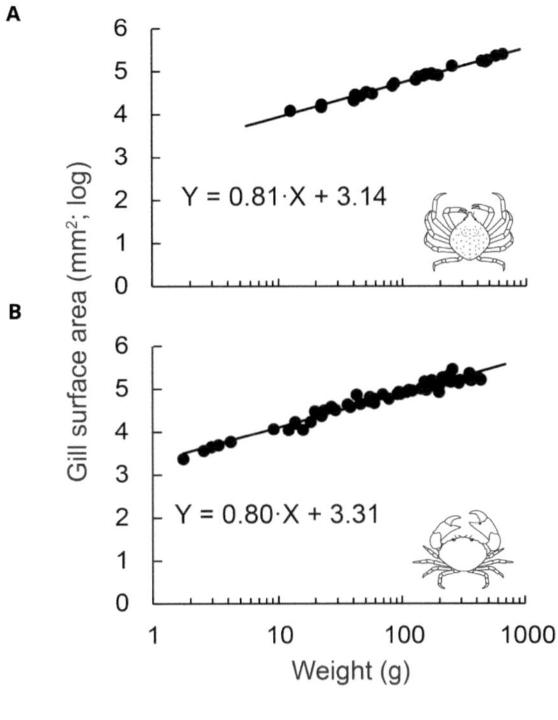

Figure 12.2. Relationship between the gill surface area of crabs and their live (or wet) weight, adapted from Gray (1957). **A:** Portly spider crab (*Libinia emarginata*). **B**: Stone crab (*Menippe mercenaria*).

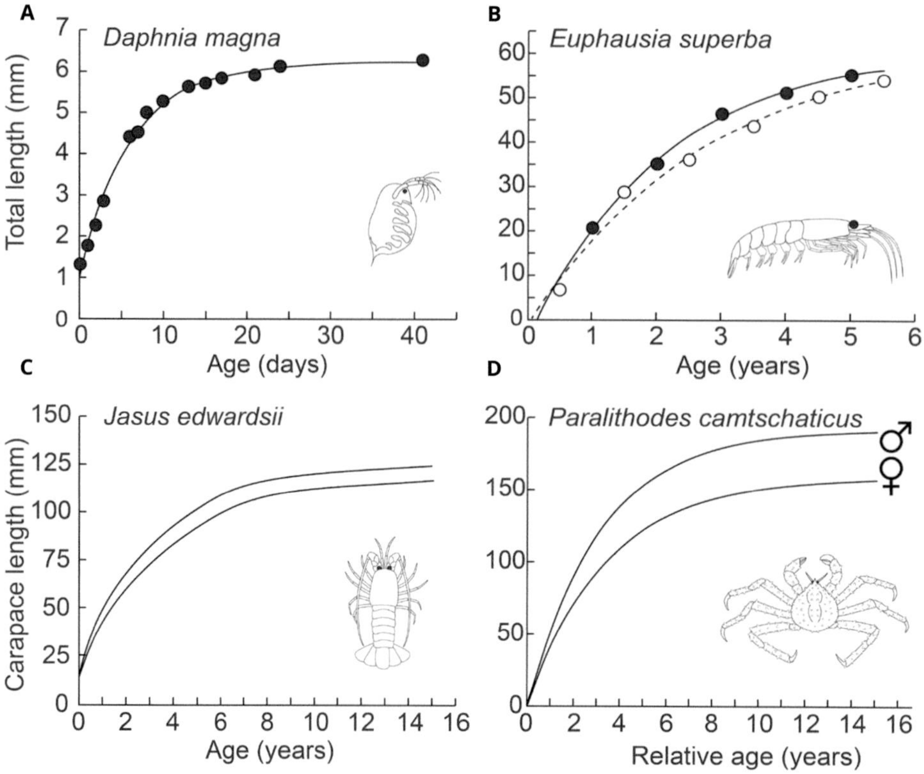
A
Daphnia magna
Total length (mm)
Age (days)
B
Euphausia superba
Age (years)
C
Jasus edwardsii
Carapace length (mm)
Age (years)
D
Paralithodes camtschaticus
♂
♀
Relative age (years)

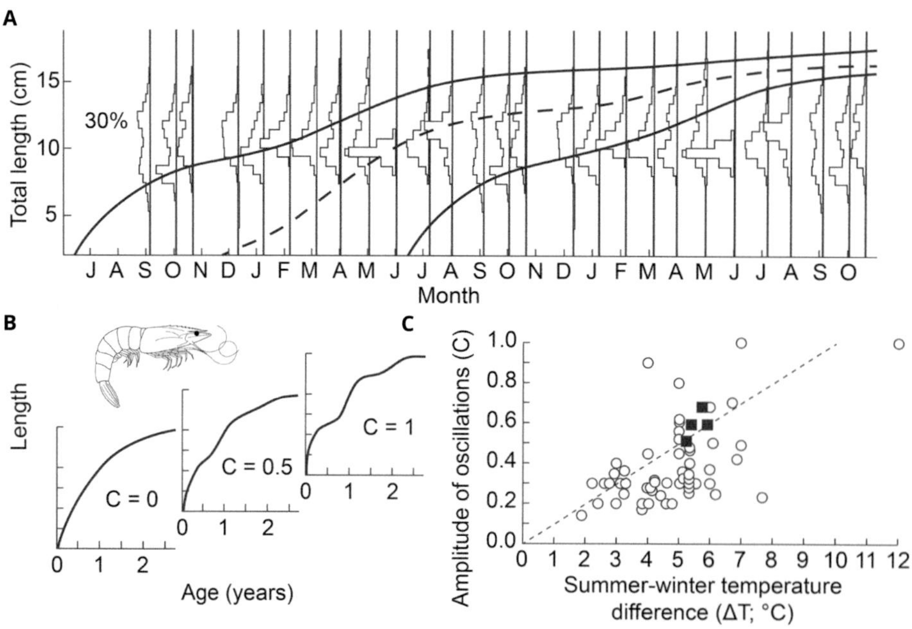
A
Total length (cm)
30%
J A S O N D J F M A M J J A S O N D J F M A M J J A S O
Month
B
Length
C = 0
C = 0.5
C = 1
Age (years)
C
Amplitude of oscillations (C)
Summer-winter temperature difference (ΔT; °C)

Figure 12.3 (opposite page, top). Asymptotic growth (not accounting for seasonal oscillations) in four crustaceans. **A**: The cladoceran *Daphnia magna* at 20 °C (modified from Martínez-Jerónimo 2012). **B**: Antarctic krill (*Euphausia superba*) sampled in 1977 (black dots) and 1981 (circles); modified from Siegel (1987). **C**: The red rock lobster (*Jasus edwardsii*) at two localities in New Zealand; adapted from McKoy and Esterman (1981). **D:** Red king crab (*Paralithodes camtschaticus*), adapted from Windsland *et al.* (2013).

Figure 12.4 (opposite page, bottom). Seasonal growth in crustaceans. **A:** Seasonal growth of male pink shrimp (*Farfantepenaeus duorarum*) from Tortugas, Florida, 1957-1958, with L_∞ = 17.6 cm, $K \approx$ 1.3 year^{-1}, and C = 0.6 (from Pauly *et al.* 1984, based on length-frequency data in Iversen *et al.* 1960). **B:** Effect of different values of the parameter C on growth curves. **C**: Relationship between the summer–winter temperature differences (in °C) and estimates in fishes and invertebrates, including four penaeid shrimps (black squares) species. Modified from Pauly *et al.* (1984) and Pauly (2019).

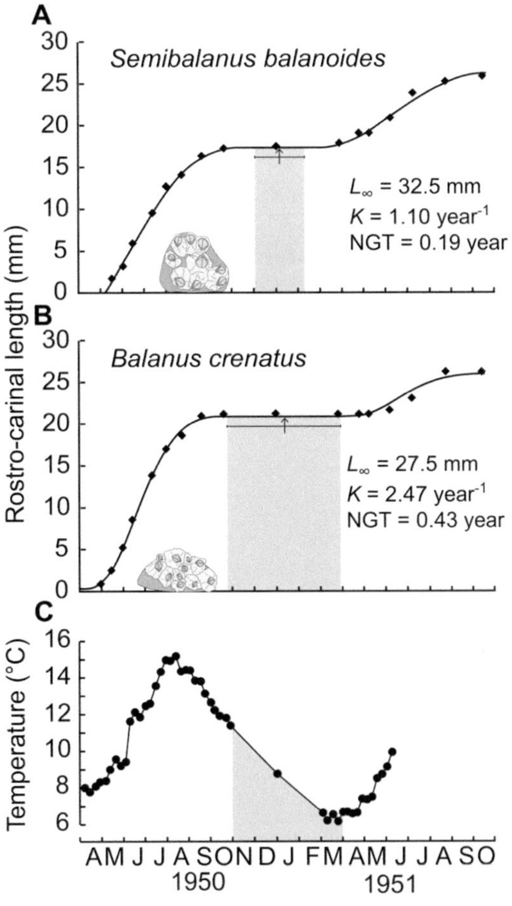

Figure 12.5. Seasonal growth of the Northern rock barnacle (*Semibalanus balanoides*) (**A**) and crenate barnacle (*Balanus crenatus*) (**B**) based on growth curves in figure 1 in Barnes and Powell (1953), fitted with the software of Ogle (2017) allowing for no-growth periods in winter (*NGT*; in grey) when temperatures (same source) are dropping (**C**, in grey).

Equation (12.2) requires two more parameters than the standard VBGF; C and t_s, of these, the former expresses the amplitude of growth oscillations; when $C = 0$, Equation (12.1) turns back into standard VBGF. When $C = 0.5$, the seasonal growth oscillations increase growth by 50% at the height of summer and reduce it by 50% in the depth of winter. When $C = 1$, growth rates finally increase by 100%, doubling at the height of summer and becoming zero in the depth of winter.

The other new parameter, t_s, expresses the time between $t = 0$ and the start of a growth oscillation. Here, we use a Winter Point, $WP = t_s + 0.5$, expressed as a fraction of the year, to represent the time of the year when growth is the slowest. WP is usually near 0.1 (early February) in the Northern Hemisphere and 0.6 (early August) in the Southern Hemisphere.[217] Equation (12.1) can be used when a WBE maintains a positive growth rate (in length) throughout the year except at WP when $C = 1$. **Figure 12.4** illustrates seasonal growth in *Farfantepenaeus duorarum* and other penaeid shrimps.

When growth in length ceases entirely for an extended period, the no-growth time (*NGT*) can be estimated using a modification of Equation (12.1), developed by Pauly *et al.* (1992), and now implemented in R (Ogle 2017). This is here illustrated by the growth of barnacles, which, because they are stuck to one place, cannot avoid long-lasting cold temperatures (**Figure 12.5**).

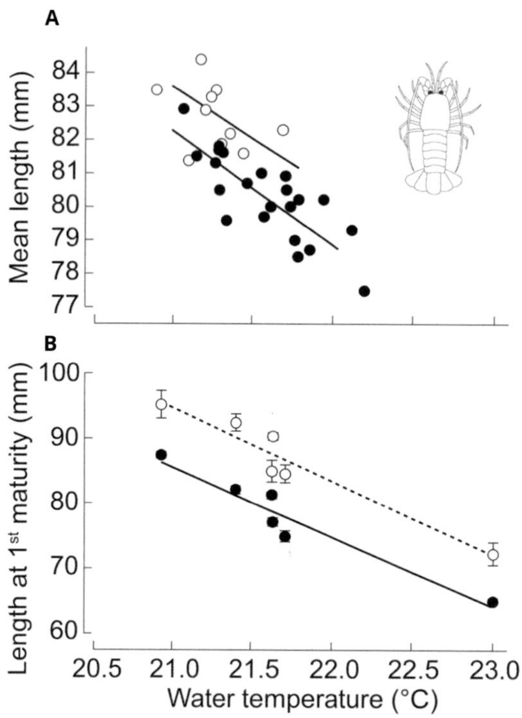

Figure 12.6. Effect of temperature on growth and reproduction of rock lobster (*Palinurus cygnus*). **A:** Decline of the standardized mean carapace length of migrating adults vs. the mean water temperature 4 years earlier (modified from Caputi *et al.* 2010). **B:** Length at which 50% of rock lobster reach first maturity vs. mean temperature (modified from Melville-Smith and De Lestang 2006).

The effect of temperature increases on the growth of crustaceans can then be straightforwardly interpreted, given Pütter's model, whose anabolic term increases with temperature more slowly than the catabolic term. Given scarce respiratory oxygen, this has the effect that the WBE approach occurs at a faster rate, a weight at which $HW^d = kW$. Adult size in crustaceans declines with temperature, as illustrated in **Figure 12.6A** for the West Australian rock lobster *Palinurus cygnus*, along with their size at first maturity (**Figure 12.6B**).

In terms of the VBGF's parameters, this response results in asymptotic length (L_∞) being reduced and K increasing, and the VBGF parameters for the various populations of the same or closely related species forming ellipsoid clusters on 'auximetric' plots (Pauly 1998a, 2019; see **Figure 12.7**). Such plots can be extended to cover disparate taxa and used, for example, to compare their growth and identify outliers.

The growth of crustaceans and other WBE with widely different shapes (e.g., copepods vs. lobsters) cannot be reliably compared in terms of length and should be performed using weights. In fishes, the length-weight relationships used for this purpose usually have the form $W = a{\cdot}L^b$, where b, past the larval stage, usually ranges between 2.5 and 3.5 (Froese 2006). In crustaceans, more species fall outside this range (Pauly *et al.* 2022a), some being interesting extreme cases. For the stalked barnacle, *Pollicipes pollicipes*, a length-weight relationship of the form

$$\log(W) = 1.81 + 0.1195{\cdot}F + 1.484{\cdot}\log(L) \qquad \ldots(13.5)$$

was estimated, where W is the wet weight in g, and L the total length (from 2.4 to 8.4 cm), based on 57 fresh specimens of stalked barnacle, or *percebes*, caught in Baiona (NW of Spain; 42° 7.64' N and 8° 52.04' W) in January 2022, for which the dummy variable F was 0,

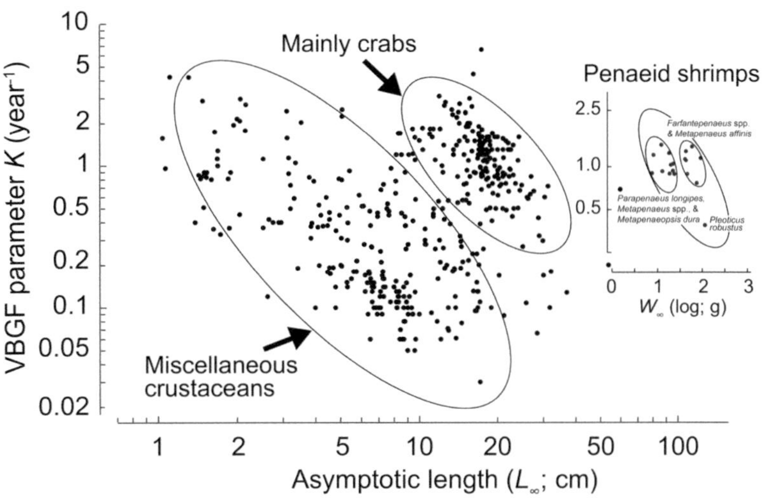

Figure 12.7. Auximetric plots for crustaceans. i.e., double-logarithmic plots of the von Bertalanffy parameter K vs. the corresponding asymptotic length in the larger plot, with each dot (i.e., K-L_∞ pair) describing a growth curve documented in SeaLifeBase. The insert (modified from Pauly *et al.* 1984) documents the tighter fit obtained when the W_∞ values for different groups of penaeid shrimps are used instead of their L_∞ values.

while F =1 for 40 specimens caught and frozen earlier.[218] This allowed for a significant effect of freezing on the weight of stalked barnacles to be identified and for two length-weight relationships to be estimated, one for fresh specimens ($W = 0.015 \cdot L^{1.484}$), the other for frozen specimens ($W = 0.020 \cdot L^{1.484}$), both of which relate weight to length via an exponent << 3.

Aspects of the reproduction of crustaceans

Reproduction in crustaceans is a set of complex biological phenomena that differ between families and genera, and to which no single account can do justice. At least in long-lived crustaceans (and we will take lobsters as an example), reproduction usually occurs after several years. The question then arises: what is so peculiar about the spawning season in the year that spawning occurs?

A moment of reflection will suffice to realize that, barring exceptional circumstances, any year and spawning season are like the others, and that the trigger for spawning must be found in the individual lobster in question. It is likely not its age (though computer models of lobster populations frequently use 'age' as the trigger) nor a critical size, because size is temperature-dependent (see above); and spawning should be possible at all tolerated temperatures.

Work on fishes (see Chapter 7) suggests that the triggering mechanism for maturation and spawning (i.e., for even perceiving the environmental stimuli that are conventionally stated to trigger maturation and spawning) is, in any individual, the ratio between its current oxygen supply (Q_m) and the oxygen supply corresponding to its maintenance metabolism (Q_{maint}).

As lobsters grow, the Q/Q_{maint} ratio declines (because $d < 1$), and they experience more frequent periods of internal hypoxia (e.g., after a strenuous activity). At some point, this ratio, as it drops, reaches a threshold informing the individual in question that, given the local conditions, it should begin to look for environmental stimuli, which will then push it further toward the behavioural and hormonal processes leading to maturation and spawning. Maturation and spawning occur at a size corresponding to an oxygen consumption $Q_m > Q_{maint}$, with Q_m declining as Q_{maint} increases, as occurs at higher temperatures.

In bony fishes, this is ensured by a fixed Q_m/Q_{maint} ratio of ~ 1.35 with a 95% confidence interval of 1.23 – 1.53. The Q_m/Q_{maint} ratio can be shown to be mathematically equivalent to $(L_{max}/L_m)^D$, where $D = b \cdot (1 - d)$, b is the exponent of the length-weight relationship, and d is the species-specific scaling factor between the respiratory area and weight (see above). Here, L_m, L_{max}, b, and d values were assembled from palinurid lobsters and crabs (**Figure 12.8**). They suggested a mean Q_m/Q_{maint} ratio of 1.46 and 1.41, with confidence intervals that overlap with those estimated for bony fishes.

These results suggest that not only the heuristic (Budaev *et al.* 2019), providing the trigger for maturation and spawning, may not only occur in crustaceans as it does in fish, but that the spawning threshold Q_m/Q_{maint} ratio in crustaceans is roughly similar to that occurring in bony fishes. Studies on other groups of crustaceans should test these results. They suggest that the life-history constraints on (large) crustaceans are similar enough to those of fish to have led to the emergence of the same heuristics (Morbey and Pauly 2022).

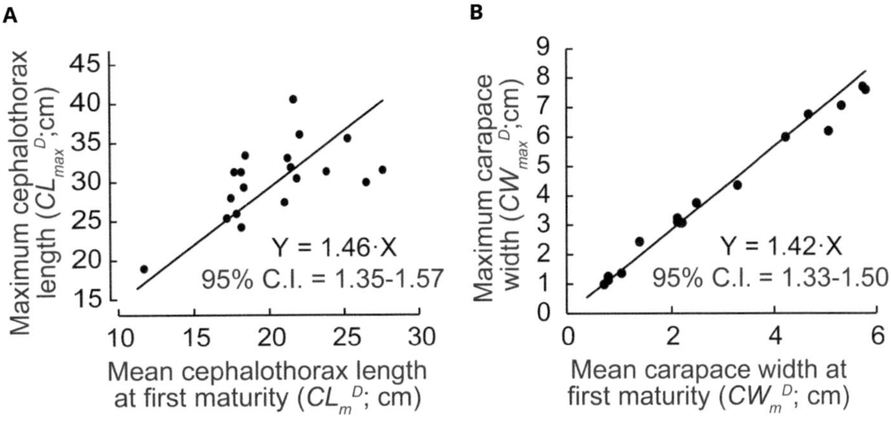

Figure 12.8. Relationships between size at first maturity and maximum size in 2 groups of crustaceans. **A**: In 10 lobster species of the genus *Panulirus* (n = 19). **B:** Size at first maturity and maximum size in 14 species of crabs (see table S3 in Pauly *et al.* 2022a for data sources).

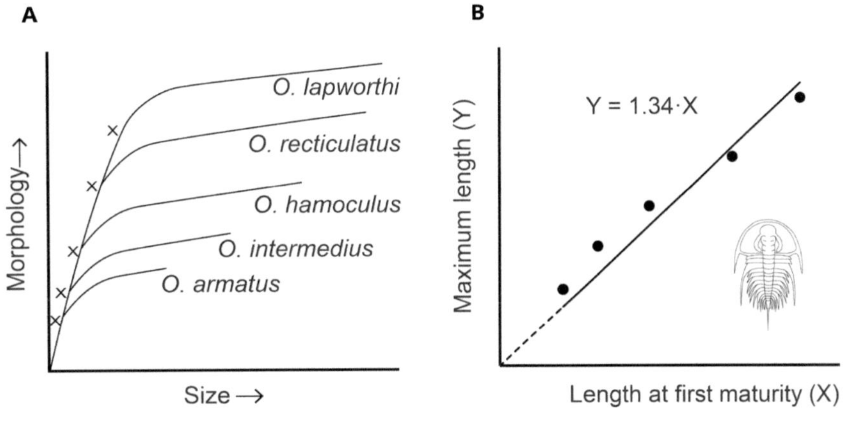

Figure 12.9. Tentative evidence suggests that a Q_m/Q_{maint} (= $L_{max}{}^{D}/L_m{}^{D}$) ratio similar to recent WBE may already have occurred in Cambrian trilobites (*Olenellus* spp.). **A**: Reproduction of figure 6 in McNamara (1978), with non-specified axis labels. **B**: assuming that the end of the (growth?) curves correspond to the maximum length in each *Olenellus* species, the crosses to the lengths at first maturity, and $d \sim 2/3$ (i.e., $D = 1$), leads to a ratio of 1.34 (95% C.I. = 1.14 – 1.54), i.e., the same for recent fish and various arthropod species.

Generalizing to other arthropod classes

Alternatively, and this is speculative and should be further tested by future evolutionary models, the heuristic described above may have emerged very early in the evolution of metazoans and was conserved in various phyla. This aligns with recent demonstrations that the growth of at least one Ordovician trilobite species, *Triarthrus eatoni,* can be described by the VBGF (Pauly and Holmes 2022) and that a Q_m/Q_{maint} ratio of 1.34 (for $d = 2/3$ and $D = 1$) can be computed from the length at first maturity and the maximum lengths of trilobites in figure 6 of McNamara (1978), about five *Olenellus* species from the Early Cambrian (**Figure 12.9**)[219].

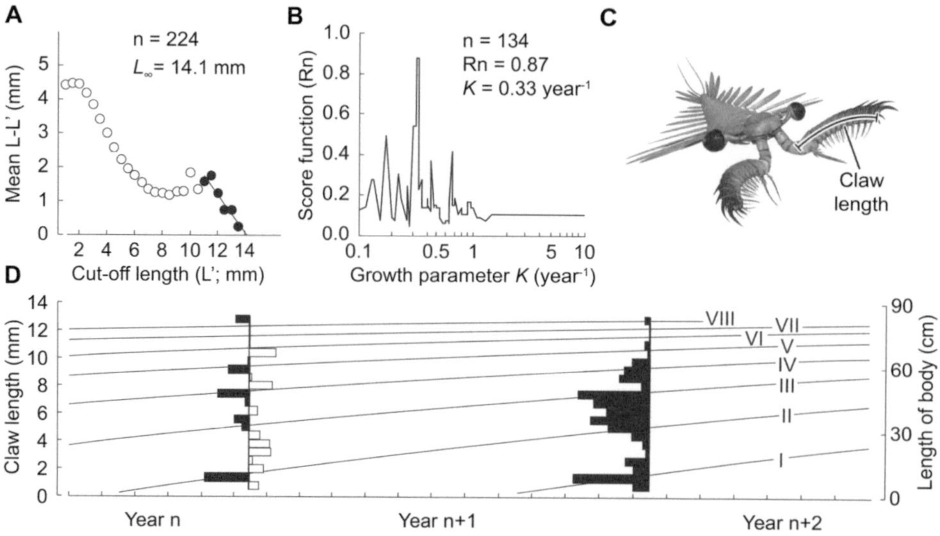

Figure 12.10. Result of the length-frequency (L/F) analyses for the Cambrian radiodont *Amplectobelua symbrachiata*. **A**: Modified Wetherall plot to obtain a preliminary estimate of its asymptotic (claw) length (L_∞) using data from several sites documented in Wu *et al.* (2024). **B**: Estimating the best *K* value (associated with the highest Rn value) for L_∞ = 14 mm, given the L/F data from Jianshan locality (n = 134). **C:** Representation of *A. symbrachiata* (modified from Wu *et al.* 2024), with the claw in darker grey. **D**: Growth of *A. symbrachiata* as inferred by the ELEFAN software, with L_∞ = 14 mm and K = 0.33 year^{-1} from L/F data (in Year n+1), also shown in 'restructured' form (Year n). The length scale on the right of panel **D** shows the actual body length of the individuals, as opposed to the length of their claw, as analysed by ELEFAN (modified from Wu *et al.* 2024).

Another ancient arthropod species, the Cambrian radiodont *Amplectobelua symbrachiata*, exhibited asymptotic growth, as also inferred from a length-frequency analysis using the ELEFAN approach (Pauly and David 1980) and software (Gayanilo *et al.* 2005). This confirmed the rapid growth inferred from fossilized instars of *A. symbrachiata* (Wu *et al.* 2024), which reached a length of 90 cm and was the largest predator of its time (see **Figure 12.10**).

We also note that in the largest known radiodont ever to exist, *Aegirocassis benmoulai*, which reached over 2 m (van Roy *et al.* 2015), the prehensile 'arms' were modified for filter feeding like the baleens in the largest whales. This suggests that large sizes, even in the Cambrian, and even with multiple side flaps which may have aided their respiration,[220] could be only reached by filter feeding, as also occurs in whale shark (*Rhincodon typus*), basking shark (*Cetorhinus maximus*) or the enormous extinct bony fish *Leedsichthys problematicus* (Pauly 2019, p. 84-89 and 156).

For another important group of water-breathing arthropods, the horseshoe crabs (Xiphosura), evidence exists that the operational surface of their 'book gills' and their respiration scales with body weight such that $d < 1$. In the well-studied Atlantic horseshoe crab *Limulus Polyphemus*, gill surface area was shown to relate to body weight via an exponent d = 0.776 (Suzuki *et al.* 2008), which should lead, as in crustaceans, and despite their frequent molting, to asymptotic growth.

Indeed, there is evidence for asymptotic growth in two horseshoe crab species, *L. polyphemus* and the Chinese horseshoe crab *Tachypleus tridentatus*, as inferred by a rearing

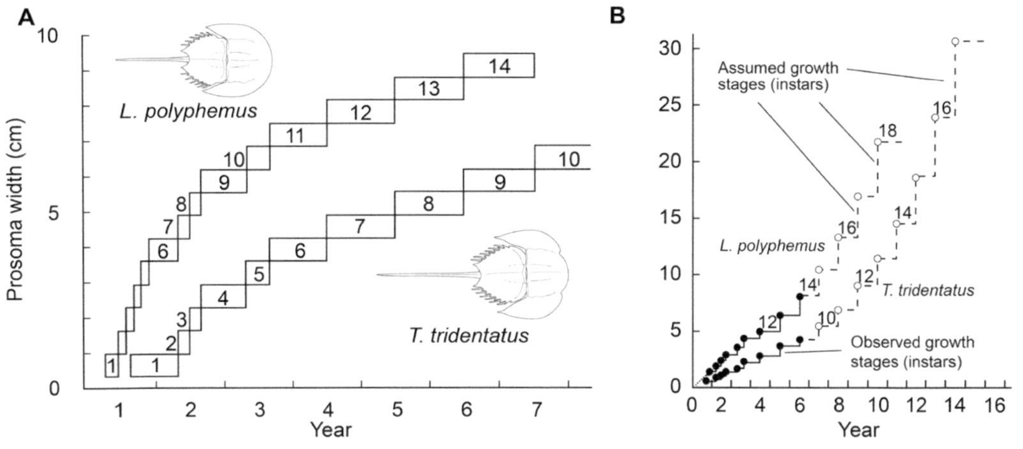

Figure 12.11. Growth curves for the Atlantic horseshoe crab, *Limulus polyphemus*, and the Chinese horseshoe crab *Tachypleus tridentatus* (from Sekiguchi *et al.* 1988). **A**: Growth curves based on specimens held in captivity under suboptimal conditions, resulting in asymptotic, but stunted growth that included fewer instars than occur in the wild. **B**: Imaginary growth curves consisting of the growth in **A**, onto which missing instars were grafted, reflecting presumed (higher) growth in the wild. The results are two exponential and unrealistic 'growth curves' (see text).

experiment (**Figure 12.11A**). This figure, adapted from Sekiguchi *et al.* (1998), depicts the growth of these two species in captivity. It suggests asymptotic sizes (i.e., the width of their 'prosoma,' or helmet-like carapace), which are much lower than the maximum sizes reported from the wild for these two species, while fully compatible with growth in captivity, i.e., under stressful conditions.[221]

Because these growth curves did not include all instars known for these two species (18 in *L. polyphemus* and 16 in *T. tridentatus*), Sekiguchi *et al.* (1988) grafted the missing instars on top of the curves in **Figure 12.11A,** along with the size increase between instars known from the wild (see also figure 5.10 in Shuster and Sekiguchi 2003). The resulting hybrid 'growth' curves are exponential (**Figure 12.11B**). Such growth patterns are improbable, as they would require that a critical *adult* size exists at which, following a phase of exponential growth, length and weight would reach a plateau, just like the size of a house fly plateaus following the exponential growth of its maggoty larva.[222] In insects, this is the critical size at which they metamorphose; adult horseshoe crabs do not metamorphose, and this conception of the growth of horseshoe crabs will have to be revised.[223]

Except for some authors perhaps unwittingly proposing exponential growth in adult horseshoe crabs, the data on the growth, respiration, and reproduction of various species of water-breathing arthropods presented above suggest that critical features of their life-histories can be explained by the effect of the "*dimensional tension*" (see Pauly and Cheung 2017, and Chapter 1 in this book) between the growth of their gills and that of their bodies.

Despite the claims of the generality of the GOLT as a unified theory, it does not explain everything crustaceans and other water-breathing arthropods do. Rather, it predicts the limits of what these animals can do under constraints such as hypoxia and/or elevated temperatures. The applicability of the GOLT suggests that fish, crustaceans, and other WBE are subjected to the same constraints — scarce oxygen in the medium they inhabit and a limited oxygen supply via their gills relative to the demands of their body tissues and their (potential) activity levels.

The available oxygen levels in air and water varied dramatically in geologic time (Ward 2006). Crustaceans first appeared during the Cambrian species explosion (~ 450 million years ago) after oxygen concentrations in the atmosphere and the sea rose from ~5% to ~15% of the current level (Krause *et al.* 2018). Large decapod crustaceans only emerged later, during the Devonian, once atmospheric oxygen concentrations reached near present levels (Wolfe *et al.* 2019).

Recognizing the constraints the GOLT poses on crustaceans may also help interpret observed patterns in their behavior and reproduction. For example, when approached by a predator, large lobsters are typically less apt to flee than smaller individuals (Bouwma 2006). If they do, they cannot match the endurance of small individuals (i.e., the number of tail-flips) (Vermeer 1986). Indeed, large decapod crustaceans typically take around 8 hours to return to their pre-exercise oxygen consumption levels after a period of exercise (McMahon *et al.* 1979; Field *et al.* 1991). This size-specific difference in escape behavior and limits to aerobic endurance is consistent with the GOLT, as is the limiting role of lactate-induced fatigue among hermit crabs (*Pagurus bernhardus*) fighting over shells of optimal sizes (Doake and Elwood 2010).

Activity levels of WBE after feeding are also influenced by oxygen availability, as an increase in oxygen consumption is associated with the extra energy required for transportation of food in the alimentary tract, its digestion, absorption, and post-absorptive metabolic processes collectively referred to as specific dynamic action (Jobling 1981). This is also true of large crustaceans whose aerobic scope for activity is reduced, often for hours, after feeding (Crear and Forteath 2000).[224]

Limitations imposed by the GOLT may also explain the remarkable feature that female crustaceans – ranging from the smallest copepods to the largest lobsters – carry fertilized eggs outside their bodies, as do horseshoe crabs. Their eggs are supplied with oxygen from the surrounding water, rather than from within their bodies, where they would have to obtain oxygen through their gills. Similar behavior occurs only in a few bony fishes, the Syngnathidae, i.e., seadragons, seahorses, and pipefishes (which have small, inefficient gills; Pauly *et al.* 2022c), and in 5 species of 'pelvic brooders' in the Family Adrianichthyidae (Gani *et al.* 2022). For large, cold-water decapod crustaceans (e.g., clawed lobster), the developmental period for eggs is nearly a year and, therefore, would represent a considerable oxygen demand on the female if eggs developed internally.

Seasonal aspects of crustacean reproductive phenology driven by temperature may also be tied to the GOLT. Most crustaceans extrude their eggs in the spring as water temperatures rise; therefore, the metabolic cost of internal egg provisioning of oxygen increases. Within the spring spawning season, it is common among the largest crustaceans (e.g., lobsters, crabs) for the largest females to spawn first, as do fish. This is consistent with predictions of the GOLT as rising seawater temperatures challenge larger individuals more than smaller ones.

Admittedly, much of the behavioral and reproductive evidence from crustaceans and other arthropods that can be marshaled to evaluate the relevance of the GOLT is based on studies performed for other purposes. We are aware of no comprehensive review of data or direct experimental test of the GOLT for crustaceans or other arthropods, but many physiologically driven aspects of crustacean biology are consistent with the theory.

Additionally, it may be possible to understand better the remarkable feature that female horseshoe crabs and crustaceans hold their eggs externally.

While viviparity and ovoviviparity in aquatic crustaceans are rare (Subramoniam 2016), they do occur. A comparison between gill surface area and reproductive mode may allow for inferences on possible correlations between these traits. Recent work on the live-bearing marine isopod *Cirolana harfordi* shows that higher temperatures correlate with increased offspring numbers and shorter gestation periods (Gavel 2022). An increase in temperature of only 4 °C halved the gestation period, which may reflect a limiting role of oxygen as a result of increased maintenance costs.

While this hypothesis requires further investigation, it has been established that the GOLT applies to crustaceans, and probably to ancient arthropods such as horseshoe crabs, trilobites and radiodonts as well. Despite the fundamental differences in the body plans of vertebrates and invertebrates, the parallels in the Q_m/Q_{maint} ratio between bony fishes and (at least several) aquatic arthropods illustrate that constraints of aquatic respiration have the same or similar implications across different phyla and body plans. These relationships need confirmation in a larger number of aquatic crustaceans and other arthropods. Still, the comparison reflects that the GOLT relies on fundamental constraints rather than specific historical contingencies that occurred in the evolution of a given phylum. The two following chapters will explore the application of the GOLT to even more phyla and probe the theory's scope across phyla and classes as diverse as molluscs, sponges, jellyfish, and arrow worms. Even if the GOLT's claims are restricted to what we earlier called 'constrained generality,' due to a specific mode of respiration, its generality across metazoic phyla, however, can claim universal applicability.

13

Molluscs and the GOLT

Synopsis: The respiration and growth of bivalves, and the frequent hypoxic event they experience, inspired earlier authors to publish a theory that anticipated several elements of the Gill-Oxygen Limitation Theory (GOLT). After briefly reviewing this early theory, and the growth of bivalves and some other smaller molluscs, this chapter is devoted to counter a mythology which has long held back research on cephalopods – squid and octopus, but not cuttlefish – whose major tenets are that *all* cephalopods (*i*) breathe through their skin and are therefore not oxygen-limited, (*ii*) grow exponentially, (*iii*) for a year, then (*iv*) spawn and die, (*v*) even when they reach monstrous sizes, as in giant squids. Counter-examples are presented which suggest that the GOLT provides more credible interpretations of the available data on cephalopod biology, notably on their growth and reproduction. This suggests that the GOLT applies to them as well, though the applicability of the theory to this group remains a challenge, notably due to problems in age determination, e.g., through their statoliths.

Markers of internal hypoxia

Molluscs are another group of invertebrates with singular features, to which, nevertheless, the GOLT appears to apply. Due to the commercial importance of many molluscan taxa, such as bivalves and cephalopods, which play essential roles in mariculture and fisheries, respectively, data for many species are readily available, and the group as a whole has received considerable scientific attention. The availability of data offers opportunities for our theory to potentially explain important aspects of the biology of these animals, especially their growth in relation to temperature. As is the case in other water-breathing phyla, the GOLT argues that these aspects of mollusc biology can best be understood in the context of their respiration, and the mechanisms and physical constraints involved in extracting oxygen from a fluid medium.

To illustrate the relationship between respiration and their most important physiological traits, we briefly explore the existing knowledge on the impact of internal hypoxia in this group.

Almost half a century ago, Lutz and Rhoads (1977) published a paper that anticipated various aspects of the GOLT. The first part of its summary reads: "*Microstructural growth increments within the shells of numerous Recent and fossil molluscs are interpreted as reflections of alternating periods of shell deposition and dissolution, occurring during aerobic and anaerobic respiration, respectively. The acidic end products of anaerobic metabolism are neutralized by calcium carbonate from the shell, leaving a relatively insoluble organic residue*

Table 13.1. Estimates of d in the relationships $S \propto W^d$ or $R \propto W^d$, where S is the gill surface area of bivalves, R their respiratory rate, and W their body weight (from Vakily 1992).

# - Species	d	S or R	Sources and Remarks
1 - *Argopecten irradians*	0.87	R	Jorgensen (1976); data from Kirby-Smith (1970)
2 - *Aulacomya ater*	0.66	R	Griffith and King (1979)
3 - *Cardium edule*[a)]	0.63	S	Foster-Smith (1975, figure 3), with $W \propto L^3$.
4 - *Chlamys islandica*	0.78	R	Vahl (1978); mean of experiments 2-4, table 4.
5 - *Chlamys opercularis*	0.65	R	McLusky (1973); from relative O_2 consumption.
6 - *Modiolus demissus*	0.80	R	Read (1962)
7 - *Mya arenaria*	0.58	S	Hughes (1969, figure 8B), with $W \propto L^3$.
8 - *Mytilus edulis*[b)]	0.68	R	See Vakily (1992, p. 44)
9 - *Scrobicularia plana*	0.70	S	Hughes (1969, figure 8A), with $W \propto L^3$.
10 - *Venerupis pullastra*	0.65	S	Foster-Smith (1975, figure 3), with $W \propto L^3$.
11 - *Venus mercenaria*	0.65	S	Hughes (1969, figure 8A), with $W \propto L^3$.

Note in # 3: a) Median of 3 values; # 8: b) Median of 9 values ranging from 0.51 to 0.84.

at the mantle-shell interface. With the return of oxygenated conditions and resumption of aerobic respiration, this organic material is reincorporated within the shell."

Hypoxia within bivalve shells is frequent because bivalves close their shells under suboptimal environmental conditions (Mitchell 1914), as may occur during a neap tide or when they sense the presence of a predator. This does not obviate the dimensional tension between the gills and their bodies. Indeed, from both measurements of gill surface area and respiratory data, there is strong evidence that bivalves, as expected, have values of $d < 1$ (**Table 13.1**).

This implies that bivalves must switch more frequently between aerobic and anaerobic metabolism as they grow larger and gradually shift from a general reliance on aerobic to anaerobic metabolism.[225] This pattern has been demonstrated in various bivalve species, and, in general, larger animals and adults rely more heavily on anaerobic metabolism than juveniles (see, e.g., Cueto-Vega *et al.* 2022; Brokordt *et al.* 2013).[226]

The metabolites involved in anaerobic metabolism are different from those in fish; however, one result is the same: the pH in the body of a mollusc relying on anaerobic metabolism is low, which, among other things, impacts the deposition of minerals onto hard structures. Given nychthemeral variation in activity levels, this results in the formation of the very readable daily 'rings' in the statoliths of young squids (Morris 1991; Balgos and Pauly 1998), similar to those in young fish (Panella 1971). Later, in adult squids, the statoliths (and the otoliths of fish) become largely unreadable because the above-mentioned dimensional tension causes semi-permanent hypoxia in their bodies, which masks the diel or 'nychthemeral' variations of pH levels that lead to markings (Pauly 2019, pp. 113-117).

Asymptotic growth in molluscs

Given the markings on their shells, bivalves can be 'aged' relatively easily, which allowed for the establishment of their growth being asymptotic very early on. Vakily (1992) presented the most extensive compilation of VBGF parameters for bivalves published at the time

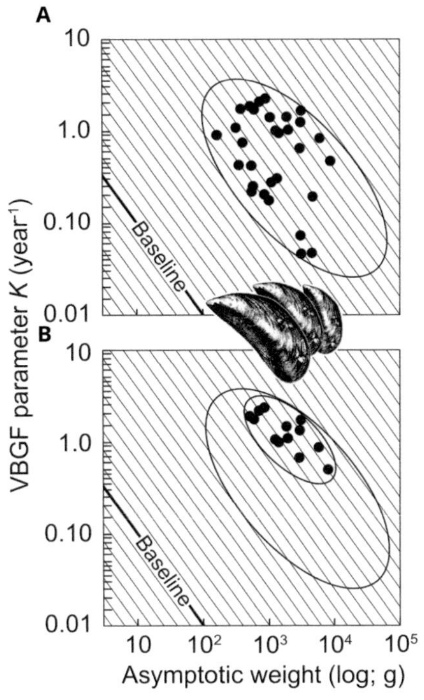

Figure 13.1. Illustrating that the VBGF parameters of bivalves take predictable positions on auximetric plots, suggesting that the different populations and species share traits that constrain their growth and size similarly and predictably. **A**: Position of 12 species (and subspecies) in 30 populations of Mytilidae. **B**: The 3 species and 13 populations of *Perna* spp., a mytilid genus, occupy only a part of the ellipsoid for Mytilidae as a whole (data from Vakily 1992, table 5.11).

(84 species and 190 populations), which he then compared with each other using auximetric plots. As can be seen, for example, in mussels (Family Mytilidae), their different growth curves, each represented by a dot, form an ellipsoid pattern, similar to that of fish and other WBE (**Figure 13.1A**). Moreover, an ellipse drawn around the data points representing a family will includes sub-patterns representing genera (**Figure 13.1B**) and so on for species and populations.

These patterns suggest that the different populations, species, genera, and higher taxa share traits that constrain the ellipsoid formed by their growth parameters to a given area of an auximetric plot. The most important of these traits is that their gill surface area determines how large they can be and how fast they will approach their asymptotic size. Local conditions, especially the quantity and quality of the food they can access, determine the variance about the main axis of the ellipsoids, i.e., how 'fat' the ellipsoid will be, while their main axes (ranging from high K and low W_∞ on the upper left to low K and high L_∞ on the lower right) largely reflect temperature-size effects that we know (as TSR) from other taxa and phyla.

Similar findings were made for gastropods, at least among the few aquatic species whose growth has been studied, which also exhibit asymptotic growth patterns (see **Figure 13.2** for an example).

Asymptotic growth patterns also occur in planktonic pteropods, whose seasonal growth does not appear to have been studied except for the contribution by Wang *et al.* (2017). As

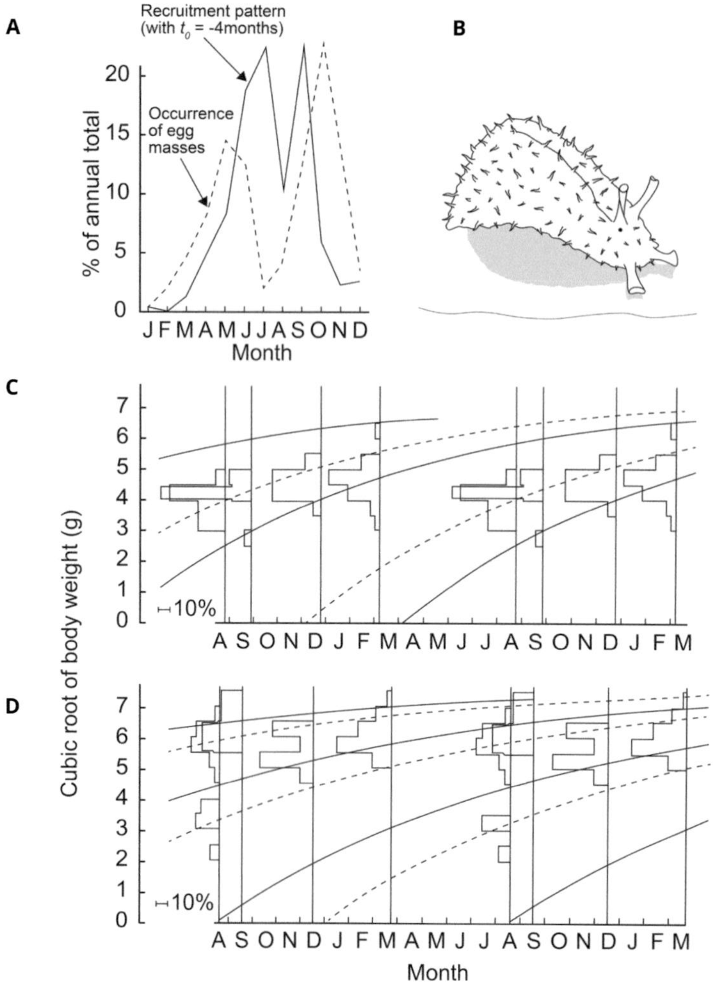

Figure 13.2. Growth of the sea slug *Dolabella auricularia* (Gastropoda) in the Central Visayas, Philippines, using ELEFAN and the cubic root of body weight as 'length equivalent,' as its body length cannot be measured precisely. **A**: Mean seasonal pattern of recruitment into the 'length'-frequency data in **C** and **D**, compared with the seasonal occurrence of egg masses in Bais and Syit Bays, confirming the existence of two cohorts per year (one solid, the other dotted in **C** and **D**), and allowing the estimation to $t_0 \approx -0.33$ years (see Pauly and Calumpong 1984). **B**: Representation of *D. auricularia* by E. Chu. **C**: Bais Bay, with $W_\infty = 512$ g and $K = 0.8$ year^{-1}. **D:** Syit Bay, with $W_\infty = 475$ g and $K = 1.0$ year^{-1}.

Observation	Remarks
The largest specimens occur in the coldest part of their range.	See text and **Figure 13.3** for other squids.
Maturation occurs at a smaller size in the warmer part of their range.	As in fish, see Chapter 7.
Gill filaments grow relatively faster in warm parts of their range.	This is required to increase O_2 supply in warmer water with less dissolved O_2.
Males have more gill filaments than females.	Males reach 50 cm (ML), females 40 cm off New England (Roper *et al.* 1984).
D. pealeii has relatively larger gills than *Doryteuthis roperi*.	*D. pealeii* is the larger species wherever they co-occur.

Table 13.2. Evidence of adaptive compensation for oxygen limitation in longfin inshore squid *Doryteuthis pealeii* (extracted from Cohen 1976).

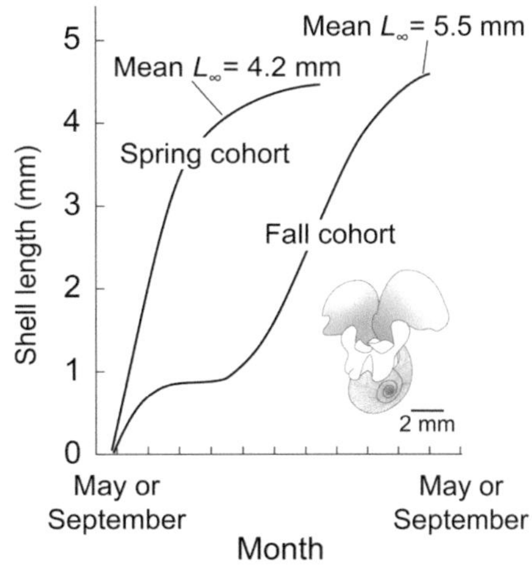

Figure 13.3. Seasonal growth curves of the North Pacific pteropod *Limacina helicina* as estimated by ELEFAN from shell length-frequency data (adapted from Wang *et al.* 2017). Note the large difference between the Spring and the Fall cohorts.

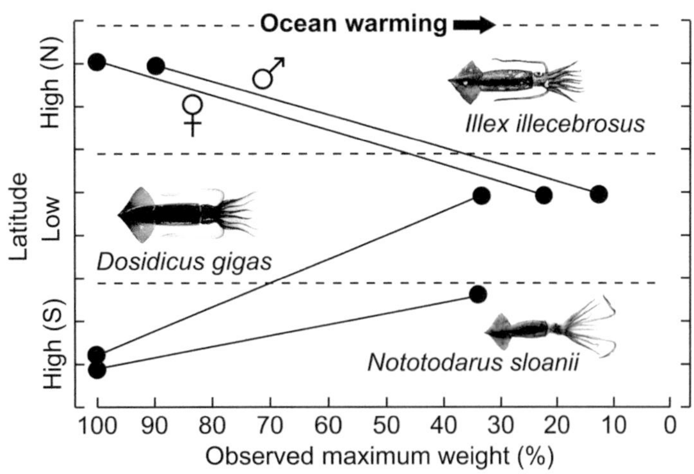

Figure 13.4 (opposite page, bottom). Maximum weight in 3 species of squid in relation to latitude, a proxy for temperature, in percent of the maximum weight reported for a species, or females and males in *I. illecebrosus,* and both sexes combined for *D. gigas* and *N. sloanii* (based on data in Ehrhardt *et al.* 1983 and Roper *et al.* 1984). As expected, squid are smaller at low latitudes (and higher temperatures), a trend likely to intensify with global warming.

these authors found, the short-lived North Pacific pteropod *Limacina helicina* exhibits growth curves whose shape depends on the time of hatching: when hatched in spring, most of their short lives unfold when temperatures are relatively high; hence their initial growth is fast, but they remain smaller (mean asymptotic shell length, i.e., L_∞ = 4.2 mm, K = 5.1 year^{-1}). On the other hand, individuals hatched in the fall tend not to grow much through the winter but reach a larger size (L_∞ = 5.3 mm; K = 4.1 year^{-1}) in the spring (**Figure 13.3**), in line with the temperature-size rule and the GOLT, as presented in the earlier chapters.[227]

Similarly, **Table 13.2** documents respiration-related differences between the longfin inshore squid *Doryteuthis pealeii* from the coldest and warmest ends of their range.

Figure 13.4, on the other hand, illustrates that 3 squid species are larger in higher (and colder) latitudes, while Lavin *et al.* (2022) confirmed these results for the Wellington flying squid, *Nototodarus sloanii*.[228] This implies that there should be no problem applying the GOLT to cephalopods. But there are.

Are cephalopods really that different?

The semelparity and rapid growth of some smaller species of squid have led to numerous misunderstandings regarding the growth performance and life span of all species of cephalopods, which include the largest[229] and heaviest[230] of all invertebrates. Or, in plain language: there are several specialists on squid – to be gradually introduced in the text below or in the endnotes – who believe that *all* cephalopods (*i*) breathe through their skin and therefore (!) are not oxygen-limited; also, (*ii*) they grow exponentially, (*iii*) for a year, then (*iv*) spawn and die – and yes, (*v*) this applies even to the biggest ones (see Pauly 1998c, 2019).[231]

The first of these five tenets, held, e.g., by O'Dor and Hoar (2000) – but which never made much sense in the first place, given that the surface area of the body of cephalopods (even in tubular squid) is much smaller than their gill surface area (Pauly 2021a) – had been thoroughly refuted by the actual measurements of Birk *et al.* (2018). The second part of (*i*) is also easily disposed of, considering that the metabolism of squid increases with a power of weight $d < 1$ (O'Dor and Wells 1987; Seibel 2007), which implies that oxygen supply does indeed limit their growth, at least in adults. If oxygen were not limiting the growth of cephalopods, they would indeed grow exponentially, according to tenet (*ii*). However, most large squid can be shown to grow asymptotically if an objective method of curve fitting explicitly considers seasonal growth oscillations (see, e.g., **Figure 13.5** and Jarre *et al.* 1991 for more cases).

In a direct challenge to the suggestion that squid grow asymptotically and that this form of growth has profound implications for their biology, Lipinski and Roeleveld (1990) showed that a Growing Sealife™ plastic squid, when placed in water, displays asymptotic growth. This was supposed to show that "*if a fit to growth data by the von Bertalanffy curve looks satisfactory, it does not follow that this is evidence for anabolic and catabolic changes.*" This is a "*judo argument*" (Asimov 1977), easily turned against its very proponents, as this Growing Sealife™ plastic squid exhibited as much 'anabolism' and 'catabolism' as can be expected from a piece of porous plastic.

We first note that the plastic squid in Lipinski and Roeleveld (1990, figure 2) did not grow 'exponentially' as, e.g., fish larvae do. This would have implied that growth was limited only by food intake (i.e., water intake in the case of a plastic squid) and that absolute growth rate (e.g., g·day^{-1}) increased until an external limit was reached (metamorphosis in the case of fish larvae). Instead, the plastic squid displayed a form of asymptotic growth

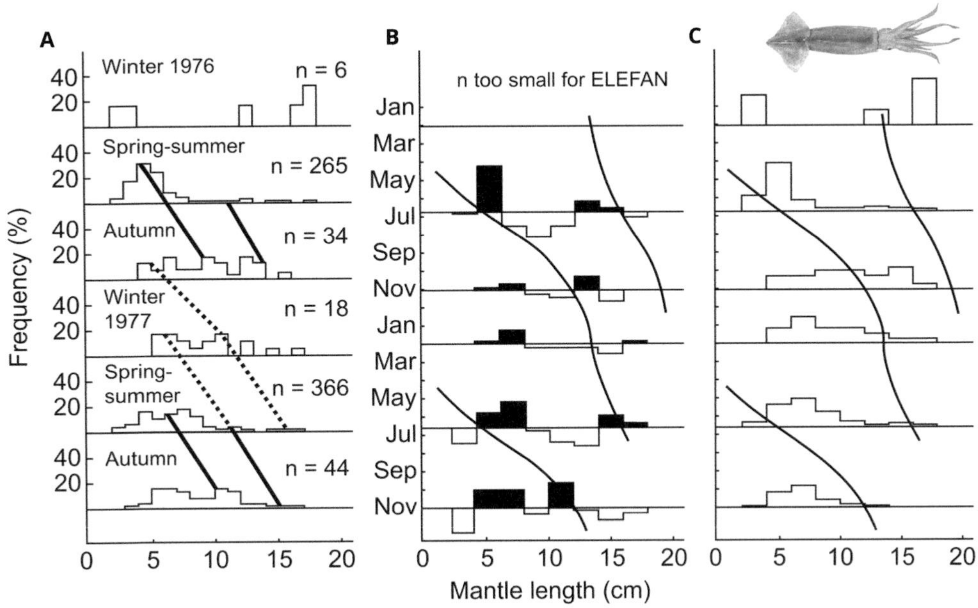

Figure 13.5. Demonstrating that females *Doryteuthis pealeii* from the northwestern Gulf of Mexico grow asymptotically once subjective methods of analysis – as in **A**, redrawn from Hixon *et al.* 1981 – are replaced by an objective interpretation of length-frequency (L/F) data, as provided by ELEFAN (Pauly and David 1981). **B**: 'Restructured' L/F data used to select the best combination of growth parameters, i.e., those generating a growth curve which, by 'hitting' the peaks of the original L/F data (the positive black histograms) while avoiding the troughs in between (the negative white histogram), collects the most 'points.' **C**: The objective procedure in B yielded the parameters ML_∞ = 21.5 cm and K = 0.285 year^{-1}, with C = 0.65 quantifying the amplitude of seasonal growth oscillations (adapted from Pauly 1985).

well represented by the special VBGF, i.e., with a growth rate that declined linearly with length. This is evidence that something was working, from the outset, to prevent water from seeping into the plastic.

This 'something' is the cohesive force of the plastic itself, whose magnitude (k) should be proportional to the volume (or W) of the plastic whose pores were filled with water. As a force acting against the expansion of the plastic squid, we have a negative term, $-kW$. The force that pulls water into the plastic body of the squid is capillarity, which is by definition proportional to a surface, i.e., $HW^{2/3}$ (Thompson 1917). This implies that the growth in weight (plastic + water) of a Growing Sealife™ plastic squid should conform to

$$dW/dt = HW^{2/3} - kW \qquad ...(13.1)$$

which we recognize as Pütter's equation, as presented in earlier chapters, and whose integral is the VBGF, which describes asymptotic growth.

Several cephalopod specialists have suggested that the VBGF cannot duplicate the growth patterns reported from squid or octopuses. The seasonally oscillating VBGF is versatile enough to duplicate cephalopod growth curves, including what appears to be the 'exponential' growth of squid. But before this can be elaborated upon, the evolution of cephalopods must be briefly mentioned.

The ancestors of today's cephalopods were shelled nautiloids; their extant descendants in the genus *Nautilus* have growth rates that are much lower than those of squid and longevities ranging from 10 to 20 years (Saunders 1983; Cochran and Landman 1984; Landman *et al.* 1988, 1989; Pauly and Ward 2024), in contrast to 1– 4 years in octopus and squid of similar sizes (see SeaLifeBase). Rodhouse (1998) suggested that the differences between the growth of *Nautilus* species and that of squid were due to the latter having evolved from nautiloids through a neoteny-like process he called *"physiological progenesis,"* which corresponds to the concept of *"paedomorphosis"* as proposed by Gould (1977). This suggestion would explain their high metabolic rate and rapid growth, similar to that occurring among the juveniles of most WBE.

Also, in many squid and other cephalopods, substantial post-spawning mortality eliminates many of the older individuals, leaving us with the initial 'linear' or 'exponential' segments of growth curves, to which any function can be fitted, whether this makes physiological sense or not (see, e.g., Bigelow 1993). The disadvantage of this approach is that when purely empirical curves are fitted to actual observations (e.g., age-length pairs, length-frequency data, or tagging-recovery data; see **Figure 1.3**), no theory is being tested and/or enriched, and potential errors cannot be detected.

For example, and this deals with tenets (*ii*) to (*v*), the near impossibly fast growth estimate of Clarke (1980b) for the oceanic squid *Kondakovia longimana* stood unchallenged for years, presumably because they coincided with the widespread belief among specialists that squid – including huge ones, living in Antarctica's frigid waters – always have an annual life span.

Re-establishing plausible cephalopod growth patterns

The implausibly low longevity estimated for *K. longimana* prompted a re-examination of the length-frequency data, which had led to it. As agreed by Malcolm Clarke, who co-authored the paper in which the rectification was presented (Jarre *et al.* 1991), there was no evidence to justify the assumption of a one-year lifespan in *K. longimana*. Indeed, based on the annual markings of the chitinous pen ('gladius') inside this species, its longevity is likely 8-10 years (Bizikov 1991). Again, length-frequency data may be used to infer growth patterns, but they must be analyzed with an objective method.

Another re-assessment occurred with the jumbo flying (or Humboldt) squid *Dosidicus gigas*, which ranges from California to Chile and currently supports lucrative fisheries in the Eastern Pacific, including around the Galapagos Islands. At least since the publication of Erhardt *et al.* (1983), the growth parameters – and hence the life-history – of this squid were a matter of controversy, notably because different authors (*i*) continue to use subjective methods for length-frequency analyses, (*ii*) usually do not consider that high temperatures impact the sizes that WBE can reach, and (*iii*) nor consider that the hypoxic water which Humboldt squid often inhabit tend to suppress the formation of rings assumed to be laid daily on their statoliths, leading to an underestimation of their age.[232]

Nevertheless, a coherent account of their life-history is beginning to emerge, based on their *"extreme"* phenotypic plasticity, which identifies two forms: a 'large' form, with a mean mantle length of about 60 cm and reaching maturity at 1.5 years (or possibly 2 years or more if their age is underestimated), and a 'small' form less than 40 cm and maturity at 6 months (Hoving *et al.* 2013), but which may also require more time if ages are underestimated. These authors write, under the subheading "*What controls size-at-maturity,*" that "*elevated temperatures and limited nutrition typically result in smaller size at maturity for squid*

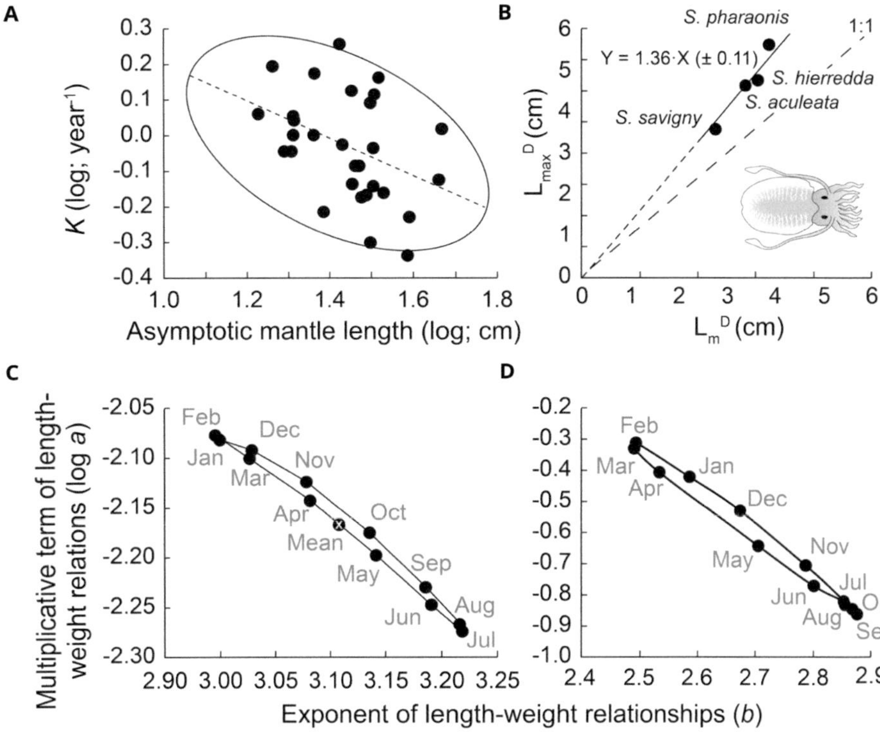

Figure 13.6. Basic features of the growth of *Sepia* species. **A**: Auximetric plot of different populations of *Sepia aculeata, S. dollfusi, S. hierradda,* and *S. officinalis*, where each dot represents a K-L_∞ pair, i.e., a growth curve. **B**: Plots of L_{max}^D vs. L_m^D in 4 *Sepia* spp. allowing the estimation of a mean L_{max}^D/L_m^D ratio = 1.36. **C**: Seasonal cycles of the parameters *a* and *b* of length-weight relationships, from a simulation in Pauly (2019), based on Equation (12.2) and a similar equation for seasonal growth in weight. **D**: Seasonal cycle based on monthly length and weight measurements in 2016 of *Sepia hierredda* caught along the coast of Ghana (Sakyi-Djan 2017), smoothed with 3-month running means (adapted from Pauly *et al.* 2024b).

(Boucher-Rodoni, 1973; Rowe and Mangold, 1975; Mangold, 1987; Forsythe and Hanlon, 1989; Arkhipkin and Mikheev, 1992; Jackson, 1995; Jackson and Domeier, 2003; Moreno et al. *2005). This generalization regarding temperature is consistent with the smaller size of mature squid that would have been born in the Guaymas Basin* [of the Gulf of California] *during the 2009 – 2010 El Niño… .*"

All of this is in line with the GOLT because the large form occurs during cold La Niña periods while the small form appears during and following the warm El Niño events, and because their size at first maturity is reduced when their maximum size also is (see Chapter 7 for a mechanism that explains why).

The plasticity of *D. gigas* is likely due to this species having evolved and inhabiting ecosystems in which the deeper but resource-rich waters are hypoxic[233] and in which the cold surface water of La Niña years, with enough dissolved oxygen, was, for millions of years, periodically replaced by the warm waters of El Niño events. This required its ancestral populations to either evolve the capability to generate a progeny consisting of individuals tolerant of hypoxic and/or warm water (but which remain smaller when exposed to such conditions) or go extinct.

Our last case is dozens of cuttlefish species, close relatives of squid, which not only exhibit asymptotic growth, as reported by multiple authors (see, e.g., Ragonese and Jereb 1991; Mehanna and Amin 2005; Sakyi-Djan *et al.* 2022; Yesilyurt *et al.* 2019; **Figure 13.6**) but reach ages of up to 4 years. Also, the parameters *a* and *b* of their length-weight relationships exhibit, in a year, cyclical patterns that match precisely the results of simulations based on Equations (12.2) and a related version of the VBGF for seasonal growth in weight (see Pauly 2019, p. 94-95; **Figure 13.6C** and **D**). Moreover, they appear to use a maturation-triggering threshold value $L_{max}^D/L_m^D = 1.36$ (**Figure 13.6B**), which is the same as the threshold estimated in bony fishes (Chapter 7) and similar to that in various invertebrates (Chapters 12 and 14). Overall, this suggests that the GOLT may apply to all species of the Family Sepiidae and, by extension, to cephalopods as a whole.

Overall, we can only reiterate the hope expressed in Pauly (1998c) that the community of scientists who study cephalopods abandon the 5 tenets mentioned above and concentrate, instead, on identifying what the animals they study – which indeed *look* different from fish – have in common with them, given that they are subjected to the same physical and dimensional constraints, as emphasized by the GOLT.

14

Growth and reproduction in sponges, cnidarians and chaetognaths

Synopsis: This chapter attempts to apply the Gill-Oxygen Limitation Theory (GOLT) to three groups of water-breathing invertebrates, i.e., near spherical sponges, here represented mainly by the commercial wool sponge (*Hippospongia lachne*), medusoid cnidarians, and chaetognaths, or arrow worm, neither of which possesses gills and whose bodies contain more water (≥ 90%) than fish (75-80%). In all three groups, oxygen consumption rates scale with body weight via an exponent d <1, resulting in a good fit of asymptotic growth curves, including variants that account for seasonal temperature fluctuations. Also, *Hippospongia lachne* and the chaetognaths exhibited strong body size reductions with increasing temperature. Indeed, the GOLT appears to explain well major traits of near-spherical sponges and chaetognaths, including their relatively slow growth of the latter, which is attributable to their apparent lack of a dedicated respiratory organ, while some features of medusoid cnidarians remain to be investigated.

Biological theory and asymmetrical knowledge

Our current knowledge of marine species and ecosystems is still dominated by a focus on animals that are of commercial interest, which leads to an asymmetry in the available data and studies that allow for a more comprehensive understanding of their life-histories and ecologies. As mentioned earlier in this book, the first elements of the GOLT were developed for fish, many of which are exploited by fisheries and studied intensely (Pauly 1979). Chapter 12, devoted mainly to crustaceans, and Chapter 13, covering molluscs, also dealt predominantly with species exploited by fisheries or raised in aquaculture, and for which considerable literature exists, notably on their growth.

Many other invertebrate taxa have been less intensively studied because fisheries do not exploit them. One example is chaetognaths (Pauly *et al.* 2021a), but also the sponges (Pauly *et al.* 2022b) and jellyfish (Pauly *et al.* 2009), of which only a few species are targeted by fisheries and have been intensively studied. Here, based mainly on the publications just mentioned, key aspects of the biology of some species representing their respective phyla are presented and discussed in relation to the GOLT. As we demonstrate in this chapter, these animals also experience limitations of their oxygen supply and the 'dimensional tension' between body mass and respiratory surfaces that shape the most important life-history parameters of bony fishes, Chondrichthyes, crustaceans, molluscs, and other aquatic ectotherms. We will first focus on Porifera (sponges), then cnidarians, and finally on chaetognaths. As the examples of these different taxa show, the central tenets of the GOLT are applicable across different body plans and reveal fundamental features of water-breathing ectotherms in all metazoan phyla and classes.

Growth, respiration, and reproduction in near-spherical sponges

This section addresses issues related to the growth, reproduction, and respiration of sponges, emphasizing the sheepswool sponge (*Hippospongia lachne*), better known as the 'wool sponge.' Emphasis is given to the relationship between growth and respiration in the wool and other sponges because this issue is often overlooked in studies of their biology, even though an appropriate supply of oxygen is as essential to their survival and growth as is an appropriate amount of food (Pauly 2021a).

The work presented in this section of Chapter 14 is based on the pre-submission version of Pauly *et al.* (2022b), which represented the first attempt to fit a sponge to the standard version of the VBGF. As this equation was derived from first-order physiological principles involving the interplay of surfaces and volumes leading to asymptotic growth (see Chapters 2 and 3), it should therefore be enlightening to test whether these principles, first derived for fish, also apply to sponges, the oldest multicellular animal taxon. This section does not address the growth of encrusting or tubular sponges that may not suffer from the respiratory constraints emphasized here because of their laminar bodies. Instead, it focuses on nearly spherical species such as the wool sponge. The choice of this species as an exemplary case is also pertinent because it is one of the most valuable and abundant commercially exploited species (Martinangeli *et al.* 2022) and, therefore, has been better studied than many other sponge species (Stevely *et al.* 1978; Witzell 1998; Butler *et al.* 2017, 2018, 2021).

Pauly *et al.* (2022b) fitted the VBGF to the diameters-at-(relative)-age published by Storr (1964), which were based on wool sponges tagged in the upper Gulf of Mexico; this included specimens of diameters up to 18 cm, beyond which he inferred their growth indirectly (**Table 14.1**). He also noted that *"the rate of growth in diameter indicated by* [his study] *appears to be valid for the first 5 or 6 years. Beyond this point, the growth rate must be assumed to be somewhat less* [...]*, the increase in diameter gradually approaching a uniform rate as the sponge assumes the doughnut shape."*

Table 14.1. Diameter (inches), volume (cubic inches), and age of wool sponges (*Hippospongia lachne*) tagged off Florida in the Gulf of Mexico, as read off the graph in figure 7b in Storr (1964). The last 4 years of data (in parentheses) are as indirectly inferred by Storr (1964).

Age (years)	Diameter (in)	Volume (in^3)	Age (years)	Diameter (in)	Volume (in^3)
1	2.6	8.6	7	8.9	388.7
2	4.3	44.2	(8)	(9.5)	(434.8)
3	5.5	88.8	(9)	(10.0)	(480.8)
4	6.6	181.9	(10)	(10.7)	(609.8)
5	7.5	269.4	(11)	(11.3)	(716.3)
6	8.2	333.3	--	--	--

The VBGF derived from the 11 age-diameter data pairs in **Table 14.1** was

$$D_t = 31.5(1 - e^{-(0.191t)}) \quad ...(14.1)$$

where D_t is the predicted diameter at age t in centimeters, and D_∞ is the asymptotic diameter in cm. The parameter t_0 is not included in this equation because it was set at zero, forcing the growth curve to start at the origin of the age scale in **Figure 14.1**. As for the value of 0.191 year^{-1} for the parameter K, its 95% CI, estimated by bootstrapping the data in **Table 14.1**, is 0.185 – 0.199 year^{-1}. This confidence interval is likely underestimated because the age-at-length data in **Table 14.1** are means, i.e., they do not account for the variability of individual growth. Still, the estimate of asymptotic diameter is close to the maximum reported by Storr (1964), 12 inches (or ~30.5 cm), the approximate size at which the sponge's spherical shape is maintained.

Given the conversion between centimeters and inches, the relationship between the age in years and the volume in cm³ becomes

$$V_t = 1046(1 - e^{-(0.191 \cdot t)})^3 \quad ...(14.2).$$

Note that in the wool sponge, the dry weight of the skeleton (e.g., as used as a bath 'sponge') is about 4% of its wet weight (after draining its internal channels and removing the commensals and/or parasites[234] living therein, see Pauly *et al.* 2022b).

As sponges lack specialized respiratory surfaces, which can grow hyper-allometrically, as do the gills of fish, surface-to-volume relationships play an even more immediate role. This implies that the cumulative cross-section of the pores through which water can enter their bodies is a surface and cannot be increased indefinitely. This (2-D) surface, at least in compact species such as the wool sponge, cannot keep up with the growth of their (3-D) bodies, and, as they grow, their interior will become increasingly hypoxic (Hoffmann *et al.* 2005; Lavy *et al.* 2016).

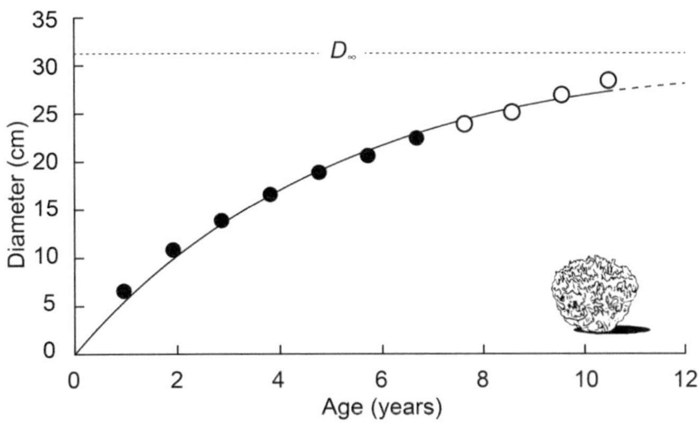

Figure 14.1. Growth of the wool sponge (*Hippospongia lachne*) in the upper Gulf of Mexico as indicated by a von Bertalanffy growth curve fitted to diameter-at-age data from Storr (1964), with the data for the last 4 years (open dots) inferred indirectly (see **Table 14.1**). The estimates of asymptotic length (i.e., diameter) and growth coefficient are 31.5 cm and 0.191 year^{-1}, respectively. The horizontal dotted line indicates the asymptotic diameter (D_∞).

Sponges have evolved a complex symbiotic relationship with microbes to deal with the geometric constraints imposed on their physiology. The sponge canal system operates as subdivided units among which pumping rates (i.e., oxygen acquisition) vary, creating spatio-temporally variable zones of hypoxia and anoxia that serve to support the respiratory activity of anaerobic symbionts (Lavy *et al.* 2016). Nevertheless, in some species, including the wool sponge, the development of internal hypoxic zones leads to tissue necrosis, as suggested by Storr (1964), who attributed central tissue "*death*" in large wool sponges to lack of food, rather than to lack of oxygen.

Concerning 12 inches being the maximum diameter attained by near-spherical wool sponges, Storr (1964) wrote the following: "*This is confirmed in the observed growth of wool sponges by the death of the central portion of the sponge when this diameter is reached. The form of the sponge from then on becomes more and more doughnut in shape. Continued growth beyond a 12-inch diameter suggests that one other growth factor operates as the sponge approaches the limit of growth indicated* [by Storr's model, not included here]. *Since the sponge is uniform in structure and the intake of water carrying the food is through the sides, the greatest amount of food uptake is in the periphery of the sponge. This area, therefore, continues to grow vigorously, but the rate of food intake is not sufficient, nor is the rate of food transfer through the sponge efficient enough to support active metabolism in the central portion of the sponge when the diameter of the sponge is 12 inches or over. The growth formula obtained would probably be directly applicable to the rate of sponge growth except for this phenomenon of sponge physiology.*"

Therefore, the growth trajectory documented in Storr's study pertains to the phase where wool sponges remain near-spherical. Beyond this phase, they change shape, resulting in different ratios between respiratory surfaces and the mass of oxygen-consuming tissues. This transition could be modeled using a biphasic version of the VBGF (Soriano *et al.* 1992); this is not attempted here, mainly because the second-phase growth of this sponge species is not well-documented and of little interest to the sponge fishing or farming industries.

The respiratory area of the wool sponge is not known. Measurements of the respiratory area for the European coastal sponge *Suberites massa*, taken over a century ago, have been used to estimate the area at 100 $cm^2 \cdot g^{-1}$ (Pütter 1914; von Ledebur 1939). Gatti *et al.* (2002) used oxygen microprobes to measure oxygen concentrations within the tissue of the Mediterranean bacteriosponge *Suberites domuncula* and detected oxygen concentrations of approximately 50% of the surrounding ambient seawater. Weisz (2006) employed oxygen microprobes and tetrazolium salts to examine oxygen concentrations and the occurrence of anoxic zones in 4 species of low-microbial-abundance and high-microbial-abundance sponges. Oxygen concentrations within the ectoderm and endoderm of low-microbial-abundance sponges were similar to ambient seawater concentrations, whereas endoderm tissues in high-microbial-abundance sponges were all hypoxic or anoxic.[235]

Measurements of oxygen levels in the interior of a near-spherical sponge, the Barrett's horny sponge (*Geodia barretti*), performed by Hoffmann *et al.* (2005) were read off their figure 2 by Pauly *et al.* (2022b), re-expressed as fractions of ambient oxygen levels, and fitted with a logistic curve, which yielded the model:

$$\textit{Fraction of ambient oxygen level} = 1/(1 + e^{-0.675(2.76 - r)}) \qquad ....(14.3)$$

where r is the radius, measured as the depth within the sponge; 0.675 is the slope of the logistic curve at 50% of the ambient O_2 level; and 2.26 is the depth within the sponge

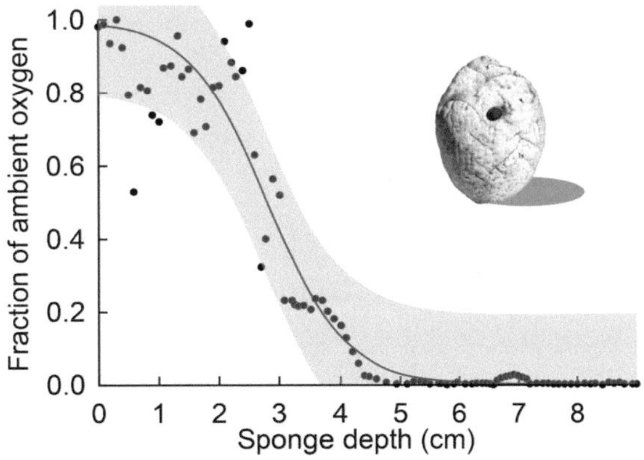

Figure 14.2. Decline of oxygen content within a near-spherical sponge, the Barrett's horny sponge (*Geodia barretti*), based on a specimen sampled between 100 and 200 m on the hard bottom slope of the Korsfjord, near Bergen, Norway, as read off figure 2 in Hoffmann *et al.* (2005) and re-expressed as a fraction of ambient oxygen levels. The relationship between oxygen content and sponge depth is fitted with a logistic curve, and the gray shaded area represents the 95% confidence interval (see text).

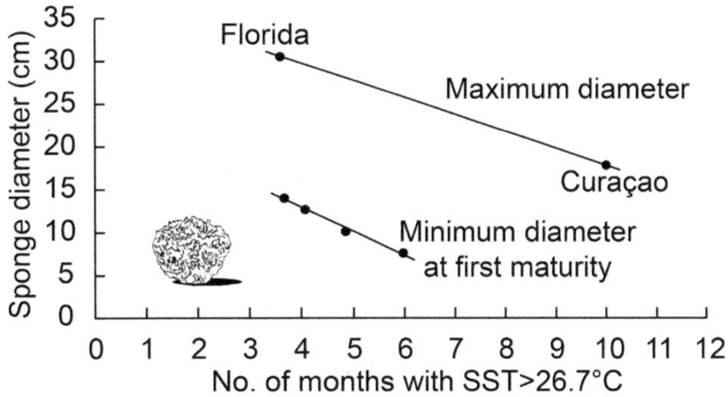

Figure 14.3. Temperature dependence of size at first maturity (solid line) and maximum size (dotted line) of wool sponge (*Hippospongia lachne*) off Florida in the Gulf of Mexico (based on data from Storr 1964) and in the waters of Curaçao (based on data from van Soest 1978). SST = sea surface temperature.

at which O_2 is 50% of the ambient level. The 95% confidence interval of the slope at the inflexion point was estimated using bootstrapping (Fieberg *et al.* 2020) and found to be 2.52 to 2.98 (**Figure 14.2**).

The oxygen level decline is most rapid at the 50% inflexion point of 2.76 cm (95% CI: 2.52 – 2.98), where the slope (0.675) has a 95% CI of -0.998 to -0.548. The generality of the numbers estimated using Equation (14.3) and illustrated in **Figure 14.2** cannot be currently assessed, as they were derived from an experiment on only one species at temperatures from 6 to 15 °C (Hoffmann *et al.* 2003). The shape of this relationship is likely realistic and can be adapted to different species by increasing or decreasing the two parameters of Equation (14.3). This equation enables the computation of oxygen levels in near-spherical sponges by adding the dissolved oxygen in successive depth layers and the oxygen level of each depth layer, as the mean of the oxygen levels at its outer and inner limits.

Van Soest (1978) concluded that wool sponges reach a size up to 18 cm in diameter in the waters of Curaçao, where the waters are warmer than in the upper Gulf of Mexico, where wool sponges reach diameters of over 30 cm (Storr 1964). These numbers are illustrated in **Figure 14.3**, which shows that the maximum sizes of wool sponges decrease with increasing temperature. Moreover, this figure also illustrates the inverse relationship between minimum length at first maturity and water temperature, as has been reported for other WBE.

The mean ratio of 0.4 for the relationship of L_m and L_∞, shown in **Figure 14.3**, cannot be used to test, in wool sponges, whether $L_{max}{}^D/L_m{}^D$ = 1.3 - 1.4, as predicted for most WBE (see Chapters 7 and 12). The reason is that the data in Storr (1964) do not pertain to the *mean* size at which wool sponges mature (i.e., the length at which 50% of a cohort reaches maturity) but rather indicate the *minimum* size at which some do.

Sponges, which originated in the Pre-Cambrian, had ample time to evolve sophisticated defenses in the form of spicules embedded in a tough matrix, which partly protects them against spongivores (Wulff 2006), a strategy resembling that of thorny plants on land. Similarly, like plants, sponges cannot evade predators and, as a result, have evolved the ability to synthesize a dazzling array of complex organic molecules to protect themselves against potential consumers and pathogens (Pawlik 2011; Loh and Pawlik 2014; Rohde *et al.* 2015). This array of compounds is the reason they are sought after as a source of potentially beneficial medical compounds (Blunt *et al.* 2003).

Although very ancient, sponges did not evolve the many differentiated cells, tissues, organs, and appendages characteristic of other phyla, such as molluscs, arthropods, and chordates. Therefore, sponges allow for easier detection of the constraints that have shaped the transition to multicellularity in the first animals and the emergence of the first body plans. As Ward (2006) argued, "*respiration was perhaps the most important driver of animal body plans,*" a statement that reflects some of the most fundamental assumptions on which the GOLT is built. Indeed, the appearance of sponges on Earth, predating oxygen-producing multicellular plants by 300 million years (Brocks *et al.* 2017), occurred during a period when a day on Earth was only 21 hours and atmospheric oxygen levels were only about 50% of what they are today (Klatt *et al.* 2021). Although global oxygen stores later increased rapidly with the evolution of multicellular plants, for approximately 500 million years, sponges evolved in a world depauperate in oxygen. Without tissues or organs specialized for oxygen acquisition, sponge morphology was strongly constrained by physiological oxygen demand.

That oxygen constrains sponge energetics and morphology contradicts the suggestion in Storr (1964) that the supply of nutrients (i.e., food) is the limiting factor to the survival and multiplication of the cells in the interior of a cell mass. In their review of food competition and limitation for coral-reef-dwelling sponges, Pawlik *et al.* (2015) found no evidence for food limitation of sponge growth. The reef sponge community is dominated by sponges with low microbial abundance and high filtration rates, characterized by large canal volumes, and they dwell in an oxygen-enriched environment. The opposite is true for shallow-water-dwelling coastal sponges, for which growth appears to be limited at high densities of sponges, which may limit available oxygen (Valentine 2019).

The effects of a limited capacity to deliver oxygen to tissues are evidenced across many animal phyla (Heim *et al.* 2020). For humans and many other mammals, effects are manifested in heart insufficiencies and attacks (Weber and Janicki 1979) and in ischemic strokes (Janardhan and Qureshi 2004). This phenomenon also applies to cancerous tumors whose cells shift to glycolytic metabolism when their interior is deprived of oxygen or when a temperature increase drives their oxygen demand beyond the capacity of the blood supply that they have commandeered (Warburg 1930; Al Tameemi *et al.* 2019).

The fundamental reason for the limiting role of oxygen in metazoan metabolism is that, because of its toxicity, molecular oxygen cannot be stored in living animal tissue (Lane 2002); the oxygen that is stored in the mesoglea of jellyfish (Thuesen *et al.* 2005) is no exception because mesoglea is not living tissue. The only way oxygen can be stored is through chemical bonding as in blood (Lenfant *et al.* 1970; Noren *et al.* 2002), and oxygen storage requires an internal circulatory system that is absent in many phyla (Heim *et al.* 2020), including the Porifera.

As a result, the growth of poriferans, and especially those with a spherical shape, is constrained by the aforementioned dimensional tension (Pauly and Cheung 2017) inherent in having to supply a growing (3-D) body with oxygen in water whose flow is determined by the cumulative (2-D) cross-section of the inhalant pores. This constraint remains even if a sponge's pumping rate increases with body size. This is why sphere-shaped demosponges, when fully grown, experience hypoxia and even anoxia in the central part of their bodies, as is commonly observed. Yet, this oxygen limitation also permits the presence of anaerobic bacteria and archaea within the commensal community dwelling inside sponges, not too dissimilar to the microbiome in the human gut (Cénit *et al.* 2014), and allowing sponges to turn a liability, at least partially, into an asset.

Not all sponges are near-spherical; their various morphologies are tube-shaped, vase-shaped, encrusting, and irregularly shaped globular. Their internal structure is equally variable; some are dense, whereas the extensive canal system of others makes them full of voids. There are also striking differences in the skeletal structure and composition of spicules, which largely define sponge families. These characteristics play a role in water flow through sponges and influence oxygen availability.

The contention, illustrated here with the wool sponge, is that the oxygen supply may limit the growth of large, near-spherical sponges to their interior, allowing several inferences about their likely response to ocean warming and deoxygenation. These inferences are also important for sponge fisheries and aquaculture, as well as the study of sponges within their ecosystem context.

Moreover, we demonstrated that the GOLT, developed mainly with data for vertebrate, molluscan, and arthropod WBE, also applies to compact, near-spherical sponges. This demonstration of new avenues for their study should allow for the formulation of various hypotheses about their biology, notably on their shape-shifting past specific sizes, that should be straightforward to test.

The growth of medusoid cnidarians

The constraints relating to oxygen uptake discussed previously can also be illustrated in another class of animals: cnidarians. Many cnidarians, e.g., the polyps of coral reefs, are probably too small to experience a dimensional tension that would inhibit their growth individually. Still, coral reefs, i.e., huge aggregates of individual polyps, can create nightly hypoxia and then suffer from it, particularly when exposed to high temperatures (Nelson and Altieri 2019; Perzer *et al.* 2023). Indeed, we can assume that it is nightly hypoxic events, which, by lowering the pH in the living tissue of coral colonies, cause the daily rings that can be used to age coral heads (see Nothdurft and Webb 2007 and references therein). This

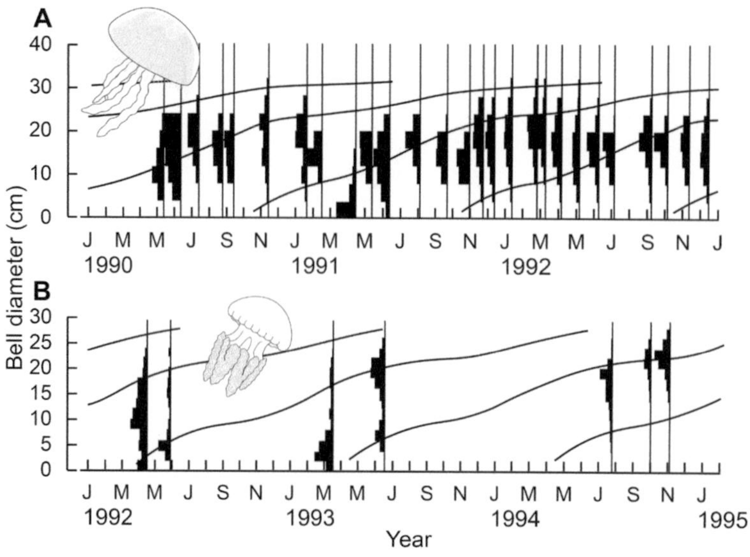

Figure 14.4. Seasonally oscillating growth curve fitted with ELEFAN to diameter (= 'length') frequency data of the medusae of 2 species of jellyfish. **A**: *Aurelia aurita* from Tokyo Bay, Japan, in 1990 – 1992 (Omori *et al.* 1995), with L_∞ = 35.5 cm and K = 0.86 year^{-1} for set values of C = 0.5 and WP = 0.1. **B**: First 3 of the 6-year samples of *Catostylus mosaicus* from Botany Bay, Australia, sampled between March 1990 and February 1998 by Pitt and Kingsford (2003), with L_∞ = 37.0 cm and K = 0.60 year^{-1} for set values of C = 0.5 and WP = 0.7.

process is analogous to the nychthemeral fluctuations of pH in the endolymph surrounding the otoliths of fish (Mugiya *et al.* 1981) or the statolith of squid (Morris 1991; Lipinski 1993), which also lead to 'daily rings,' particularly in larvae (Pauly 2019, pp. 113-117).

Another aspect of the biology of cnidarians relevant to the GOLT appears to be asymptotic growth (**Figure 14.4**), although their respiration can scale linearly with their weight (Larson 1987). This is an anomaly in terms of the theory presented here, which may be resolved by a meta-analysis of respiration in medusoid jellyfish, emphasizing how different classes respond to experimental stress, and possibly by accounting for their ability to store oxygen in their mesoglea (Thuesen *et al.* 2005).

For cnidarian species with suitable data on the frequency distributions of the diameter of their medusoid stage, Palomares and Pauly (2009) were able to fit the VBGF to all species so examined (**Table 14.2**). This was also the case when previous authors, having observed that medusae sometimes shrink when their food is scarce, claimed that the interpretation of the growth curves of jellyfish as asymptotic is inadequate. What turned out to be far more important is that their growth occurs in waters whose temperatures strongly differed seasonally, which requires the use of a seasonally oscillating variant of the VBGF, as presented in Chapter 12 and illustrated in **Figure 14.4**.

The water content of the medusoid stage of cnidarians, at approximately 96% on average (Larson 1986), compared to 75% in fish (Bykov 1973), makes it difficult to compare their growth performance to that of fish and other WBE. **Figure 14.5** illustrates a comparison between the growth of fish and jellyfish after the asymptotic weights of the latter were adjusted to a water content resembling that of fish. As can be seen, when standardized for

Table 14.2. Von Bertalanffy growth parameters (seasonal or not) of the medusoid stage of 9 species of jellyfish (Scyphozoa and Hydrozoa), as estimated with ELEFAN (extracted from table 2 in Palomares and Pauly 2009); the asymptotic size L_∞ refers to the diameter of the jellyfish 'bells' and W_∞ is in wet weight, while *K*, *C* and *WP* are parameters of a seasonally oscillating version of the VBGF (see Chapter 12).

Species	Location	L_∞ (cm)	W_∞ (g)	*K* (year-1)	*C* & *WP*	Reference
Aequorea aequorea	Namibian Shelf	11.1	141	0.87	0.5 & 0.7	Buecher *et al.* (2001) and Bierley *et al.* (2001)
Aurelia aurita	Tokyo Bay	35.5	1161	0.86	0.5 & 0.1	Omori *et al.* (1995)
Catostylus mosaicus	Botany Bay, Australia	37.0	3083	0.60	0.5 & 0.7	Pitt and Kingsford (2000)
Chiropsalmus sp.	4 Mile Beach, Australia	14.8	210	1.50	--	Gordon *et al.* (n 2004)
Chrysaora hysocella	Namibian Shelf	68.8	10991	1.10	0.25 & 0.7	Buecher *et al.* (2001)
Chrysaora tuberculata	Ionian Sea, Greece	39.0	6845	0.73	--	Kikinger (1992)
Cyanea sp.	New England, USA	14.2	214	2.30	0.5 & 0.1	Brewer (1989)

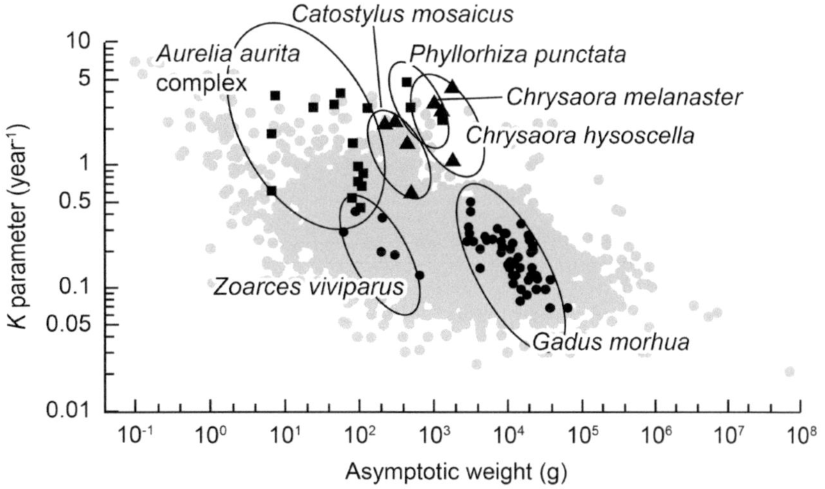

Figure 14.5. Plot of log(*K*) against log(W_∞) in jellyfishes (re-scaled for their different water contents) on an auximetric background of grey dots representing the growth curves of fish in general (in addition to two fish species for comparison, *Zoarces viviparus,* and *Gadus morhua,* both represented by black dots). As can be seen, the *Aurelia aurita* complex (black squares) and *Catostylus mosaicus* (black triangles) resemble small fishes in their growth pattern. Still, *Phryllorhiza punctata* (black squares) and *Chrysaora* spp. (black triangles) may grow faster (= higher *K* for a given W_∞) than most fishes. These results are very sensitive to the assumed ratio between the water contents of jellyfish and fish (Palomares and Pauly 2009).

their water content and positioned on an auximetric plot, medusoid jellyfish lose some of their alienness, and one can conceive that they might be subjected to the same dimensional tension that shapes the biology of other WBE.

Chaetognaths and the GOLT

Another group of invertebrates to which the central principles of the GOLT can be applied is Chaetognatha, or 'arrow worms' (Pauly *et al.* 2021a), a marine invertebrate phylum including 132 extant, carnivorous species in 9 families and 2 orders (see **Figure 14.6** for a few species). The chaetognaths are a group about which Darwin (1844) noted the "*obscurity of their affinities,*" which are still being debated today (Harzsch *et al.* 2015). The carnivorous arrow worms are understudied relative to their importance in the marine zooplankton, where they rank second in abundance after the herbivorous copepods, their major prey. Although arrow worms lack gills or other dedicated respiratory organs, we show that the Gill-Oxygen Limitation Theory (GOLT) can explain how temperature and respiration affect their growth and related life-history traits.

The maximum recorded body lengths (BL, in mm) for the 132 species of chaetognaths currently recognized in WoRMS (see www.marinespecies.org) and SeaLifeBase were linked by Pauly *et al.* (2021) to the mean latitude of their distribution range (*LAT*), in degrees north or south (as a proxy for temperature; Lindsey 1966; Fisher *et al.* 2010) and their habitat through the multiple regression

$$\log(BL) = 0.616 + 0.445 \cdot \log(LAT) - 0.580 \cdot B - 0.354 \cdot BP \qquad \ldots(14.4)$$

where *B* and *BP* are dummy variables (*B* = 1 for benthic and 0 for pelagic and benthopelagic species; *BP* = 1 for benthopelagic and 0 for pelagic and benthic species). The multiple correlation coefficient R = 0.573 implies that this model, with $R^2 = 0.329$, explains almost 1/3 of the variance in length among these 132 species. Additionally, its partial slopes have the expected signs and are significantly ≠ 0 ($p < 0.001$).

Equation (14.4) implies that chaetognath species tend to be smaller the closer they are to the Equator. Also, it allows for the calculation that the 14 benthic species are, on average, 3.8 times smaller than the 77 pelagic species. Finally, it also allows inferring that the 41 benthopelagic species (which, along with 'benthic' and 'pelagic' species, may have been misclassified) tend to be 2.3 times smaller than the pelagic species, i.e., intermediate, as may be expected (**Figure 14.6**).

The temperature-size relationship among chaetognath species presented in **Figure 14.6** likely has the same cause as that within the 3 species in **Figure 14.7**.

In **Figure 14.6,** this relationship represents a pattern on an evolutionary scale, while the correlation between temperature and size in **Figure 14.7** occurs on an ontogenetic scale. The respiratory surfaces of chaetognaths, which grow approximately with length squared, cannot keep up with the oxygen demand of their tissues, which grow approximately with length cubed, and, like in other WBE, must increase when temperatures are high. Hence, the smaller sizes of chaetognaths where and/or when the temperature is high.

In water-breathing ectotherms (WBE) that have hard parts — such as otoliths, other bones, or scales in fish, or statoliths in squid — the ages at different sizes can be estimated by counting annual rings in long-lived individuals exposed to marked summer–winter temperature differences in their habitats, or by counting daily rings in otoliths or statoliths. (Panella 1971; Daryanani *et al.* 2021). Unfortunately, arrow worms do not have statoliths (Bullock 1965). The growth of chaetognaths must be inferred from laboratory specimens or the analysis of length-frequency (L/F) data sampled in the wild. However, the correct way to draw inferences about growth from L/F data is often misunderstood; many authors

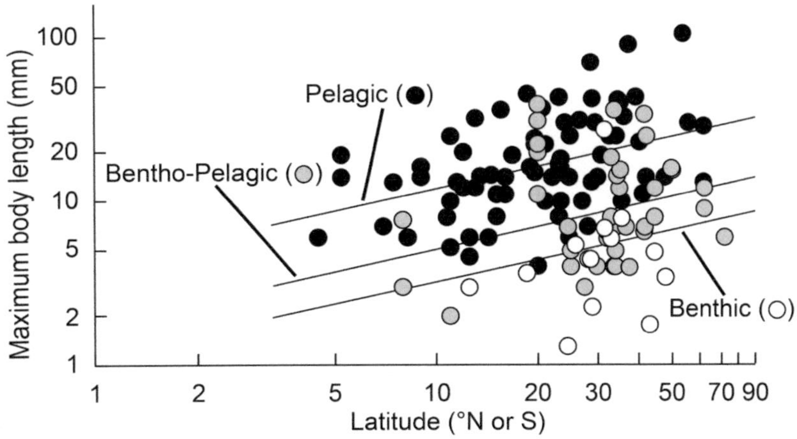

Figure 14.6. Relationship between the body lengths of the 132 extant chaetognath species and the mean latitudes of their distribution ranges (as a proxy for sea surface temperature). On average, pelagic species (n = 77) are 2.3 times longer than benthopelagic species (n = 41) and 3.8 times longer than benthic species (n = 14). Modified from Pauly *et al.* (2021).

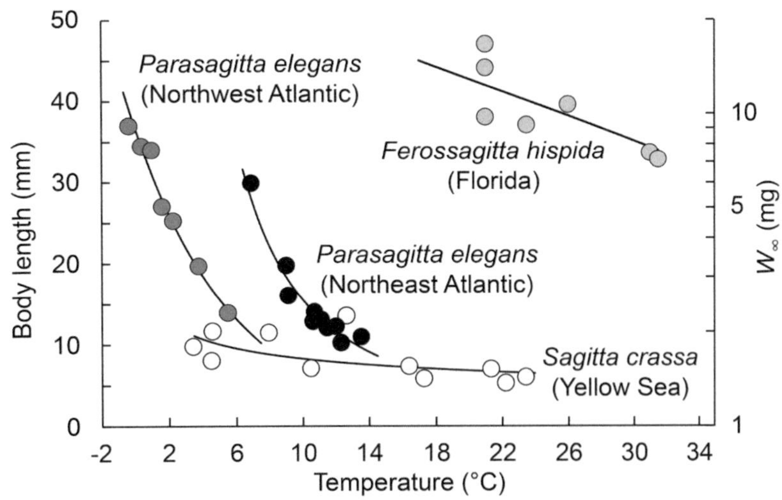

Figure 14.7. Relationship between the size (body length or W_∞, i.e., asymptotic weight) of 3 species of chaetognaths and the mean temperature of the water in which they live or were raised. The body lengths and the curves for *Parasagitta elegans* in the North Atlantic were redrawn from figure 1 in McLaren (1966), and the points for *Aidanosagitta crassa* in the Yellow Sea are from Huo *et al.* (2010) with a curve replacing their straight line. The grey data points from Reeve (1970), Reeve and Baker (1975), and Reeve and Walter (1972) for *Ferosagitta hispida* in Florida are documented in **Table 14.3** (note the logarithmic scale on the right).

emphasize identifying the peaks of their various L/F samples, assumed to represent cohorts (see **Figure 13.4**). On the other hand, they place less emphasis on the criteria they (should) use for linking the peaks of successive L/F samples; however, it is decisions based on these criteria which generate the growth increments from which the parameters of growth curves can be derived (Pauly 1998b).

As a result, the literature on the growth of chaetognaths includes many examples of growth curves whose author(s) presupposed slow growth. This has resulted in 'cohorts' being subjectively created whose growth curves extend to well over a year, sometimes two (Grigor *et al.* 2017; Welch *et al.* 1996) and including long periods of near-zero growth at minuscule sizes (see, e.g., Zo 1973).

Here, we present a set of published "*mass-at-age*" data presented by Hirst and Foster (2013), based on aquarium experiments by various authors. The standard VBGF was fitted after the data point representing larval growth was removed (**Figure 14.8**).

For comparisons of growth performance within and between different chaetognath species and between chaetognaths and other WBE, one can use the index

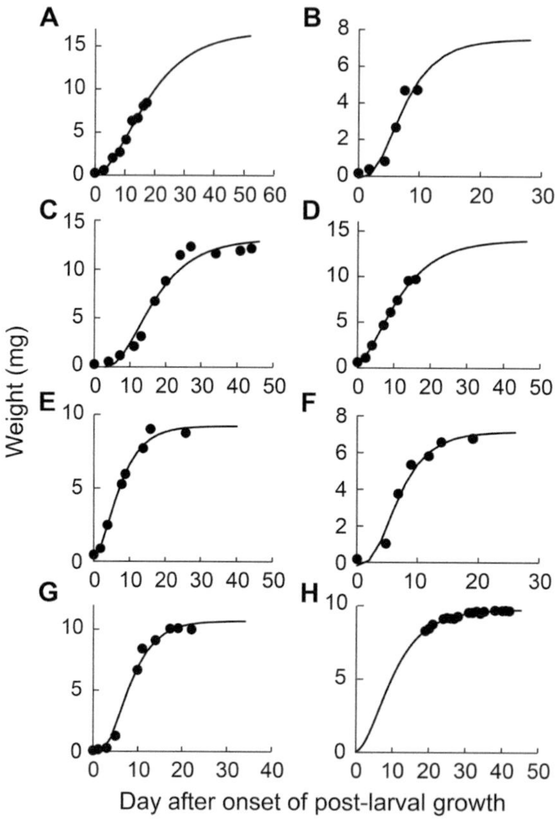

Figure 14.8. Weight growth curves of *Ferosagitta hispida* based on the "*mass-at-age*" data. The origins of these data, the corresponding growth parameters, and the pertinent length-weight relationships are documented in **Table 14.3**.[237]

$$Ø' = \log(K) + 2 \cdot \log(L_\infty) \quad ...(14.5a)$$

which is relatively constant within species (and higher taxa with similar shapes), as assessed by studying and relating hundreds of K-L_∞ data pairs (see Longhurst and Pauly 1987; Binohlan and Pauly 2000; Pauly 2019). To allow for comparisons that consider the different shapes that different clades of WBE can have, we use the index

$$Ø' = \log(K) + 2/3 \cdot \log(W_\infty) \quad ...(14.5b)$$

which assumes a length-weight relationship with slope $b = 3$, as commonly occurs in WBE.

In captivity, fish and other WBE usually grow rapidly to a smaller size than they would in nature, resulting in lower asymptotic sizes (length and weight) and higher values of the parameter K when von Bertalanffy growth curves are fitted to their size-at-age data (Saikkonen *et al.* 2011). This also appears to be the case for chaetognaths, whose growth curves are shown in **Figure 14.8** (see also **Table 14.3**). In any case, these growth curves provide a realistic alternative to many of the hand-traced growth curves found in the literature, and some suggest longevity of up to 2-3 years.

Indeed, the growth curves in **Figure 14.8** may be realistic, as some of the omitted points may correspond to the period between hatching and first feeding, characteristic of chaetognath larvae that can last up to 10 days (Pearre 1991).

Even if preceded by a larval period of slow growth, **Figure 14.8** suggests that at least for *Ferosagitta hispida*, the emergence and disappearance of a cohort would occur in a matter of about 2 months, which is possible given the small body length they reach (≈ 1 cm) and the high temperatures to which they were exposed. These results are compatible with Russell (1932), who suggested 4-5 generations per year near Plymouth, and with Murakami (1966), who reported 4 generations per year from areas in the U.K. and Japan, all of which are cooler than the waters of Florida.

Table 14.3. Growth parameters of *Ferosagitta hispida* raised in aquaria and estimated from "*mass-at age*" data of Hirst and Foster (2013), as illustrated and documented in panels **A**-**G** of **Figure 14.8**[a)] and size-at-age data in figure 2 of Reeve (1970), as illustrated in panel **H** of **Figure 14.8b)**. All samples originated from coastal water in Florida, U.S.

Panel	Temp (oC)	W_∞ (mg)	K (year-1)	t_0 (year)	Øc)	Data sources
A	21	16.7	31.09	0.0045	0.308	Reeve and Baker (1975; figure 1)
B	31	7.5	88.07	0.0024	0.528	Reeve and Baker (1975; figure 1)
C	17	13.1	41.75	0.0072	0.367	Reeve and Walter (1972; figure 3)
D	21	14	43.79	0.0077	0.406	Reeve and Walter (1972; figure 3)
E	23.5	9.2	72.51	0.0037	0.504	Reeve and Walter (1972; figure 3)
F	26	10.7	83.67	0.0035	0.609	Reeve and Walter (1972; figure 3)
G	31.5	7.1	97.00	0.0027	0.556	Reeve and Walter (1972; figure 3)
H	21	9.8	53.37	0.0033	0.387	Reeve (1970; figure 2)

a) Converted to wet weight by multiplying by 10 (see table 2 Pauly *et al.* 2021a); b) Converted from lengths via $W = 0.001 \cdot L^3$; c) Ø (= log K+2/3·log W_∞) was computed with W_∞ in g, facilitating comparison with other groups; mean Ø = 0.458.

Table 14.4. Comparison of the growth performance of 6 species of fish, 2 species of chaetognaths and 1 species of chironomid using Ø = log(*K*) + 2/3log(W_∞).

# and Speciesa)	W_∞ (g)	*K* (year-1)	Ø
1 - *Thunnus albacares*	198,940	0.250	2.93
2 - *Morone saxatilis*	17,543	0.186	2.10
3 - *Mugil cephalus*	13,800	0.110	1.80
4 - *Trigla gurnardus*	534	0.312	1.31
5 - *Callionymus lyra*	53	0.490	0.84
6 - *Taurulus bubalis*	102	0.230	0.70
7 - *Pseudosagitta gazellae*[b)]	≈ 1	2.87	(0.458)
8 - *Ferrosagita hispida*	≈ 0.01	55	0.458
9 - *Chironomus anthracinus*[c)]	0.0125	2.88	-0.81

See **Species** header for: a) Pauly (1981) for the authors of the W_∞ and *K* estimates of the 6 fish species; # 7: b) Assuming the Ø as in *F. hispida* (see **Table 14.3** and text) and # 9: c) Bsed on weights-at-age in figure 15b of Jónasson and Kristiansen (1967).

The Ø-values in **Table 14.3** have a mean of Ø = 0.458. Given this Ø-value, the largest extant chaetognath, *Pseudosagitta gazellae*, with an asymptotic weight of $W_\infty \approx 1$ g, would have a *K* value of 4-5 year^{-1} (see **Table 14.4**), which is compatible with having a single cohort per year, as reported by Pearre (1991), based on David (1955).

As **Table 14.4** clearly illustrates, chaetognaths have low growth performance compared to fish, even when considering only the most sluggish fish species in the table are considered. These slow growth rates are probably due to chaetognaths' respiration, which limits their growth potential compared to animals with elaborate breathing organs — our next topic.

Parry (1944) suggested that the long, flagella-like cilia in the gut of chaetognaths "*maintain a current in the gut lumen. This may be concerned with respiration or with the removal of dissolved excretory matter liberated by the intestinal cells. Neither circulation of body fluids nor excretory organs have been described in* Spadella, *so it may be concluded that the intake of oxygen and the excretion of katabolites takes place at the surface of the body. If these processes occur at the surface of the gut, then some means would be necessary to renew the water in the lumen. The cilia would provide this means.*" The last suggestion is not realistic because it would require either a rapid flow (through the vent?) of ingested water or nearly constant swallowing and regurgitation of oxygen-depleted water, neither of which has ever been observed.

This suggests that it is unavoidable to infer that the integument of chaetognaths serves as their respiratory organ. Indeed, the integument appears to be thinner, and given Equation (4.3), more efficient for respiratory purposes near the 'corona,' a group of "*cilia arranged around a depression on the dorsal side of the head and/or neck*" (Bleich *et al.* 2017, citing Kapp 1991). The corona is where the cilia generate a flow that contributes "*2 to 3 times more to the oxygen transport than diffusion does*" (Bleich *et al.* 2017).

Bleich *et al.* (2017) conclude that "*while the corona* [...] *may incidentally support respiration, especially for the oxygen supply of brain, eyes and head muscles, this is most probably not its main function.*" They also note that "*the continuous activity of the cilia of*

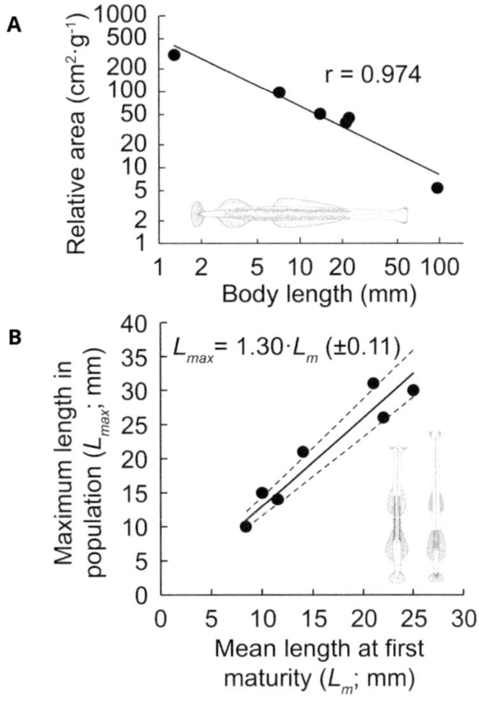

Figure 14.9. Basic information on the respiration and reproduction of chaetognaths (modified from Pauly *et al.* 2021a, table S3 and 4, respectively). **A**: Illustrating the relationships between the body length of 6 individual chaetognaths representing 6 species and their relative surface area/volume. **B**: Maximum lengths (L_{max}) vs. lengths at first maturity (L_m) of 4 species of chaetognaths (n = 7; see also **Table 14.5**), fitted with a linear regression with a zero intercept. The estimated slope is 1.30, with a 95 % confidence interval (C.I.) of 1.19 - 1.41, which overlaps with the slope of L_{max}^D vs. L_m^D in fishes (1.35; C.I. = 1.22 - 1.53).

the corona means a considerable energy investment (Katsu-Kimura 2009). It lets us conclude that the corona and the generated flow are important to chaetognaths."

No function is more critical for a small organism, which must acquire all the oxygen it needs by diffusion from the boundary layer of water adjacent to its integument, than to induce a flow that would renew the water surrounding it.[236] Indeed, this is necessary if a high oxygen gradient is to be maintained between the surrounding water and its integument.

Despite the extreme differences between the body plans of bony fishes and arrow worms, the underlying constraints arising from aquatic respiration central to the GOLT reveal a surprising parallel between the life-histories of these different phyla. As demonstrated in Chapter 7, mean length at first maturity in bony fishes (L_m, i.e., the length at which 50 % of the individuals are mature) occurs at $L_{max}^D/L_m^D \approx 1.35$, with $D = 3(1\text{-}d)$. Here, we also tested whether chaetognaths reach maturity and spawn at a fraction of their maximum length (L_{max}), similar to fishes. Since post-larval chaetognaths maintain their basic shape as they grow, their integument (i.e., their respiratory surface) should grow proportionately to 2/3 of their weight, i.e., $d \sim 0.67$ (**Figure 14.9A**). This means that for chaetognaths, $D = 1$, and hence here, $L_{max}^D/L_m^D = L_{max}/L_m$.

Table 14.5. Approximate length at fist maturity (L_m) and maximum length (L_{max}) of chaetognaths; *P* = *Pseudosagitta; F* = *Ferosagitta*, and *E* = *Eukrohnia*; all length in mm.

Species	Location/time (if indicated)	L_m	L_{max}	References
P. elegans	Bedford Basin, May-June '68	21	31	Zo (1973; figure 2)
P. elegans	Plymouth, September 1930	10	15	Russell (1932, plate I)
P. elegans	Plymouth, May 1930; 1931	14	21	Russell (1932, plate I)
F. hispida	Laboratory, at ~ 30 °C	8.4	10	Reeves and Walters (1972; figures 3 and 4A)
F. hispida	Laboratory, at ~ 17 °C	11.5	14	
E. bathypelagica	Lazarev Sea, Antarctica	22	26	Kruse (2010; fig. 1, p. 55)
E. bathyantarctica	Lazarev Sea, Antarctica	25	30	Kruse (2010; fig. 2, p. 56)

Figure 14.9B, based on **Table 14.5**, shows that in chaetognaths, L_{max}, when plotted vs. L_m in a regression with zero intercept, leads to a slope of 1.30 (C.I. = ± 0.11), which is close to the estimate of 1.35 for bony fishes, and well within their 95% confidence interval of 1.22 - 1.53 (see Chapter 11). This suggests that maturity in chaetognaths is triggered by a decrease in oxygen supply to their tissues as they grow, and their surface area to volume ratio declines to a level near 1.35 times their maintenance metabolism, as also occurs in bony fish.

Chaetognaths have long posed challenges regarding their identity and biology, primarily due to their bizarre anatomy, which is distinct from that of other invertebrate phyla. We hope to have shown that they behave like other, better-studied WBE regarding some of their physiological traits. We could demonstrate that chaetognaths behave as hypothesized, i.e.,

1. species occurring in colder waters are generally larger than those in warmer waters;
2. individual chaetognaths reach larger sizes in colder rather than warmer waters;
3. the VBGF can describe the growth of chaetognaths;
4. they lack a dedicated respiratory organ and exhibit low growth performances; and
5. they mature at a fraction of their maximum size which is similar to that occurring in bony fishes.

Therefore, we conclude that the chaetognaths conform to the Gill-Oxygen Limitations Theory (GOLT) despite lacking gills.

While numerous studies have been conducted on the metabolic rate (i.e., O_2 consumption) of chaetognaths (see references in Pauly *et al.* 2021a), very little thought has been devoted to how they breathe, i.e., how they transfer oxygen dissolved in the water surrounding them into their bodies. For example, there is no mention of breathing, respiration, or oxygen in the otherwise wonderfully detailed 12 chapters in the book "*The Biology of Chaetognaths*" (Bone *et al.* 1991).

Finally, we should also mention that the particular 'dart-and-sink' mode of locomotion exhibited by chaetognaths described a "*swimming pattern* [which] *consists of repetitive head-upwards darting motions, each followed by a period of passive descent*" (Sweatt and Forward 1985; see also Bone and Duvert 1991) may be linked to their lack of an efficient respiratory organ. Here, we propose the testable hypothesis that the oxygen in their muscle tissue is exhausted at the end of a 'darting' event, with a high oxygen level recovered at the end of the subsequent 'sinking' event.

The GOLT and breathing water without gills

The GOLT is built around the assumption that animals that must extract the oxygen they require from the water surrounding them will have increasing difficulties breathing as their size increases because the volume of their bodies grows in three dimensions, i.e., faster than the surface through which the required O_2 enters the body. Here, in the three groups we documented – near-spherical sponges, medusoid cnidarians, and arrow worms – the respiratory surfaces differed from gill lamellae as we find them in fish.

A 'dimensional tension' involving their oxygen supply occurs in 2 of the 3 groups of metazoans examined here (near-spherical sponges and chaetognaths). This tension not only affects their growth but also causes their maximum size to be strongly dependent on temperature, limiting the O_2 content of water and increasing their O_2 requirements (Pauly and Cheung 2017; see also Chapter 4). As for medusoid cnidarians, their growth does appear to be asymptotic, but the oxygen supply to their watery bodies may not be the factor that ultimately causes their growth to slow down and eventually stop. This is an anomaly that needs to be resolved.

Demonstrating the limiting effect of size on growth and the association of higher temperatures with lower maximum sizes in WBE are both tests of the GOLT's generality. The near-spherical sponges and the chaetognaths passed these tests, which suggests that the GOLT applies to these WBE. Questions remain regarding the medusoid cnidarians, which will require their respiratory physiology to be examined in detail. This may also be true for another gelatinous phylum, the Ctenophora or comb jellies (see Salihoglu *et al.* 2011 and references therein).

15

Limits of Adaptation

"So if you see that a creature possesses a certain advantage,
Put the question at once: What is the fault that afflicts it
Elsewhere?—and seek to discover the defect, always inquiring;
Then at once you will find the key to the world of formation."
J.W. Goethe, *The Metamorphosis of Animals* (1798) 40-43; 50-53.

Synopsis: This chapter situates the GOLT within the framework of evolutionary theory and engages with critics who argue that the theory neglects the adaptive potential of biological organisms to overcome physical and physiological constraints. It discusses the role of organisms' *Bauplan* (body plan) in determining the extent to which adaptation can overcome physiological limitations and how biological constraints shape evolutionary trajectories. We then address the tension between teleological perspectives – viewing evolution as goal-directed – and the reality of adaptation as an immediate, often suboptimal response to environmental pressures. This introduces the concept of 'lazy' adaptation, where natural selection favours energetically efficient but seemingly 'imperfect' solutions. By integrating these themes, the chapter highlights how the constraints on oxygen supply to metabolism and growth do not contradict evolutionary principles but rather shape the evolutionary trajectories of aquatic organisms.

Adaptation and the *Bauplan* of biological organisms

Because *"nothing in biology makes sense except in the light of evolution,"* every biological theory claiming scientific status must explain its central mechanisms in relation to evolutionary processes (Dobzhansky 1973). The mechanism we present in this book to explain the asymptotic growth in fishes and other water-breathing ectotherms – surface-limited oxygen supply – may at first appear counterintuitive in the light of evolutionary history, which is largely driven by adaptive processes. The idea that physiological functions are constrained by innate geometrical and morphological traits that cannot be overcome by natural selection may even suggest that the GOLT is a theory of *mal*-adaptation. This seems to be the understanding of colleagues who are troubled by the idea of *"an intrinsic inability to develop mechanisms for adequate oxygen supply,"* which may appear unconvincing *"from an evolutionary perspective"* (Audzijonyte *et al.* 2019). As some physiologists put it, *"oxygen demand determines the size of their respiratory surface area, not the other way around."* (Lefevre *et al.* 2017), and others echo the same views. This seems to be based on the underlying assumption that supply always meets demand and that adaptive processes can overcome every constraint that hinders an organism from achieving what would be advantageous regarding survival, growth, or reproduction. The apotheosis of this understanding of evolutionary adaptation is the idea that "*the invocation of constraints is unnecessary to explain the ontogenetic trajectories of metabolism and growth*" and that adaptive optimization is sufficient to explain every feature of an organism (White *et al.* 2022).

In this chapter, we discuss the central tenets of the GOLT in relation to evolutionary and adaptive processes, and we argue that its central explanatory mechanism does not constitute a maladaptation but can be explained as a result of adaptive processes. Many geometrical factors and physiological trade-offs constrain these adaptive processes. As summarized above, the opposing views reflect assumptions about evolution, of which some seem hardly compatible with Darwinian theory in its historical and current state. While many contemporary physiologists have difficulties in accepting the idea that adaptation does not always lead to outcomes that entirely overcome intrinsic constraints, this form of adaptationism has philosophical foundations whose implicit intellectual baggage is seldom reflected upon, and pays little attention to debates on the relationships between phylogenetic constraints and their optimization concerning specific physiological traits (see, e.g., Godfrey-Smith 2001; Orzack and Forber 2010).

If we set contemporary notions of adaptationism – some less radical than sketched above – in a historical context, they appear as idiosyncratic interpretations of evolutionary theory. The initial reactions to Darwin's original outline of an evolutionary theory based on variation and selection often reflect bewilderment that not all evolved life forms develop their full functional potential. As Darwin retrospectively writes in the first edition of *On the Origin of Species*, "[...] *a distinguished German naturalist has asserted that the weakest part of my theory is, that I consider all organic beings as imperfect: what I have really said is, that all are not as perfect as they might have been in relation to their conditions*" (Darwin 1859).[237] The lack of assumed perfection in nature puzzled Darwin's contemporaries. While evolutionary explanations were later invoked to explain unconstrained optimization, this was not the case when early proposals of evolutionary theories were first presented.

An awareness of this problem can be traced back to the earliest morphological studies, which describe homologies between organisms sharing the same *Bauplan*. Early morphological theorists like Étienne Geoffroy St. Hilaire (1772–1844) or Johann Wolfgang von Goethe already realized that if the homology of organs in the vertebrate *Bauplan* was explained historically, i.e., as a consequence of phylogeny and descending from a common ancestor, it was impossible to say that forelimbs in vertebrates evolved *for* walking, flying, or swimming since they could attain all these functions (Richards 2017; Le Guyader 2004). Since none of these functions was original, their 'perfection' could not be explained further than their ancestral morphological construction allowed.

The idea that not every trait was originally 'designed' for its current function in an organism challenged notions about the assumed perfection of life on Earth. If functions and traits were not 'made' for each other, this opened the gate for the possibility of imperfections and deficiencies in what was earlier believed to be a perfect Divine creation.[238] For Darwin, this was no immediate problem: optimization did not mean perfection in an absolute sense, but it could only be understood in relative terms: "*Natural selection tends only to make each organic being as perfect as, or slightly more perfect than, the other inhabitants of the same country with which it comes into competition. And we see that this is the standard of perfection attained under nature. The endemic productions of New Zealand, for instance, are perfect one compared with another; but they are now rapidly yielding before the advancing legions of plants and animals introduced from Europe. Natural selection will not produce absolute perfection, nor do we always meet, as far as we can judge, with this high standard under nature.*" (Darwin 1872, p. 163)

Consequently, the notion of the "*survival of the fittest,*" inspired by Spencer and adopted by Darwin in later editions of the "*Origin of Species*" only makes sense when understood in relative terms: the "*fittest*" trait is never an absolute, universal optimum or the best of all imaginable designs but only the 'fitter' within in variety of less fit traits, whose range is limited by the ancestral *Bauplan* that is available and that preconditions the range of possibilities from which natural selection can favor certain traits over others.

Hence, optimization can only occur within the constraints of the given *Bauplan* in which the traits emerged and further evolved. The structures of this plan are not the only constraints that determine the functional ranges of a biological trait or organ. Gills evolved several times (even though they only emerged once in vertebrates; see Gillis and Tidswell 2017), and their evolution can be understood as the result of convergent evolutionary processes. Gills and their highly efficient system of countercurrent gas exchange are an impressive example of physiological optimization. As we have seen in Chapter 3, the very 'paradox of optimization' sets a constraint on expanding the length of their lamellae. Once the water has moved along its highly vascularized surface, it is quickly exhausted of oxygen, and a further extension would be pointless.

Despite their different ancestral body plans, vertebrate and invertebrate gills share another feature that constrains their maximum size: they have multiple functions. Their surface is relative to the animal's body size, a tradeoff between the need to take up sufficient oxygen and the risk of ionic or osmotic imbalances, which can result in organ failure. As Rombough (2007) puts it, the functional design of fish gills is not "*rational,*" adding "*that evolution does not act rationally in the human sense*" since "*neither freshwater nor marine fish are in ionic equilibrium with their environments and a large surface area compounds the problems of trying to maintain ionic homeostasis under such circumstances.*" Such notions of "*rationality*" in evolution (*pace* Rombough, whose argument illustrates this) assume the possibility of constraint-free design in which features can be arranged such that tradeoffs with other processes do not impair the effectiveness of a given organ or trait.

This is not how evolutionary processes optimize traits, and the GOLT's model of the relationship between gills and the metabolic performance and growth of the organisms they supply with oxygen can be described as a model of optimization. In the introduction, we distinguished between constraints and limitations. As the previous sections and chapters argued, adaptation and optimization can only be defined *within* the constraints that limit the range of the phenotypical options that can be realized. But what about the limitations that prevent an individual animal from reaching a specific size, for example, under conditions under which the final sizes typical for the species or the population in question cannot be realized? As critics of the GOLT suggest, a mechanism that reduces growth as a result of surface-limited oxygen supply would be implausible in the context of adaptive optimization, which implies that water-breathing animals would be able to increase their gill surface infinitely if the circumstances (e.g., hypoxia or higher temperatures) were to require it (Lefevre *et al.* 2017). While gill remodeling is indeed a well-described response to hypoxic and thermal stress in a vast number of species (see Chapter 3), this resource is not endless, and its effects are far less effective than often assumed (see, e.g., Rutjes 2006; Bowden *et al.* 2014). The inability to overcome the limitations is not a maladaptation. Rather, temperature-induced growth reductions can also be understood as a form of phenotypical plasticity that results from a limitation and the outcome of an adaptive strategy that protects animals from reaching sizes at which they risk asphyxiation at higher temperatures or hypoxia. This illustrates what has been described as the *"physiological benefits of being small in a changing world"* (Clark *et al.* 2012).

Teleological aporias vs. life in the here and now

The idea that a water-breathing organism could 'evolve out of' the limitation of oxygen uptake through its two-dimensional gills to supply its three-dimensional body seems to be based on an inherently teleological understanding of evolution, according to which supply is always shaped by demand. The teleological problem underlying this argument is not superficial but inherent in the logic of symmorphosist theories of evolution. This understanding of evolutionary processes centers on the idea that biological units always develop to produce optimal (potentially unlimited) output. This optimization of biological functions is also considered economical: "*the formation of structural elements is regulated to satisfy but not exceed the requirements of the functional system*" (Taylor and Weibel 1981). As we have seen, the relationship between the growth of gill surface areas and body volume has evolved to allow a given population to reach a given size within a given temperature range. The limiting factor – two-dimensional gill surface areas in relation to three-dimensional body volume – does not push the organism to a constant limit where metabolic need and oxygen supply intersect at a critical point. Instead, it leads to a redistribution of energy (or oxygen) allocation to other processes. In contrast, the relationship between gill surface area and body volume in young organisms allows for enough resources to be devoted to growth; these resources are unavailable in older individuals. The resulting proportionality between gill surface area and metabolic rates may then appear as a harmonious balance between supply and demand, but only at the *cost* of available resources for growth. This apparent harmony leads physiologists to assume a perfect balance between available resources and a growing organism's energetic demand. As one recent paper symptomatically puts it: "*as whole-animal metabolic rate increases with size, the oxygen supply capacity increases to match it*" (Seibel and Deutsch 2020).

This sentence is correct but does not support the argument it is meant to, as it overlooks the crucial point that larger (older) WBE have to allocate their energetic budgets differently from smaller (younger) WBE. The statement above is only valid if we subtract the cost of growth from the energy budgets of larger, non-growing adults. Suppose we imagine that adult growth would continue at the same rate as in juveniles; it becomes clear that no gill shape could exist (within the constraints of fish anatomy) to support a growing body infinitely.

What is less obvious is why temperature-induced 'shrinking' cannot easily be overcome through evolutionary mechanisms. If we look at global biogeographical patterns and paleontological data, body sizes almost universally follow thermal and climatic trends.[239] Why do surfaces not keep up with higher metabolism and resist reduced body size trends? The assumed 'perfect harmony' between energetic supply and metabolic demand is a fragile and highly contingent balancing act that always requires tradeoffs and compromises. Assuming a 'balanced harmony' between supply and demand requires a symmorphosist understanding of evolution in which life is expected to solve whatever problem it has been 'assigned' and where it will develop proper or even perfect solutions. This view has been refuted in earlier critiques of symmorphosis and does not need to be repeated here (but see, e.g., Garland 1998; Dudley and Gans 1991).

When such views on adaptive optimization are applied to evolutionary adaptation to quickly changing environmental conditions, such as rising temperatures, it becomes clear that adaptation is severely constrained by the short timescales available to adjust physiological and morphological features to the new situation. As Niles Eldredge has reminded us in a review paper on the contribution of Theodosius Dobzhansky to modern evolutionary thought, it is an obvious fact that *"evolution has no 'eyes' for the future. And life is about living, not evolving"* (Eldredge 2008).[240] First and foremost, life – understood as the interplay of physiological functions that allow for an organism's survival, growth, and reproduction – is always about the *"here and now"* and shaped by the selective pressures that exist in the moment, and hence selection will typically result in short-term 'solutions.' To expect that it could somehow anticipate future situations and challenges would involve notions of design or teleology. Species or populations subjected to quickly changing environmental circumstances can develop traits that allow for some degree of adaptive flexibility or plasticity (or they will disappear, which is always the most likely outcome from a long-term perspective). Sudden, unique, and critical change requires different responses for organisms and populations to survive and adjust.

Adaptive responses in the face of sudden change may be characterized as *lazy*; in the sense that they always take the path of least resistance. In the face of the current climate crisis, the limitations of assumptions about the adaptability of animals to new circumstances become increasingly clear. The expectations that fish and other water animals would come up with 'adaptive solutions' are problematic insofar as they fail to distinguish between the actual challenge and the adaptation and, as a result, fail to explain the mechanisms that lie underneath the current trends of 'shrinking' fish.

The global and widely acknowledged trend that water-breathing animals reach smaller sizes in warming water begs the question of how adaptive mechanisms can become effective in situations of sudden change. Some of these primary responses to sudden change are generally agreed upon, and it is in their nature that they inhibit selection mechanisms that might provide a long-term solution to a particular adaptive challenge. The first response to a sudden rise in temperatures involves new forms of energy allocation on the level of individual organisms, which means that evolution cannot yet play a role. This new strategy of allocating energy to different functions within the organism is why fish spend fewer resources on growth. This means that individuals do not grow as previously and can be colloquially described as 'shrinking.' The reason is apparent: since metabolic rates are elevated due to higher temperatures, more energy is invested into maintenance metabolism and activity, and less energy is available for growth.

This response does not involve evolution yet, since energy allocation works on the level of ontogeny. As studies on long-term size developments with fish taxa have shown, size reductions also correlate with higher temperatures at a phylogenetic level (Troyer *et al.* 2022). The transition from ontogenetic growth patterns to phylogenetic levels is discussed in Chapter 10 with respect to increasing sizes, but similar mechanisms should also inform opposite trends. In the case of increasing temperatures, the push towards smaller sizes is fast and will quickly put larger animals at a disadvantage. Since fish reach sexual maturity earlier and are smaller in warmer water (see Chapter 7), this response will favor the reproduction of smaller individuals that are also better adapted to the new temperatures. A transformation of sexual selection mechanisms is, therefore, a result that will, in the long term, shape growth patterns. Smaller sizes are not the constraint itself, but the resulting accommodation strategy. This implies that there is no further need to replace this simple and low-cost adaptation with a new, more complex one. This mechanism can be described as *lazy evolution*, a form of adaptation that does not favor 'perfect,' complex, and costly adaptive strategies, but rather easy, inexpensive, and imperfect ones that often emerge as assimilation and accommodation on an ontogenetic level.

'Lazy' adaptation in nature

As long as feeding strategies, predation schemes, or other size-dependent features of a population are not critically pressured through selection for smaller sizes, shrinking is the least costly and, thereby, most immediate accommodative response of organisms that experience this specific constraint (Passow *et al.* 2015). The potential negative consequences of size reduction, such as lower fecundity and productivity, are not immediately felt by the population whose individuals phenotypically adjust to the changing situation, but rather only after multiple generations. Accommodation through size reduction does not have to 'wait' for specific mutations that will accidentally appear at some unforeseeable point and then enter the gene pool. Still, since growth constraints are an immediate response to a physiological imbalance, they already become effective in ontogeny. That means it can become effective before genetically adaptive and evolutionary mechanisms come into play.

As the growing body of literature on the role of phenotypic plasticity in evolution suggests, ontogenetic adjustments to changing situations may also have consequences for adaptive processes (Levis and Pfennig 2016, 2020). While smaller final sizes in response to higher temperatures are primarily an ontogenetic response that can be explained as a result of a limitation, the ability to cope with this problem through adjusted reproduction patterns can then be acted upon by selection. Selection for phenotypes that exhibit fast growth, small final size, and early reproduction can then occur and dominate in subsequent generations. At the same time, greater final size and later reproduction would be suppressed due to the pressure exerted on ontogenetic development. The range of future adaptations is shaped by the immediate – plastic – ontogenetic responses to the new situation. The outcome might not be the selection of the most ideal phenotypes in the long term, but a mode of adaptation that reflects the *ad hoc* adjustments to critical and threatening changes.

Simply put, our concept of *lazy adaptation* implies that selective pressure typically favors the least costly and, thereby, the least '*perfect*' option. Since the respiratory systems of fish are involved in multiple trade-offs, such as the 'osmorespiratory compromise,' and do not have safety margins that would allow them to slowly adapt to changing environments by developing more complex and efficient breathing organs, they must adapt through size reduction. Shrinking can therefore be understood as an accommodation through *lazy adaptation*. This perspective on evolutionary processes provides an important insight: if it is true that adaptation to critical situations is typically *lazy* and sloppy – like putting a bucket under a leaking sink instead of fixing the pipe – it will eliminate the urgency for more perfect solutions (or the urgency will occur once no further adaptation is possible).

This is not to say that features that have not (yet) acquired a specific function will be eliminated (*lazy* is not the same as *ascetic* – one could even argue that it is the opposite).

Useless features can remain and even develop certain levels of complexity without acquiring certain functions – as long as they are not *too* disadvantageous.[241] Neutral traits may later become functional, but this would be just a secondary effect, and they can be inherited over generations without any beneficial function. It is indeed possible that they will reach considerable degrees of 'perfection' later, but they can also remain stuck in the form of a *bricolage* that still performs some functions. This phenomenon has sometimes been addressed in the context of a similar concept: "*tinkering evolution*" (Jacob 1977). "*Tinkering*" implies that life will present imperfect solutions by developing half-baked or rudimentary mechanisms without a clear prior function. The idea of "*tinkering*" is largely compatible with the theory presented in this book, but it does not go far enough.

The heuristic value of *lazy adaptation* lies in its ability to explain why adaptation is often sloppy and why perfection is avoided: *lazy* forms of coping work on a pre-adaptive and pre-evolutionary level. That means that organisms and populations do not have to 'wait' for random mutations when they can adjust during ontogeny. Once this assimilative mechanism becomes adaptive and is then transferred to phylogeny (and becomes evolutionary in a proper sense), it becomes increasingly unlikely that other alternatives will still have a chance to intervene. Despite its apparent inertia (or actually *because* of it), *lazy adaptation* is a force that tends to block other evolutionary mechanisms that might provide 'better,' more elaborate, and more 'perfect' solutions – terms that are problematic in the language of evolution but may indicate evolutionary developments that increase the long-term viability of certain traits.

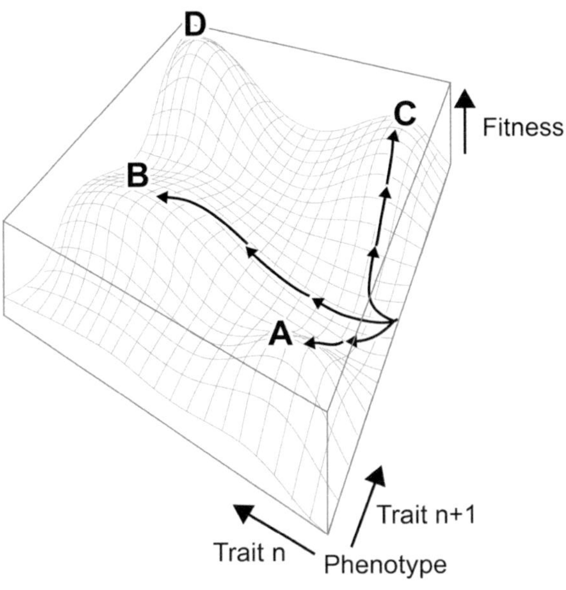

Figure 15.1. Schematic representation of phenotypical accommodation and selective adaptation. In the case of temperature-induced size reduction in water-breathing animals, the lower accommodative peaks in A, B or C, which can all be climbed from the starting point, will, once they are reached, hinder the possibility of populations from climbing the highest adaptive peak (D), because they would have to reduce their fitness initially. Any adaptation will involve more than the two traits illustrated here, but the principle remains the same if multiple traits are considered.

If we express this in the metaphor of a fitness landscape where the vertical axis represents fitness and the two horizontal axes represent the combination of specific phenotypic traits, the different responses to environmental change, in our case, higher temperatures, can be marked as different hills in the landscape. Assumptions of a 'perfect' solution would presuppose the quick modification of respiratory surfaces, which would then allow for larger final sizes – and more offspring, which is often a correlate of greater size – constituting the highest hill in the landscape. A more realistic scenario would be the selection for earlier maturation and faster juvenile growth, with reductions in final size and, ultimately, fewer offspring (**Figure 15.1**). This strategy has costs and can be considered a tradeoff at best, but it constitutes an adaptive solution that allows a population to survive in scenarios of rapid change.

As discussed above, constraints limit the degrees of freedom of an organism while simultaneously creating certain opportunities. The limitations from these constraints structure the life-histories so that their most important traits – metabolic performance, growth, maturation, and reproduction – become predictable. Drastic reductions in the degrees of freedom can limit the available phase space of possibilities, such that adaptation is only possible within very close ranges. The current climatic transition we are experiencing may be at such a moment, and the physiological limitations it represents for water-breathing should not be underestimated.

'Lazy adaptation' in the laboratory

Recent research on the multigenerational impact of increased temperatures has produced results that illustrate how such forms of "*lazy adaptation*" concretely manifest themselves in the context of temperature-induced size reduction in fish. In an exciting contribution to the present debate on temperature-size effects in ectotherms, Wootton *et al.* (2022) explored the adaptive effects of higher temperatures on six different generations of zebrafish (*Danio rerio*) and the impact of these conditions on growth and final size. As expected, elevated temperature correlated with size reduction, and the experiment confirmed typical temperature-size patterns as described elsewhere in the literature.

Wootton *et al.* (2022) explicitly tested two hypotheses that are commonly used to explain the phenomenon of reduced growth in fish (and other water-breathing animals) under elevated temperatures: (*i*) the role of resource limitation because of elevated metabolic rates, and (*ii*) the idea that elevated temperatures result in a trade-off between growth and reproduction. The authors studied six generations of zebrafish that were gradually adapted to warmer water (from 26°C to 30°C), then presented two general conclusions:

1. Standard metabolic rates (*SMR*) in warm-adapted fish (of the F6 generation) were lower in warmer water than in the previous generations, but all generations showed the same temperature-induced growth reduction;
2. Elevated maximum metabolic rates (*MMR*) remained the same in all generations: the F6 generation could not reduce them, even though they seemed fully adapted to warmer water.

Wootton *et al.* (2022) do not present an explicit answer to the question of adult size reduction. Still, they conclude that *"smaller adult fish size in warmer water is not explained by elevated metabolism."* The findings of this study are a perfect example of *lazy adaptation*.

The fact that maximum metabolic rates (*MMR*) remained high among all six generations (while standard metabolic rates were lowered in later generations) is important here, and it reveals the different mechanisms that are at play here: while the measurements of *SMR* included overhead costs of growth (as the authors acknowledge; but see also Rosenfeld *et al.* 2015), MMR mainly reflects the costs of activity. As the experiment shows, adaptation to higher temperatures does not imply that energetic expenditures become 'cheaper' – activity comes at the same price for adapted and non-adapted individuals. What changes are growth and reproduction, and a reduction of *SMR* must come at the cost of one of those two. The data generated in the experiment indicate that adaptation comes at the expense of growth – indeed, the experiment's results strongly support the limitation hypothesis that it sought to challenge.[242]

This experiment corroborates the differential prioritization of physiological processes in changing environments. From a teleological perspective, it may have been beneficial not to reduce metabolic rates and instead invest more energy in growth. As the findings show, the organism's response is an economization on its total energy budget. The fact that adult sizes are reduced among warm-adapted and non-adapted generations is therefore not surprising: the adaptation to increasing temperatures does not consist of complex mechanisms to increase oxygen uptake but to a reduction of its growth budget, a form of *lazy adaptation* that may be typical for responses to quickly changing environmental conditions everywhere in the biosphere.

16

Inferring causality in metabolic scaling

Synopsis: To a large extent, the current debate around the GOLT revolves around the question of whether it is reduced demand or supply that shapes the metabolic and growth patterns in water-breathing organisms. In this chapter, we discuss how causal inferences can be drawn to determine whether limited supply or reduced demand dictates metabolic scaling and the decreasing growth rates with age and size. Building on literature priorities in energy allocation, we argue that such inferences are possible if we consider different degrees of plasticity in both resource supply and demand.

Why does metabolic rate scale with body size?

The negative allometry between respiratory surfaces and body volume in water-breathing ectotherms – the central morphological pattern described in this book and the proposed mechanism for asymptotic growth – is not limited to water-breathers. In aquatic and terrestrial animals, both respiratory surfaces and metabolic rates increase in proportion to each other. While earlier theoretical accounts postulated more or less fixed scaling exponents of either 2/3 (Sarrus and Rameaux 1839; Pütter 1920) or 3/4 (Kleiber 1932), the GOLT offers an adaptive explanation for the diversity of values, which typically range from 0.6 to 0.9.[243]

If negative allometric scaling of metabolic rate with body mass is a universal pattern in both terrestrial and aquatic animals, the question arises whether surface or transport limitation can indeed be invoked as a valid explanation for declining growth rates at greater body size: intuitively, observed scaling patterns might also suggest that metabolic scaling may not be a result of diminishing supply, but rather a consequence of a shift in required consumption, i.e., a declining demand. Indeed, this hypothesis has inspired several theoretical models focusing on metabolic demand rather than supply to explain metabolic scaling trends (e.g., Wels 1925; Kleiber 1961; Harrison 2017).[244] While the data that inform the current debates around metabolic scaling are typically not disputed, the central problem revolves around inferring causality.[245] This chapter sets this debate in perspective and evaluates the arguments invoked by proponents of supply- and demand-based theories. A central problem in this debate seems to be the ambiguous definition of 'demand' in physiological and ecological terms. To enrich this term with content and make it biologically relevant, we present a simple model that relates supply and demand in relation to the different physiological functions. Since organisms are never in stasis, demand cannot be objectively defined unless it is related to specific functions and expenditures, some of which can only be performed as parts of energetic tradeoffs.

The idea that ontogenetic metabolic scaling results from declining maintenance demand is at the heart of many criticisms of the GOLT (see, e.g., Lefevre *et al.* 2017; 2021; Scheuffele *et al.* 2021). As these critiques argue, the maintenance costs of larger animals are lower on a per-unit-of-body-weight basis. To support this assumption, Lefevre *et al.* (2017) refer to a study on nine species of mammals and conclude that "*cells from small animals have higher rates of oxygen uptake than those from large ones.*" The study that suggests this did not evaluate the differing energy consumption of growing animals at different life stages, but only examined interspecific differences in cellular oxygen consumption among mice, rats, ferrets, horses, and other large species (Porter and Brand 1995). However, comparing the relative metabolic rates of animals across four orders of magnitude pertains to a very different problem and is unrelated to intraspecific metabolic scaling. As outlined in Chapter 10, the evolution of large body sizes coincides with a lowering of maintenance costs – to illustrate this with the example chosen by Lefevre *et al.* (2017): if a horse had the same energy metabolism as a mouse, it would immediately overheat and collapse, aside from the fact that horse-size animals could not find (and metabolize) enough food to support their extraordinary (mouse-like) demand.

As this example illustrates, the debate surrounding supply and demand in metabolic scaling often neglects the fundamental differences between inter- and intraspecific scaling mechanisms. As should have become apparent (see, e.g., Chapters 2 and 10), these differences are crucial. While interspecific scaling patterns can be explained by evolutionary optimization, intraspecific scaling is mainly a consequence of reduced growth, i.e., a result of reducing the energetic costs of protein synthesis and related processes. Extrapolating interspecific scaling patterns to ontogenetic growth trajectories leads to the problematic assumption "*that the scaling exponents of total metabolic rate and maintenance metabolic rate* [...] *are identical*," as postulated by White and Marshall (2023a), and that the fraction of energy devoted to growth and reproduction remains constant and is "*independent of size.*" This claim lacks empirical evidence, and the only source cited as support for this assumption is a study on juvenile brown trout, which used modeled estimates of standard and maintenance metabolic rates rather than measured oxygen uptake (Elliot 1976).[246] For demand-based explanations to hold, it would be necessary to demonstrate that maintenance costs themselves – i.e., excluding costs of growth – would be diminishing with size on an intra-species level. The lack of evidence for this notion suggests that the proposed shift from supply- to demand-based models is a step that obscures the mechanisms behind scaling patterns, both in evolutionary and ontogenetic terms.

What does 'demand' represent for whole organisms?

The GOLT's assumption that the scaling of metabolic rate and gill surface area is essentially the same is confirmed by the theory's critics (Scheuffele *et al.* 2021; Lonthair *et al.* 2024a). It follows that the gill area does not exceed metabolic demand and that the entire available respiratory surface is utilized for either activity + growth or activity alone. As outlined in Chapter 3, surfaces that would exceed organismic demand would needlessly increase the risk of parasitic infestation and ion loss. It is challenging to define 'demand' in physiological terms since organisms use consumed resources – here: oxygen – for various functions, most importantly maintenance, growth, and reproduction. Demand can only be meaningfully conceptualized when defined as demand *for* a given function. Physiologists typically define demand as the amount of uptake required for the optimal function of a given organ (Weber and Janicki 1979).

While this definition is helpful and viable at the cellular and organ level, it is unclear what 'optimal' function represents at the level of whole organisms. At this organizational scale, 'optimality' requires a physiological and ecological definition since maintenance, growth, and reproduction are always related to specific environments and life-history strategies. As demonstrated by a wide array of studies, energy budgets must be allocated among different functions, which often compete with one another (Priede 1985; Korsmeyer *et al.* 1996; Zhang *et al.* 2010). The balance between supply and demand can be explained as a result of selection processes, but only in specific ecological and life-historical contexts. Demand can then be defined as the energy needed to reach a size that allows sustainable reproduction in a specific environment. This demand can drastically change under different thermal circumstances or conditions, requiring higher activity levels for survival.

As we have seen in Chapter 10, selection for greater body size correlates with and is often dependent on lower maintenance costs. If we pose the question of supply-demand balance at the level of ontogeny, a different relationship between metabolic supply and demand emerges. Demand cannot be understood as a constant that takes fixed and predictable values; it is, in fact, a highly variable parameter. The problem of how to conceptualize metabolic demand may be solved if we consider a general principle that should be assumed to underlie physiological and ecological processes: to infer causation in supply-demand relationships, it seems reasonable to assume that the factor with the lesser degree of plasticity is the cause of the observed balance between supply and demand. A review of the degrees of plasticity of supply and demand reveals substantial differences between the two factors (**Table 16.1**): the supply side is far less plastic than the demand side. Gill surface areas (supply) show a certain degree of plasticity in some species (e.g., through gill remodeling in some cyprinids, killifish, and eels – see Nilsson *et al.* 2012), while ventilation rate is more plastic. Ventilation is energetically costly, and high ventilation rates are unlikely to offset the overhead costs of growth or reproduction. In contrast, energetic demand is extremely plastic above the level of maintenance metabolism. At the same time, growth and reproduction cannot be understood as presenting an intrinsically fixed demand but as a feature that adjusts itself to the available supply of resources.

This suggests that demand-based models fail to explain the proportional scaling of respiratory surface growth and metabolic rate during ontogeny. As the supply components of metabolism have a lower plasticity, the demand side is adjusted to the available resource supply.[247] Conversely, adjustments to supply are far more limited, and aside from increases in ventilation rate and an inherently limited enlargement of gill surface area, an increase in supply capacity occurs mainly through evolutionary adaptation, and at an interspecific level. More importantly, it is impossible to identify a limitation on a demand level from an ontogenetic perspective. Since the different components of metabolic demand typically compete with each other (Priede 1985; Whitledge *et al.* 2002), the regulation of each component occurs on the supply side (see **Figure 2.1**). Oxygen must be partitioned between various functions since an organism cannot simultaneously invest maximal energy into activity, growth, and reproduction.

Table 16.1. Components of metabolic supply and demand and their respective degree of plasticity.

Supply	Demand
Gill surface area: Low degree of plasticity, despite documented remodeling in several freshwater species (several cyprinids, killifish, and eels; see Nilsson *et al.* 2012)	**Metabolic demand of maintenance:** Low degree of plasticity as fish are dependent on a steady O_2 supply to support vital functions. They can delay feeding, but (with some exceptions), but not breathing.
Ventilation rate: Short-term plasticity, but limited due to high energetic costs of increased ventilation rates (see, e.g., Xiong *et al.* 2020)	**Metabolic demand of growth expenditures:** High degree of plasticity; growth ceases if the energetic supply is cut off.
Cardiovascular transport: Some degree of plasticity via the regulation of heart rate, vasoconstriction, and related mechanisms	**Metabolic demand of reproduction:** High degree of plasticity; reproduction stops if the energetic supply is cut off.
Cell size: Low degree of (developmental) plasticity.	**Metabolic demand of activity:** High degree of plasticity; fish will typically become sluggish if energy supply declines.

Optimizing the partition of energy represents a balance in dividing resources rather than overcoming the supply constraints. While organisms can regulate demand so that energy partitioning allows them to survive, grow, and reproduce, the mechanisms that shape these partition strategies are ecological and evolutionary. The question of which energetic investment is prioritized at a given time is a question of evolutionary fitness, thereby determining the level of trait selection and optimization. At the level of intraspecific ontogenies, it would be implausible to explain metabolic scaling in terms of reduced demand, even if mechanisms behind reduced maintenance demand at greater sizes could be identified.[248]

Hierarchies and priorities in energy allocation

As outlined in Chapter 2, evolutionary accounts of phenotype optimization are only convincing if they successfully define the boundary conditions and the constraints within which traits can be optimized. Since, in nature, all organisms are vulnerable to predators or parasites and limited in their reproduction and scope of activity, optimization is always relative to constraints and limitations (Maynard Smith 1978). The options are, therefore, not endless, and virtually every form of optimization is, to some degree, a tradeoff.

To increase their fitness, organisms have evolved different forms of metabolic prioritization that enable them to function in optimal, suboptimal, and critical circumstances (Jourdan-Pineau *et al.* 2010; Duskey 2023). The metabolic priorities directly align with the different components of metabolic demand in **Table 16.1**, whose degree of plasticity negatively correlates with their expected metabolic prioritization.[249] While maintenance metabolism is critical for survival and must always be prioritized over other expenditures, the prioritization of growth, activity, and reproduction depends on the respective life stage, ecology, and environmental stressors (Kozłowski *et al.* 2020).[250]

Considering the different degrees of metabolic priorities can be helpful in extrapolating causal patterns in metabolic scaling on an intraspecific level. Even though supply capacities and basic demands can be optimized through adaptive processes, every individual organism can function within the (evolved) constraints of its respective phenotype. This makes it even more surprising that demand- or optimization-based models of metabolic scaling often coincide with the assumption of a 'reproductive drain' as an explanation for reduced growth at larger sizes (e.g., White *et al.* 2022, 2023). If metabolic scaling *"can be explained without the need to invoke any of the assumed constraints traditionally imposed by metabolic theories"* (White *et al.* 2022) and declining oxygen consumption results from reduced maintenance demand, it is unclear why growth and reproduction would have to compete with each other. According to this argument, the competition between growth and reproduction should not occur if geometrical or physiological constraints do not limit supply. A crucial problem with demand-based theories is that they must incorporate diminishing maintenance as an artificial factor into their model to relate their explanation to real-world data.[251] While the reduction of maintenance costs can be seen as an evolutionary pathway to larger body sizes (as we argue in Chapters 10 and 15), it is not a mechanism that explains ontogenetic scaling. The confusion between inter- and intraspecific scaling appears to be a central problem in the current theory formation on this topic (e.g., Lefevre *et al.* 2017; Harrison 2017).[252] If we separate the two levels of observed scaling patterns, it becomes clear that different mechanisms are at play, and demand-based explanations cannot explain ontogenetic scaling.

Gills and the whole organism

As the introduction outlines, the GOLT is a theory of whole organisms, which relies on causal mechanisms to explain why young and growing organisms consume (relatively) more energy than fully grown adults. Therefore, for a coherent explanation of ontogenetic metabolic scaling, it is essential to look for answers at the level of whole organisms. As argued in this chapter, explanations must be sought at the energetic (i.e., oxygen) supply level, and neither speculative accounts of lower maintenance demand at greater size nor investment in reproductive activities can provide solutions here. While supply limitations can occur throughout an organism and at different levels of organization, the GOLT's focus on gills is not arbitrary. As Bigman *et al.* (2023 argued, *"oxygen acquisition at the gills is the first of many steps in the 'oxygen cascade' as oxygen is removed from the environment and transported through the body to the cells where it is ultimately utilized."* This is correct, but it does not adequately address the challenges involved in the oxygen cascade's initial steps and the three-dimensional organic tissue's capacity to absorb oxygen from a liquid medium such as water. Since the surface-volume ratio of these gas exchanges is the greatest at the boundaries between the organism and the environment (see also our analogy of 'Russian dolls' in Chapter 3), it is the first step in the oxygen cascade that is the most limiting one. As Priede (1985) put it concisely: *"gill (fish-water interface) area may be the ultimate limitation, but the whole of the respiratory transport chain from diffusion across the gill epithelium through the circulatory system, cardiac performance and mechanisms of tissue gas exchange must be matched to this gas-exchange capacity."* While biological organisms are highly complex systems that rely on many fine-tuned functions and processes as well as reciprocal networks of causality, causal mechanisms in whole-organism metabolic processes are directional. They can be inferred by detecting the differential degrees of plasticity at the demand and supply levels.[253]

17

Epilog – Size and shape in a three-dimensional universe

Synopsis: The final chapter serves as an epilog, exploring unresolved questions and future research directions – some of which are still in their early stages and may even be currently rather speculative. We examine the relevance of the GOLT to fundamental biological processes, such as egg hatching in aquatic environments, as well as broader ecological phenomena like viviparity, parental care, and hermaphroditism in fish. The chapter concludes with a call for further research, emphasizing the vast potential for future discoveries in aquatic life and the need for a deeper understanding of how the GOLT can enhance our understanding of organismic adaptations and evolutionary constraints.

What it takes to become big

The vast majority of life on our planet is small.[254] Growing from a single-cell scale requires multiple morphological and physiological modifications that allow organisms to take up and metabolize energy to satisfy the needs of a larger body. While a common correlate of larger body size (in interspecies comparisons) is a lower metabolic intensity, this alone is not enough to enable life at greater body size. In addition to this, larger organisms rely on modifications of the scaling relationships between their surfaces and volumes to allow for food uptake, gas exchange, and thermoregulation. The challenges associated with these modifications are easier to overcome in terrestrial organisms since gases are far more diffusible in air than in water.[255] The previous chapter of this book described the specific metabolic constraints that WBEs face and explained the morphological adaptations that allow the evolution of body sizes beyond the smallest size scales. As we have shown, the hyper-allometric growth of gills relative to body volume facilitates the evolution of water-breathing animals into megafauna-scale organisms.

Pathways to larger sizes must navigate not only in phylogenies but also in the ontogenies of individuals. Since all organisms start small, ontogenetic growth trajectories towards greater size must involve morphological changes that enable the exchange of resources and nutrients between the organism and the environment. The ontogenies of WBE illustrate this fact impressively, and to reach the respective next developmental stage, they need to rearrange their morphologies drastically.

The transition from egg to larva

At very young life-stages, all water-breathing animals obey simple surface-volume laws, and the only interface between the organism and its environment is the outward body surface. This is sufficient only at very small sizes: while the surface of spherical eggs is only four times their squared radius times π, the only species that can afford larger egg sizes are fishes and crustaceans that perform elaborate strategies of parental care to oxygenate their eggs effectively (Kranenbarg *et al.* 2000; Sargent *et al.* 1987) and/or deposit their eggs in cold, well-oxygenated water. Even under optimal oxygenation conditions, the spherical shape of the egg presents an insurmountable constraint.

This constraint manifests itself in two different ways. The first is that the incubation time in the spherical fish eggs of bony fishes is a function of their size and ambient temperature. Based on data from 27 sources, Pauly and Pullin (1988) estimated the parameters of the multiple regression

$$\log(Inc) = 7.1 + 0.608 \cdot \log(\varnothing) + 4.09 \cdot \log(T+26) \quad \text{...(17.1)}$$

where *Inc* is the incubation time in days, Ø the egg diameter in mm, and *T* the ambient temperature in °C, which explained 82.3 % of the variance in a data set consisting of 84 marine species representing 35 families of marine fishes, with Ø ranging from 0.6 to 3.4 mm, *T* ranging from 2.8 to 29 °C and *Inc* ranging from 9 hours to 12 days. The addition of 26 °C to the rightmost term was a purely empirical device to improve the fit of Equation (17.1). Here, a new relationship was re-estimated from the same dataset, which has a more appropriate temperature term:

$$\log(Inc) = -11.5 + 0.597 \cdot \log(\varnothing) + 3477 \cdot (1/K) \quad \text{...(17.2)}$$

where *K* is Kelvin, i.e., °C + 293, which explains 85.5% of the variance in the dataset. Equation (17.2) has a better fit because it corresponds to an Arrhenius plot (Regier *et al.* 1990).[256] Note that Equations (17.1) and (17.2) do not apply to the eggs of polar fish, i.e., markedly lower than 4 °C, for reasons outlined in Chapter 4.

The other way that the spherical shape of the egg manifests itself as a constraint is the fact that the embryos inside hatch when their respiratory surface (which they can deploy only upon hatching) is ~ 1.5 times that of the egg itself (Teletchea and Pauly 2024).[257] Or, put differently, the response to the question 'Why do fish larvae hatch when they do?' is that they hatch when the surface of the eggs within which they are confined becomes unable to satisfy their growing oxygen requirements. They must be able to start relying on their own, larger respiratory surfaces or asphyxiate.[258] Note here that the 1.5 ratio between the egg surface area and the respiratory surface area of newly hatched larvae appears to be independent of egg size and temperature, as ascertained using eggs from 34 species of marine and freshwater fishes, ranging from 0.63 to 5.7 mm in diameter, with the corresponding larvae ranging from 1.13 to 22 mm in length and 3.5 to 432 mm^2 in respiratory area (i.e., the area of their bodies, as newly hatched larvae do not have functional gills).

As previously known, the eggs of fish, and possibly of other aquatic ectotherms, hatch earlier under hypoxic conditions (Oppen-Berntsen *et al.* 1990; Warkentin 2007; Polymeropoulos *et al.* 2016). While hypoxia-induced hatching is a well-described phenomenon, oxygen limitation inside the egg is likely to be the mechanism behind

Figure 17.1: Different life stages of amphibians (upper series) and fish (lower series) as represented in Romanes (1892), based on Ernst Haeckel's *Anthropogenie* (1874).[269]

hatching even under normoxic conditions, as the geometrical constraints of their spherical shape will ultimately result in internal hypoxia (Kranenbarg *et al.* 2000).

Breaking free of the oxygen constraints of an egg's spherical shape offers new scope for growth for the larvae at the post-hatching stages. Their bodies are now more elongated and show different surface-volume ratios, which enable them to take in oxygen more effectively through their skin than before.[259] This next life stage only guarantees sufficient oxygen uptake capacity up to a certain size. Once this size is reached, the organism's oxygen demand for maintenance and growth exceeds the cutaneous oxygen supply. Developing a respiratory system with higher surface-volume ratios becomes necessary (see **Figure 17.1**).

Except for slow-growing and sluggish aquatic species, such as the Lake Titicaca frog (*Telmatobius culeus*), which has draped, loose skin that allows for substantial cutaneous oxygen uptake, all larger water-breathers must develop filamentous structures like gills to survive, grow, and reproduce underwater. Overcoming the larval oxygen constraint by developing gills with surface areas beyond the surface-cube or surface-sphere laws does not mean that surface limitation ceases to constrain energy uptake, growth, and size. As we demonstrate in this book, water-breathing animals from all phyla can be assumed to be subjected to this constraint[260] and, therefore, after a short larval and early juvenile phase, when the food they ingest is the main factor affecting their (exponential) growth, they grow asymptotically.

Alien oceans and the GOLT

The aforementioned patterns should apply everywhere in the universe where gas exchange occurs in fluid media and where Fick's law holds (Denny 1993). They should also apply to the oceans that appear to exist on several of the moons around Jupiter (Europa, Ganymede, Callisto) and Saturn (Enceladus, Mimas), not to mention on other possible oceans within the Solar system (Hand 2010; Lainey *et al.* 2024) and obviously on some exoplanets and their moons (Cooper 2015).

In the oceans covered by ice, such as those on Europa or Enceladus, water remains liquid due to the heat generated by tidal forces caused by the gravity of nearby Jupiter and Saturn, respectively. If these oceans harbor life, it may depend on other sources of energy than that provided by processes requiring oxygen (see, e.g., Oro *et al.* 1992; Irwin and Schulze-Makuch 2003). Such processes do not need to be imagined: they exist on Earth, or rather in the depths of our oceans. They are based on sulfur and hydrogen compounds metabolized by chemoautotrophic or chemolithotrophic bacteria, which form the basis of the ecosystems along the flanks of hydrothermal vents (Hand 2010).[261]

Whatever the chemical basis of life in alien oceans, the argument can be made that physical and dimensional constraints such as those explored in this book would not only 'work' in extraterrestrial systems but generate anatomical and physiological convergences (Conway Morris 2003) – as would Darwin's natural selection. We suggest that the GOLT may help focus research on the mechanisms that structure the lives of animals we may (or will?) discover in these alien oceans.

Ecological implications of the GOLT

Back on Earth, the mechanisms described in this book have important ecological implications, not only in quickly changing environments (as described in Chapter 5) but also in the long-term evolution of ecosystems and trophic networks. Oxygen availability and evolved oxygen uptake capacity shape how species coexist, interact, form symbiotic or parasitic relationships, or prey on each other. While evolutionary accounts of body size evolution have described mechanisms such as 'Cope's Rule' or an evolutionary selection for larger body sizes (Chapter 10), not everyone benefits from getting bigger. Growing to a larger size also increases maintenance costs and increases the risk of higher vulnerability to hypoxia and/or thermal stress (Rubalcaba *et al.* 2020; Müller *et al.* 2023). On the other hand, this disadvantage of larger species and individuals creates opportunities for smaller organisms to occupy niches inaccessible to larger predators or competitors.

Larger organisms have an advantage in aquatic environments due to reduced frictional drag. This advantage favors predators, who can swim faster and more efficiently than their prey. Despite this advantage, predators cannot continually follow prey schools due to the risk of oxygen depletion. Large predators face an energetic constraint, which can become dangerous if their oxygen reserves are squandered. While larger sizes offer hydrodynamic benefits, size-dependent constraints on oxygen uptake give smaller species and individuals an advantage, and prevent predators from maintaining continuous, direct contact with their prey (Bakun 2019, 2022; see also Robb and Abrahams 2003).

Predators must adapt their feeding behaviors to balance energy expenditure with prey availability and oxygen conservation. Most large predatory fish employ preying strategies that minimize energy expenditures with maximum efficiency. Large individuals of freshwater species, such as perch and pike, and marine fish such as groupers and pollack, become ambush predators that lurk sluggishly behind plants or rocks. In marine environments, predator groups developed cooperative prey-herding behavior to counteract the oxygen advantage of smaller prey. By encircling prey schools and launching occasional attacks, predators induce oxygen-

Table 17.1. Biomass distribution of life on earth in gigatons of carbon, adapted from Bar-On *et al.* (2018).

Taxa	Biomass
All plants	450
All animals	2
Arthropods	1
Fish	0.7
Molluscs	0.2
Annelids	0.2
Cnidarians	0.1
Livestock	0.1
Humans	0.06
Nematodes	0.02
Wild mammals	0.007
Wild birds	0.002

consuming evasive behaviors, rendering prey immobile and facilitating consumption with minimal oxygen expenditure. When prey abundance is too low for effective cooperation, the prey species may regain an advantage, making cooperative predator activity less efficient (Bakun 2019, 2022). Another solution is to become a passive filter feeder, as do whale and basking sharks.

The differential dynamics of oxygen demand play a crucial role in preserving both prey and predator populations and maintaining the structure and function of marine and freshwater ecosystems. Oxygen availability may play a vital role in preventing the collapse of these ecosystems and ensuring the continuation of diverse biological structures.[262] Without the constraints imposed by oxygen demand, the evolution of the diversity of the planet's aquatic ecosystems and their inhabitants might not have been possible. The oxygen constraint on the growth and metabolism of water-breathing ectotherms may even have ecological implications above the level of local ecosystems and might be relevant on a global level. In contrast to plant biomass, of which 99% occurs on land, 75% of animal biomass is aquatic (Bar-On *et al.* 2018), and ectotherms like fish, aquatic arthropods, and molluscs constitute the largest share of all carbon stored in animals (see **Table 17.1**).

Recent studies have suggested that dredging up this "*blue carbon*" may impact the global carbon cycle (Mariani *et al.* 2020). While carbon stored in marine animals would normally sink to the bottom of the sea in the form of excrement or corpses and feed other marine organisms, fisheries incorporate this carbon into terrestrial cycles. The aquatic lifestyles

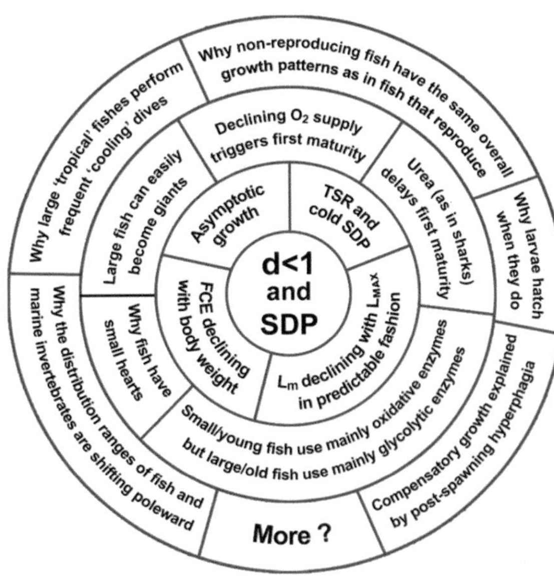

Figure 17.2. Inspired by Crete's enigmatic Phaistos Disk, the GOLT-disk is an attempt to summarize the content of this book via a summary statement of what the GOLT explains. At the center are its two basic components, i.e., that $d < 1$ in adult fish and other WBE (Chapter 3), and that all living organisms are affected by spontaneous denaturation of proteins (SDP; Chapter 4). The first concentric ring features asymptotic growth as a consequence of the centers' contents (Chapter 2), and other consequences are outlined in Chapters 4, 7 and 9. The two outer rings refer to further items explicitly covered in various parts of this book, with space left open for 'More?' Further rings may be added, as new applications of the GOLT are identified, for example dealing with hypoxia-induced pain in fish (see below).

and respiratory modes of fishes and WBE may be relevant for the global carbon cycle in yet another way: respiratory constraints only allow for metabolic levels that are very slow compared to the metabolic level of endotherms. Carbon is stored much longer in WBE and excreted slower than in marine mammals or seabirds, implying yet another way that relative metabolic inertia of WBE may affect the global carbon cycle.

In this book, except for brief visits to putative extraterrestrial oceans, we have limited ourselves overwhelmingly to mechanisms and implications of GOLT that can be supported by existing literature or published data on various organisms (**Figure 17.2**). Still, we are convinced that the theory presented here will have more and wider implications for the physiology and ecology of water-breathing animals,[263] and we encourage others to test our hypotheses and extend them beyond the themes we have discussed here, and beyond the taxa that we have used to illustrate these themes (Clark *et al.* 2023).[264] To conclude this book, we outline 5 research topics – viviparity, hermaphroditism, hypoxia-induced pain, parental care, and bioenergetics – that readers may explore in the future in addition to the various untested hypotheses presented in earlier chapters.

The GOLT and viviparity in fish

One subject area where we believe that applications of the GOLT can lead to novel insights is the reproductive biology and ecology of fishes and other water-breathing ectotherms. Reproductive investment is among the most highly prioritized energetic expenditures and comes at great costs, especially in the context of the oxygen-constrained energy budgets of water-breathing taxa. While most oviparous fishes and crustaceans 'outsource' the oxygenation of their embryos by storing their eggs outside their bodies (often by broadcast spawning without any further parental care), viviparous and ovoviviparous taxa pay the full price of internal oxygenation. The resulting differences are substantial, with oxygen demand increasing from 27% to 117% relative to males or non-gestating females (Skov *et al.* 2010; Timmerman and Chapman 2003; Webb and Brett 1972; Boehlert *et al.* 1991; Hopkins *et al.* 1995).

Such drastic increases in oxygen demand force pregnant fish to seek cooler areas or develop other strategies to cope with oxygen limitation. As Banet *et al.* (2016) have shown, gestating guppies in flowing waters move to quieter areas where they can save energy, and pregnant rockfish retreat to cooler spots to lower their metabolism (Boehlert *et al.* 1991). Auer *et al.* (2021) showed that gestation lowered the thermal tolerance of poeciliid females. Energetic costs increase strongly with body size: in gestating Korean rockfish (*Sebastes schlegeli*), the oxygen consumption of gestating females of 30 cm standard length increases by only 35%, whereas individuals of 50 cm consumed 117% more oxygen (Boehlert *et al.* 1991).

This raises the question of how oxygen constraints shaped the biogeography and the occupation of ecological niches of live-bearing fishes. As Goodwin *et al.* (2005) pointed out, live-bearers are relatively uncommon at the lowest latitudes (most species inhabit subtropical climates). The two smallest live-bearing families, ovoviviparous poeciliids and viviparous goodeids, both exhibit one remarkable exception: one poeciliid species (*Tomeurus gracilis*), and the entire goodeid subfamily Empetrichthyinae, are egg-layers. An interesting parallel emerges if we look at the biogeographies of the Poeciliidae and Goodeidae. While the thermal niches of goodeids are rather narrow, and most species inhabit the cool Mexican highlands, the Empetrichthyinae subfamily only occurs in hot

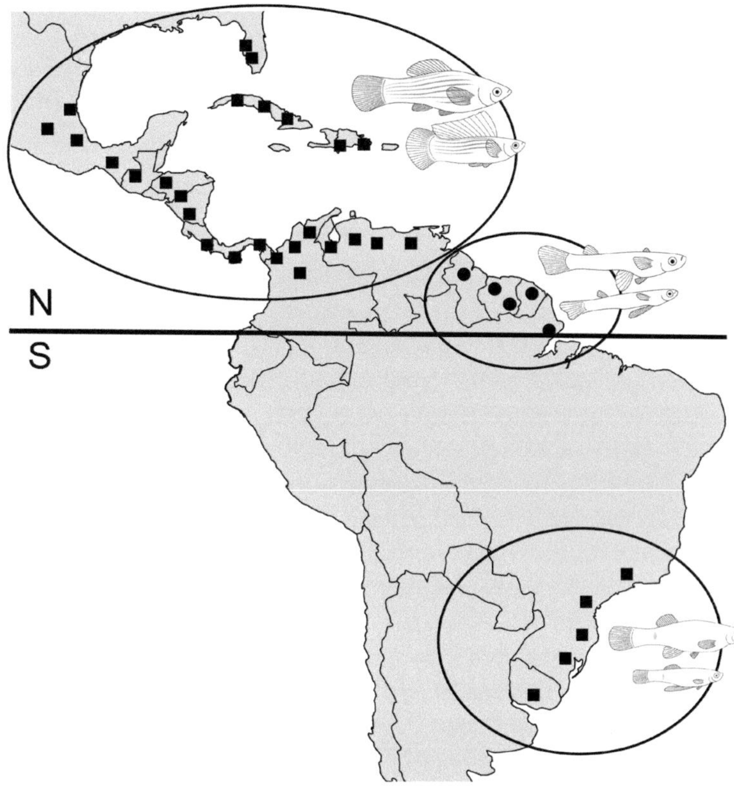

Figure 17.3. Biogeographical map of poeciliid fishes (modified from Hrbek *et al.* 2007). Black squares = live-bearing species; black dots = egg-laying species. As in goodeids, the egg-laying species inhabit the warmest climates.

springs in Nevada. The only extant species of this subfamily, *Crenichthys baileyi*, is the only fish that can enter the hottest part of these water bodies and has been recorded at temperatures of up to 37.8 °C (Baugh and Brown 1980). It remains unclear if the reproductive mode of Empetrichthyinae represents the ancestral state of the Goodeidae and if viviparity is a later development. Still, no live-bearing fish tolerates such extreme values, especially not during gestation.

Similar to the notable exception in goodeids, the only egg-laying poeciliid, *Tomeurus gracilis*, is also a geographical outlier. While live-bearing poeciliids are concentrated in the Caribbean and the south of Brazil, Uruguay, and Argentina, *Tomeurus* is one of the few (non-invasive) members of this family in the Guyanas and northern Brazil (see **Figure 17.3**).

Both the assumption of facultative viviparity and the idea that egg-laying re-evolved in *Tomeurus gracilis* (see Reznick *et al.* 2007) remain hypothetical, but it would be in line with the hypothesis that live-bearing is constrained by thermal tolerance. Under higher temperatures, the energetic costs of viviparity might be so high that the organism must be either very small or switch (back) to egg-laying. Even though maintenance metabolism in poecilids is lower than in related species (Nordlie 2014), their thermal tolerance has limits beyond which viviparity is no longer advantageous or sustainable.

According to Wourms's (1991) "*temperature hypothesis*," the distribution range of the entire genus *Sebastes* may be shaped by temperature and oxygen. This genus is present in both hemispheres but only in the respective temperate and subtropical regions (Wourms 1991), thus, the distribution ranges of the *Sebastes* species are effectively separated by the equator. Future research on distributions of this sort should benefit from GOLT-derived hypotheses.

The GOLT and parental care

Another pattern that the GOLT may explain is the distribution of parental care in fishes and other aquatic ectotherms. Compared to terrestrial vertebrates, especially birds and mammals, fishes with parental care show an exceptional pattern: in contrast to most other animal groups with parental care, the caring parent in fish is primarily the father.[265] While mammals exhibit strong bonds between offspring and their mother, birds are typically biparental caregivers; reptiles and amphibians show a more diverse array of care, but paternal care is rare except for some anurans. About 30% of fish families show some form of protection of eggs or larvae, which is often combined with the provision of oxygen and sometimes food, and the vast majority of these taxa show paternal care (Balshine and Sloman 2011). While about 6% of known bony fish families have species where both parents provide care, monoparental maternal care is rare.

This pattern requires an explanation. Older studies often focused on the differences between internal and external fertilization that might predict the investment of male parents in the offspring of their female partners. In species with external fertilization, it was either assumed that males would show more paternal involvement because their chance of fatherhood would be higher (Ridley 1978; Perrone and Zaret 1979; Blumer 1979) or that external fertilization would increase the likelihood that females would desert their eggs (and partners) because the male gametes were only released later (Dawkins and Carlisle 1976). Both hypotheses were abandoned later because empirical support for higher chances of fatherhood in species with external fertilization could not be found, and because eggs and sperm are often released simultaneously (Gross and Shine 1981; Balshine and Sloman 2011).

A different explanation for the prevalence of paternal care was sought in the energetic asymmetries in reproductive investment between the sexes and the degree of 'association' between the eggs and the respective parent. Proponents of this hypothesis argued that the energetic cost of parental care would be lower for the parent already associated with the eggs or to which the eggs are attached at the moment of fertilization. This hypothesis may explain why maternal care is more common in species with internal fertilization (because the female already holds the eggs while the male might no longer be around); however, its explanatory power for paternal care remains limited. Even in cases where the male is the last parent present at the fertilized egg site, it is unclear why this should decrease the energetic cost of care. In short, none of these hypotheses explain the striking overrepresentation of piscine paternal care compared to other vertebrates.

If we accept the fact that the ripe gonads of females can make up a quarter or a third of their total body weight and that they lose most of this weight during spawning (see Chapter 9), then energetic factors become far more plausible as an explanation for the prevalence of paternal care in fish. As females invest more energy in the production of gonadal material, this energy needs to be compensated for after spawning. Large amounts of energy are needed to support post-spawning growth, but acquiring this energy requires time. Providing care limits the ability to feed, and in extreme cases, for example, in mouthbrooding labyrinth fishes, the caring parent must fast for two weeks or even longer after spawning. Recent research shows that mouthbrooding species in this family are paternal brooders, and reports of maternal care are likely based on incorrect observations (Zworykin *et al.* 2024). Fasting would dramatically increase the energetic cost of reproduction for females that have already lost 25-30% of their weight and now depend on quick growth for the next spawning event or season.

This hypothesis is supported by experiments on species with plastic parental care systems in which both parents can be mouthbrooders. In the Galilee St. Peter's fish, *Sarotherodon galilaeus*, the energetic costs differ significantly between the mouthbrooding sexes. As Balshine-Earn (1995) showed, mouthbrooding females lost more weight than males, and maternal care also increased the intervals between spawning events. Males, by contrast, were able to spawn more than 10 days earlier after mouthbrooding than females. Finally, caring and non-caring females showed strong differences in brood weight, which illustrates the high cost of producing gonadal materials in combination with providing care. While this study did not measure the number of eggs in the spawning events that followed mouthbrooding, a reduction in hatch sizes should be an inevitable outcome. This suggests that the differential energetic costs of parental care in fishes favor male care. Since females typically feed heavily after spawning and invest a lot of energy (i.e., food and oxygen) into post-spawning growth spurts (which allows them to produce more offspring in the following breeding season), the job of caregiving is outsourced to the males. More research is needed in this area, but the GOLT's model seems well-suited to solve this problem.

The GOLT and hermaphroditism

An issue not explained in this book is the growth of hermaphroditic fish, a phenomenon for which the GOLT might have the potential to shed new light on an unresolved question. As many authors who report on the growth of protogynous hermaphroditic fishes suggest, the individuals who have become males seem to grow "*better*" than the females or their growth is "*accelerated*" (see, e.g., **Figure 17.4A**, adapted from Garratt *et al.* 1993), which would come handy in species that engage in some form of male-to-male contests (Parker 1992). Looking at **Figure 17.4B**, one quickly realizes that something is profoundly wrong with interpreting length-at-age data, such that females (i.e., individuals who did not change sex) are supposed to *shrink*. What occurs is, as is always the case, that there is much variability in the growth of a cohort of fish, even if it initially consists only of one sex (here: females), as shown by the high variance of length-at-age data. Also, this sex change appears to occur at some fraction of the maximum length they can reach, i.e., after reaching a threshold that may be similar to that which triggers first maturity in dioecious species (see Chapter 7).

Once a fast-growing female turns into a male, it is subtracted from the still-female members of her cohort, which 'reduces' their mean lengths. This process is repeated until all females capable of reaching the threshold in question have done so, leaving only the slow-growing females to be identified as such. This gives the impression that the males grow "*better*," while what happens is that only the better-growing females become males. While the GOLT has not yet examined this phenomenon, the hypothesis could be tested that sex change occurs at a point that could be determined with a similar regularity as the first moment of maturation or spawning (see Chapters 7 and 8). A possible explanation could be

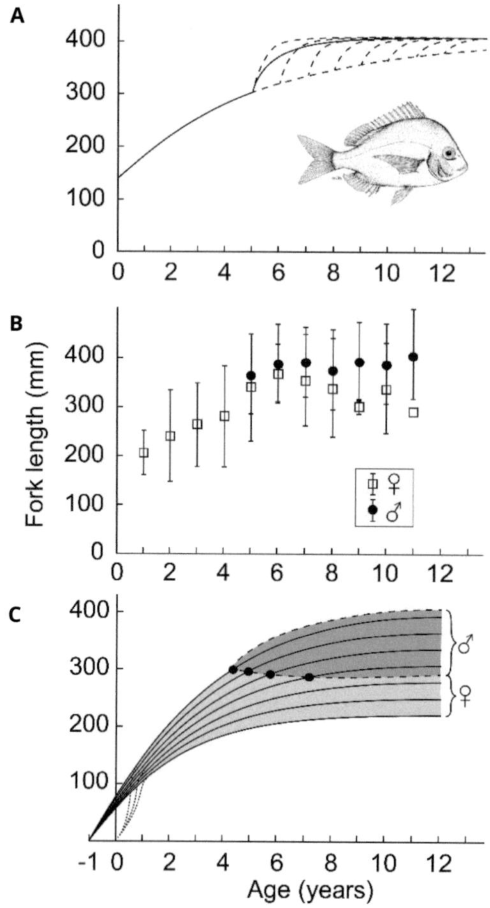

Figure 17.4. Two conceptions of protogynous hermaphroditism in fish. **A**: 'Standard' version, which sees (and models) females turning into males, and *then* (therefore?) experiencing an acceleration of their growth (adapted from figure 3 in Garratt *et al.* 1993). **B**: Empirical age-at-length data on slinger seabream *Chrysoblephus puniceus* along the southeast coast of Africa, (inadvertently) suggesting that females who do not change sex ultimately 'shrink' (adapted from figure 2 in Garratt *et al.* 1993). **C**: Alternative explanation, proposing that it is only the fast-growing females that change sex (e.g., at the size suggested by the dots), which explains both why the resulting males seem to 'grow faster' and why the mean lengths of slower-growing females (which fail to change sex) appear to 'shrink.' The dotted lines near the origin suggest the actual length trajectories of the larvae and early juveniles, whose growth is usually not well represented by the VBGF.

that the observed regularities are connected to social hierarchies within the group, i.e., that the subdominant individuals (still 'females') experience more stress, which is induced by the dominant males (see also Todd *et al.* 2016). Once a recently transformed individual has the chance to pass the threshold, its maintenance costs decrease, and it can grow further. If this hypothesis can be confirmed, oxygen is expected to mediate this further growth, even though it should also not be neglected that social dominance also results in advantages in food competition.

Without more evidence, this is not the place to elaborate on this issue; however, the conceptual confusion that seems to underlie the work cited above and several related studies suggests the need for clarification.

Oxygen, hypoxia and pain in fish

To our knowledge, Bakun (2010, 2011, 2019) was the first colleague who noted the close relationship between the oxygen limitation emphasized by the GOLT and pain in fish. As he wrote under the subheading "*Oxygen Deficit is a Pain*" in Bakun (2011): "*Evidently, falling into a state of oxygen deficit in a medium where replenishment is a slow, activity-restricting process presents a distinct set of problems and could even constitute a grave danger to the survival of a fish. Thus, it would seem important to evolutionary fitness that a state of oxygen deficit would continually be strongly avoided. It might be expected that, as in all animals, a fish's nervous system would generate an urgent sensory signal (i.e., physical pain) that would have turned, over evolutionary time, a voluntary aversion into a compelling involuntary response. Accordingly, lower limits of acceptable dissolved oxygen levels as often measured by laboratory fish physiologists (e.g., concentration levels at which a fish may begin to exhibit active distress symptoms or behaviors) may not be the relevant ones in provoking evident responses. Rather, a fish endowed by evolution with a pain-like aversion to depleting its oxygen reserves may be very hesitant to engage in vigorous physical exertion, even in situations where there would be entirely sufficient oxygen available in the water to support respiration and replenishment while a fish is in a resting, or cruising, state.*"

This line of argument makes sense, but it would have been good for supporting evidence to be cited, especially given that there are authors who argue that fish don't feel pain (some of whom funded by the 'sport' fishing industry, and need not be cited here, as it is biologically implausible for fish to have evolved nociceptors if they were incapable of experiencing pain).

In the meantime, however, an important contribution has appeared (Schuck-Paim *et al.* 2025) which summarizes numerous quantitative neurochemical studies of pain in fish, all showing that internal hypoxia, which leads to hypercapnia and acidosis, induces pain ranging from 'annoying' to 'excruciating'.[266]

We lack the expertise to follow up up on this relationship, which may also apply to other WBE, and which seems important in an age when animal welfare has become an issue of increased concern.

A final plea

Many of the ideas in this chapter and other parts of this book could be followed up by ourselves and/or through our students, and perhaps at least some of them will be. Still, there are two issues with the GOLT that we would not know how to tackle, if only because we do not master the required, mostly mathematical tools.

One is the reformulation of the GOLT as a "*causal network*," as defined by Pearl and MacKenzie (2018). This book attempts to explore causal linkages by proposing mechanisms, drawing on a hefty dose of science, history, and philosophy. However, we are aware that formal logico-mathematical methods exist to identify the causal links within a web of complex interactions (see Pearl and MacKenzie 2018). This could settle the issue of whether gill surface areas not keeping up with the mass of the bodies to which they supply oxygen is the *cause* of the ultimate cessation of the growth of WBE, as stated by the GOLT.

The other issue is the GOLT's relationship to the body of literature dealing with what is often called 'bioenergetics,' i.e., how animals allocate the energy they acquire through food to their various organs to support the processes that sustain their different activities.[267] The GOLT is incomplete as a theory for two reasons: (*i*) because it is largely qualitative, i.e., its shows the likely direction of changes, but not their magnitude and (*ii*) because its formulation is based on a simplification: most of its hypotheses and mechanisms assume that the WBE in question have constant access to all the food they need. This is not likely to be the case in the real world. However, this is the flip side of studies in 'bioenergetics,' which focus on the fate of ingested proteins and fats in agonizing detail but assume that the oxygen required to produce ATP will magically appear where and when needed.[268] A quantitative version of the GOLT, integrated into a version of bioenergetics (or *vice versa*), is needed; however, we are currently unable to deliver it. We believe it would be valuable and rewarding to follow up on this.

References

Aaskov, M.L., R.J. Jensen, P.V. Skov, C.M. Wood, T. Wang, H. Malte and M. Bayley. 2022. *Arapaima gigas* maintains gas exchange separation in severe aquatic hypoxia but does not suffer branchial oxygen loss. *Journal of Experimental Biology*, 225(6), jeb243672.

Aaskov, M.L., D. Nelson, H. Lauridsen, D.T.T. Huong, A. Ishimatsu, D.A. Crossley, H. Malte and M. Bayley. 2023. Do air-breathing fish suffer branchial oxygen loss in hypoxic water? *Proceedings of the Royal Society B*, 290(2006): 20231353.

Adhikari, S. 2015. *World-wide body size patterns in freshwater fish by geography, size class, trophic level, and taxonomy*. Doctoral dissertation, Wright State University, Dayton, OH, 94 p. https://corescholar.libraries.wright.edu/etd_all/1436

Ahas, R. and A. Aasa. 2006. The effects of climate change on the phenology of selected Estonian plant, bird and fish populations. *International Journal of Biometeorology*, 51(1): 17–26. DOI:10.1007/s00484-006-0041-z

Ahluwalia, A. 2017. Allometric scaling *in vitro*. *Scientific Reports* 7(1): 42113.

Akaike, H. 1973. Information theory and an extension of the maximum likelihood principle, p. 267–281 *In:* B.N. Petrov and F. Csáki (eds) *2nd International Symposium on Information Theory*. Akadémia Kiadó, Budapest.

Al Tameemi, W., T.P. Dale, R.M.K. Al-Jumaily and N.R. Forsyth. 2019. Hypoxia-modified cancer cell metabolism. *Frontiers in Cell Development Biology*, 7: 4. DOI:10.3389/fcell.2019.00004

Alajmi, A., A. Ben-Hasan, T. Alrushaid, A. Vahabnezhad and D. Pauly. 2024. Mean temperature of the catch index can be masked by changes in catch composition unrelated to ocean warming. *Fisheries Management and Ecology*, 31(1): 1–6. DOI:10.1111/fme.12659

Ali, M., A. Nicieza and R.J. Wootton. 2003. Compensatory growth in fishes: a response to growth depression. *Fish and Fisheries*, 4(2): 147-190.

Alix, M., O.S. Kjesbu and K.C. Anderson. 2020. From gametogenesis to spawning: How climate-driven warming affects teleost reproductive biology. *Journal of Fish Biology*, 97: 607–632.

Al-Kadhomiy, N. 1985. *Gill development, growth and respiration of the flounder, Platichthys flesus (L.)*. Doctoral dissertation, University of Bristol, 114 p. https://research-information.bris.ac.uk/ws/portalfiles/portal/34494339/354444.pdf

Allen, A. 1877. The influence of physical conditions in the genesis of species. *Radical Review*, 1: 108–140.

Allen, K.R. 1935. The food and migration of the perch (*Perca fluviatilis*) in Windermere. *Journal of Animal Ecology*, 4(2): 264-273. www.jstor.org/stable/1016

Amarasinghe, U.S. and D. Pauly. 2021. The relationship between size at maturity and maximum size in cichlid populations corroborates the Gill-Oxygen Limitation Theory (GOLT). *Asian Fisheries Science*, 34: 14–22. DOI:10.33997/j.afs.2021.34.1.002

Amitai, Y., E. Strobach, D.P. Hamilton, S. Assouline, A. Nishri and T. Zohary. 2024. Seiche-Induced Fish Kills in the Sea of Galilee – A Possible Explanation for Biblical Miracles? *Water Resources Research*, 60 (10): e2024WR037894.

Aminur Rahman, M., A. Arsha and S.M. Nurul Amin. 2012. Growth and production performance of threatened snakehead fish, *Channa striatus* (Bloch), at different stocking densities in earthen ponds. *Aquaculture Research*, 43: 297–302.

Anderson, A.W. 1954. A 152-year-old lake sturgeon caught in Ontario. *Commercial Fisheries Review*, 18: 28.

Andrews, A.H., D.M. Tracey and M.R. Dunn. 2009. Lead-radium dating of orange roughy (*Hoplostethus atlanticus*): Validation of a centenarian life span. *Canadian Journal of Fisheries and Aquatic Science*, 66: 1130–1140. DOI:10.1139/F09-059

Angilletta Jr, M.J., T.D. Steury and M.W. Sears. 2004. Temperature, growth rate, and body size in ectotherms: fitting pieces of a life-history puzzle. *Integrative and Comparative Biology*, 44(6): 498–509.

Angilletta, M.J. and A.E. Dunham. 2003. The temperature-size rule in ectotherms: Simple evolutionary explanations may not be general. *American Naturalist*, 162: 332-342.

Arantes, C., L. Castello and Garcez, D.S. 2007. Variaçôes entre contagens de *Arapaima gigas* (Schinz) (Osteoglossomorpha, Osteoglossidae) feitas por pescadores individualmente em Mamirauá, Brasil. *Pan-American Journal of Aquatic Sciences*, 2: 263–269.

Arkhipkin, A.I. 2004. Diversity in growth and longevity in short-lived animals: squid of the suborder Oegopsina. *Marine and Freshwater Research*, 55: 341–355.

Arranz, I., T. Mehner, L. Benejam, C. Argillier, K. Holmgren, E. Jeppesen and S. Brucet. 2015. Density-dependent effects as key drivers of intraspecific size structure of six abundant fish species in lakes across Europe. *Canadian Journal of Fisheries and Aquatic Sciences*, 73(4): 519-534.

Asano, T., K. Hashimoto and R. Craig Everroad. 2023. Eco-evolutionary implications for a possible contribution of cuticle hardening system in insect evolution and terrestrialisation. *Physiological Entomology*, 48: 55-60. DOI:10.1111/hhen.12406

Asch, R.G., C.A. Stock and J.L. Sarmiento. 2019. Climate change impacts on mismatches between phytoplankton blooms and fish spawning phenology. *Global Change Biology*, 25: 2544–2559. DOI:10.1111/gcb.14650

Asch, R.G. 2015. Climate change and decadal shifts in the phenology of larval fishes in the California Current ecosystem. *Proceedings of the National Academy of Sciences*, 112(30): E4065-E4074. DOI:10.1073/pnas.1421946112

Asimov, I. 1977. *The Planet that wasn't: A Mind-dazzling Excursion into the Realm of Myth, Science and Speculation.* Discus Books, New York, 206 p.

Atkinson, T.P. 2005 Arthropod body fossils from the Union Chapel Mine, p. 69–176 *In:* R.J. Buta, A.K. Rindsberg and D.C. Kopaska-Merkel (eds) Pennsylvanian footprints in the Black Warrior basin of Alabama. *Alabama Paleontological Society Monographs* No. 1.

Atkinson, D. 1994. Temperature and organism size: a biological law for ectotherms? *Advances in Ecological Research,* 25: 1–58.

Atkinson, D., Leighton, G. and M. Berenbrink. 2022. Controversial roles of oxygen in organismal responses to climate warming. *Biological Bulletin,* 243(2): 207-219.

Audzijonyte, A., D.R. Barneche, A.R. Baudron, J. Belmaker, T.D. Clark, C.T. Marshall, J.R. Morrongiello and I. van Rijn. 2019. Is oxygen limitation in warming waters a valid mechanism to explain decreased body sizes in aquatic ectotherms? *Global Ecology and Biogeography,* 28(2): 64–77.

Audzijonyte, A., S.A. Richards, R.D. Stuart-Smith, G. Pecl, G.J. Edgar, N.S. Barrett, N. Payne and J.L. Blanchard. 2020. Fish body sizes change with temperature but not all species shrink with warming. *Nature Ecology & Evolution,* 4: 809– 814. DOI:10.1038/s41559-020-1171-0

Auer, S.K., E. Agreda, A.H. Chen, M. Irshad and J. Solowey. 2021. Late-stage pregnancy reduces upper thermal tolerance in a live-bearing fish. *Journal of Thermal Biology,* 99: 103022.

Babák, E. 1910. Über die Oberflächenentwicklung bei Organismen und ihre Anpassungsfähigkeit. *Biologisches Centralblatt,* 30: 225-239.

Bahtiar, B., Y.E. Jiwani, M.N. Findra and E. Ishak. 2024. Temporal variation of peanut worm (Siphonosoma australe-australe) reproduction in Toronipa Beach of Southeast Sulawesi, Indonesia. *Biodiversitas Journal of Biological Diversity,* 25(8): 2533-2540.

Baird, I.G., V. Inthaphaisy, P. Kisouvannalath, B. Phylavanh and B. Mounsouphom. 1999. *The fishes of southern Lao.* Lao Community Fisheries and Dolphin Protection Project. Ministry of Agriculture and Forestry, Lao PDR. 161 p.

Baker, D.W., B. Sardella, J.L. Rummer, M. Sackville and C.J. Brauner. 2015. Hagfish: Champions of CO_2 tolerance question the origins of vertebrate gill function. *Scientific Reports,* 5(1): 11182.

Bakun, A. 2010. Afterword, p. 137–140 *In:* D. Pauly, *Gasping Fish and Panting Squids: Oxygen, Temperature and the Growth of Water-Breathing Animals.* International Ecology Institute, Oldendorf/Luhe, Germany.

Bakun, A. 2011. The Oxygen Constraint, p. 11–23 *In:* V. Christensen and J. MacLean (eds) *Ecosystem Approaches to Fisheries.* Cambridge University Press, New York.

Bakun, A. 2019. The oxygen constraint, p. 143–146 *In:* D. Pauly. 2019. *Gasping Fish and Panting Squids: Oxygen, Temperature and the Growth of Water-Breathing Animals – 2nd Edition,* International Ecology Institute, Oldendorf/Luhe, Germany.

Bakun, A. 2022. Adjusting intuitions as to the role of oxygen constraints in shaping the ecology and dynamics of ocean predator-prey systems. *Environmental Biology of Fishes,* 105(10): 1287–1299.

Bakun, A. 1990. Global climate change and intensification of coastal ocean upwelling. *Science,* 247: 198–201.

Balgos, M. and D. Pauly. 1998. Age and growth of the squid *Sepioteuthis lessoniana* in N.W. Luzon, Philippines. *South African Journal of Marine Science,* 20(1): 449-452.

Ballantyne, J.S. 2016. Some of the most interesting things we know, and don't know, about the biochemistry and physiology of elasmobranch fishes (sharks, skates and rays). *Comparative Biochemistry and Physiology B,* 199: 21–28. DOI:10.1016/j.cbpb.2016.03.005

Ballantyne, J.S and D.I. Fraser. 2012. Euryhaline elasmobranchs. *Fish Physiology,* 32: 125–198. DOI:10.1016/B978-0-12-396951-4.00004-9

Ballantyne, J.S. and J.W. Robinson. 2010. Freshwater elasmobranchs: A review of their physiology and biochemistry. *Journal of Comparative Physiology B,* 180: 475–493. DOI:10.1007/s00360- 010-0447-0

Balshine, S. and K.A. Sloman. 2011. Parental care in fishes, p. 670-677 *In:* A.P. Farrell (ed.) *Encyclopedia of Fish Physiology: from Genome to Environment,* Vol. 1. Academic Press, San Diego.

Balshine-Earn, S. 1995. The costs of parental care in Galilee St Peter's fish, *Sarotherodon galilaeus. Animal Behaviour,* 50(1): 1-7.

Banavar, J.R., J. Damuth, A. Maritan and A. Rinaldo. 2002. Supply-demand balance and metabolic scaling. *Proceedings of the National Academy of Sciences,* 99(16): 10506-10509.

Banet, A.I., J.C. Svendsen, K.J. Eng and D.N. Reznick. 2016 Linking reproduction, locomotion, and habitat use in the Trinidadian guppy (*Poecilia reticulata*). *Oecologia,* 181: 87-96.

Baran, E., I.G. Baird and G. Cans. 2005. *Fisheries Bioecology at the Khone Falls (Mekong River, Southern Laos).* WorldFish Centre, Penang, Malaysia, 80 p.

Barber, I. 2005. Parasites grow larger in faster-growing fish hosts. *International Journal for Parasitology,* 35(2): 137-143.

Barber, I. and P.A. Svensson. 2003. Effects of experimental *Schistocephalus solidus* infections on growth, morphology and sexual development of female three-spined sticklebacks, *Gasterosteus aculeatus. Parasitology,* 126(4): 359-367.

Bardi, U. 2017. *The Seneca Effect: Why Growth is Slow but Collapse is Rapid.* Springer International Publishing, 209 p.

Barnes, H. and H.T. Powell. 1953. The growth of *Balanus balanoides* (L.) and *B. crenatus* Brug. under varying conditions of submersion. *Journal of the Marine Biological Association of the United Kingdom,* 32(1): 107–127.

Bar-On, Y.M., R. Phillips and R. Milo. 2018. The biomass distribution on Earth. *Proceedings of the National Academy of Sciences,* 115(25): 6506–6511.

Barry- Gérard, M. 1994. Migration des poissons le long du littoral sénégalais, p. 215-234 *In:* M. Barry-Gerard, T. Diouf and A. Fonteneau (eds) *L'évaluation des ressources exploitables par la pêche sénégalaise.* Edition ORSTOM, Paris.

Bassi, M., H. Giusti, G.S. da Silva, J. Amin-Naves and M.L. Glass. 2010. Blood gases and cardiovascular shunt in the South American lungfish (*Lepidosiren paradoxa*) during normoxia and hyperoxia. *Respiratory Physiology & Neurobiology,* 173(1): 47–50.

Baugh, T.M. and B.G. Brown. 1980. Field observations on the response of the Railroad Valley springfish (*Crenichthys nevadae*) to temperature. *Great Basin Naturalist,* 40(4): 5.

Bayley, M., C. Damsgaard, N.V. Cong and N.T. Phuong. 2020. Aquaculture of air-breathing fishes, p. 315–535 *In:* A.P. Farrell, C.J. Brauner and T.J. Benfey (eds) *Aquaculture.* Elsevier Science & Technology.

Beauregard, D., E. Enders and D. Boisclair. 2013. Consequences of circadian fluctuations in water temperature on the standard metabolic rate of Atlantic salmon parr (*Salmo salar*). *Canadian Journal of Fisheries and Aquatic Sciences,* 70: 1072-1081.

Bechtel, W. 2011. Mechanism and biological explanation. *Philosophy of Science*, 78(4): 533–557.

Bechtel, W. 2017. Top-down causation in biology and neuroscience: control hierarchies, p. 203–224 *In:* M. Paolini Paoletti and F. Orilia (eds) *Philosophical and Scientific Perspectives on Downward Causation.* Routledge, London.

Bechtel, W. and A. Abrahamsen. 2005. Explanation: A mechanist alternative. *Studies in History and Philosophy of Science Part C: Studies in History and Philosophy of Biological and Biomedical Sciences*, 36(2): 421–441.

Bednarsek, N., J. Mozina, M. Vogt, C. O'Brien and G.A. Tarling. 2012. The global distribution of pteropods and their contribution to carbonate and carbon biomass in the modern ocean. *Earth System Science Data*, 4: 167–186. DOI:10.5194/essd-4-167-2012

Beekes, R.S.P. 2010. *Etymological Dictionary of Greek.* (Leiden Indo-European Etymological Dictionary Series, Vol. 10), with the assistance of Lucien van Beek, Leiden & Boston, Brill, 930 p.

Belkin, I.M. 2009. Rapid warming of Large Marine Ecosystems. *Progress in Oceanography*, 81: 207–213. DOI:10.1016/j.pocean.2009.04.011

Belton, B. and S.H. Thilsted. 2014. Fisheries in transition: Food and nutrition security implications for the global South. *Global Food Security*, 3(1): 59–66.

Bergmann, C. 1847. Über die Verhältnisse der Wärmeökonomie der Thiere zu ihrer Grösse. *Göttinger Studien*, 3(1): 595–708.

Bertalanffy, L. von. 1932. *Theoretische Biologie, Band. I. Allgemeine Theorie, Physikochemie, Aufbau und Entwicklung des Organismus*, Vol. I. Borntraeger, Berlin, 349 p.

Bertalanffy, L. von. 1934. Untersuchungen über die Gesetzlichkeit des Wachstums: I. Teil: Allgemeine Grundlagen der Theorie; Mathematische und physiologische Gesetzlichkeiten des Wachstums bei Wassertieren. *Wilhelm Roux' Archiv für Entwicklungsmechanik der Organismen*, 131: 613–652.

Bertalanffy, L. von. 1960. Principles and theory of growth, p. 137–259 *In:* W.W. Nowinski (ed.) *Fundamental Aspects of Normal and Malignant Growth.* Elsevier, Amsterdam and New York.

Bertalanffy, L. von. 1938. A quantitative theory of organic growth (inquiries on growth laws. II). *Human Biology*, 10(2): 181–213.

Bertalanffy, L. von. 1950. An outline of general system theory. *The British Journal for the Philosophy of Science*, 1(2): 134-165.

Bertalanffy, L. von. 1951. *Theoretische Biologie. Zweiter Band: Stoffwechsel, Wachstum.* A. Francke Verlag, Bern, 418 p.

Bertalanffy, L. von. 1952. *Problems of Life: An Evaluation of Modern Biological Thought.* John Wiley and Sons, New York, 216 p.

Bertalanffy, L., von. 1964. Basic concepts in quantitative biology of metabolism. *Helgoland Marine Research*, 9(1): 5-37.

Bertalanffy, L., von and J. Krywienczyk. 1953. The surface rule in crustaceans. *American Naturalist*, 87: 107–110.

Beverton, R.J.H. 1963. Maturation, growth and mortality of clupeid and engraulid stocks in relation to fishing. *Rapports et Procès-Verbaux des Réunions du Conseil International pour l'Exploration de la Mer*, 154: 44–67.

Beverton, R.J.H. and S.J. Holt. 1957. *On the Dynamics of Exploited Fish Populations.* Fisheries Investigations 19, Series 2. HM Stationary Office, London, 533 p.

Beverton, R.J.H. and S.J. Holt. 1959. A review of the lifespans and mortality rates of fish in nature and their relation to growth and other physiological characteristics, p. 142–180 *In:* G.E.W. Wolstenholme and M. O'Connor (eds) *The Lifespan of Animals. Ciba Foundation Colloquia on Ageing*, Vol. 5. Churchill, London. DOI:10.1002/9780470715253.ch10

Bhattacharya, S. 1992. Endocrine control of fish reproduction. *Current Science,* 63(3): 135–139.

Bhattacharya, S. 1999. Recent advances in the hormonal regulation of gonadal maturation and spawning in fish. *Current Science,* 76: 342–349.

Bianucci, L., K. Fennel, D. Chabot, N. Shackell and D. Lavoie. 2016. Ocean biogeochemical models as management tools: a case study for Atlantic wolffish and declining oxygen. *ICES Journal of Marine Science,* 73(2): 263–274.

Brierley, A. S., B. E. Axelsen, E. Buecher, C. A. J. Sparks, H. Boyer and M. J. Gibbons. 2001. Acoustic observations of jellyfish in the Namibian Benguela. *Marine Ecology and Progress Series,* 210: 55–66.

Bigelow, K.A. 1993. Age and growth of the oceanic squid *Onychoteuthis borealjaponica* in the North Pacific. *Fisheries Bulletin,* 92: 13–25.

Bigman, J.S., N.C. Wegner and N.K. Dulvy. 2023. Gills, growth and activity across fishes. *Fish and Fisheries,* 24(5): 730–743.

Bigman, J.S., S.A. Pardo, T.S. Prinzing, M. Dando, N.C. Wegner and N.K. Dulvy. 2018. Ecological lifestyles and the scaling of shark gill surface area. *Journal of Morphology,* 279: 1716-1724. DOI:10.1002/jmor.20879

Binohlan, C. and D. Pauly. 2000. The POPGROWTH table, p. 138–145 *In:* R. Froese and D. Pauly (ed.) *FishBase 2000 Concepts, Design and Data Sources.* ICLARM, Los Baños, Philippines.

Bird, A.F. 1971. *The Stucture of Nematodes.* Academic Press, New York and London, 309 p.

Birk, M.A., A.K. Dymowka and B.A. Seibel. 2018. Do squid breathe though their skin? *Journal of Experimental Biology,* 221(19): jeb185553.

Bizikov, V.A. 1991. A new method of squid age determination using the gladius, p. 39–51 *In:* P. Jereb, S. Ragonese and S. Boletzky (eds) *Squid age determination using statolith.* Proceedings of an International Workshop, 9-14 October 1989, Mazara del Vallo, Italy. NTR-ITPP Special Publication No 1.

Blanchet, S., G. Grenouillet, O. Beauchard, P.A. Tedesco, F. Leprieur, H.H. Dürr, F. Busson, T. Oberdorff and S. Brosse. 2010. Non-native species disrupt the worldwide patterns of freshwater fish body size: implications for Bergmann's rule. *Ecology Letters,* 13(4): 421–431.

Blanckenhorn, W.U. and M. Demon. 2004. Bergmann and converse Bergmann latitudinal clines in arthropods: two ends of a continuum? *Integrative and Comparative Biology,* 44(6): 413-424.

Blasco, F.R., E.W. Taylor, C.A. Leite, D.A. Monteiro, F.T. Rantin and D.J. McKenzie. 2022. Tolerance of an acute warming challenge declines with body mass in Nile tilapia: Evidence of a link to capacity for oxygen uptake. *Journal of Experimental Biology,* 225(16): jeb244287.

Bleich, S., C.H.G. Müller, G. Graf and W. Hanke. 2017. Flow Generation by the Corona Ciliata in Chaetognatha – Quantification and Implications for Current Functional Hypotheses. *Zoology,* 125: 79–86. DOI:10.1016/j.zool.2017.09.001

Blumer, L.S. 1979. Male parental care in the bony fishes. *The Quarterly Review of Biology,* 54(2): 149–161.

Blumer, L.S. 1986. Parental care sex differences in the brown bullhead, *Ictalurus nebulosus* (Pisces, Ictaluridae). *Behavioral Ecology and Sociobiology,* 19: 97–104.

Blunt, J. W., B.R. Copp, M.H.G. Munro, P.T. Northcote and M.R. Prinsep. 2003. Marine natural products. *Natural Product Report,* 20:1–48. DOI:10.1039/B207130B

Bochdansky, A.B. and W. Leggett. 2001. Winberg revisited: convergence of routine metabolism in larval and juvenile fish. *Canadian Journal of Fisheries and Aquatic Sciences,* 58: 220–230.

Boehlert, G.W., M. Kusakari and J. Yamada. 1991. Oxygen consumption of gestating female *Sebastes schlegeli*: estimating the reproductive costs of live-bearing, p. 81–90 *In:* G.W. Boehlert and J. Yamada (eds) *Rockfishes of the genus* Sebastes *their reproduction and early life history.* Springer, Dordrecht.

Boëly, T. 1979. Biologie de deux espèces de sardinelles (*Sardinella aurita* Valenciennes 1887 et *Sardinella maderensis* Lowe 1841) des côtes sénégalaises. *La Pêche maritime.* Juillet 1979: 426–430.

Boëly, T., J. Chabanne and P. Fréon. 1978. Schéma migratoire de poissons pélagiques côtiers dans la zone sénégalo-mauritanienne. p. 63–70 *In: Rapport du groupe de travail ad hoc sur les poissons pélagiques côtiers ouest africains de la Mauritanie au Liberia* (26° N à 5° N). FAO/Comité des pêches pour l'Atlantique. Centre-Est — COPACE/ RACE, Sér. 78/10.

Bojkowska, K., de Sio, F.S., Barde, I., Offner, S., Verp, S., Heinis, C., Johnsson, K. and D. Trono. 2011. Measuring in vivo protein half-life. *Chemistry & Biology*, 18(6): 805-815.

Boltzmann, L. 1884. Ableitung des Stefan'schen Gesetzes, betreffend die Abhängikeit der Wärmestrahlung von der Temperatur aus der elektromagnetischen Licht-theorie. *Annalen der Physik und Chemie,* 258: 291-294.

Bonds, M.H. 2006. Host life-history strategy explains pathogen-induced sterility. *American Naturalist,* 168: 281–293.

Bone, Q., H. Kapp and A.C. Pierrot-Bults (Editors). 1991. *The Biology of Chaetognaths.* Oxford University Press, Oxford, 184 p.

Bone, Q. and M. Duvert. 1991. Locomotion and buoyancy, p. 32–44 *In:* G. Bone, H. Kapp and A.C. Pierrot-Bults (eds) *The Biology of Chaetognaths.* Oxford University Press, Oxford.

Bos, A. and R. Ferwerda. 2008. *Aristotle, On the Life-Bearing Spirit (De spiritu): A Discussion with Plato and his Predecessors on Pneuma as the Instrumental Body of the Soul.* Introduction, Translation, and Commentary by Abraham P. Bos and Rein Ferwerda. Brill, Leiden and Boston, 209.

Botton, M.L. and C.N. Shuster. 2003. Horseshoe crab in a food web: who eats whom? p. 133-151 *In:* C.N. Shuster, R.B. Barlow and H.J. Brockman (eds) *The American Horseshoe Crab*. Harvard University Press, Cambridge.

Botton, M.L. 1984. Diet and food preferences of the adult horseshoe crab *Limulus polyphemus* in Delaware Bay, New Jersey, USA. *Marine Biology,* 81(2): 199-207.

Boucher-Rodoni, R. 1973. Nutrition, digestion et transfert énergétique chez les céphalopodes *Eledone cirrosa* (Lamarck) et *Illex illecebrosus* (Lesueur). Doctoral dissertation, Université de Genève, 96 p.

Bouwma, P.E. 2006. *Aspects of antipredation in* Panulirus argus *and* Panulirus guttatus*: behaviour, morphology, and ontogeny.* Doctoral dissertation, Florida State University, Tallahassee, Florida.

Bowden, A.J., S.J. Andrewartha, N.G. Elliott, P.B. Frappell and T.D. Clark. 2018. Negligible differences in metabolism and thermal tolerance between diploid and triploid Atlantic salmon (*Salmo salar*). *Journal of Experimental Biology,* 221, jeb166975. DOI:10.1242/jeb.166975

Bowden, A.J., N.M. Gardiner, C.S. Couturier, J.A.W. Stecyk, G.E. Nilsson, P.L. Munday and J.L. Rummer. 2014. Alterations in gill structure in tropical reef fishes as a result of elevated temperatures. *Comparative Biochemistry and Physiology A,* 175: 64–71.

Boyle, R. 1662. *New Experiments Physico-Mechanical, Touching the Spring of the Air, and its Effects*. Oxford.

Brandts, T.F. 1967. Heat effects on proteins and enzymes, p. 25–72 *In:* A.E. Rose (ed.) *Thermobiology*. Academic Press, London and New York.

Brauner, C.J., V. Matey, J.M. Wilson, N.J. Bernier and A.L. Val. 2004. Transition in organ function during the evolution of air-breathing; insights from *Arapaima gigas*, an obligate air-breathing teleost from the Amazon. *Journal of Experimental Biology,* 207(9): 1433–1438.

Breen, P.A. 1994. Population dynamics and stock assessment of lobsters: a review. *Crustaceana,* 67(2): 239–255. DOI:10.1163/156854094X00602

Breder, C.M. and D.E. Rosen 1966. *Modes of reproduction in fishes*. T.F.H. Publications, Neptune City, New Jersey. 941 p.

Brett, J.R. 1972. The metabolic demand for oxygen in fish, particularly salmonids, and a comparison with other vertebrates. *Respiration Physiology,* 14: 151- 170.

Brett, J.R. 1964. The respiratory metabolism and swimming performance of young sockeye salmon. *Journal of the Fisheries Research Board of Canada,* 21: 1183–1226.

Brett, J.R. and T.D.D. Groves. 1979. Physiological energetics. *Fish Physiology,* 8(6): 280–352.

Bridges, C.R. and A.R. Brand. 1980. Oxygen consumption and oxygen-independence in marine crustaceans. *Marine Ecology Progress Series,* 2: 133–141.

Bringloe, T.T., S. Starko, R.M. Wade, C. Vieira, H. Kawai, O. De Clerck, J.M. Cock, S.M. Coelho, C. Destombe, M. Valero and J. Neiva. 2020. Phylogeny and evolution of the brown algae. *Critical Reviews in Plant Sciences* 39(4): 281-321.

Britz, R. and K.W. Conway. 2009. Osteology of *Paedocypris*, a miniature and highly developmentally truncated fish (Teleostei: Ostariophysi: Cyprinidae). *Journal of Morphology,* 270(4): 389–412. DOI:10.1002/jmor.10698

Brocks, J.J., A.J.M. Jarrett, E. Sirantoine, C. Hallmann, Y. Hoshino and T. Liyanage. 2017. The rise of algae in Cryogenian oceans and the emergence of animals. *Nature,* 548: 578–581. DOI:10.1038/nature23457

Brody, S. 1945. *Bioenergetics and Growth*. Hafner, New York. 1023 p.

Brody, S. and R.C. Procter. 1932. Relation between basal metabolism and mature body weight in different species of mammals and birds. *Universiy of Missouri College of Agriculture Agricultural Experimental Station Research Bulletin* 116: 89–101.

Broertjes, J.J.S., J. Jens, D. van Oudheusden and P. Voogt. 1982. Demonstration of nutrient flow in the starfish *Asterias rubens* (L.) with 1251 labelled proteins. *Netherlands Journal of Zoology,* 32: 472–478.

Brokordt, K., H. Pérez and F. Campos. 2013. Environmental hypoxia reduces the escape response capacity of juvenile and adult scallops *Argopecten purpuratus*. *Journal of Shellfish Research,* 32(2): 39–376.

Brooks, G., J. Uyeda, N. Bone, H. Conrad, C. Mull and H. Kindsvater. 2025. Fundamental constraints on the evolution of vertebrate life histories. *Nature Ecology & Evolution 9:* 857–866.

Broussonet, P.M.A. 1785. Mémoire pour servir à l'histoire de la respiration des Poissons. *Mémoires de l'Académie Royale des Sciences,* p. 174–196.

Brown, M.F., T.P. Gratton and J.A. Stuart. 2007. Metabolic rate does not scale with body mass in cultured mammalian cells. *American Journal of Physiology-Regulatory, Integrative and Comparative Physiology,* 292(6): R2115-R2121.

Bruneaux, M., M. Nikinmaa, V.N. Laine, K. Lindström, C.R. Primmer and A. Vasemägi. 2014. Differences in the metabolic response to temperature acclimation in nine-spined stickleback (*Pungitius pungitius*) populations from contrasting thermal environments. *Journal of Experimental Zoology Part A: Ecological Genetics and Physiology,* 321: 550–565.

Brunner, N. and M. Kühleitner. 2020. The growth of domestic goats and sheep: A meta study with Bertalanffy-Pütter models. *Veterinary and Animal Science,* 10: 100135.

Brunner, N., M. Kühleitner and K. Renner-Martin. 2021. Bertalanffy-Pütter models for avian growth. *PloS ONE,* 16(4): e0250515.

Brunnschweiler, J.M., H. Baensch, S.J. Pierce and D.W. Sims. 2009. Deep-diving behaviour of a whale shark *Rhincodon typus* during long-distance movement in the western Indian Ocean. *Journal of Fish Biology,* 74: 706–714

Brewer, R. H. 1989. The annual pattern of feeding, growth, and sexual reproduction in *Cyanea (Cnidaria*: Scyphozoa) in the Niantic River Estuary, Connecticut. *Biological Bulletin,*176: 272–281.

Buchmann, K. 1989. Relationship between host size of *Anguilla anguilla* and the infection level of the monogeneans *Pseudodactylogyrus* spp. *Journal of Fish Biology,* 35(4): 599–601.

Budaev S., C. Jørgensen, M. Mangel, S. Eliassen and J. Giske. 2019. Decision-making from the animal perspective: bridging ecology and subjective cognition. *Frontiers in Ecology and Evolution,* 7: 164. DOI:10.3389/fevo.2019.00164

Buecher, E., C. Sparks, A. Brierley, H. Boyer and M. Gibbons. 2001. Biometry and size distribution of *Chrysaora hysoscella* (Cnidaria, Scyphozoa) and *Aequorea aequorea* (Cnidaria, Hydrozoa) off Namibia with some notes on their parasite Hyperia medusarum. *Journal of Plankton Research,* 23: 1073–1080.

Bullock, T.H. 1965. Chaetognatha, Pogonophora, Hemichordata, and Chordata Tunicata, p. 1559–1592 *In:* T.H. Bullock and G.A. Horridge (eds) *Structure and Function in the Nervous Systems of Invertebrates,* Vol. 2. W.H. Freeman, San Francisco.

Bullock, R.W., C. Foster and J.S. Lea. 2024. Just keep swimming? Observations of resting behavior in gray reef sharks *Carcharhinus amblyrhynchos* (Bleeker, 1856). *Journal of Fish Biology*, 104(3): 898-900.

Burness, G.P., S.C. Leary, P.W. Hochachka and C.D. Moyes. 1999. Allometric scaling of RNA, DNA, and enzyme levels: an intraspecific study. *American Journal of Physiology – Regulatory, Integrative and Comparative Physiology,* 277: R1164–R1170. DOI:10.1152/ajpregu.1999.277.4.R1164

Butler, J. 2002. What is critique? An essay on Foucault's virtue, p. 301–321 *In:* S. Salih and J. Butler (eds) *The Judith Butler Reader*. Blackwell. Malden, MA.

Butler, M.J., D.C. Behringer and M.M. Valentine. 2017. Commercial sponge fishery impacts on the population dynamics of sponges in the Florida Keys, FL (USA). *Fisheries Research,* 190: 113–121.

Butler, J., W.C. Sharp, J.H. Hunt and M.J. Butler IV. 2021. Setting the foundation for renewal: restoring sponge communities aids the ecological recovery of Florida Bay. *Ecosphere,* 12(12): e03876. DOI:10.1002/ecs2.3876

Butler, M. J., IV, J. B. Weisz and J. Butler. 2018. The effects of water quality on back-reef sponge survival and distribution in the Florida Keys, Florida (USA). *Journal of Experimental Marine Biology and Ecology,* 503: 92–99.

Bykov. V.P. 1983. *Marine Fishes: Chemical Composition and Processing Properties.* Amerind Publishing, New Delhi, 322 p.

Caldwell, L.K., A.L. Pierce and J.J. Nagler. 2013. Metabolic endocrine factors involved in spawning recovery and rematuration of iteroparous female rainbow trout (*Oncorhynchus mykiss*). *General and Comparative Endocrinology,*194: 124–132.

Campbell, T., K. Pin, P.B. Ngor and Z. Hogan. 2020. Conserving Mekong megafishes: Current status and critical threats in Cambodia. *Water* 12(6): 1820.

Cannell, B.L., P.G. Thomson, V. Schoepf, C.B. Pattiaratchi and M.W. Fraser. 2019. Impacts of marine heatwaves, p. 123–140 *In:* E. Tetchera and G. Winter (eds) *Marine Extremes.* Routledge, London.

Caputi, N., R. Melville-Smith, S. de Lestang, Pearce, A. and M. Feng. 2010. The effect of climate change on the western rock lobster (*Panulirus cygnus*) fishery of Western Australia. *Canadian Journal of Fisheries and Aquatic Sciences,* 67(1): 85–96. DOI:10.1139/F09-167

Cardillo, M., G.M. Mace, G.M. Jones, J. Bielby, O.R. Bininda-Emonds, W. Sechrest, C.D.L. Orme and A. Purvis. 2005. Multiple causes of high extinction risk in large mammal species. *Science,* 309: 1239–1241.

Carpenter, S.R., S.G. Fisher, N.B. Grimm and J.F. Kitchell. 1992. Global change and freshwater ecosystems. *Annual Review of Ecology and Systematics,* 23(1): 119–139.

Carter, A.M. and H. Soma. 2020. Viviparity in the longest-living vertebrate, the Greenland shark (*Somniosus microcephalus*). *Placenta,* 97: 26–28. DOI:10.1016/j.placenta.2020.05.014

Cavalier-Smith, T. 1991. Coevolution of vertebrate genome, cell and nuclear sizes, p. 51-86 *In:* G. Ghiara, F. Angelin, E. Olmo and L. Varano (eds) *Symposium on the Evolution of Terrestrial Vertebrates*. Selected Symposia and Monographs - Collana U.Z.I. 4. Mucchi Editore, Modena.

Cashion, T., F. Le Manach, D. Zeller and D. Pauly. 2017. Most fish destined for fishmeal production are food-grade fish. *Fish and Fisheries,* 18(5): 837-844.

Cavender, T.M. and R.R. Miller. 1972. *Smilodonichthys rastrosus*: a new Pliocece salmonid fish from western United States. *University of Oregon Natural History Bulletin,* 18:1-44.

Cavero, B.A.S., M. Pereira-Filho, R. Roubach, D.R. Ituassú, A.L. Gandra and R. Crescêncio. 2003. Stocking density effect on alimentary efficiency in juveniles pirarucu (*Arapaima gigas*) in a confined environment. *Acta Amazonica,* 33: 631–636.

Cénit, M.C., V. Matzaraki, E.F. Tigchelaar and A. Zhernakova. 2014. Rapidly expanding knowledge on the role of the gut microbiome in health and disease. *Biochemica et Biophysica Acta,* 1842: 1981-1992. DOI:10.1016/j.bbadis.2014.05.023

Chabot, D., J.F. Steffensen and A.P. Farrell. 2016. The determination of standard metabolic rate in fishes. *Journal of Fish Biology,* 88(1): 81–121.

Champagnat, C. and F. Domain. 1978. Migrations des poissons démersaux le long des côtes ouest-africaines de 10 à 24° de latitude nord. *Cahier ORSTOM, Série Océanographie,* 16(3/4): 239–261.

Chapelle, G. and L.S. Peck. 1999. Polar gigantism dictated by oxygen availability. *Nature,* 399: 114–115.

Chapelle, G. and L.S. Peck. 2004. Amphipod crustacean size spectra: new insights in the relationship between size and oxygen. *Oikos,* 106: 167–175. DOI:10.1111/j.0030-1299.2004.1294.x

Charnov, E.L. 1993. *Life History Invariants: Some Explorations of Symmetry in Evolutionary Ecology.* Oxford University Press, New York, 182 p.

Charnov, E.L. 2008. Fish growth: Bertalanffy *k* is proportional to reproductive effort. *Environmental Biology of Fishes,* 83: 185–187.

Chatterji, A, J.K. Mishra and A.H. Parulekar. 1992. Feeding behaviour and food selection in the horseshoe crab, *Tachypleus gigas* (Müller). *Hydrobiologia,* 246(1): 41-48.

Chen, Z., J. Bigman, W. Xian, C. Liang, E. Chu and D. Pauly. 2021. The ratio of length at first maturity to maximum length in marine and freshwater fish. *Journal of Fish Biology,* 101(2): 401-407. DOI:10.1111/jfb.14970

Chen, Z., M. Snow, C.S. Lawrence, A.R. Church, S.R. Narum, R.H. Devlin and A.P. Farrell. 2015. Selection for upper thermal tolerance in rainbow trout (*Oncorhynchus mykiss* Walbaum). *Journal of Experimental Biology,* 218: 803-812.

Cheung, W.W.L., C. Close, V.W.Y. Lam, R. Watson and D. Pauly. 2008a. Application of macroecological theory to predict effects of climate change on global fisheries potential. *Marine Ecology Progress Series,* 365: 187-193.

Cheung, W.W.L., V.W.Y. Lam and D. Pauly. 2008b. Dynamic bioclimate envelope model to predict climate-induced changes in distributions of marine fishes and invertebrates, p. 5–50 *In* W.W.L. Cheung, V.W.Y. Lam and D. Pauly (eds) *Modelling Present and Climate-shifted Distribution of Marine Fishes and Invertebrates.* Fisheries Centre Research Reports 16(3). University of British Columbia, Vancouver.

Cheung, W.W.L., V.W.Y. Lam, J.L. Sarmiento, K. Kearney, R. Watson and D. Pauly. 2009. Projecting global marine biodiversity impacts under climate change scenarios. *Fish and Fisheries,* 10: 235–251.

Cheung, W.W.L., V.W.Y. Lam, J.L. Sarmiento, K. Kearney, R. Watson, D. Zeller and D. Pauly. 2010. Large-scale redistribution of maximum fisheries catch potential in the global ocean under climate change. *Global Change Biology,* 16: 24-35.

Cheung, W.W.L., D. Pauly and J. Sarmiento. 2013a. How to make scientific progress in projecting climate change impacts. *ICES Journal of Marine Sciences,* 70(6): 1069-1074.

Cheung, W.W.L., R. Watson and D. Pauly. 2013c. Signature of ocean warming in global fisheries catch. *Nature,* 497: 365–368.

Cheung, W.W.L., J.L. Sarmiento, J. Dunne, T.L. Frölicher, V.W.Y. Lam, M.L.D. Palomares, R. Watson and D. Pauly. 2013b. Shrinking of fishes exacerbates impacts of global ocean changes on marine ecosystems. *Nature Climate Change,* 3: 254-258. DOI:10.1038/nclimate1691

Chu, E. and D. Pauly. 2024a. A revision of 'Key information on 25 species of sturgeon; Family Acipenseridae,' p. 11-22 *In:* D. Pauly and V. Ruiz-Leotaud (eds) *Marine and Freshwater Miscellanea V.* Fisheries Centre Research Reports 30(4). University of British Columbia, Vancouver.

Chu, E. and D. Pauly. 2024b. The relationship between mean length at maturity and maximum length in coral reef fishes. *Fishes,* 9(4): 130. DOI:10.3390/fishes9040130

Claeson, K.M., B.L. Sidlauskas, R. Troll, Z.M. Prescott and E.B. Davis. 2024. From sabers to spikes: A newfangled reconstruction of the ancient, giant, sexually dimorphic Pacific salmon,† *Oncorhynchus rastrosus* (Salmoninae: Salmonini). *PLoS ONE,* 19(4), e0300252.

Claessen, D., A.M. de Roos and L. Persson. 2000. Dwarfs and giants: Cannibalism and competition in size-structured populations. *American Naturalist,* 155(2): 219-237.

Claessen, D., C. Van Oss, A.M. de Roos and L. Persson. 2002. The impact of size-dependent predation on population dynamics and individual life history. *Ecology,* 83(6): 1660–1675.

Claireaux, G., C. Couturierand and A.L. Groison. 2006. Effect of temperature on maximum swimming speed and cost of transport in juvenile European sea bass (*Dicentrarchus labrax*). *Journal of Experimental Biology,* 209: 3420–3428.

Claireaux, G., D.M.Webber, J.P. Lagardère and S.R. Kerr. 2000. Influence of water temperature and oxygenation on the aerobic metabolic scope of Atlantic cod (*Gadus morhua*). *Journal of Sea Research,* 44: 257–265.

Clark, T.D., E. Sandblom, G.K. Cox, S.J. Hinch and A.P. Farrell. 2008. Circulatory limits to oxygen supply during an acute temperature increase in the Chinook salmon (*Oncorhynchus tshawytscha*). *American Journal of Physiology,* 295: R1631–R1639.

Clark, CJ., J.R. Hutchinson and T. Garland. 2023. The inverse Krogh Principle: all organisms are worthy of study. *Physiological and Biochemical Zoology,* 96(1): 1-16.

Clark, T.D., W.T. Brandt, J. Nogueira, L.E. Rodriguez, M. Price, C.J. Farwell and B.A. Block. 2010. Postprandial metabolism of Pacific bluefin tuna (*Thunnus orientalis*). *Journal of Experimental Biology,* 213: 2379–2385.

Clark, T.D., M.R. Donaldson, S. Pieperhoff, S.M. Drenner, A. Lotto, S.J. Cooke, S.J. Hinch, D.A. Patterson and A.P. Farrell. 2012. Physiological benefits of being small in a changing world: responses of coho salmon (*Oncorhynchus kisutch*) to an acute thermal challenge and a simulated capture event. *PLoS ONE,* 7(6): e39079.

Clark, T.D., C.J. Farwell, L.E. Rodriguez, W.T. Brandt and B.A. Block. 2013. Heart rate responses to temperature in free-swimming Pacific bluefin tuna (*Thunnus orientalis*). *Journal of Experimental Biology,* 216: 3208–3214.

Clark, T.D., K.M. Jeffries, S.G. Hinch and A.P. Farrell. 2011. Exceptional aerobic scope and cardiovascular performance of pink salmon (*Oncorhynchus gorbuscha*) may underlie resilience in a warming climate. *Journal of Experimental Biology,* 214: 3074–3081

Clarke, A. 1980a. A reappraisal of the concept of metabolic cold adaptation in polar marine invertebrates. *Biological Journal of the Linnean Society,* 14: 77–92.

Clarke, A. 1983. Life in cold water: the physiological ecology of polar marine ectotherms. *Oceanography and Marine Biology Annual Review,* 21: 351–453.

Clarke, A. 1991. What is cold adaptation and how should we measure it? *American Zoologist,* 31: 81–92.

Clarke, A. 1993. Seasonal acclimatization and latitudinal compensation in metabolism: do they exist? *Functional Ecology,* 7: 139–149.

Clarke, A. and N. Johnson. 1999. Scaling of metabolic rate with body mass and temperature in teleost fish. *Journal of Animal Ecology,* 69:893–905.

Clarke, M.R. 1980b. *Cephalopods in the Diet of Sperm Whales of the Southern Hemisphere and their Bearing on Sperm Whale Biology.* Discovery Reports 37, Cambridge University Press, London, 324 p.

Clauset, A. and D.H. Erwin. 2008. The evolution and distribution of species body size. *Science,* 321: 399-401.

Cochran, J.E.M., R.S. Hardenstine, C.D. Braun, G.B. Skomal, S.R. Thorrold, K. Xu, M. G. Genton and M.L. Berumen. 2016. Population structure of a whale shark *Rhincodon typus* aggregation in the Red Sea. *Journal of Fish Biology,* 89: 1570–1582. DOI:10.1111/jfb.13054

Cochran, J.K. and N.H. Landman. 1984. Radiometric determination of the growth rate of *Nautilus* in nature. *Nature,* 308: 725–727.

Cohen, J.M., M.J. Lajeunesse and J.R. Rohr. 2018. A global synthesis of animal phenological, responses to climate change. *Nature Climate Change,* 8: 224–228. DOI:10.1038/s41558-018-0067-3

Cohen, A.C. 1976. The systematics and distribution of *Loligo* in the Western North Atlantic, with description of two new species. *Malacologica,* 15: 299–367.

Constance, J.M., E.A. Garcia, R.D. Pillans, V. Udyawer and P.M. Kyne. 2023. A review of the life history and ecology of euryhaline and estuarine sharks and rays. *Rev Fish Biol Fisheries,* 1-25. DOI:10.1007/s11160-023-09807-1

Conway Morris, S. 2003. *Life's solution: inevitable humans in a lonely universe.* Cambridge University Press, Cambridge, 464 p.

Cooper, K. 2025. *Amazing Worlds of Science Fiction and Science Fact.* Reaktion Books, London, 248 p.

Correa-González, J.C., M. del Carmen, Chávez-Parga, J.A. Cortés and R.M. Pérez-Munguía. 2014. Photosynthesis, respiration, and reaeration in a stream with complex dissolved oxygen patterns and temperature dependence. *Ecological Modelling,* 273: 220–227.

Cortés, E. 2000. Life History Patterns and Correlations in Sharks. *Reviews in Fisheries Science* 8: 299–344. DOI:10.1080/10408340308951115

Costello, M.J., R. Corkrey, A.E. Bates, M.T. Burrows, C. Chaudhary, G.E. Edgar, R.D. Stuart-Smith, M. Yasuhara and C.L. Wei. 2023. The universal evolutionary and ecological significance of 20 °C. *Frontiers of Biogeography,* 15(4), e61673.

Coutant, C.C. 1985. Striped bass, temperature, and dissolved oxygen: a speculative hypothesis for environmental risk. *Transactions of the American Fisheries Society,* 114: 31– 61.

Coutant, C.C. 1986. Thermal niches of striped bass. *Scientific American,* 255(2): 98–105.

Coutant, C.C. 1990. Temperature-oxygen habitat for freshwater and coastal striped bass in a changing climate. *Transactions of the American Fisheries Society,* 119: 240–253.

Coutant, C.C. and D.S. Carroll. 1980. Temperatures occupied by ten ultrasonic-tagged striped bass in freshwater lakes. *Transactions of the American Fisheries Society,* 109: 195–202.

Cramer, J.L., R.M. Nakamura, A.E. Dizon and W.N. Ikehara. 1981. Burnt tuna: conditions leading to rapid deterioration in the quality of raw tuna. *Marine Fisheries Review,* 43: 12–16.

Cramp, R.L., M.J. Hansen and C.E. Franklin. 2015. Osmoregulation by juvenile brown-banded bamboo sharks, *Chiloscyllium punctatum*, in hypo- and hyper-saline waters. *Comparative Biochemistry and Physiology,* A 185: 107–114. DOI:10.1016/j.cbpa.2015.04.001

Crear, B.J. and G.N.R. Forteath. 2000. The effect of extrinsic and intrinsic factors on oxygen consumption by the southern rock lobster, *Jasus edwardsii*. *Journal of Experimental Marine Biology and Ecology,* 252: 129–147.

Crête-Lafrenière, A., L.K. Weir and L. Bernatchez. 2012. Framing the Salmonidae family phylogenetic portrait: a more complete picture from increased taxon sampling. *PLoS ONE*, e46662.

Crespi, B.J. and R. Teo. 2002. Comparative phylogenetic analysis of the evolution of semelparity and life history in salmonid fishes. *Evolution* 56(5): 1008-1020.

Cueto-Vega, R., J. Flye-Sainte-Marie, A. Aguirre-Velarde, F. Jean, P. Gil-Kodaka and G. Thouzeau. 2022. Size-based survival of cultured *Argopecten purpuratus* (L, 1819) under severe hypoxia. *Journal of the World Aquaculture Society,* 53(1): 151-173.

Cury, P. 1994. Obstinate nature: an ecology of individuals. Thoughts on reproductive behavior and biodiversity. *Canadian Journal of Fisheries and Aquatic Sciences,* 51(7): 1664–1673.

Cushing, D.H. 1990. Plankton production and year-class strength in fish populations: An update of the Match/Mismatch Hypothesis. *Advances in Marine Biology,* 26: 249–293. DOI:10.1016/S0065-2881(08)60202–3

Cushing, D.H. 1981. *Fisheries Biology: A Study in Population Dynamics,* 2nd edition. University of Wisconsin Press, Madison, 295 p.

Cutts, C.J., N.B. Metcalfe and A.C. Taylor. 2002. Juvenile Atlantic salmon (*Salmo salar*) with relatively high standard metabolic rates have small metabolic scopes. *Functional Ecology,* 16: 73–78.

Dabrowski, K.R. 1985. Energy budget of coregonid (*Coregonus* spp.) fish growth, metabolism and reproduction. *Oikos,* 45: 358–364.

Dahlke, F.T., S. Wohlrab, M. Butzin and H.O. Pörtner. 2020. Thermal bottlenecks in the life cycle define climate vulnerability of fish. *Science,* 369(6499): 65–70.

Darwin, C. 1859. *On the Origin of Species by Means of Natural Selection, or the Preservation of Favoured Races in the Struggle for Life.* John Murray, London, 502 p.

Darwin, C. 1844. Observations on the structure and propagation of the genus *Sagitta. Annals and Magazine of Natural History,* 13: 1–6. DOI:10.1080/03745484409442559

Darwin, C.R. 1861. *On the Origin of Species by Means of Natural Selection, or the Preservation of Favoured Races in the Struggle for Life,* 3rd edition. John Murray, London, 538 p.

Darwin, C.R. 1872. *The Origin of Species,* 6th edition, with additions and corrections, John Murray, London, 458 p.

Daryanani, D.S., J.C. Martino and Z.A. Doubleday. 2021. Statolith chemistry: a new tool to understand the ecology and provenance of octopus. *Reviews in Fish Biology and Fisheries,* 31: 923–934.

Daufresne, M., K. Lengfellner and U. Sommer. 2009. Global warming benefits the small in aquatic ecosystems. *Proceedings of the National Academy of Sciences,* 106: 12788 – 12793.

David, P.M. 1955. The Distribution of *Sagitta gazellae* Ritter-Záhony. *Discovery Reports,* 27: 235–278.

Davies, R. and C.D. Moyes. 2007. Allometric scaling in centrarchid fish: origins of intra- and inter-specific variation in oxidative and glycolytic enzyme levels in muscle. *Journal of Experimental Biology,* 210: 3798–3804. DOI:10.1242/jeb.00389.

Davies, R.W., M.A. Monita, E. Dratnal and L.R. Linton. 1992. The effect of different dissolved oxygen regime on the growth of a freshwater leech. *Ecography,* 15: 190–194. DOI:10.1111/j.1600-0587.1992.tb00023.x

Davison, I.R., R.M. Greene and E.J. Podolak. 1991. Temperature acclimation of respiration and photosynthesis in the brown alga *Laminaria saccharina. Marine Biology,* 110: 449-454.

Dawes, J. 1998. *Aquaculture: the Claims Story*. MMI General Insurance Ltd, Sydney, 9 p.

Dawkins, R. and T.R. Carlisle. 1976. Parental investment, mate desertion and a fallacy. *Nature,* 262: 131–133.

Day, T. and P.D. Taylor. 1997. Von Bertalanffy's growth equation should not be used to model age and size at maturity. *American Naturalist,* 149(2): 381–393.

Debelius, B., A. Gómez-Parra and J.M. Forja. 2009. Oxygen solubility in evaporated seawater as a function of temperature and salinity. *Hydrobiologia,* 632: 157-165.

De Jager, S. and W.J. Dekkers. 1974. Relations between gill structure and activity in fish. *Netherlands Journal of Zoology,* 25: 276–308. DOI:10.1163/002829675X00290

De Oliveira, E.G., F.J. de Seixas Santos, V.Q. de Oliveira, P.E.C. Mesquista, M.G. de Moraes, R.R. de Sousa, I.G.R.F. de Gomes, L.F.G. de Albuquerque, I.R.C.B. de Rocha and F.H.F. Costa. 2020. Influence of stocking density on water quality and growth performance in production of juvenile pirarucu, *Arapaima gigas*, in irrigation canals. *Brazilian Journal of Development,* 6(2): 8725–8743.

De Sousa Rangel, B., N.E. Hussey, Y. Niella, L.A. Martinelli, A.D.P. Gomes and R.G. Moreira. 2020. Neonatal nutritional strategy of a viviparous elasmobranch with extremely low reproductive output. *Marine Ecology Progress Series,* 638: 107–121. DOI:10.3354/meps13261

De Silva, D.P. 1974. Development of the Respiratory System in Herring and Plaice Larvae, p. 465–485 *In:* J.H.S. Blaxter (ed.). *The Early Life History of Fishes.* Springer, Berlin and New York.

Degani, G., M.L. Gallagher and A. Meltzer. 1989. The influence of body size and temperature on oxygen consumption of the European eel, *Anguilla anguilla. Journal of Fish Biology,* 34: 19–24.

Denny, M.W. 1993. *Air and Water: the Biology and Physics of Life's Media.* Princeton University Press, Princeton, 342 p.

Deutsch, C., J.L. Penn and N. Lucey. 2024. Climate, oxygen, and the future of marine biodiversity. *Annual Review of Marine Science,* 16: 217–245.

Dewdney, A.K. 1984. *The Planiverse. Computer Contacts with a Two-dimensional World.* Picador - Pan Books, London, 267 p.

Diaz Pauli, B., J. Kolding, G. Jeyakanth and M. Heino. 2017. Effects of ambient oxygen and size-selective mortality on growth and maturation in guppies. *Conservation Physiology*, 5(1). DOI:10.1093/conphys/cox010

Dice, J.F. 1987. Molecular determinants of protein half-lives in eukaryotic cells. *The FASEB Journal*, 1(5): 349-357.

Dickson, I.W. and R.H. Kramer. 1971. Factors influencing scope for activity and active and standard metabolism of rainbow trout (*Salmo gairdneri*). *Journal of the Fisheries Research Board of Canada*, 28: 587–596.

Dimarchopoulou, D. and A.C. Tsikliras. 2022. Linking growth patterns to sea temperature and oxygen levels across European sardine (*Sardina pilchardus*) populations. *Environmental Biology of Fishes*, 105: 1335-1345. DOI:10.1007/s10641-022-01229-5

Dimarchopoulou, D., M. Makino, M.R. Prayoga, D. Zeller, G.M.S. Vianna and Humphries, A.T. 2022. Responses in fisheries catch data to a warming ocean along a latitudinal gradient in the western Pacific Ocean. *Environmental Biology of Fishes*, 1347-1362. DOI:10.1007/s10641-021-01162-z

Distasio, R., V. Gobre and Tkatchenko, A. 2014. Many-body van der Waals interactions in molecules and condensed matter. *Journal of Physics: Condensed Matter*, 26: 213202. DOI:10.1088/0953-8984/26/21/213202

Doake, S. and R.W. Elwood. 2011. How resource quality differentially affects motivation and ability to fight in hermit crabs. *Proceedings of the Royal Society B: Biological Sciences*, 278(1705): 567-573.

Dobzhansky, T. 1973. Nothing in biology makes sense except in the light of evolution. *The American Biology Teacher*, 75(2): 87–91.

Donelson, J.M. and P.L. Munday. 2012. Thermal sensitivity does not determine acclimation capacity for a tropical reef fish. *Journal of Animal Ecology*, 81: 1126–1131.

Donelson, J.M. 2015. Development in a warm future ocean may enhance performance in some species. *Journal of Experimental Marine Biology and Ecology*, 472: 119–125.

Doonan, I.J. 1994. *Life history parameters of orange roughy: Estimates for 1994*. New Zealand Fisheries Assessment Research Document 94/19. Unpublished report held in NIWA Library, Wellington, New Zealand.

Doroshov, S.I., G.P. Moberg and J.P. Van Eenennaam. 1997. Observations on the reproductive cycle of cultured white sturgeon, *Acipenser transmontanus*. *Environmental Biology of Fishes*, 48(1): 265–278. DOI:10.1023/A:1007336802423

Drack, M. 2015. Ludwig von Bertalanffy's organismic view on the theory of evolution. *Journal of Experimental Zoology Part B: Molecular and Developmental Evolution*, 324(2): 77–90.

Drew, M.M., S. Harzsch, M. Stensmyr, S. Erland and B.S. Hansson. 2010. A review of the biology and ecology of the robber crab, *Birgus latro* (Linnaeus, 1767) (Anomura: Coenobitidae). *Zoologischer Anzeiger - A Journal of Comparative Zoology*, 249(1): 45–67. DOI:10.1016/j.jcz.2010.03.001

Dudley, R. and C. Gans. 1991. A critique of symmorphosis and optimality models in physiology. *Physiological Zoology* 64(3): 627–637.

Duponchelle, F. and J. Panfili. 1998. Variations in age and size at maturity of female Nile tilapia, *Oreochromis niloticus*, populations from man-made lakes of Côte d'Ivoire. *Environmental Biology of Fishes*, 52(4): 453–465. DOI:10.1023/A:1007453731509

Duskey, E. 2023. Metabolic prioritization of fish in hypoxic waters: an integrative modeling approach. *Frontiers in Marine Science*, 10, 1206506.

Duthie, G.G. and G.M. Hughes. 1982. Some effects of gill damage on the swimming performance of rainbow trout (*Salmo gairdneri*). *Journal of Physiology (London)*, 327: 21–22.

Duthie, G.G. 1982. The respiratory metabolism of temperature-adapted flatfish at rest and during swimming activity and the use of anaerobic metabolism at moderate swimming speeds. *Journal of Experimental Biology*, 97: 359–373.

Dutilloy, A. and M.R. Dunn. 2020. Observations of sperm storage in some deep-sea elasmobranchs. *Deep Sea Research Part I: Oceanographic Research Papers*, 166:103405. DOI:10.1016/j.dsr.2020.103405

Dwyer, R.G., H.A. Campbell, R.L. Cramp, C.L. Burke, M.A. Micheli-Campbell, R.D. Pillans, B.J. Lyon and C.E. Franklin. 2020. Niche partitioning between river shark species is driven by seasonal fluctuations in environmental salinity. *Functional Ecology*, 34: 2170–2185. DOI:10.1111/1365-2435.13626

Dwyer, W.P. and R.H. Kramer. 1975. The influence of temperature on scope for activity in cutthroat trout, *Salmo clarkii*. *Transactions of the American Fisheries Society*, 104: 552–554.

Earhart, M.L., T.S. Blanchard, A.A. Harman and P.M. Schulte. 2022. Hypoxia and high temperature as interacting stressors: will plasticity promote resilience of fishes in a changing world? *Biological Bulletin*, 243(2): 149-170.

Ebert, D., H.J. Carius, T. Little and E. Decaestecker, E. 2004. The evolution of virulence when parasites cause host castration and gigantism. *American Naturalist*, 164: 19–32.

Ebert, D.A., L.J.V. Compagno and P.D. Cowley. 2006. Reproductive biology of catsharks (Chondrichthyes: Scyliorhinidae) off the west coast of southern Africa. *ICES Journal of Marine Science*, 63: 1053–1065. DOI:10.1016/j.icesjms.2006.04.016

Ebert, D.A., M. Dando and S. Fowler. 2021. *Sharks of the World: a Complete Guide.* Princeton University Press, Princeton, 608 p.

Ebert, T.A. 1975. Growth and mortality of post-larval echinoids. *Americam Zoologist*. 15: 755–775. DOI:10.1093/icb/15.3.755

Edwards, R.R.C., D.M. Finlayson and J.H. Steele. 1972. An experimental study of the oxygen consumption, growth and metabolism of the cod (*Gadus morhua* L.). *Journal of Experimental Marine Biology and Ecology*, 8: 299–309.

Ege, R. and A. Krogh. 1914. On the relation between the temperature and the respiratory exchange in fishes. *Internationale Revue der Gesamten Hydrobiologie und Hydrologie*, 7: 48–55.

Ehrhardt, N.M., P.S. Jacquemin, B.F. Garcia, D.G. Gonzalez, B.M.J. Lopez, B.M.J. Ortiz and N.A.I. Solis. 1983. On the fishery and biology of the giant squid, *Dosidocus gigas*, in the Gulf of California, Mexico, 306–340 *In:* J.F. Caddy (ed.) *Advances in Assessment of World Cephalopod Resources*. FAO Fisheries Technical Papers 231.

Einstein, A. 1905. Über die von der molekularkinetischen Theorie der Wärme geförderte Bewegung von in ruhenden Flüssigkeiten suspendierten Teilchen. *Annalen der Physik*, 322(8): 549–560.

Eldredge, N. 2008. Some Thoughts on 'Adaptive Peaks,' 'Dobzhansky's Dilemma' – and How to Think About Evolution. *Evolution: Education and Outreach*, 1(3): 243–246.

Eliason, E.J., T.D. Clark, M.J. Hague, L.M. Hanson, Z.S. Gallagher, K.M. Jeffries, M.K. Gale, D.A. Patterson, S.G. Hinch and A.P. Farrell. 2011. Differences in thermal tolerance among sockeye salmon populations. *Science* 332: 109–112.

Elliott, J.M. 1976. The energetics of feeding, metabolism and growth of brown trout (*Salmo trutta* L.) in relation to body weight, water temperature and ration size. *Journal of Animal Ecology,* 45: 923–948.

Erman, P. 1808. Untersuchungen über das Gas in der Schwimmblase der Fische, und über die Mitwirkung des Darmkanals zum Respirationsgeschäfte bei der Fischart *Cobitis fossilis* (Schlammspeitzger). *Annalen der Physik,* 10: 113–160.

Eva, B., P. Harmony, G. Thomas, G. Francois, V. Alice, M. Claude and D. Tony. 2016. Trails of river monsters: detecting critically endangered Mekong giant catfish *Pangasianodon gigas* using environmental DNA. *Global Ecology and Conservation,* 7: 148–156

Evans, D.H., P.M. Piermarini and K.P. Choe. 2005. The multifunctional fish gill: dominant site of gas exchange, osmoregulation, acid-base regulation, and excretion of nitrogenous waste. *Physiological Reviews,* 85(1): 97–177.

Farhat, E., E.D. Turenne, K. Choi and J.M. Weber. 2019. Hypoxia-induced remodelling of goldfish membranes. *Comparative Biochemistry and Physiology B,* 237: 110326.

Farmer, C.G. 1999. Evolution of the vertebrate cardio-pulmonary system. *Annual Review of Physiology,* 61: 573-592.

Farrell, A.P., E.J. Eliason, E. Sandblom and T.D. Clark. 2009. Fish cardiorespiratory physiology in an era of climate change. *Canadian Journal of Zoology,* 87(10): 835–851.

Farrell, C.J., B.M. Johnson, A.G. Hansen, B.W. Avila and C.A. , B.W. and C.A. Myrick. 2025. Induced sterility illuminates the effects of reproduction on growth. *Canadian Journal of Fisheries and Aquatic Sciences,* 82: 1-13.

Felline, L. 2016. Mechanistic Causality and the bottoming-out problem, p. 257–266 *In:* L. Felline, A. Ledda, F. Paoli and E. Rossanece (eds) *New Directions in Logic and the Philosophy of Science,* SILFS Series 3, College Publications: London. http://www.college-publications.co.uk

Ferguson, R.A., J.D. Kieffer and B.L. Tufts. 1993. The effects of body size on the acid-base and metabolite status in the white muscle of rainbow trout before and after exhaustive exercise. *Journal of Experimental Biolog,y* 180(1): 195–207.

Fernandes, F.A., T.J. Fernandes, A.A. Pereira, S.L.C. Meirelles and A.C. Costa. 2019. Growth curves of meat-producing mammals by von Bertalanffy's model. *Pesquisa Agropecuária Brasileira,* 54: e01162. DOI:10.1590/S1678-3921.pab2019.v54.01162

Fernie, A.R. and E. Pichersky. 2015. Focus issue on metabolism: Metabolites, metabolites everywhere. *Plant Physiology,* 169(3): 1421–1423.

Ferreira, E.O., K. Anttila and A.P. Farrell. 2014. Thermal optima and tolerance in the eurythermic goldfish (*Carassius auratus*): Relationships between whole-animal aerobic capacity and maximum heart rate. *Physiological and Biochemical Zoology,* 87: 599–611.

Finke, P. and M. Hallmann. 2013. *Prachtguramis: Juwelen des Urwalds in der Natur und im Aquarium.* Aqualog, Rodgau, Germany. 200 p.

Fick, A. 1855. On liquid diffusion. *The London, Edinburgh, and Dublin Philosophical Magazine and Journal of Science.* 10(63): 30–39. [Also: Fick, A. 1855. Ueber Diffusion. *Annalen der Physik,* 94(1): 59–86].

Fieberg, J.R., K. Vitense and D. Johnson. 2020. Resampling-based methods for biologists. *PeerJ*, 8:e9089. DOI:10.7717/peerj.9089

Field, R.H., A.C. Taylor, D.M. Neil and C.J. Chapman. 1991. Factors affecting swimming ability and its recovery in the Norway lobster (*Nephrops norvegicus*). *Journal of the Marine Biological Association of the United Kingdom* 71: 735–736.

Fieller, E.C. 1940. The biological standardization of insulin. Supplement to the *Journal of the Royal Statistical Society*, 7: 1–64. DOI:10.2307/2983630

Fiksen, Ø. and J. Jørgensen. 2011. Model of optimal behaviour in fish larvae predicts that food availability determines survival, but not growth. *Marine Ecology Progress Series*, 32: 207-219.

Filbee-Dexter, K. and T. Wernberg. 2018. Rise of turfs: a new battlefront for globally declining kelp forests. *Bioscience*, 68(2): 64-76.

Fincham, J.I., Rijnsdorp, A.D. and G.H. Engelhard. 2013. Shifts in the timing of spawning in sole linked to warming sea temperatures. *Journal of Sea Research*, 75: 69-76. DOI:10.1016/j.seares.2012.07.004

Fisher, J.A.D., K.T. Frank and W.C. Leggett. 2010. Global Variation in Marine Fish Body Size and its Role in Biodiversity – Ecosystem Functioning. *Marine Ecology Progress Serie,s* 405: 1–13. DOI:10.3354/meps08601

Fitting, J.W. 2015. From breathing to respiration. *Respiration* 89(1): 82–87.

Flüchter, J. and T.J. Pandian. 1968. Rate and efficiency of yolk utilization in developing eggs of the sole *Solea solea*. *Helgoländer Wissenschaftliche Meeresuntersuchungen*, 18: 53–50.

Foo, C.H. 2019. Crustacean growth estimation from tag-recapture data. *Journal of Advanced Research in Dynamical & Control Systems*, 11(Special issue 6): 1859–1867.

Foo, C.H. 2020. Parameter estimation of laboratory-reared *Panulirus ornatus*. *SN Applied Sciences*, 2: 805. DOI:10.1007/s42452-020-2612-8

Forbes, T.L. and G.R. Lopez. 1990. The effect of food concentration, body size and environmental oxygen tension on the growth of the deposit-feeding polychaete, *Capitella* species 1. *Limnology and Oceanography*, 35(7): 1535–1544. DOI:10.4319/lo.1990.35.7.1535

Forster, M.E., C.E. Franklin, H.H. Taylor and W. Davison. 1987. The aerobic scope of an Antarctic fish, *Pagothenia borchgrevinki*, and its significance for metabolic cold adaptation. *Polar Biology*, 8: 155–159.

Forster, R.P. and L. Goldstein. 1969. Formation of excretory products, p. 313–350, *In* W.S. Hoar and D.J. Randall (eds) *Fish Physiology*, Vol. 1. Academic Press, New York, USA.

Forsythe J.W. Hanlon R.T. 1989. Growth of the Eastern Atlantic squid, *Loligo forbesi* Steenstrup (Mollusca: Cephalopoda). *Aquaculture and Fisheries Managemen,t* 20: 1–14.

Fortey, R.A. and L.R.M. Cocks. 2003. Palaeontological evidence bearing on global Ordovician-Silurian continental reconstructions. *Earth-Science Reviews*, 61: 245–307. DOI:10.1016/S0012-8252(02)00115-0

Foster-Smith, R.L. 1975. The effect of concentration of suspension on the filtration rates and pseudofaecal production for *Mytilus edulis* L., *Cerastoderma edule* (L.) and *Venerupis pullastra* (Montagu). *Journal of Experimental Marine Biology and Ecology*, 17: 1-22.

Frandsen, F., H.J. Malmquist and S.S. Snorrason. 1989. Ecological parasitology of polymorphic Arctic charr, *Salvelinus alpinus* (L.), in Thingvallavatn, Iceland. *Journal of Fish Biology,* 34(2): 281–297.

Frank, K.T., B. Petrie, W.C. Leggett and D.G. Boyce. 2018. Exploitation drives an ontogenetic-like deepening in marine fish. *Proceedings of the National Academy of Sciences,* 115(25): 6422–6427.

Franks, F. 2002. Protein stability: the value of 'old literature.' *Biochemical Chemistry,* 96(2-3): 117–127.

Fraser, K.P., L.S. Peck, M.S. Clark, A. Clarke and S.L. Hill. 2022. Life in the freezer: protein metabolism in Antarctic fish. *Royal Society Open Science* 9(3): 211272.

Freedman, J.A. and D.L. Noakes. 2002. Why are there no really big bony fishes? A point-of-view on maximum body size in teleosts and elasmobranchs. *Reviews in Fish Biology and Fisheries,* 12(4): 403–416.

Friedman, M., K. Shimada, L.D. Martin, A. Maltese and M. Triebold. 2010. 100-million-year dynasty of giant planktivorous bony fishes in the Mesozoic seas. *Science,* 327: 990–993. pmid:20167784

Froese, R. and M.L.D. Palomares. 2000. Growth, Natural Mortality, Length-weight Relationship, Maximum Length and Length-at-first-maturity of the Coelacanth *Latimeria chalumnae. Environmental Biology of Fishes,* 58: 45–52. DOI:10.1023/A:1007602613607

Froese, R. 2006. Cube law, condition factor and weight-length relationships: History, meta-analysis and recommendations. *Journal of Applied Ichthyology,* 22: 241–253. DOI:10.1111/j.1439- 0426.2006.00805.x

Froese, R. and C. Binohlan. 2000. Empirical relationships to estimate asymptotic length, length at first maturity and length at maximum yield per recruit in fishes, with a simple method to evaluate length frequency data. *Journal of Fish Biology,* 56: 758–773. DOI:10.1111/j.1095-8649.2000.tb00870.x

Froese, R. and D. Pauly. 2023. Comment on "Metabolic scaling is the product of life-history optimization." *Science,* 380(6643): eade6084.

Froese, R. and D. Pauly. 2024. FishBase. World Wide Web electronic publication. Retrieved from www.fishbase.org, version (02/2024).

Froese, R., F. Flindt, E. Meyer, J. Meyer and O. Egerland. 2020. Untersuchung zum Laichverhalten des Dorsches in der Kieler Bucht im Frühjahr 2020. Manuscript; available from https://oceanrep.geomar.de/id/eprint/50614

From, J. and G. Rasmussen. 1968. A growth model, gastric evacuation and body composition in rainbow trout, *Salmo gairdneri* Richardson 1836. *Dana,* 3: 61–139.

Fry, F.E.J. 1947. Effects of the environment on animal activity. *Ontario Fisheries Research Laboratory Publication, Biological Series 55,* 68: 1–62.

Fu, C., J.M. Wilson, P.J. Rombough and C.J. Brauner. 2010. Ions first: Na+ uptake shifts from the skin to the gills before O_2 uptake in developing rainbow trout, *Oncorhynchus mykiss. Proceedings of the Royal Society B: Biological Sciences,* 277(1687): 1553–1560.

Gajbhiye, D.S., G.L. Fernandes, I. Oz, Y. Nahmias and M. Golan. 2024. A transient neurohormonal circuit controls hatching in fish. *Science,* 386: 1173-1178.

Galbreath, P.F., W. St. Jean, V. Anderson and G.H. Thorgaard. 1994. Freshwater performance of all female diploid and triploid Atlantic salmon. *Aquaculture,* 128: 41–49.

Galilei, G. 1638. *Discorsi e dimostrazioni matematiche intorno a due nuove scienze.* Elzevir, Leiden.

Gani, A., N. Suhendra, F. Herder, J. Schwarzer, J. Mohring, J. Montenegro, M. Herjayanto and D.F. Mokodongan. 2022. A new endemic species of pelvic-brooding ricefish (Beloniformes: Adrianichthyidae: *Oryzias*) from Lake Kalimpa'a, Sulawesi, Indonesia. *Bonn Zoological Bulletin*, 71(1): 77-85.

Garcia-Gómez, G. 2023. *The role of growth in metabolic scaling: a case study across habitats and life histories.* Doctoral dissertation, University of Liverpool, U.K., 168 p.

Gardiner, N.M., P.L. Munday and G.E. Nilsson. 2010. Counter-gradient variation in respiratory performance of coral reef fishes at elevated temperatures. *PLoS ONE*, 5, e13299.

Garland, T. 1998. Conceptual and methodological issues in testing the predictions of symmorphosis, p. 40 - 47 *In:* E.R. Weibel, C.R. Taylor and L. Bolis (eds) *Principles of Animal Design – the Optimization and Symmorphosis Debate.* Cambridge University Press, Cambridge.

Garratt, P.A., A. Govender and A.E. Punt. 1993. Growth acceleration at sex change in the protogynous hermaphrodite *Chrysoblephus puniceus* (Pisces: Sparidae). *South African Journal of Marine Science*, 13(1): 187-193. DOI:10.2989/025776193784287176

Garstang, W. 1909. The distribution of the plaice in the North Sea, Skagerrak and Kattegat, according to size, age and frequency. *Rapports et Procès-Verbaux des Réunions du Conseil permanent international pour l'Exploration de la Mer*, 1: 136–138.

Gascuel, D., A. Fonteneau and C. Capisona. 1992. Modélisation d'une croissance en deux stances chez l'albacore (*Thunnus alalunga*). *Aquatic Living Resources*, 5(3): 155-172.

Gatti, S., T. Brey, W. Müller, O. Heilmayer and G. Holst. 2002. Oxygen microoptodes: a new tool for oxygen measurements in aquatic animal ecology. *Marine Biology*, 140:1075–1085. DOI:10.1007/s00227-002-0786-9

Gavel, I., M. Byrne and M. Thomson. 2022. Effects of raised temperature on viviparous reproduction in the marine isopod *Cirolana harfordi*. *Journal of Experimental Marine Biology and Ecology*, 546: 151648.

Gayanilo, F.C., P. Sparre and D. Pauly. 2005. *FAO-ICLARM Sock Assessment Tool II (FiSAT II) – User Guide (Revised version).* FAO Computerized Information Series (Fisheries), No 8, 168 p.

Geddes, P. and J.A. Thomson. 1889. *The Evolution of Sex.* Water Scott, London, 322 p.

Gehrke, P.C. 1987. Cardio-respiratory morphometrics of spangled perch, *Leiopotherapon unicolor* (Günther, 1859), (Percoidei, Teraponidae). *Journal of Fish Biology*, 31: 617–623.

Genner, M.J., N.C. Halliday, S.D. Simpson, A.J. Southward, S.J. Hawkins and D.W. Sims. 2010. Temperature-driven phenological changes within a marine larval fish assemblage. *Journal of Plankton Research*, 32(5): 699–708. DOI:10.1093/plankt/fbp082

Geoffroy-Saint-Hilaire, E. 1822. *Philosophie anatomique*, Vol. 1. Méquignon-Marvis, Paris, 551 p.

Gerking, S.D. 1952. The protein metabolism of sunfishes of different ages. *Physiological Zoology*, 25: 358–371.

Gerking, S.D. 1971. Influence of rate of feeding and body weight on protein metabolism of bluegill sunfish. *Physiological Zoology*, 44: 9–19.

Giareta, E.P., R.A. Hauser-Davis, V. Abilhoa and N. Wosnick. 2023. Carbonic anhydrase in elasmobranchs and implications of the current climate change scenario. *Comparative Biochememistry and Physiology A*, 281: 111435. DOI:10.1016/j.cbpa.2023.111435

Gillet, C. and P. Quétin. 2006. Effect of temperature changes on the reproductive cycle of roach in Lake Geneva from 1983 to 2001. *Journal of Fish Biology*, 69(2): 518–534. DOI:10.1111/j.1095-8649.2006.01123.x

Gillis, J.A. and O.R. Tidswell. 2017. The origin of vertebrate gills. *Current Biology*, 27(5): 729-732.

Gingerich, A.J. and C.D. Suski. 2012. The effect of body size on post-exercise physiology in largemouth bass. *Fish Physiology and Biochemistry*, 38(2): 329–340.

Ginther, S.C., H. Cameron, C.R. White and D.J. Marshall. 2024. Metabolic loads and the costs of metazoan reproduction. *Science*, 384: 763–767.

Giomi, F., M. Fusi, A. Barausse, B. Mostert, H.O. Pörtner and S. Cannicci. 2014. Improved heat tolerance in air drives the recurrent evolution of air-breathing. *Proceedings of the Royal Society B*. 281: 20132927 DOI:10.1098/rspb.2013.2927

Glazier, D.S. 2002. Resource-allocation rules and the heritability of traits. *Evolution*, 56:1698–700.

Glazier, D.S. 2005. Beyond the '3/4-power law': variation in the intra-and interspecific scaling of metabolic rate in animals. *Biological Reviews*, 80(4): 611–662.

Glazier, D.S. 2010. A unifying explanation for diverse metabolic scaling in animals and plants. *Biological Reviews*, 85(1): 111–138.

Glazier, D.S. 2022. Variable metabolic scaling breaks the law: from 'Newtonian' to 'Darwinian' approaches. *Proceedings of the Royal Society B*, 289 (1985): 20221605.

Glazier, D.S. 2024. Multiphasic allometry: the reality and significance of ontogenetic shifts in the body-mass scaling of metabolic rate. *Academia Biology*, 2. DOI:10.20935/AcadBiol7411

Glennan, S. 2002. Rethinking mechanistic explanation. *Philosophy of Science*, 69(S3): S342-S353.

Glennan, S. 2017. *The New Mechanical Philosophy*. Oxford University Press, Oxford, 256 p.

Glennan, S.S. 1996. Mechanisms and the nature of causation. *Erkenntnis* 44(1): 49-71.

Godfrey-Smith, P. 2001. Three kinds of adaptationism, p. 335–357 *In:* S.H. Orzack and E. Sober (eds) *Adaptationism and Optimality*, Cambridge University Press.

Goethe, J.W. von. 1798 (1994). *Selected Poems*, transl. Christopher Middleton, Princeton University Press, 298 p.

Gompertz, B. 1825. On the nature of the function expressive of the law of human mortality, and on a new mode of determining the value of life contingencies. *Philosophical Transactions of the Royal Society of London B*, 182: 513-585.

Gonzalez, R.J. and D.G. McDonald. 1992. The relationship between oxygen consumption and ion loss in a freshwater fish. *Journal of Experimental Biology*, 163(1): 317-332.

González, A., J. Beltrán, L. Hiriart-Bertrand, V. Flores, B. de Reviers, J.A. Correa and B. Santelices. 2012. Identification of cryptic species in the *Lessonia nigrescens* complex (Phaeophyceae, Laminariales). *Journal of Phycology*, 48(5): 1153-1165.

Goodwin, N.B., N.K. Dulvy and J.D. Reynolds. 2002. Life-history correlates of the evolution of live bearing in fishes. *Philosophical Transactions of the Royal Society of London. Series B*, 357(1419): 259–267.

Goolish, E.M. 1989a. A comparison of oxygen debt in small and large rainbow trout, *Salmo gairdneri* Richardson. *Journal of Fish Biology*, 35: 597–598.

Goolish, E.M. 1989b, The scaling of aerobic and anaerobic muscle power in rainbow trout. (*Salmo gairdneri*). *Journal of Experimental Biology*, 147(1):493–505.

Goolish, E.M. 1991. Aerobic and anaerobic scaling in fish. *Biology Review,* 66: 33–56.

Goolish, E.M. 1995. The metabolic consequences of body size, p. 335–366 *In:* P.W. Hochachka and T.P. Mommsen (eds) *Biochemistry and Molecular Biology of Fishes,* Vol. 4. Elsevier, Amsterdam.

Gordon, M., Hatcher, C. and J. Seymour. 2004. Growth and age determination of the tropical Australia cubozoan *Chiropsalmus* sp. *Hydrobiologia,* 530/531: 339-345.

Gould, S.J. 1997. Cope's rule as a psychological artifact. *Nature,* 385: 199–200.

Gould, S.J. 1977. *Ontogeny and Phylogeny.* Harvard University Press, Cambridge, 520 p.

Gould, S. J. 1989. A developmental constraint in *Cerion,* with comments on the definition and interpretation of constraint in evolution. *Evolution,* 43(3): 516-539.

Gould, S J. and R.C. Lewontin. 1979. The spandrels of San Marco and the Panglossian paradigm: a critique of the adaptationist programme. *Conceptual Issues in Evolutionary Biology,* 2: 73–90.

Graham, J.B. 1997. *Air-breathing Fishes: Evolution, Diversity, and Adaptation.* Elsevier, San Diego, 299 p.

Graham, J.B., R.H. Rosenblatt and C. Gans. 1978. Vertebrate air-breathing arose in fresh waters and not in the oceans. *Evolution,* 32(2): 459-463.

Gräns, A., F. Jutfelt, E. Sandblom, E Jönsson, K. Wiklander, H. Seth and M. Axelsson. 2014. Aerobic scope fails to explain the detrimental effects on growth resulting from warming and elevated CO_2 in Atlantic halibut. *Journal of Experimental Biology,* 217: 711-717.

Graham, J. 2024. *Sharks don't Sink – Adventures of a Rogue Shark Scientist.* Pantheon Books, New York, 211 p.

Gray, I.E. 1957. A comparative study of the gill area of crabs. *Biological Bulletin,* 112(1): 34–42. DOI:10.2307/1538877

Green, S. and N. Jones. 2016. Constraint-based reasoning for search and explanation: Strategies for understanding variation and patterns in biology. *Dialectica,* 70(3): 343–374.

Greenwell, M.G., J. Sherrill and L.A. Clayton. 2003. Osmoregulation in fish: mechanisms and clinical implications. *Veterinary Clinics: Exotic Animal Practice,* 6: 169–189. DOI:10.1016/S1094- 9194(02)00021-X

Grey, J. 2001. Ontogeny and dietary specialization in brown trout (*Salmo trutta* L.) from Loch Ness, Scotland, examined using stable isotopes of carbon and nitrogen. *Ecology of Freshwater Fish,* 10(3): 168–176.

Grey, J., S.J. Thackeray, R.I. Jones and A. Shine. 2002. Ferox trout (*Salmo trutta*) as 'Russian dolls:' complementary gut content and stable isotope analyses of the Loch Ness food web. *Freshwater Biology,* 47(7): 1235–1243.

Griffiths, D. 1994. The size structure of lacustrine Arctic charr (Pisces: Salmonidae) populations. *Biological Journal of the Linnean Society,* 51(3): 337–357.

Griffiths, C.L. and J.A. King. 1979. Some relationships between size, food availability and energy balance in the ribbed mussel *Aulacomya ater. Marine Biology,* 51: 141–149.

Griffith, R.W. 1991. Guppies, toadfish, lungfish, coelacanths and frogs: a scenario for the evolution of urea retention in fishes, p. 199-218 *In:* J.A. Musick, M.N. Bruton and E.K. Balon (eds) The biology of *Latimeria chalumnae* and evolution of coelacanths. Spinger-Science+Business Media.

Grigor, J.J., M.S. Schmid and L. Fortier. 2017. Growth and reproduction of the chaetognaths *Eukrohnia hamata* and *Parasagitta elegans* in the Canadian Arctic Ocean: Capital breeding versus income breeding. *Journal of Plankton Research,* 39: 910–929. DOI:10.1093/plankt/fbx045

Grosberg, R.K., G.J. Vermeij and P.C. Wainwright. 2012. Biodiversity in water and on land. *Current Biology* 22(21), R900-R903.

Gross, M. R. and R. Shine. 1981. Parental care and mode of fertilization in ectothermic vertebrates. *Evolution* 35: 775–793.

Grutter, A.S. 1994. Spatial and temporal variations of the ectoparasites of seven reef fish species from Lizard Island and Heron Island, Australia. *Marine Ecology Progress,* Series 115: 21–30.

Guan, X.H. and W.X. Cao. 2007. Study on the hatchdate and growth of juvenile grass carp from middle reaches of the Yangtze River using daily increment technology. *Acta Hydrobiologica Sinica,* 31(1): 18-23.

Guerra-Marrero, A., A. Bartolomé, L. Couce-Montero, A. Espino-Ruano, D. Jiménez-Alvarado, J.J. Castro and C. Perales-Raya. 2023. Age, growth and population structure of the African cuttlefish *Sepia bertheloti* based on beak microstructure. *Marine Biology,* 170(10): 118.

Gunwardane, N.D.P., M.D.S.T. De Croos and U.S. Amarasinghe. 2024. Recent declining trends in pelagic fish catches in the Indian Ocean off Sri Lanka: Is gill oxygen limitation theory (GOLT) a possible explanation? *Asian Fisheries Science,* 37: 69–80. DOI:10.33997/j.afs.2024.37.1.005

Gupta, A. and S.C. Liew. 2007. The Mekong from satellite imagery: A quick look at a large river. *Geomorphology,* 85(3-4): 259–274.

Gutiérrez-Marco, J.C., A.A. Sá, D.C. García-Bellido, I. Rábano and M. Valério. 2009. Giant trilobites and trilobite clusters from the Ordovician of Portugal. *Geology,* 37(5): 443–446. DOI:10.1130/G25513A.1

Gvoždík, L. 2018. Just what is the thermal niche? *Oikos,* 127(12): 1701-1710.

Haeckel, E.H.P.A. 1877. *Anthropogenie oder, Entwickelungsgeschichte des Menschen, Keimes- und Stammesgeschichte.* W. Engelmann, Leipzig, 732 p.

Haldane, J.B.S. 1926. On being the right size. *Harper's Magazine,* 152: 424–427.

Hall, S.R., C. Becker and C.E. Cáceres. 2007. Parasitic castration: a perspective from a model of dynamic energy budgets. *Integrative and Comparative Biology,* 47: 295-309.

Hamlett, W.C. and M.K. Hysell. 1998. Uterine specializations in elasmobranchs. *Journal of Experimental Zoology,* 282: 438–459.

Hand, K.P. 2010. *Alien Oceans: the Search for Life in the Depth of Space.* Princeton University Press: Princeton and Oxford, 281 p.

Hansen, M.J., D. Boisclair, S.B. Brandt, S.W. Hewett, J.F. Kitchell, M.C. Lucas and Ney, J.J. 1993. Applications of bioenergetics models to fish ecology and management: where do we go from here? *Transactions of the American Fisheries Society,* 122(5): 1019-1030.

Harden-Jones, F.R.1968. *Fish Migrations.* Edward Arnold, London. 325 p.

Hariharan, I.K., D.B. Wake and M.H. Wake. 2016. Indeterminate growth: could it represent the ancestral condition? *Cold Spring Harbor Perspectives in Biology,* 8(2): a019174.

Harrington, M.E. 1993. Aggression in damselfish: adult-juvenile interactions. *Copeia,* 1993(1): 67–74.

Harris, J.O., G.B. Maguire, S.J. Edwards and D.R. Johns. 1999. Low dissolved oxygen reduces growth rate of juvenile greenlip abalone, *Haliotis laevigata* Donovan. *Aquaculture,* 174: 265–278.

Harrison, J.F. 2017. Do performance–safety tradeoffs cause hypometric metabolic scaling in animals? *Trends in Ecology & Evolution,* 32(9): 653–664.

Hartman, K.J. and S.B. Brandt. 1995. Comparative energetics and the development of bioenergetics models for sympatric estuarine predators. *Canadian Journal of Fisheries and Aquatic Sciences,* 52: 1647–1666.

Harvey, B.C. and A.J. Stewart. 1991. Fish size and habitat depth relationships in headwater streams. *Oecologia,* 87: 336-342.

Harzsch, S., C.H. Müller and Y. Perez. 2015. Chaetognatha, p. 215-240 *In:* A. Wanninger (ed.). *Evolutionary Developmental Biology of Invertebrates,* Vol. 1: *Introduction, Non-Bilateria, Acoelomorpha, Xenoturbellida, Chaetognatha.* Springer-Verlag, Vienna.

Hay, A., W. Xian, N. Bailly, C. Liang and D. Pauly. 2020. The why and how of determining length-weight relationships of fish from preserved museum specimens. *Journal of Applied Ichthyology,* 36: 373-379. DOI:10.1111/jai.14014

Healy, T.M. and P.M. Schulte. 2012. Thermal acclimation is not necessary to maintain a wide thermal breadth of aerobic scope in the common killifish (*Fundulus heteroclitus*). *Physiological and Biochemical Zoology,* 85: 107-119

Heibo, E., C. Magnhagen and L.A. Vøllestad. 2005. Latitudinal variation in life-history traits in Eurasian perch. *Ecology,* 86(12): 3377-3386.

Heim, N.A., S.H. Bakshi, L. Buu, S. Chen, S. Heh, A. Jain, C. Noll, A. Patkar, N. Rizk, S. Sundararajan, I. Villante, M.L. Knope and J.L. Payne 2020. Respiratory medium and circulatory anatomy constrain size evolution in marine macrofauna. *Paleobiology,* 46: 288–303. DOI:10.1017/pab.2020.16

Heincke, F. 1913. Investigations on the plaice. General Rapport I. Plaice fishery and protective regulations. Part I. *Rapports et Procès-Verbaux des Réunions du Conseil permanent international pour l'Exploration de la Mer,* 17A: 1-153.

Heinisch, G., A. Corriero, A. Medina, F.J. Abascal, J.M. de la Serna, J. R. Vassallo-Agius, A.B. Ríos, A. García, F. de la Gándara, C. Fauvel, C.R. Bridges, C.C. Mylonas, S.F. Karakulak, I. Oray, D. De Metrio, H. Rosenfeld and H. Gordin. 2008. Spatial-temporal pattern of bluefin tuna (*Thunnus thynnus* L. 1758) gonad maturation across the Mediterranean Sea. *Marine Biology,* 154(4): 623-630.

Heins, D.C., J.A. Baker, M.A. Toups and E.L. Birden. 2010. Evolutionary significance of fecundity reduction in threespine stickleback infected by the diphyllobothriidean cestode *Schistocephalus solidus. Biological Journal of the Linnean Society,* 100(4): 835-846.

Hempel, C. 1962. Deductive-nomological vs. statistical explanation, p. 98-169 *In:* H. Feigl and G. Maxwell (eds) *Scientific Explanation, Space & Time.* Minnesota Studies in the Philosophy of Science, Vol. 3. University of Minnesota Press, Minneapolis.

Henderson, P. 2005. The growth of tropical fishes. *Fish Physiology,* 21: 85–100.

Henderson, P. and I. Walker. 1986. On the leaf litter community of the Amazonian blackwater stream Tarumazinho. *Journal of Tropical Ecology,* 2(1): 1–16.

Henry, J.-L. 1989. Paléoenvironnements et dynamique de faunes de Trilobites dans l'Ordovicien (Llanvirn Supérieur-Caradoc basal) du Massif Armoricain (France): *Palaeogeography, Palaeoclimatology, Palaeoecology,* 73: 139–153. DOI:10.1016/0031-0182(89)90049-7

Hentschel, U., K.M. Usher and M.W. Taylor. 2006. Marine sponges as microbial fermenters. *FEMS Microbiolology Ecology*, 55: 167–177. DOI:10.1111/j.1574-6941.2005.00046.x

Herczeg, G., A. Gonda and J. Merilä. 2009. Evolution of gigantism in nine-spined sticklebacks. *Evolution: International Journal of Organic Evolution,* 63(12): 3190-3200.

Herder, J.G. 1772 [2001] *Abhandlung über den Ursprung der Sprache.* Reclam, Stuttgart, 175 p.

Hermaniuk, A., L.L. van de Pol and W.C. Verberk. 2021. Are acute and acclimated thermal effects on metabolic rate modulated by cell size? A comparison between diploid and triploid zebrafish larvae. *Journal of Experimental Biology,* 224(1): jeb227124.

Herrmann, J.P. and Enders, E. 2000. Effect of body size on the standard metabolism of horse mackerel. *Journal of Fish Biology,* 57: 746-760.

Hesse, R. 1913. Die ökologischen Grundlagen der Tierverbreitung. *Geographische Zeitschrift,* 19(5): 241-259.

Hetz, S.K. and T.J. Bradley. 2005. Insects breathe discontinuously to avoid oxygen toxicity. *Nature* 433: 516–519.

Hicks, C.C., P.J. Cohen, N.A. Graham, K.L. Nash, E.H. Allison, C. D'Lima, D.J. Mills, M. Roscher, S.H. Thilsted, A.L. Thorne-Lyman and M.A. MacNeil. 2019. Harnessing global fisheries to tackle micronutrient deficiencies. *Nature,* 574: 95–98.

Hirst, A.G. and J. Forster. 2013. When growth models are not universal: Evidence from marine invertebrates. *Proceedings of the Royal Society B,* 280. DOI:10.1098/rspb.2013.1546

Hislop, J.R.G. 1988. The influence of maternal length and age on the size and weight of the eggs and the relative fecundity of the haddock, *Melanogrammus aeglefinus,* in British waters. *Journal of Fish Biology,* 32(6): 923–930.

Hixon, R.F., R.T. Hanlon and W.H. Hulet. 1981. Growth and maximum size of the long-finned squid *Loligo pealei* in the northwestern Gulf of Mexico. *Journal of Shellfish Research,* 1: 181-185.

Hoar, W.S., D.J. Randall and E.M. Donaldson. 1983. Reproduction, Part A: endocrine tissues and hormones, p. 229–231 *In:* W.S. Hoar, D.J. Randall and E.M. Donaldson (eds) *Fish Physiology*. Academic Press, New York.

Hochachka, P.W. 1969. Intermediary metabolism in fishes, p. 351-389 *In:* W.S. Hoar and D.J. Randall (eds) *Fish Physiology,* Vol. 1. Academic Press, New York.

Hoffmann, F., H.T. Rapp, T. Zöller and J. Reitner. 2003. Growth and regeneration in cultivated fragments of the boreal deep-water sponge *Geodia barretti* Bowerbank, 1858 (Geodiidae, Tetractinellida, Demospongiae). *Journal of Biotechnology,* 100:109–118. DOI:10.1016/s0168-1656(02)00258-4

Hoffmann, F., O. Larsen, V. Thiel, H.T. Rapp, T. Pape, W. Michaelis and J. Reitner. 2005. An anaerobic world in sponges. *Geomicrobiology Journal,* 22:1–10. DOI:10.1080/01490450590922505

Hogan, Z.S. 2004. Threatened fishes of the world: *Pangasianodon gigas* Chevey, 1931 (Pangasiidae). *Environmental Biology of Fishes,* 70: 210.

Hogan, Z. and S. Lovgren. 2023. *Chasing Giants: In Search of the World's Largest Freshwater Fish.* University of Nevada Press, Reno, 257 p.

Holeton, G.F. 1973. Respiration of arctic char (*Salvelinus alpinus*) from a high Arctic lake. *Journal of the Fisheries Research Board of Canada,* 30: 717–723.

Holeton, G.F. 1976. Respiratory morphometrics of white and red-blooded Antarctic fish. *Comparative Biochemistry and Physiology,* 54(A): 215–220.

Holeton, G.F. 1974. Metabolic cold adaptation of polar fish: fact or artifact? *Physiological Zoology*, 47(3): 137–152.

Hone, D.W. and M.J. Benton. 2005. The evolution of large size: how does Cope's Rule work? *Trends in Ecology & Evolution,* 20(1): 4–6.

Hopkins, T.E., M.B. Eldridge and J.J. Cech. 1995. Metabolic costs of viviparity in yellowtail rockfish, *Sebastes flavidus. Environmental Biology of Fishes,* 43(1): 77–84.

Hordijk, W. 2016. Evolution: limited and predictable or unbounded and lawless? *Biological Theory,* 11(4): 187–191. DOI:10.1126/science.155.3762.568

Hossain, M.Y., R.L. Vadas Jr., R. Ruiz-Carus and S.M. Galib. 2018. Amazon sailfin catfish *Pterygoplichthys pardalis* (Loricariidae) in Bangladesh: a critical review of its invasive threat to native and endemic aquatic species. *Fishes,* 3(1): 14.

Hoving, H.J.T., W.F. Gilly, U. Markaida, K.J. Benoit-Bird, Z.W. Brown, P. Daniel, J.C. Field, L. Parassenti, B. Liu and B. Campos. 2013. Extreme plasticity in life-history strategy allows a migratory predator (jumbo squid) to cope with a changing climate. *Global Change Biology,* 19(7): 2089-2103.

Hrbek, T., J. Seckinger and A. Meyer 2007. A phylogenetic and biogeographic perspective on the evolution of poeciliid fishes. *Molecular Phylogenetics and Evolution,* 43(3): 986–998.

Huang, Q., Y. Zhang, S. Liu, W. Wang and Y. Luo. 2013. Intraspecific scaling of the resting and maximum metabolic rates of the Crucian carp (*Carassius auratus*). *PLoS ONE, e* 8, e82837.

Hubert, N., Kadarusman, P. D., Wibowo, A., Busson, F., Caruso, D., Sulandari, S., Nafiqoh, N., Pouyaud, L. Rüber, L. Avarre, J.C., Herder, F., Hanner, R., Philippe Keith, P. and R.K. Hadiaty. 2015. DNA barcoding Indonesian freshwater fishes: challenges and prospects. *DNA Barcodes* 3(1): 144-169.

Hubbs, C.L. 1926. The structural consequences of modifications of the developmental rate in fishes, considered in reference to certain problems of evolution. *American Naturalist,* 60: 57–81. DOI:10.1086/280071

Hughes, G.M. 1966. The dimensions of fish gills in relation to their function. *Journal of Experimental Biology,* 45: 177-195.

Hughes, G.M. 1972. Morphometrics of fish gills. *Respiration Physiology,* 14: 1-25.

Hughes, G.M. 1983. Allometry of gill dimensions in some British and American decapod Crustacea. *Journal of Zoology,* 200: 83–97. DOI:10.1111/j1469-7998.1983.tb06110.x.

Hughes, G.M. 1979. Morphometry of fish gas exchange in relation to their respiratory function, p. 33–26 *In:* M.E. Ali (ed.) *Environmental Physiology of Fishes*. Plenum, New York, USA.

Hughes, G.M. 1978. On the respiration of *Torpedo marmorata. Journal of Experimental Biology,* 73: 85–105.

Hughes, G.M. 1984. Measurement of gill area in fishes: practices and problems. *Journal of the Marine Biological Association of the United Kingdom,* 64: 637–655.

Hughes, G.M. 1972. Morphometrics of fish gills. *Respiration Physiology* 14: 1-25.

Hughes, G.M. and M. Morgan. 1973. The structure of fish gills in relation to their respiratory function. *Biological Review,* 48:419–475. DOI:10.1111/j.1469-185X.1973.tb01009.x

Hughes, M. R., O.E. Hooker, T.E. Van Leeuwen, A. Kettle-White, A. Thorne, P. Prodöhl and C.E. Adams. 2019. Alternative routes to piscivory: contrasting growth trajectories in brown trout (*Salmo trutta*) ecotypes exhibiting contrasting life history strategies. *Ecology of Freshwater Fish,* 28(1): 4-10.

Hughes, R.N. 1969. A study of feeding in *Scrobicularia plana. Journal of the Marine Ecological Association of the United Kingdom,* 49: 805-823.

Hui, T.H. and P.K.L. Ng. 2005. The labyrinth fishes (Teleostei: Anabantoidei, Channoidei) of Sumatra, Indonesia. *The Raffles Bulletin of Zoology,* 13: 115-138.

Humboldt, A.V. and J.M. Provençal. 1809. Recherches sur la respiration des poissons. *Mémoire de Physique et de Chimie de la Societé d'Arcueil,* 2: 359-404.

Humphries, N.E., D.W. Fuller, K.M. Schaefer and D.W. Sims. 2024. Highly active fish in low oxygen environments: vertical movements and behavioural responses of bigeye and yellowfin tunas to oxygen minimum zones in the eastern Pacific Ocean. *Marine Biology,* 171: 55. DOI:10.1007/s00227-023-04366-2

Huo, Y.Z., S. Sun and B. Yang. 2010. Life cycle of the Chaetognath *Sagitta crassa* in the southern Yellow Sea, China. *Oceanologia et Limnologia Sinica,* 41: 181–185.

Hurley, D. 1968. Transition from water to land in amphipod crustaceans. *American Zoologist,* 8(3): 327-353.

Huxley, J.S. and G. Teissier. 1936. Terminology of relative growth. *Nature,* 137: 780-781.

Ibn al-Nafīs. 1968. *The Theologus Autodidactus of Ibn al-Nafīs.* Translated by J. Schacht and M. Meyerhof. Clarendon Press, Oxford [83 p. in English and 53 p. in Arabic].

Iles, T.D. 1974. The tactics and strategy of growth in fishes, p. 331–345 *In:* E. R. Harden-Jones (ed.) *Sea Fisheries Research.* Elek Science, London. https://www.seaaroundus.org/wp-content/uploads/2023/12/Iles-D.-1974-The-Tactics-and-Stategy-of-Fish-Growth.pdf

Imsland, A.K. 1999. Sexual maturation in turbot (*Scophthalmus maximus*) is related to genotypic oxygen affinity: Experimental support for Pauly's juvenile-to-adult transition hypothesis. *ICES Journal of Marine Science,* 56(3): 320–325. DOI:10.1006/jmsc.1999.0470

Irwin, L.N. and D. Schulze-Makuch. 2003. Europa: Strategy for modeling putative multilevel ecosystems on Europa. *Astrobiology,* 3(4): 813-821.

Ivantsov, A., M. Zakrevskaya, A.H. Knoll, M.A. Fedonkin and D. Pauly. 2025. The growth of the enigmatic Ediacaran *Parvancorina minchami. Paleontology,* 1-8, DOI:10.1017/pab.2024.55

Iversen, E.S., A. Jones and C.P. Idyll. 1960. *Size distribution of pink shrimp* Penaeus duorarum *and fleet concentration on the Tortugas fishing grounds.* U.S. Fish and Wildlife Service Special Science Report on Fish no. 356, 62 p.

Jablonski, D. 1996. Body size and macroevolution, p. 256-289 *In:* D. Jablonski, D.H. Erwin and J.H. Lipps (eds) *Evolutionary Paleobiology*. University of Chicago Press, Chicago.

Jacob, F. 1977. Evolution and tinkering. *Science,* 196: 1161–1166.

Jäger, G. 1884. *Lehrbuch der allgemeinen Zoologie: ein Leitfaden für Vorträge und zum Selbstunterricht,* Vol. 3. Ernst Günther Verlag, Leipzig, 437 p.

Jackson G.D. 1995. Seasonal influences on statolith growth in the tropical nearshore loliginid squid *Loligo chinensis* (Cephalopoda, Loliginidae) off Townsville, North-Queensland, Australia. *Fishery Bulletin,* 93: 749–752.

Jackson G.D. and M.L. Domeier. 2003. The effects of an extraordinary El Niño/La Niña event on the size and growth of the squid *Loligo opalescens* off Southern California. *Marine Biology*, 142: 925–935.

Janardhan, V. and A. I. Qureshi. 2004. Mechanisms of ischemic brain injury. *Current Cardiolology Reports,* 6: 117–123. DOI:10.1007/s11886-004-0009-8

Jansen, T. and H. Gislason. 2011. Temperature affects the timing of spawning and migration of North Sea mackerel. *Continental Shelf Research,* 31(1): 64–72. DOI:10.1016/j.csr.2010.11.003

Jarre, A., M.R. Clarke and D. Pauly. 1991. Re-examination of growth estimates in oceanic squids: the case of *Kondakovia longimana* (Onychotheutidae). *ICES Journal of Marine Science,* 48: 195–200.

Jarrett, M.L., A.L. Smith, G.J. Langford and P.M. Gravinese. 2024. Lower seawater pH reduces the foraging activity of the Florida stone crab, *Menippe mercenaria* (Say, 1818) (Decapoda: Brachyura: Menippidae). *Journal of Crustacean Biology,* 44: ruae024. DOI:10.1093/jcbiol/ruae024

Jarrett, M.L. A. Avedo, J. Newland, D. Pauly and M.J. Butler. 2025. Balancing body size and respiratory surface: a morphometric analysis of the Caribbean king crab (*Maguimithrax spinosissimus*) and Caribbean spiny lobster (*Panulirus argus*). *Journal of Morphology* 286(10): e70098.

Jerde, C.L., K. Kraskura, E.J. Eliason, S.R. Csik, A.C. Stier and M.L. Taper. 2019. Strong evidence for an intraspecific metabolic scaling coefficient near 0.89 in fish. *Frontiers in Physiology,* 10: 1166.

Jiwyam, W. 2011. The effect of stocking density on yield, growth, and survival of Asian river catfish (*Pangasius bocourti* Sauvage, 1880) cultured in cages. *Aquaculture International,* 19(5): 987–997.

Jobling, M. 2010. Are compensatory growth and catch-up growth two sides of the same coin? *Aquaculture International,* 18: 501-510.

Jobling, M. 1993. Bioenergetics: feed intake and energy partitioning, p. 1–44 *In* J.C. Rankin and F.B. Jensen (eds) *Fish Ecophysiology*. Chapman and Hall, London.

Jobling, M. 1981. The influences of feeding on the metabolic rate of fishes: a short review. *Journal of Fish Biology,* 18: 385–400.

Jobling, M., O.H. Meløy, J. Dos Santos and B.J.A.I. Christiansen. 1994. The compensatory growth response of the Atlantic cod: effects of nutritional history. *Aquaculture International,* 2(2): 75–90.

Johansen, K. 1970. Air breathing in fishes. *Fish Physiology,* 4: 361–411.

Johansen, J.L., M.D. Mitchell, G.O. Vaughan, D.M. Ripley, H.A. Shiels, H.A. and J.A. Burt. 2024. Impacts of ocean warming on fish size reductions on the world's hottest coral reefs. *Nature Communications,* 15(1): 5457.

Jónasson, P.M. and Kristiansen, J. 1967. Primary and secondary production in Lake Esrom: Growth of *Chironomus anthracinus* in relation to seasonal cycles of phytoplankton and dissolved oxygen. *Internationale Revue der gesamten Hydrobiologie und Hydrographie,* 52(2): 163-217.

Jones, R. 1976. Growth of fishes, p. 251–279 *In:* D.H. Cushing and J.J. Walsh (eds) *The Ecology of the Seas*. Blackwell Scientific Publications, London.

Jørgensen, C., B. Ernande, O. Fiksen and U. Dieckman. 2006. The logic of skipped spawning in fish. *Canadian Journal of Fisheries and Aquatic Science,* 63: 200–211. DOI:10.1139/f05-210

Jorgensen, C.B. 1976. Growth efficiencies and factors controlling size in some mytilid bivalves, especially *Mytilus edulis* L.: review and interpretation. *Ophelia,* 15(2): 175-192.

Jourdan-Pineau, H., A. Dupont-Prinet, G. Claireaux and D.J. McKenzie. 2010. An investigation of metabolic prioritization in the European sea bass, *Dicentrarchus labrax*. *Physiological and Biochemical Zoology,* 83(1): 68-77.

Jurado Jurado-Molina, J., O. Gutiérrez-Benítez and A. Roldan-Heredia. 2018. Model uncertainty and Bayesian estimation of growth parameters of Yellowtail Snapper (*Ocyurus chrysurus*) from Veracruz, Mexico. *Hidrobiológica,* 28(2): 191-199.

Jutfelt, F., R. Ern, R.H.J. Leeuwis and T.D. Clark. 2024. Effects of climate warming, p. 14-31 *In:* S.L. Alderman and T. E. Gillis (eds) *Encyclopedia of Fish Physiology, 2nd Edition,* Vol. 3. Academic Press. DOI:10.1016/B978-0-323-90801-6.00183-X

Jutfelt, F., T. Norin, E.R. Åsheim, L.E. Rowsey, A.H. Andreassen, R. Morgan, T.D. Clark and B. Speers-Roesch. 2021. 'Aerobic scope protection' reduces ectotherm growth under warming. *Functional Ecology,* 35(7): 1397–1407.

Jutfelt, F., T. Norin, R. Ern, J. Overgaard, T. Wang, D.J. McKenzie, S. Lefevre, G.E. Nilsson, N.B. Metcalfe, A.J.R. Hickey, J. Brijs, B. Speers-Roesch, D.G. Roche, A.K. Gamperl, G.D. Raby, R. Morgan, A.J. Esbaugh, A. Gräns, M. Axelsson, A. Ekström, E. Sandbloman, S.A. Binning, J.W. Hicks, F. Seebacher, C. Jørgensen, S.S. Killen, P.M. Schulte and T.D. Clark. 2018. Oxygen- and capacity-limited thermal tolerance: blurring ecology and physiology. *Journal of Experimental Biology,* 221(1): jeb169615.

Kaiser, A., C.J. Klok, J.J. Socha, W.K. Lee, M.C. Quinlan and J.F. Harrison. 2007. Increase in tracheal investment with beetle size supports hypothesis of oxygen limitation on insect gigantism. *Proceedings of the National Academy of Sciences,*104(32): 13198-13203.

Kajimura, M., S.J. Cooke, C.N. Glover and C.M. Wood. 2004. Dogmas and controversies in the handling of nitrogenous wastes: the effect of feeding and fasting on the excretion of ammonia, urea and other nitrogenous waste products in rainbow trout. *Journal of Experimental Biology,* 207: 1993-2002. DOI:10.1242/jeb.00901

Kangur, K., E. Tammiksaar and D. Pauly. 2021. A reflection of global warming in the long-term fisheries catch of a large north-temperate lake: Insights from the 'Mean Temperature of the Catch' concept. *Environmental Biology of Fishes,* 105: 1405–1413. DOI:10.1007/s10641-021-01114-7

Kant, I. 1790. *Kritik der Urteilskraft.* Lagarde und Friedrich, Berlin, 480 p. [Also Kant, I. 1951. *Critique of Judgment.* Translated by J. H. Bernard, Hafner Publishing. New York, 477 p.].

Kapp, H. 1991. Morphology and anatomy, p. 5–17 *In:* Q. Bone, H. Kapp and A.C. Pierrot-Bults (eds) *The Biology of Chaetognaths.* Oxford University Press, Oxford.

Karpouzi, V.S. and D. Pauly. 2008. Life-history patterns in marine birds. p. 27-53 *In:* M.L.D. Palomares and D. Pauly (eds) *Von Bertalanffy Growth Parameters of Non-Fish Marine Organisms.* Fisheries Centre Research Report 16(10).

Kaspersson, R., Höjesjö, J. and T. Bohlin. 2012. Habitat exclusion and reduced growth: a field experiment on the effects of inter-cohort competition in young-of-the-year brown trout. *Oecologia*, 169: 733-742.

Kassowitz, M. 1899. *Allgemeine Biologie, Band 1. Aufbau und Zerfall des Protoplasmas*. Perles, Wien, 440 p.

Kassowitz, M. 1905. Vitalismus und Teleologie. *Biologisches Centralblatt*, XXV: 753-777.

Katsu-Kimura, Y., F. Nakaya, S.A. Baba and Y. Mogami. 2009. Substantial Energy Expenditure for Locomotion in Ciliates Verified by Means of Simultaneous Measurement of Oxygen Consumption Rate and Swimming Speed. *Journal of Experimental Biology*,, 212: 1819–1824. DOI:10.1242/jeb.028894

Kearney, M.R. 2021. What is the status of metabolic theory one century after Pütter invented the von Bertalanffy growth curve? *Biological Reviews* 96(2): 557–575.

Keas, M.N. 2018. Systematizing the theoretical virtues. *Synthese*, 195: 2761–2793. DOI:10.1007/s11229-017-1355-6

Keller, E.F. 1983. *A Feeling for the Organism: The Life and Work of Barbara McClintock*. W.H. Freeman, New York. 235 p.

Ken-Ichi, Y. 1992. Relationship of respiration to body weight in the tilapia *Oreochromis niloticus* under resting and normoxic conditions. *Comparative Biochemistry and Physiology A*, 103: 81–83.

Keskin, Ç. and D. Pauly. 2014. Changes in the 'Mean Temperature of the Catch': application of a new concept to the North-eastern Aegean Sea. *Acta Adriatica*, 55(2): 213-218.

Keskin, Ç. and D. Pauly. 2018. Reconciling Trends of Mean Trophic Level and Mean Temperature of the Catch in the Eastern Mediterranean and Black Seas. *Mediterranean Marine Science*, 19(1): 79-83.

Keskin, Ç. and D. Pauly. Testing predictions of length at first maturity of teleostean fishes, given their maximum length. *Cybium*, 47(3): 249-257. DOI:10.26028/cybium/2023-001

Khalfallah, M., H.H. Mahmoud, R.M. Fahim and D. Pauly. 2023. Once upon a century, the Mediterranean Egyptian fisheries (1920-2017) as affected by 'fishing down' and climate change. *Ocean and Coastal* , 245, p. 106831. DOI:10.1016/j.ocecoaman.2023.106831

Khan, J.R., F.I. Iftikar, N.A. Herbert, E. Gnaiger and A.J.R. Hickey. 2014a. Thermal plasticity of skeletal muscle mitochondrial activity and whole animal respiration in a common intertidal triplefin fish, *Forsterygion lapillum* (Family: Tripterygiidae). *Journal of Comparative Physiology B*, 184: 991–1001.

Khan, J.R., S. Pether, M. Bruce, S.P. Walker and N.A. Herbert. 2014b. Optimum temperatures for growth and feed conversion in cultured hapuku (*Polyprion oxygeneios*) — is there a link to aerobic metabolic scope and final temperature preference? *Aquaculture*, 430: 107–113.

Kieffer, J.D. 1995. *The role of body size and temperature in the physiological response to exercise in fish*. Doctoral dissertation, Queen's University, Kingston, Ontario.

Kieffer, J.D. 2000. Limits to exhaustive exercise in fish. *Comparative Biochemistry and Physiolology A*, 126(2): 161–179.

Kieffer, J.D. 2010. Perspective—exercise in fish: 50+ years and going strong. *Comparative Biochemistry and Physiology A*, 156(2): 163–168.

Kieffer, J.D., R.A. Ferguson, J.E. Tompa and B.L. Tufts. 1996. Relationship between body size and anaerobic metabolism in brook trout and largemouth bass. *Transactions of the American Fisheries Society,* 125(5): 760–767.

Kieffer, J.D. and B.L. Tufts. 1998. Effects of food deprivation on white muscle energy reserves in rainbow trout (*Oncorhynchus mykiss*): the relationships with body size and temperature. *Fish Physiology and Biochemistry,* 19(3): 239–245.

Kikinger, R. 1992. *Cotylorhiza tuberculata* (Cnidaria: Scyphozoa) – life history of a stationary population. *Marine Ecology* (Berlin), 13: 333–362.

Kingsolver, J.G. and D.W. Pfennig. 2004. Individual-level selection as a cause of Cope's rule of phyletic size increase. *Evolution,* 58(7): 1608-1612.

Kinne, O. 1960. Growth, foods intake, and food conversion in a euryplastic fish exposed to different temperatures and salinities. *Physiological Zoology,* 33: 288–317.

Kirby-Smith, W.W. 1970. Growth of the scallops, *Argopecten irradians concentricus* (Say) and *Argopecten gibbus* (Linné), as influenced by food and temperature. Duke University, Durham, North Carolina, Doctoral dissertation, 126 p.

Kisia S. and G.M. Hughes. 1992. Estimation of oxygen-diffusing capacity in the gills of different sizes of a tilapia, *Oreochromis niloticus. Journal of Zoology,* 227: 405-415.

Klatt, J. M., A. Chennu, B.K. Arbic, B.A. Biddanda and G.J. Dick. 2021. Possible link between Earth's rotation rate and oxygenation. *Nature Geoscience,* 14:564–570. DOI:10.1038/s41561-021-00784-3

Kleiber, M. 1932. Body size and metabolism. *Hilgardia,* 6(11): 315-353.

Kleiber, M. 1961. *The fire of life: An introduction to animal energetics.* Wiley, New York. 454 p.

Klumpp, D.W., B.L. Bayne and A.J.S. Hawkins. 1992. Nutrition of the giant clam *Tridacna gigas* (L.) I. Contribution of filter feeding and photosynthates to respiration and growth. *Journal of Experimental Marine Biology and Ecology,* 155(1): 105–122.

Knight, J.M. 2010. Invasive ornamental fish: a potential threat to aquatic biodiversity in peninsular India. *Journal of Threatened Taxa,* 2(2): 700–704.

Knouft, J.H. and L.M. Page. 2003. The evolution of body size in extant groups of North American freshwater fishes: speciation, size distributions, and Cope's rule. *American Naturalist,* 161(3): 413-421.

Koch, F. and W. Wieser. 1983. Partitioning of energy in fish: can reduction of swimming activity compensate for the cost of production? *Journal of Experimental Biology,* 107(1): 141-146.

Koehler, N., M. Goubeaud, O. Hildebrandt, A.K. Sohrabi and U. Koehler. 2011. Die Geschichte des Sauerstoffs – von der Entdeckung bis zur medizinischen Anwendung. *Pneumologie,* 65(12): 736-741.

Koenigbauer, S.T., M.L. Cubbage, L.D. Warren, J.M. Tellier, O.M. Selz, G.G. Sass and and T.O. Höök. 2025. Fish reproductive phenology shifts with increasing temperature and year. *Biology Letters,* 21(1): 20240240.

Kolachana, A., K. Mahesh and K. Ramasubramanian (Editors). 2019. *Studies in Indian mathematics and astronomy: selected articles of Kripa Shankar Shukla.* Springer Singapore, 742 p.

Kolding, J., L. Haug and S. Stefansson. 2008. Effect of ambient oxygen on growth and reproduction in Nile tilapia (*Oreochromis niloticus*). *Canadian Journal of Fisheries and Aquatic Sciences,* 65: 1413–1424. DOI:10.1139/F08-059

Kooijman, S.A.L.M. 2000. *Dynamic Energy and Mass Budgets in Biological Systems.* Cambridge University Press, Cambridge, 424 p.

Korsmeyer, K. E., H. Dewar, N.C. Lai and J.B. Graham. 1996. The aerobic capacity of tunas: adaptation for multiple metabolic demands. *Comparative Biochemistry and Physiology A,* 113(1): 17–24.

Koshelev, V.N. and G.I. Ruban. 2012. Maturation and fecundity of kaluga *Acipenser dauricus. Journal of Ichthyology,* 52: 528–536. DOI:10.1134/S0032945212040054

Kottelat, M. 2001. *Fishes of Laos.* WHT Publications, Colombo, Sri Lanka. 198 p.

Kozłowski, J., M. Czarnołęski and M. Dańko. 2004. Can optimal resource allocation models explain why ectotherms grow larger in cold? *Integrative and Comparative Biology,* 44(6): 480–493. DOI:10.1093/icb/44.6.480

Kozłowski, J., M. Konarzewski and M. Czarnoleski. 2020. Coevolution of body size and metabolic rate in vertebrates: a life-history perspective. *Biological Reviews,* 95(5): 1393-1417.

Kramer, D.L. 1983. The evolutionary ecology of respiratory mode in fishes: an analysis based on the costs of breathing. *Environmental Biology of Fishes,* 9: 145-158.

Kramer, D.L. 1987. Dissolved oxygen and fish behavior. *Environmental Biology of Fishes,* 7: 47-55.

Kramer, D. and J. Graham. 1976. Synchronous air-breathing is a social component of respiration in fishes. *Copeia,* 1976(4): 689-697.

Kramer, D. L. and E.A. Braun. 1983. Short-term effects of food availability on air-breathing frequency in the fish *Corydoras aeneus* (Callichthyidae). *Canadian Journal of Zoology,* 61(9): 1964-1967.

Kranenbarg, S., M. Muller, J.L.M. Gielen and J.H.G. Verhagen. 2000. Physical constraints on body size in teleost embryos. *Journal of Theoretical Biology,* 204(1): 113–133.

Krause, A.J., B.J.W. Mills, S. Zhang, N.J. Planavsky, T.M. Lenton and S.W. Poulton. 2018. Stepwise oxygenation of the Paleozoic atmosphere. *Nature Communications,* 9: 4081. DOI:10.1038/s41467-018-06383-y

Krogh, A. 1937. Osmotic regulation in freshwater fishes by active absorption of chloride ions. *Zeitschrift für vergleichende Physiologie,* 24(5): 656-666.

Krogh, A. 1939. *Osmotic Regulation in Aquatic Animals.* Cambridge University Press, Cambridge, 252 p.

Krogh, A. 1941. *The Comparative Physiology of Respiratory Mechanisms.* The University of Pennsylvania Press, Philadelphia, 184 p.

Kruse, S. 2010. *Biology of Meso- and Bathypelagic Chaetognaths in the Southern Ocean.* Berichte zur Polar and Meeresforschung 610, 145 p.

Kuhar, M.J. 2009. On the use of protein turnover and half-lives. *Neuropsychopharmacology,* 34: 1172–1173.

Kühleitner, M., N. Brunner, W.G. Nowak, K. Renner-Martin and K. Scheicher, K. 2019. Best fitting tumor growth models of the von Bertalanffy-Pütter type. *BMC Cancer* (19): 1-11.

Kuhn, T. 1962. *The Structure of Scientific Revolutions.* University of Chicago Press, Chicago, 264 p.

Kuhn, T.S. 1977. Objectivity, value judgment, and theory choice, p. 74-86 *In:* A. Bird and J. Ladyman (eds) *Arguing about Science.* Routledge, London.

Kunwar, G.K. and J.D. Munshi. 1988. Gill surface area of a giant size (weight 17 kg) riverine specimen of a major carp, *Catla catla* (Ham). *Current Science*, 57(10): 563-566.

Kuparinen, A., D. Gielewski and J.A. Hutchings 2022. Gill area explains deviations from body size–metabolic rate relationship in teleost fishes. *Journal of Fish Biology*, 101(1): 308-311.

Kutschera, U. and K.J. Niklas. 2013. Metabolic scaling theory in plant biology and the three oxygen paradoxa of aerobic life. *Theory in Biosciences*, 132(4): 277-288.

L'Abée-Lund, J.H., A. Langeland and H. Sægrov. 1992. Piscivory by brown trout *Salmo trutta* L. and Arctic charr *Salvelinus alpinus* (L.) in Norwegian lakes. *Journal of Fish Biology*, 1(1): 91-101.

Lafferty, K.D. and A.M. Kuris. 2009. Parasitic castration: the evolution and ecology of body snatchers. *Trends in Parasitology*, 25(12): 564-572.

Lagler, K.F., J.E. Bardach, R.R. Miller and D.R.M. Passino. 1977. *Ichthyology, 2nd Edition.* John Wiley and Sons, Inc., New York, 506 p.

Lainey, V., N. Rambaux, G. Tobie, N. Cooper, Q. Zhang, B. Noyelles and K. Baillié. 2024. A recently formed ocean inside Saturn's moon Mimas. *Nature*, 626: 280-282. DOI:10.1038/s41586-023-06975-9

Lam, V.W.Y., W.W.L. Cheung, C. Close and D. Pauly. 2008. Modelling seasonal distributions of pelagic marine fishes and squids, p. 51-62 *In:* W.W.L. Cheung, V.W.Y. Lam and D. Pauly (eds) *Modelling Present and Climate-shifted Distribution of Marine Fishes and Invertebrates.* Fisheries Centre Research Report 16(3), University of British Columbia, Vancouver.

Lamarck, J.B.D.M. de. 1797. *Mémoires de physique et d'histoire naturelle.* Museum d'Histoire Naturelle, Paris, 410 p.

Lamborn, C.C. and J.W. Smith. 2019. Human perceptions of, and adaptations to shifting runoff cycles: A case-study of the Yellowstone River (Montana, USA). *Fisheries Research*, 216: 96–108.

Landman, N.H., E.R.M. Druffel, J.K. Cochran, D.J. Donohue and A.J.T. Jull. 1988. Bomb produced radiocarbon in the shell of the chambered *Nautilus*: rate of growth at age at maturity. *Earth and Planetary Science Letters*, 89: 28–34.

Landman, N.H., J.K. Cochran, J.A. Chamberlain and D.J. Hirschberg. 1989. Timing of septal formation in two species of *Nautilus* based on radiometric and aquarium data. *Marine Biology*, 102: 65-72.

Lane, N. 2002. *Oxygen: the Molecule that Made the World.* Oxford University Press, Oxford, 374 p.

Larson, R.J. 1986. Water content, organic content, and carbon and nitrogen composition of medusae from the northeast Pacific. *Journal of Experimental Marine Biology and Ecology*, 99: 107–120.

Larson, R.J. 1987. Respiration and carbon turnover rates of medusae from the NE Pacific. *Comparative Biochemistry and Physiology A*, 87(1): 93–100.

Lasso, C.A., R.S. Rosa, M.A. Morales-Betancourt, D. Garrone-Neto y, M. Carvalho, Editors. 2016. XV. *Rayas de agua dulce (Potamotrygonidae) de Suramérica. Parte II: Colombia, Brasil, Perú, Bolivia, Paraguay, Uruguay y Argentina.* Serie Editorial Recursos Hidrobiológicos y Pesqueros Continentales de Colombia. Instituto de Investigación de Recursos Biológicos Alexander von Humboldt, Bogotá, Colombia, 220 p.

Lasso, C.A., R.S. Rosa, P. Sánchez-Duarte, M.A. Morales-Betancourt and E. Agudelo-Córdoba. 2013. IX. *Rayas de agua dulce (Potamotrygonidae) de Suramérica Parte I: Colombia, Venezuela, Ecuador, Perú, Brasil, Guyana, Surinam y Guayana Francesa: diversidad, bioecología, uso y conservación.* Serie Editorial Recursos Hidrobiológicos y Pesqueros Continentales de Colombia, Instituto de Investigación de Recursos Biológicos Alexander von Humboldt, Bogotá, Colombia, 370 p.

Last, P., G. Naylor, B. Séret, W. White, M. de Carvalho and M. Stehmann (Editors). 2016. *Rays of the World.* CSIRO Publishing, Clayton, Australia, 790 p.

Latham, K.E. and J.J. Just. 1989. Oxygen availability provides a signal for hatching in the rainbow trout (*Salmo gairdneri*) embryo. *Canadian Journal of Fisheries and Aquatic Sciences,* 46: 55-58.

Lauzanne, L. 1976. Régimes alimentaires et relations trophiques des poissons du Lac Tchad. *Cahier ORSTOM – Série Hydrobiologie,* 10: 267-310.

Lavaud, R., R. Filgueira and S. Augustine. 2021. The role of Dynamic Energy Budgets in conservation physiology. *Conservation Physiology,* 9(1): coab083.

Lavin, C.P., D. Pauly, D. Dimarchopoulou, C. Liang and M.J. Costello. 2023. Fishery catch is affected by geographic expansion, fishing down food webs and climate change in Aotearoa, New Zealand. *PeerJ,* 11:e16070. DOI:10.7717/peerj.16070

Lavin, C.P., C. Gordó-Vilaseca, M.J. Costello, Z. Shi, F. Stephenson, F. and A. Grüss. 2022. Warm and cold temperatures limit the maximum body length of teleost fishes across a latitudinal gradient in Norwegian waters. *Environmental Biology of Fishes,* 105: 415-1429. DOI:10.1007/s10641-022-01270-4

Lavin, C.P., C. Gordó-Vilaseca, F. Stephenson, Z. Shi and M.J. Costello. 2022. Warmer temperature decreases the maximum length of six species of marine fishes, crustaceans, and squid in New Zealand. *Environmental Biology of Fishes,* 105(10): 431-1446.

Lavoisier, A. and S. de la Place. 1780. Mémoire sur la Chaleur. *Mémoires de l'Académie des Sciences,* 283-333.

Lavy, A., R. Keren, G. Yahel and M. Ilan. 2016. Intermittent hypoxia and prolonged suboxia measured *in situ* in a marine sponge. *Frontiers in Marine Science,* 3: 263. DOI:10.3389/fmars.2016.00263

Learn, J.R. 2024. They'd give an arm (or three) to survive. *The New York Times,* June 11, p. D3.

Le Cren, E.D. 1951. The length-weight relationship and seasonal cycle in gonad weight and condition in the perch (*Perca fluviatilis*). *Journal of Animal Ecology,* 20(2): 201–219.

Le Cren, E.D. 1992. Exceptionally big individual perch (*Perca fluviatilis* L.) and their growth. *Journal of Fish Biology,* 40: 599-625.

doi.org/10.1111/j.1095-8649.1992.tb02609.x

Le Franc, G. 1970. Biologie de la morue du Sud de la mer du Nord et de la Manche Orientale. *Revue des Traveaux de l'Institut des Pêches Maritimes,* 34(3): 277–296.

Le Guyader, H. 2004. *Geoffroy Saint-Hilaire: A Visionary Naturalist.* University of Chicago Press, Chicago, 288 p.

Le Pape, O. and S. Bonhommeau. 2015. The food limitation hypothesis for juvenile marine fish. *Fish and Fisheries,* 16(3): 373-398.

Lea, E. 1913. Further studies concerning the methods of calculating the growth of herrings. *Conseil permanent international pour l'Exploration de la Mer - Publications de Circonstance,* 1(66): 1-35.

Ledebur, J.F. von. 1939. Über die Atmung der Schwämme und Coelenteraten, p. 262–291 *In:* K. v. Frisch, W. Ruhland, H. Stubbe and W. Vogt (eds) *Ergebnisse der Biologie.* Springer, Berlin, Heidelberg.

Lee, C.G., A.P. Farrell, A. Lotto, M.J. MacNutt, S.G. Hinch and M.C. Healey. 2003. The effect of temperature on swimming performance and oxygen consumption in adult sockeye (*Oncorhynchus nerka*) and coho (*O. kisutch*) salmon stocks. *Journal of Experimental Biology,* 206: 3239-3251.

Lee, D.J. and P.G. Matthews. 2021. How insects transition from water to air: Respiratory insights from dragonflies. *Comparative Biochemistry and Physiology A,* 253: 110859.

Lefevre, S., D.J, McKenzie and G.E. Nilsson. 2017. Models projecting the fate of fish populations under climate change need to be based on valid physiological mechanisms. *Global Change Biology,* 23: 3449–3459.

Lefevre, S., D.J. McKenzie and G.E. Nilsson. 2018. In modelling effects of global warming, invalid assumptions lead to unrealistic projections. *Global Change Biology,* 24(2): 553-556.

Lefevre, S., T. Wang, N.T. Phuong and M. Bayley. 2013. Partitioning the oxygen uptake and cost of surfacing during swimming in the air-breathing catfish *Pangasianodon hypothalamus. Journal of Comparative Physiology B,* 183: 215–221.

Lefevre, S., T. Wang and D.J. McKenzie. 2021. The role of mechanistic physiology in investigating impacts of global warming on fishes. *Journal of Experimental Biology,* 224, jeb238840.

Lefevre, S., Watson, S.A., P.L. Munday and G.E. Nilsson. 2015. Will jumping snails prevail? Influence of near-future CO_2, temperature and hypoxia on respiratory performance in the tropical conch *Gibberulus gibberulus gibbosus. Journal of Experimental Biology,* 218(19): 2991-3001.

Lefevre, S. 2016. Are global warming and ocean acidification conspiring against marine ectotherms? A meta-analysis of the respiratory effects of elevated temperature, high CO_2 and their interaction. *Conservation Physiology,* 4(1): cow009.

Lefrançois, C. and G. Claireaux. 2003. Influence of ambient oxygenation and temperature on metabolic scope and scope for heart rate in the common sole *Solea solea. Marine Ecology Progress Series,* 259: 273-284.

Lenfant, C., K. Johansen and J.D. Torrance. 1970. Gas transport and oxygen storage capacity in some pinnipeds and the sea otter. *Respiratory Physiology,* 9: 277–286. DOI:10.1016/0034-5687(70)90076-9

Lester, N.P., B.J. Shuter and P.A. Abrams. 2004. Interpreting the von Bertalanffy model of somatic growth in fishes: the cost of reproduction. *Proceedings of the Royal Society of London B,* 271: 1625-1631.

Levis, N.A. and D.W. Pfennig. 2016. Evaluating 'plasticity-first' evolution in nature: key criteria and empirical approaches. *Trends in Ecology & Evolution,* 31(7): 563-574.

Levis, N.A. and D.W. Pfennig. 2020. Plasticity-led evolution: A survey of developmental mechanisms and empirical tests. *Evolution & Development,* 22(1-2): 71-87.

Li, G., X. Lv, J. Zhou, C. Shen, D. Xia, D., H. Xie and Y. Luo. 2018. Are the surface areas of the gills and body involved with changing metabolic scaling with temperature? *Journal of Experimental Biology,* 221(8):jeb.174474.

Li, G., H. Xie, D. He and Y. Luo. 2016. Effects of body chemical components on the allometric scaling of the resting metabolic rate in four species of cyprinids. *Fish Physiology and Biochemistry,* 42: 295-301.

Liang, C., Xian, W. and D. Pauly. 2018. Impacts of Ocean Warming on China's Fisheries Catch: Application of the 'Mean Temperature of the Catch'. *Frontiers in Marine Science,* 5: 26. DOI:10.3389/fmars.2018.00026

Liebenberg, L. 2006. Persistence hunting by modern hunter-gatherers. *Current Anthropology,* 47(6): 1017–1026.

Lima, M.C.M. de, D.F. Campos, D. Kochhann and A.L. Val. 2025. Effects of oxygen level on thermal tolerance in Amazonian catfishes with bimodal respiration: physiological and behavioural changes. *Journal of Experimental Biology,* 228(3).

Lin, H.Y. and M.J. Costello. 2023. Body size and trophic level increase with latitude, and decrease in the deep sea and Antarctica, for marine fish species. *PeerJ,* 11:e15880. doi 10.7717/peerj.15880

Lindmark, M., J. Olberger and A. Gårdmark. 2022. Optimum growth temperature declines with body size within species. *Global Change Biology,* 28: 2259-2271 DOI:10.1111/gcb.16067

Lindsey, C.C. 1966. Body Sizes of Poikilotherm Vertebrates at Different Latitudes. *Evolution,* 20: 456–465. DOI:10.2307/2406584

Linnaeus, C. 1758. *Systema naturæ per regna tria naturæ, secundum classes, ordines, genera, species, cum characteribus, differentiis, synonymis, locis. Tomus I. Editio decima, reformata.* Impensis Laurentii Salvii. Holmiæ. Vol. 1, 824 p.

Lipinski, M.R. 1993. The deposition of statoliths: a working hypothesis, p. 241–262 *In:* T. Okutani, R.R. O'Dor and T. Kubodera (eds) *Recent Advances in Fisheries Biology.* Tokai University Press, Tokyo.

Lipinski, M.R. and M.A. Roeleveld. 1990. Minor extension of the von Bertalanffy growth theory. *Fisheries Research* 9: 367–371.

Lipschütz, A. 1915. *Allgemeine Physiologie des Todes.* Vieweg und Teubner, Leipzig, 184 p.

Liston, J.J., M. Newbrey, T. Challands and C. Adams, C. 2013. Growth, age and size of the Jurassic pachycormid *Leedsichthys problematicus* (Osteichthyes: Actinopterygii), p. 145-175 *In:* G. Arratia, H. Schultze and H. Wilson (eds) *Mesozoic Fishes 5 – Global Diversity and Evolution.* Verlag Dr. Friedrich Pfeil, München.

Lo, C.M. 1999. Mating rendezvous in monogenean gill parasites of the humbug *Dascyllus aruanus* (Pisces: Pomacentridae). *Journal of Parasitology,* 85(6): 1178-1180.

Lobel, P. S. and S. Neudecker. 1985. Diurnal periodicity of spawning activity by the hamlet fish, *Hypoplectrus guttavarius* (Serranidae), p. 71-86 *In:* M.L. Reaka (ed.) *The Ecology of Coral Reefs.* NOAA's Undersea Research Program 3(1), Washington, D.C.

Loh, T.-L. and J.R. Pawlik. 2014. Chemical defenses and resource trade-offs structure sponge communities on Caribbean coral reefs. *Proceedings of the National Academy of Sciences,* 111: 4151–4156. DOI:10.1073/pnas.1321626111

Longhurst, A. and D. Pauly. 1987. *Ecology of tropical oceans.* Academic Press, San Diego, 407 p. DOI:10.1016/C2009-0-02861-X

Longo, G. and M. Montévil. 2014. *Perspectives on organisms: Biological time, symmetries and singularities*. Springer, Berlin, 300 p.

Longo, G., M. Montévil and S. Kauffman. 2012. No entailing laws, but enablement in the evolution of the biosphere, p. 1379-1392 *In:* T. Soule (ed.) *Proceedings of the 14th annual conference companion on Genetic and evolutionary computation*. Machinery, New York.

Lonthair, J.K., N.C. Wegner, B.S. Cheng, N.A. Fangue, M.J. O'Donnell, A.M. Regish, J.D. Swenson, E. Argueta, S.D. McCormick, B.H. Letcher and L.M. Komoroske. 2024a. Smaller body size under warming is not due to gill-oxygen limitation in a cold-water salmonid. *Journal of Experimental Biology*, 227. https://10.1242/jeb.246477

Lonthair, J.K., N.C. Wegner, B.S. Cheng, N.A. Fangue, M.J. O'Donnell, A.M. Regish, J.D. Swenson, E. Argueta, S.D. McCormick, B.H. Letcher and L.M. Komoroske. 2024b. The gill oxygen limitation hypothesis remains unsupported by experimental evidence – Response to 'Testing mechanistic theories must be based on correct interpretations'. *Journal of Experimental Biology*, 227. DOI:1242/jeb.248173

López, S., M.I. Militelli, C.M. Riestra and G.J. Macchi 2025. Energy allocation strategy during the reproductive cycle of Patagonian grouper (*Acanthistius patachonicus*) from the southwestern Atlantic Ocean. *Marine and Fishery Sciences (MAFIS)*, 38(3): 375-388.

Lorenzen, K., N. Sukumasavin and Z. Hogan. 2006. Development of a conservation strategy for the critically endangered Mekong giant catfish: quantitative assessment report. *Mekong Giant Catfish. Working Group Report 3*. NACA, Bangkok, 45 p.

Loubens, G. 1974. Quelques aspects de la biologie de *Lates niloticus* du Tchad. *Cahiers ORSTOM, Série Hydrobiologie*, 8: 3-21.

Lowe, C.J. and W. Davison. 2006. Thermal sensitivity of scope for activity in *Pagothenia borchgrevinki*, a cryopelagic Antarctic nototheniid fish. *Polar Biology*, 29: 971-977.

Lutz, R.A. and D.C. Rhoads. 1977. Anaerobiosis and a theory of growth line formation. *Science*, 198: 1222-1227.

Lyell, C. 1830. *Principle of Geology, Being an Attempt to Explain the Former Changes of the Earth's Surface, by Reference to Causes Now in Operation*. 3 volumes. Murray, London.

Magnin, E. 1966. Quelques données biologiques sur la reproduction des esturgeons *Acipenser fulvescens* de la rivière Nottaway, tributaire de la baie James. *Canadian Journal of Zoology*, 44: 257–263. DOI:10.1139/z66-024

Mahé, K, B. Ernande and M. Herbin. 2021. New scale analyses reveal centenarian African coelacanths. *Current Biology*, 31:3621-3628.e4. DOI:10.1016/j.cub.2021.05.054

Majluf, P., K. Matthews, D. J. Skerritt, D. Pauly and M.L.D. Palomares. 2024. A review of the global use of fishmeal and the Fish In:Fish Out metric. *Science Advances*, 10, eadn 5650.

Makarieva, A.M., V.G. Gorshkov, B.L. Li, S.L. Chown, P.B. Reich and V.M. Gavrilov. 2008. Mean mass-specific metabolic rates are strikingly similar across life's major domains: evidence for life's metabolic optimum. *Proceedings of the National Academy of Sciences*, 105(44): 16994–16999.

Mäkinen, T., O.L.F. Weyl, K.K.A. van der Walt and E.R. Swartz. 2013. First record of an introduction of the giant pangasius, *Pangasius sanitwongsei* Smith 1931, into an African river. *African Zoology*, 48(2): 388–391.

Maki-Petäys, A., T. Motka, A. Huusko, P. Tikkanen and P. Kreivi. 1997. Seasonal changes in habitat use and preference by juvenile brown trout, *Salmo trutta*, in a northern boreal river. *Canadian Journal of Fisheries and Aquatic Sciences,* 54(3): 520-530.

Maldeniya, R. 1996. Food consumption of yellowfin tuna, *Thunnus albacares,* in Sri Lankan waters. *Environmental Biology of Fishes,* 47: 101–107.

Mallekh, R. and J.P. Lagardère. 2002. Effect of temperature and dissolved oxygen concentration on the metabolic rate of the turbot and the relationship between metabolic scope and feeding demand. *Journal of Fish Biology,* 60: 1105-1115.

Mangel, M.M.V. and Abrahams. 2001. Age and longevity in fish, with consideration of the ferox trout. *Experimental Gerontology,* 36(4-6): 765-790.

Mangold, K. 1987. Reproduction, p. 157–200 *In:* P.R. Boyle (ed.) *Cephalopod Life Cycles,* Vol. 2, Academic Press, London.

Mariani, G., W.W.L. Cheung, A. Lyet, E. Sala, J. Mayorga, L. Velez and D. Mouillot. 2020. Let more big fish sink: Fisheries prevent blue carbon sequestration—half in unprofitable areas. *Science Advances,* 6(44): eabb4848.

Marras, S., A. Cucco, F. Antognarelli, E. Azzurro, M. Milazzo, M. Bariche and P. Domenici. 2015. Predicting future thermal habitat suitability of competing native and invasive fish species: From metabolic scope to oceanographic modelling. *Conservation Physiology,* 3. DOI:10.1093/conphys/cou059

Marshall, D.J. and C.R. White. 2019. Have we outgrown the existing models of growth? *Trends in Ecology & Evolution,* 34(2): 102–111.

Martinangeli, L., M. Fourt, M. Butler, A.C. Tsikliras, N. Smith, M.L.D. Palomares, B. Derrick, E. Chu and D. Pauly. 2022. The global catch of commercial sponges (1950 to 2019), p. 85-106 *In:* D. Pauly and E. Chu (eds) *Marine and Freshwater Miscellanea IV.* Fisheries Centre Research Reports 30(4). University of British Columbia, Vancouver.

Martínez-Jerónimo, F. 2012. Description of the individual growth of *Daphnia magna* (Crustacea: Cladocera) through the von Bertalanffy growth equation: Effect of photoperiod and temperature. *Limnology,*13(1): 65–71. DOI:10.1007/s10201-011-0356-2

Martins, E.G., S.G. Hinch, D.A. Patterson, M.J. Hague, S.J. Cooke, K.M. Miller, M.F. Lapointe, K.K. English and A.P. Farrell. 2011. Effects of river temperature and climate warming on stock-specific survival of adult migrating Fraser River sockeye salmon (*Oncorhynchus nerka*). *Global Change Biology,* 17(1): 99–114.

Massimi, M. 2022. *Perspectival Realism.* Oxford University Press, Oxford, 432 p.

Mattheeuws, A., M. Genin, A. Detollenaere, J.-C. Micha, avec la collaboration technique de Y. Mine. 1981. Etude de la reproduction du gardon (*Rutilus rutilus*) et des effets d'une élévation provoquée de la température en Meuse sur cette reproduction. *Hydrobiologia,* 85(3): 271-282. DOI:10.1007/BF00017615

Maxime, V. 2008. The physiology of triploid fish: Current knowledge and comparisons with diploid fish. *Fish and Fisheries,* 9: 67–78.

May, R.M. and J. Godfrey. 1994. Biological Diversity: Differences between Land and Sea [and Discussion]. *Philosophical Transactions: Biological Sciences,* 343(1303): 105–111.

Maynard Smith, J. 1978. Optimization theory in evolution. *Annual Review of Ecology and Systematics,* 9(1): 31-56.

Mayr, E. 1961. Cause and effect in biology: kinds of causes, predictability, and teleology are viewed by a practicing biologist. *Science,* 134(3489), 1501-1506.

McArley, T.J., E. Sandblom and N.A. Herbert. 2021. Fish and hyperoxia—from cardiorespiratory and biochemical adjustments to aquaculture and ecophysiology implications. *Fish and Fisheries, 22*: 324–355.

McGhee Jr., G.R. 2015. Limits in the evolution of biological form: a theoretical morphologic perspective. *Interface Focus,* 5(6): 20150034.

McKenzie D.J., R.M. Martinez, A. Morales, J. Acosta, R. Morales, E.W. Taylor, J.F. Steffensen and M.P. Estrada. 2003. Effects of growth hormone transgenesis on metabolic rate, exercise performance and hypoxia tolerance in tilapia hybrids. *Journal of Fish Biology,* 63: 398– 409.

McKenzie, D.J., J.F. Steffensen, E.W. Taylor and A.S. Abe. 2012. The contribution of air-breathing to aerobic scope and exercise performance in the banded knifefish *Gymnotus carapo* L. *Journal of Experiental Biology,* 215: 1323–1330.

McKenzie, D. 2023. 'Review' of 'Gasping Fish and Panting Squids - Oxygen, Temperature and the Growth of Aquatic Animals.' *Fish and Fisheries,* 24(6): 907–908.

McKenzie, J. 2007. *The Architecture of Alexandria and Egypt, c. 300 BC to AD 700,* Vol. 63. Yale University Press, New Haven, 464 p.

McKinley, S., G. Van Der Kraak and G. Power. 1998. Seasonal migrations and reproductive patterns in the lake sturgeon, *Acipenser fulvescens,* in the vicinity of hydroelectric stations in northern Ontario. *Environmental Biology of Fishes,* 51(3): 245–256.

McKoy, J.L. and D.B. Esterman. 1981. Growth of rock lobsters (*Jasus edwardsii*) in the Gisborne region, New Zealand. *New Zealand Journal of Marine and Freshwater Research,* 15(2): 121–136. DOI:10.1080/00288330.1981.9515905

McLaren, I.A. 1966. Adaptive significance of large size and long life of the chaetognath *Sagitta elegans* in the Arctic. *Ecology,* 47: 852–855. DOI:10.2307/1934273

McLusky, D.S. 1973. The effect of temperature on the oxygen consumption and filtration rate of *Chlamys* (*Aequipecten*) *opercularis* (L.) (Bivalvia). *Ophelia,* 10: 141–154.

McMahon, B.R., D.G. McDonald and Wood, C.M. 1979. Ventilation, oxygen uptake and haemolymph oxygen transport, following enforced exhausting activity in the Dungeness crab *Cancer magister*. *Journal of Experimental Biology,* 80: 271–285.

McNab, B.K. 1997. On the utility of uniformity in the definition of basal rate of metabolism. *Physiological Zoology,* 70(6): 718–720.

McNamara, K.J. 1978. Paedomorphosis in Scottish olenellid trilobites (Early Cambrian). *Paleontology,* 21(3): 6535–655.

McQueen, K. and C.T. Marshall. 2017. Shifts in spawning phenology of cod linked to rising sea temperatures. *ICES Journal of Marine Science,* 74(6): 1561–1573. DOI:10.1093/icesjms/fsx025

Mehanna, S. and A. Amin. 2005. Population dynamics of the cuttlefish *Sepia dollfusi* (Adam, 1941) in the Gulf of Suez, Egypt. *Egyptian Journal of Aquatic Biology and Fisheries,* 9(1): 1-14.

Meekan, M.G., L.A., Fuiman, R. Davis, Y. Berger and M. Thums. 2015. Swimming strategy and body plan of the world's largest fish: implications for foraging efficiency and thermoregulation. *Frontiers in Marine Science* 2, p. 64. DOI:10.3389/fmars.2015.00064

Meka, J. M. and S.D. McCormick. 2005. Physiological response of wild rainbow trout to angling: impact of angling duration, fish size, body condition, and temperature. *Fisheries Research,* 72(2-3): 311–322.

Melville-Smith, R. and S. De Lestang. 2006. Spatial and temporal variation in the size at maturity of the western rock lobster *Panulirus cygnus* George. *Marine Biology,* 150(2): 183–195. DOI:10.1007/s00227-006-0349-6

Menzel, D.W. 1958. Utilization of algae for growth by the Angelfish *Holacanthus bermudensis. Journal du Conseil international pour l'Exploration de la Mer* 24: 308–313.

Menzel, D.W. 1960. Utilization of food by a Bermuda reef fish, *Epinephelus guttatus. Journal du Conseil international pour l'Exploration de la Mer,* 25: 216–222.

Merilä, J. and A.P. Eloranta. 2017. Cannibalism facilitates gigantism in a nine-spined stickleback (*Pungitius pungitius*) population. *Ecology of Freshwater Fish,* 26(4): 686-694.

Meyer, K.A., J.L. McCormick, J.R. Kozfkay and J.C. Dillon. 2023. Effects of elevated water temperature on cutthroat trout angler catch rates and catch-and-release mortality in Idaho streams. *Fisheries Management and Ecology,* 30(2): 134–141.

Meyer, K.A. and D.J. Schill. 2021. The Gill-Oxygen Limitation Theory and size at maturity/maximum size relationships for salmonid populations occupying flowing waters. *Journal of Fish Biology,* 98: 44–49. DOI:10.1111/jfb.14555

Michalsen, K., E. Johannesen and B. Bogstad. 2008. Feeding of mature cod (*Gadus morhua*) on the spawning ground in Lofoten. *ICES Journal of Marine Science,* 65: 571-580.

Missouri Department of Conservation. 1994. *Fish kills in Ponds and Lakes.* Missouri Department of Conservation, Jefferson City, MO, USA. https://mdc.mo.gov/sites/default/files/2020-06/fishkills.pdf.

Mitamura, H., N. Arai, Y. Yamagishi, Y. Kawabata, Y. Mitsunaga, M. Khachaphichat and T. Viputhanumas. 2009. Habitat use and movement of hatchery-reared F2 Mekong giant catfish in the Mae Peum reservoir, Thailand, studied by acoustic telemetry. *Fisheries Science,* 75(1): 175–182.

Mitchell, P.H. 1914. *The Oxygen Requirements of Shellfish.* US Government Printing Office, No. 787: 207–222.

Miyashita, S., N. Hattori, Y. Sawada, Y. Ishibashi, H. Nakatsukasa, T. Okada, O. Murata and H. Kumai. 1999. Ontogenetic change in oxygen consumption of bluefin tuna, *Thunnus thynnus. Aquaculture Science,* 47: 269-275.

Moncrief, T., N.J. Brown-Peterson and M.S. Peterson. 2018. Age, growth, and reproduction of Vermilion Snapper in the north-central Gulf of Mexico. *Transactions of the American Fisheries Society,* 147(5): 996–1010.

Montévil, M. and M. Mossio. 2015. Biological organisation as closure of constraints. *Journal of Theoretical Biology,* 372: 179-191.

Mooij, W., M. De Senerpont, L.N. Domis, and S. Hülsmann. 2008. The impact of climate warming on water temperature, timing of hatching and young-of-the-year growth of fish in shallow lakes in the Netherlands. *Journal of Sea Research,* 60(1): 32–43. DOI:10.1016/j.seares.2008.03.002

Morales-Nin, B. 1986a. Chemical composition of the otoliths of the sea bass (*Dicentrachus labrax* Linnaeus, 1758) (Pisces, Serranidae). *Cybium,* 10: 115-120.

Morales-Nin, B. 1986b. Structure and composition of otoliths of Cape hake *Merluccius capensis. South African Journal of Marine Science,* 4: 3-10.

Moran, D. and R.M. Wells 2007. Ontogenetic scaling of fish metabolism in the mouse-to-elephant mass magnitude range. *Comparative Biochemistry and Physiology Part A: Molecular & Integrative Physiology,* 148(3): 611-620.

Morbey, Y.E. and D. Pauly. 2022. Juvenile-to-adult transition invariances in fishes: perspectives on proximate and ultimate causation. *Journal of Fish Biology,* 101: 874–884. DOI:10.1111/jfb.15146

Moreira, J.M., A.C. Mendes, A.L. Maulvault, A. Marques, R. Rosa, P. Pousão-Ferreira, T. Sousa, P. Anacleto and G.M. Marques. 2022. Impacts of ocean warming and acidification on the energy budget of three commercially important fish species. *Conservation Physiology,* 10(1): coac048.

Moreno, A., J. Pereira and M. Cunha. 2005. Environmental influences on age and size at maturity of *Loligo vulgaris. Aquatic Living Resources*, 18: 377–384.

Morgan, M. 1971. Gill development, growth and respiration in the trout, *Salmo gairdneri* (Richardson). Doctoral dissertation, Bristol University.

Morita, K., M. Fukuwaka, N. Tanimata and O. Yamamura. 2010. Size-dependent thermal preferences in a pelagic fish. *Oikos,* 119(8): 1265–1272.

Morris, C.C. 1991. Statocyst fluid composition and its effect on calcium carbonate precipitation in the squid *Alloteuthis subulata* (Lamarck, 1798): toward a model for biomineralization. *Bulletin of Marine Science,* 49: 379-388.

Mousseau, T. A.1997. Ectotherms follow the converse to Bergmann's rule. *Evolution,* 51(2): 630-632.

Mugiya, Y., N. Watanabe, J. Yamada, J.M. Dean, D.G. Dunkeberger and M. Shimizu. 1981. Diurnal rhythms in otolith formation in the goldfish *Carassius auratus. Comparative Biochememistry and Physiology,* 68: 659662.

Muir, B.S. and G.M. Hughes. 1969. Gill dimensions for three species of tunny. *Journal of Experimental Bio*logy, 51(2): 271–285. DOI:10.1242/jeb.51.2.271b

Müller, J. 1835. *Handbuch der Physiologie des Menschen für Vorlesungen,* Vol. 1, Hölscher, Coblenz, 852 p.

Müller, J. 2021. The fish that could climb palm trees: observation, rumour and colonial hearsay in nineteenth-century ichthyology, p. 89-100 *In:* A. van de Haar and A. Schulte-Nordholt (eds) *Figurations animalières à travers les textes et l'image en Europe*. Brill, Leiden.

Müller, J. and P. Koster. 2024. August Pütter and the mechanistic origins of the temperature-size rule. *Biological Theory,* 20(1): 41-53.

Müller, J., N. Houben and D. Pauly. 2023. On being the wrong size, or the role of body mass in fish kills and hypoxia exposure. *Environmental Biology of Fishes,* 106(7): 1651-1667.

Murakami, A. 1966. Rearing experiments of a chaetognath, *Sagitta crassa. Information Bulletin on Planktology in Japan*, 13: 62–65.

Murdy, E.O., R.S. Birdsong and J.A. Musick. 1997. *Fishes of Chesapeake Bay*. Smithsonian Institution Press, Washington and London. 324 p.

Nätscher, P.S., G. Dera, C.J. Reddin, P. Rita and K. De Baets. 2021. Morphological response accompanying size reduction of belemnites during an Early Jurassic hyperthermal event modulated by life history. *Scientific Reports,* 11(1): 14480.

Ngamsiri, T., M. Nakajima, S. Sukmanomon, N. Sukumasavin, W. Kamonrat, U. Na-Nakorn and N. Taniguchi. 2007. Genetic diversity of wild Mekong giant catfish *Pangasianodon gigas* collected from Thailand and Cambodia. *Fisheries Science,* 73(4): 792–799.

Nauen, C.E. 1978. The growth of the sea star, *Asterias rubens*, and its role as benthic predator in Kiel Bay. *Kieler Meeresforschungen-Sonderheft,* 4: 68-81.

Navaluna, N.A. and D. Pauly. 1988. Seasonality in the recruitment of Philippine fishes as related to monsoon wind patterns, p. 167-179. *In:* A. Yañez- Arancibia and D. Pauly (eds) *Proceedings of the IREP/OSLR Workshop on the Recruitment of Coastal Demersal Communities, Campeche, Mexico, 21-25 April 1986*. Supplement to IOC Workshop Rep. No. 44.

Needham, J. 1959. *Science and Civilization in China: Vol. 3, Mathematics and the Sciences of the Heavens and the Earth.* Cambridge University Press, Cambridge, 926 p.

Neidetcher, S. K., T. P. Hurst, L. Ciannelli and E.A. Logerwell. 2014. Spawning phenology and geography of Aleutian Islands and eastern Bering Sea Pacific cod (*Gadus macrocephalus*). *Deep Sea Research Part II: Topical Studies in Oceanography,* 109: 204-214.

Nelson, D.L. and M.M. Cox. 2017. *Lehninger Principles of Biochemistry (7th Edition).* W.H. Freeman and Company, New York, 1119 p.

Nelson, H.R. and A.H. Altieri. 2019. Oxygen: the universal currency on coral reefs. *Coral Reefs,* 38(2): 177-198.

Niimi, A.J. and F.W.H. Beamish. 1974. Bioenergetics and growth of largemouth bass (*Micropterus salmoides*) in relation to body weight and temperature. *Canadian Journal of Zoology,* 52: 447-456.

Nilsson, G.E., J.A. Hobbs, S. Östlund-Nilsson and P.L. Munday, P.L. 2007. Hypoxia tolerance and air-breathing ability correlate with habitat preference in coral-dwelling fishes. *Coral Reefs,* 26: 241-248.

Nilsson, G.E. 2007. Gill remodeling in fish – a new fashion or an ancient secret? *Journal of Experimental Biology,* 210(14): 2403–2409.

Nilsson, G.E. and S. Östlund-Nilsson. 2008. Does size matter for hypoxia tolerance in fish? *Biological Reviews,* 83(2): 173-189.

Nilsson, G.E., N. Crawley, I.G. Lunde and P.L. Munday. 2009. Elevated temperature reduced the respiratory scope of coral reef fishes. *Global Change Biology,* 15: 1405.

Nilsson, G.E., A. Dymowska and J.A. Stecyk. 2012. New insights into the plasticity of gill structure. *Respiratory Physiology & Neurobiology,* 184(3): 214-222.

Nordenskiöld, E. 1946. *The History of Biology*. Tudor Publishing, New York. 629 p.

Nordlie, F.G. 2014. Influences of body mass, temperature, oxygen tension, and salinity on respiratory oxygen consumption of cyprinodontoid fishes of three families. *Reviews in Fish Biology and Fisheries,* 24(1): 269–315.

Noren, S.R., G. Lacave, R.S. Wells and T.M. Williams. 2002. The development of blood oxygen stores in bottlenose dolphins (*Tursiops truncatus*): implications for diving capacity. *Journal of Zoology, London,* 258: 105–113. DOI:10.1017/S095283690200124

Norin, T., H. Malte and T.D. Clark. 2014. Aerobic scope does not predict the performance of a tropical eurythermal fish at elevated temperatures. *Journal of Experimental Biology,* 217: 244–251.

Norton, S.F., Z.A. Eppley and B.D. Sidell. 2000. Allometric scaling of maximal enzyme activities in the axial musculature of striped bass, *Morone saxatilis* (Walbaum). *Physiological and Biochememical Zoology,* 73: 819–828. DOI:10.1086/318103

Nothdurft, L.D. and G.E. Webb. 2007. Microstructure of common reef-building coral genera *Acropora*, Pocillopora, *Goniastrea* and *Porites*: constraints on spatial resolution in geochemical sampling. *Facies,* 53: 1–26.

Nowell, L.B., J.W. Brownscombe, L.F. Gutowsky, K.J. Murchie, C.D. Suski, A.J. Danylchuk and S.J. Cooke. 2015. Swimming energetics and thermal ecology of adult bonefish

(*Albula vulpes*): A combined laboratory and field study in Eleuthera, The Bahamas. *Environmental Biology of Fishes,* 98: 2133–2146.

Nyhart, L. 2012. Voyaging and the scientific expedition report, 1800–1940, p. 65–86 *In:* J.A Secord (ed.) *Science in Print: Essays on the History of Science and the Culture of Print.* University of Wisconsin Press.

Nyhart, L.K. 1995. *Biology Takes Form: Animal Morphology and the German Universities, 1800-1900*. University of Chicago Press, Chicago, 428 p.

O'Dor, R.K. and M.J. Wells. 1987. Energy and nutrient flows, p. 109–133 *In:* P.R. Boyle (ed.) *Cephalopod Life Cycle,* Vol. II. Academic Press, Orlando.

O'Dor, R.K. and J.A. Hoar. 2000. Does geometry limit squid growth? *ICES Journal of Marine Science,* 57: 8–14.

Ogle, D.H. 2017. An algorithm for the von Bertalanffy seasonal cessation in growth function of Pauly *et al.* (1992). *Fisheries Research,* 185: 1–5. DOI:10.1016/j.fishres.2016.09.020

Oikawa, S. and Y. Itazawa. 1985. Gill and body surface areas of the carp in relation to body mass, with special reference to the metabolism-size relationship. *Journal of Experimental*, 117: 1–14

Oliveira, F.J.M., D.P. Lima Junior and L.M. Bini. 2022. Body size explains patterns of fish dominance in streams. *Hydrobiologia,* 849(10): 2241-2251.

Olson, K.R. and P.O. Fromm. 1971. Excretion of urea by two teleosts exposed to different concentrations of ambient ammonia. *Comparative Biochemistry and Physiology,* 40: 999–1007.

Omori, M., H. Ishii and A. Fujinaga. 1995. Life history strategy of *Aurelia aurita* (Cnidaria, Scyphomedusae) and its impact on the zooplankton community of Tokyo Bay. *ICES Journal of Marine Science,* 52: 597–603.

Oppen-Berntsen, D.O., A. Bogsnes and B.T. Walther. 1990. The effects of hypoxia, alkalinity and neurochemicals on hatching of Atlantic salmon (*Salmo salar*) eggs. *Aquaculture,* 86(4): 417–430.

Oro, J., S.W. Squyres, R.T. Reynolds and T.M. Mills. 1992. Europa: prospects for an ocean and exobiological implications, p. 103–126 *In:* G.C. Carle, D.E. Schwartz and J. Huntington (eds) *Exobiology in Solar System Exploration.* NASA-SP-512. https://ntrs.nasa.gov/api/citations/19930009363/downloads/19930009363.pdf

Orzack, S.H. and P. Forber. 2010. Adaptationism *In:* E.N. Zalta (ed.) *The Stanford Encyclopedia of Philosophy.* https://plato.stanford.edu/entries/adaptationism/

Overnell, J. and R.S. Batty. 2000. Scaling of enzyme activity in larval herring and plaice: effect of temperature and individual growth rate on aerobic and anaerobic capacity. *Journal of Fish Bio*logy, 56: 577–589.

Palomares, M.L.D. and D. Pauly, Editors. 2024. SeaLifeBase, World Wide Web electronic publication. http://www.sealifebase.org

Palomares, M.L.D., V.A. Parducho, R. Reyes and N. Bailly. 2022. The interrelationship of temperature, growth parameters and activity level in fishes. *Environmental Biology of Fishes,* 105: 1475–1479. DOI:10.1007/s10641-022-01261-5

Palomares, M.L.D. and D. Pauly. 2009. The growth of jellyfishes. *Hydrobiologia,* 616: 11–21. DOI:10.1007/s10750-008-9582-y

Palzenberger, M. and H. Pohla. 1992. Gill surface area of water-breathing freshwater fish. *Reviews in Fish Biology and Fisheries,* 2: 187–216.

Pandian, T.J. 1967. Intake, digestion, absorption and conversion of food in the fishes *Megalops cyprinoides* and *Ophiocephalus striatus*. *Marine Biology*, 1: 16–32.

Pandian, T.J. 1970. Intake and conversion of food in the fish *Limanda limanda* exposed to different temperatures. *Marine Biology*, 5: 1–17.

Panella, G. 1971. Fish otoliths: daily growth layers and periodical patterns. *Science*, 173: 1124.

Pang, X., Z.D. Cao, J.L. Peng and S.J. Fu. 2010. The effects of feeding on the swimming performance and metabolic response of juvenile southern catfish, *Silurus meridionalis*, acclimated at different temperatures. *Comparative Biochemistry and Physiology A*, 155: 253–258.

Pang, X., X.Z. Yuan, Z.D. Cao and S.J. Fu. 2013. The effects of temperature and exercise training on swimming performance in juvenile qingbo (*Spinibarbus sinensis*). *Journal of Comparative Physiology B*, 183(1): 1–10. 10.1007/s00360-012-0690-7

Pankhurst, N.W. 2016. Reproduction and development, p. 295–331 *In:* C.B. Schreck, L. Tort, A.P. Farrell and C.J. Brauner (eds) *Fish Physiology*. Vol. 35. Elsevier, Amsterdam.

Paredes-López, D., R. Robles-Huaynate, C. Rebaza-Alfaro, J. Delgado-Ramírez and U. Aldava-Pardave. 2021. Effect of stocking density of juvenile *Arapaima gigas* on rearing water quality hematological and biochemical profile, and productive performance. *Latin American Journal of Aquatic Research*, 49(2): 193–201.

Parker, A. 2003. *In the Blink of an Eye*. Perseus Publishing, Cambridge, MA, 352 p.

Parker, G.A. 1992. The evolution of sexual size dimorphism in fish. *Journal of Fish Biology*, 41: 1–20. DOI:10.1111/j.1095-8649. 1992.tb03864.x

Parmesan, C. and G. Yohe. 2003. A globally coherent fingerprint of climate change impacts across natural systems. *Nature*, 421: 37–42.

Parry, D.A. 1944. Structure and Function of the Gut in *Spadella cephaloptera* and *Sagitta setosa*. *Journal of the Marine Biological Association of the U.K.*, 26: 16–36. DOI:10.1017/S0025315400014430

Parry, G.D. 1983. The influence of the cost of growth on ectotherm metabolism. *Journal of Theoretical Biology*, 101(3): 453–477.

Partridge, L. and J.A. Coyne. 1997. Bergmann's rule in ectotherms: is it adaptive? *Evolution*,51(2): 632–635.

Passow, C.N., R. Greenway, L. Arias-Rodriguez, P.D. Jeyasingh and M. Tobler. 2015. Reduction of energetic demands through modification of body size and routine metabolic rates in extremophile fish. *Physiological and Biochemical Zoology*, 88(4): 371–383.

Pauly, D. 1976. The biology, fishery and potential for aquaculture of *Tilapia melanotheron* in a small West African lagoon. *Aquaculture*, 7(1): 33–49.

Pauly, D. 1979. *Gill size and temperature as governing factors in fish growth: a generalization of von Bertalanffy's growth formula.* Berichte aus dem Institut für Meereskunde an der Universität Kiel. 63, xv + 156 p. Doctoral dissertation, Kiel University. https://oceanrep.geomar.de/id/eprint/41323/

Pauly, D. 1980a. On the interrelationships between natural mortality, growth parameters and mean environmental temperature in 175 fish stocks. *Journal du Conseil international pour l'Exploration de la Mer*, 39(3): 175–192.

Pauly, D. 1980b. The use of a pseudo catch-curve for the estimation of mortality rates in *Leiognathus splendens* (Pisces: Leiognathidae) in Western Indonesian Waters.

Berichte der Deutschen wissenschaftlichen Kommission für Meeresforschung, 28(1): 56–60.

Pauly, D. 1981. The relationships between gill surface area and growth performance in fish: a generalization of von Bertalanffy's theory of growth. *Berichte der Deutschen wissenschaftlichen Kommission für Meeresforschung,* 28: 251–282.

Pauly, D. 1982a. Further evidence for a limiting effect of gill size on the growth of fish: the case of the Philippine goby (*Mistichthys luzonensis*). *Kalikasan: Philippine Journal of Biology,* 11: 379–383.

Pauly, D. 1982b. The fishes and their ecology, p. 15–33 *In:* D. Pauly and A.N. Mines (eds) *Small-scale fisheries of San Miguel Bay Philippines: biology and stock assessment.* ICLARM Technical Report 7. [Reprinted in 1985 as 'Ecology of coastal and estuarine fishes in Southeast Asia: a Philippine case study,' p. 499–514 *In:* A. Yañez-Arancibia (ed.) *Fish Community Ecology in Estuaries and Coastal Lagoons.* Universidad Nacional Autónoma de Mexico, Mexico, D.F.].

Pauly, D. 1984. A mechanism for the juvenile-to-adult transition in fishes. *ICES Journal of Marine Science,* 41(3): 280–284. DOI:10.1093/icesjms/41.3.280

Pauly, D. 1985. The population dynamics of short-lived species, with emphasis on squids. *NAFO Scientific Council Studies,* 9: 143–154.

Pauly, D. 1986. A simple method for estimating the food consumption of fish populations from growth data and food conversion experiments. *Fishery Bulletin,* 4(4): 827–842. https://hdl.handle.net/20.500.12348/3402

Pauly, D. 1989. A theory of fishing for a two-dimensional world. *Fishbyte, Newsletter of the Network of Tropical Fisheries Scientists* 7(3): 28–31. [Reprinted as Essay no. 12, p. 104–111 *In*: D. Pauly. 1994. *On the sex of fish and the gender of scientists: essays in fisheries science.* Chapman and Hall, London]

Pauly, D. 1994a. Quantitative analysis of published data on the growth, metabolism, food consumption and related features of the red-bellied piranha, *Serrasalmus nattereri* (Characidae). *Environmental Biology of Fishes* 41: 423–437. [Reprinted *In:* E.K. Balon, M.N. Bruton and D.I.G. Noakes (eds) *Women in ichthyology: an anthology in honour of ET, Ro and Genie.* Kluwer Academic Publishers, Dordrecht.]

Pauly, D. 1994b. Re-sharpening Ockham's razor. *Naga, the ICLARM Quarterly,* 17(2): 7–8.

Pauly, D. 1998a. Beyond our original horizons: the tropicalization of Beverton and Holt. *Reviews in Fish Biology and Fisheries,* 8: 307–334.

Pauly, D. 1998b. Tropical fishes: patterns and propensities, p. 1–17 *In:* T.E. Langford, J. Langford and J.E. Thorpe (eds) *Tropical Fish Biology. Journal of Fish Biology,* 53 (Supplement A): 1–17. DOI:10.1111/j.1095-8649.tb01014.x

Pauly, D. 1998c. Why squids, though not fish, may be better understood by pretending they are. *South African Journal of Marine Science,* 20: 47–58. DOI:10.2989/025776198784126269

Pauly, D. 2002a. Growth and mortality of basking shark *Cetorhinus maximus,* and their implications for whale shark *Rhincodon typus*, p. 199–208 *In:* S.L. Fowler, T. Reid and F.A. Dipper (eds) *Elasmobranch biodiversity: conservation and management.* Proceedings of an International Seminar and Workshop held in Sabah, Malaysia. Occasional Papers of the IUCN Survival Commission 25, Gland, Switzerland.

Pauly, D. 2002b. Consilience in oceanographic and fishery research: a concept and some digressions, p. 41–46 *In:* J. McGlade, P. Cury, K.A. Koranteng and N.J. Hardman-Mountford (eds) *The Gulf of Guinea Large Marine Ecosystem: environmental forcing and sustainable development of marine resources*. Elsevier Science, Amsterdam.

Pauly, D. 2004. *Darwin's Fishes: an Encyclopedia of Ichthyology, Ecology and Evolution.* Cambridge University Press, Cambridge, xxv + 340 p.

Pauly, D. 2010. *Gasping Fish and Panting Squids: Oxygen, Temperature and the Growth of Water-Breathing Animals.* Excellence in Ecology (22), International Ecology Institute, Oldendorf/Luhe, Germany, xxviii + 216 p.

Pauly, D. 2018a. Female fish grow bigger: Let's deal with it. *Trends in Ecology and Evolution,* 34(3): 181–182. DOI:10.1016/j.tree.2018.12.007

Pauly, D. 2018b. Learning from peer-reviews: The Gill-Oxygen Limitation Theory and the Growth of Fishes and Aquatic Invertebrates, p. 53–70 *In:* Pauly D. and V. Ruiz-Leotaud (eds) *Marine and Freshwater Miscellanea.* Fisheries Centre Research Reports 26(2). University of British Columbia.

Pauly, D. 2019. *Gasping Fish and Panting Squids: Oxygen, Temperature and the Growth of Water-Breathing Animals – 2nd Edition.* Excellence in Ecology (22), International Ecology Institute, Oldendorf/Luhe, Germany, v + 279 p.

Pauly, D. 2020. Females grow bigger – deal with it! p. 125–130 *In:* D. Pauly and V. Ruiz-Leotaud (eds) *Marine and Freshwater Miscellanea II.* Fisheries Centre Research Reports 28(2). University of British Columbia, Vancouver.

Pauly, D. 2021a. The Gill-Oxygen Limitation Theory (GOLT) and its critics. *Science Advances,* 7(2). DOI:10.1126/sciadv.abc6050

Pauly, D. 2021b. Why do fish reach first maturity when they do? *Journal of Fish Biology,* 101(2): 1–9. DOI:10.1111/jfb.14902

Pauly, D. 2021c. Pre-submission draft of 'The GOLT and its critics' p. 67–115 *In:* D. Pauly and E. Chu (eds) *Marine and Freshwater Miscellanea III.* Fisheries Centre Research Report 29(1). University of British Columbia, Vancouver.

Pauly, D. 2023. What are marine heatwaves? *Oceana Magazine*, Winter 2023, p. 22–23.

Pauly, D. 2024. *Gill size and temperature as governing factors in fish growth: a generalization of von Bertalanffy's growth formula (2nd edition).* Fisheries Centre Research Report 32(2), University of British Columbia, Vancouver, 104 p.

Pauly, D., U.S. Amarasinghe, E. Chu, K.M.F. Freire, E. Vasquez, M.J. Butler. 2022a. The growth, respiration, and reproduction of crustaceans: a synthesis through the Gill- Oxygen Limitation Theory (GOLT*). Journal of Crustacean Biology,* 42: 1–13. DOI:10.1093/jcbiol/ruac059

Pauly, D. and H. Calumpong. 1984. Growth and mortality of the sea-hare *Dolabella auricularia* (Gastropoda: Aplysiidae) in the Central Visayas, Philippines. *Marine Biology,* 79: 289–293. DOI:10.1007/BF00393260

Pauly, D., C. Casal and M.L.D. Palomares. 2000. DNA, cell size and fish swimming. Box 34, p. 254 *In:* R. Froese and D. Pauly (eds) *FishBase 2000: Concepts, Design and Data Sources*. ICLARM, Los Baños, Philippines.

Pauly, D., W. Graham, S. Libralato, L. Morissette and M.L.D. Palomares. 2009. Jellyfish in ecosystems, online databases and ecosystem models. *Hydrobiologia,* 616 (1): 67–85. doi 10.1007/s10750-008-9583-x

Pauly, D. and W.W.L. Cheung. 2017. Sound physiological knowledge and principles in modelling shrinking of fishes under climate change. *Global Change Biology,* 24(1): e15-e26. DOI:10.1111/gcb.13831

Pauly, D. and W.W.L. Cheung. 2018. On confusing cause and effect in the oxygen limitation of fish. *Global Change Biology*. doi 10.1111/gcb.14383

Pauly, D., V. Christensen, J. Dalsgaard, R. Froese and F.C. Torres. 1998. Fishing down marine food webs. *Science,* 279: 860-863.

Pauly, D. and E. Chu. 2021. Key Traits of Amphioxus Species (Cephalochordata) and the GOLT, p. 57–66 *In:* D. Pauly and E. Chu (eds) *Marine and Freshwater Miscellanea III.* Fisheries Centre Research Report 29(1). University of British Columbia, Vancouver. DOI:10.14288/1.0396376

Pauly, D., E. Chu and J. Müller. 2024a. Brobdingnagians and Goliaths: two forms of gigantism in fish. *Journal of Fish Biology,* 1–9. DOI:10.1111/jfb.15694

Pauly, D. and N. David. 1981. ELEFAN I, a BASIC program for the objective extraction of growth parameters from length-frequency data. *Berichte der Deutschen wissenschaftlichen Kommission für Meeresforschung,* 28(4): 205–211.

Pauly, D. and D. Dimarchopoulou. 2022. Fishes in a Warming and Deoxygenating World – Introduction to a Special Issue. *Environmental Biology of Fishes,* 105: 1261–1267. DOI:10.1007/s10641-022- 01357-y

Pauly, D., R. Froese, C. Liang, J. Müller and P. Sorensen. 2023. Post-spawning growth acceleration in fish as a result of reduced live weight and thus, increased food conversion efficiency. *Environmental Biology of Fishes,* 106: 2031–2043. DOI: 10.1007/s10641-023-01482-2

Pauly, D. and J.D. Holmes. 2022. Re-assessing growth and mortality estimates for the Ordovician trilobite *Triarthrus eatoni. Paleobiology,* 49: 120-130. DOI:10.1017/pab.2022.22

Pauly, D., J. Ingles and R. Neal. 1984. Application to shrimp stocks of objective methods for the estimation of growth, mortality and recruitment-related parameters from length-frequency data (ELEFAN I and II), p. 220–234 *In:* J.A. Gulland and B.I. Rothschild (eds) *Penaeid shrimps - their biology and management.* Fishing News Books, Farnham, England.

Pauly, D. and Ç. Keskin. 2017. Temperature constraints shaped the migration routes of mackerel (*Scomber scombrus*) in the Black Sea. *Acta Adriatica,* 58(2): 339–346.

Pauly, D. and M.E. Lam. 2023. Too hot or too cold: the biochemical basis of temperature-size rules for fish and other ectotherms. *Environmental Biology of Fishes,* 106: 1519–1527. https://10.1007/s10641-023-01429-7

Pauly, D. and C. Liang. 2022a. A reconceptualization of the interactions between spawning and growth in bony fish. *Scientia Marina,* 84(4): p.e044-e044. DOI:10.3989/scimar.05280.044

Pauly, D. and C. Liang, C. 2022b. Temperature and the early maturation of fish: a simple sine-wave model for predicting spring spawning. *Environmental Biology of Fishes,* 105(10): 1481–1487. DOI:10.1007/s10641-022-01212-0

Pauly, D., C. Liang, W. Xian, E. Chu and N. Bailly. 2021a. The sizes, growth and reproduction of arrow worms (Chaetognatha) in light of the Gill-Oxygen Limitation Theory (GOLT). *Journal of Marine Science and Engineering,* 9(12): 1397. DOI:10.3390/jmse9121397

Pauly, D., L. Matsushige, J. Malacane, A. Hay, E. Chu and M. Warren. 2022c. Length-weight relationships and growth parameters of common and leafy seadragons (Syngnathidae) from a public aquarium. *Fishes* 7(2), p.77. DOI:10.3390/fishes7020077

Pauly, D., J. Moreau and M.L.D. Palomares. 1988. Detritus and energy consumption and conversion efficiency of *Sarotherodon melanotheron* (Cichlidae) in a West African lagoon. *Journal of Applied Ichthyology*, 4: 150–153.

Pauly, D. and J. Müller. 2022. Georges Cuvier 'On the respiration of fishes,' p. 13–15 *In:* D. Pauly and E. Chu (eds) *Marine and Freshwater Miscellanea IV.* Fisheries Centre Research Reports 30(4). University of British Columbia, Vancouver.

Pauly, D. and J. Müller. 2024. Testing mechanistic theories must be based on correct interpretations: A response to Lonthair *et al.* (2024). *Journal of Experimental Biology*, 227(16), p.jeb247841. DOI:10.1242/jeb.247841

Pauly, D. C. Piroddi, L. Hood, N. Bailly, El. Chu, V. Lam, E. Pakhomov, L.K. Pshenichnov, V.I. Radchenko and M.L.D. Palomares. 2021b. The biology of mesopelagic fishes and their catches (1950-2018) by commercial and experimental fisheries. *Journal of Marine Science and Engineering, 9,* 1057. 9(10): 1057. DOI:10.3390/ jmse9101057

Pauly, D. and J. Müller. 2025. Fick's diffusion laws and scaling of the gill surface area and oxygen uptake in fish. *Fishes*, 10(5): 233. DOI:10.3390/Fishes10050233

Pauly, D. and R.S.V. Pullin. 1988. Hatching time in spherical, pelagic marine fish eggs in response to temperature and egg size. *Environmental Biology of Fishes*, 22(2): 261–271.

Pauly, D., P. Sakyi-Djan and M.F. Akwetey. 2024b. Patterns of growth, respiration and reproduction in species of the genus *Sepia*, p. 74–80 *In:* D. Pauly, D. and V. Ruiz-Leotaud (eds) *Marine and Freshwater Miscellanea V.* Fisheries Centre Research Reports 32(4). University of British Columbia, Vancouver.

Pauly, D., N. Smith and M. Butler. 2022b. Growth and related traits of the wool sponge *Hippospongia lachne*: practical and theoretical considerations. *Fisheries Bulletin*, 120: 99–112. DOI:10.7755/FB.120.2.1

Pauly, D., M. Soriano-Bartz, J. Moreau and A. Jarre. 1992. A new model accounting for seasonal growth cessation in fishes. *Australian Journal of Marine and Freshwater Research*, 43: 1151–1156.

Pauly, D., J. Stuart Smith, A. Trotter, V. Delpero, E. Chu and T. Bessell. 2024b. Growth and related parameters of the red handfish (*Thymichthys politus*) p. 7–10 *In:* D. Pauly, D. and V. Ruiz-Leotaud (eds) *Marine and Freshwater Miscellanea V.* Fisheries Centre Research Reports 32(4). University of British Columbia, Vancouver.

Pauly, D. and P.D. Ward. 2024. Major biological traits of *Nautilus* spp. with emphasis on *N. belauensis, N. macromphalus* and *N. pompilius*, p. 63-73 *In:* D. Pauly, D. and V. Ruiz-Leotaud (eds) *Marine and Freshwater Miscellanea V.* Fisheries Centre Research Reports 32(4). University of British Columbia, Vancouver.

Pawlik, J.R. 2011. The chemical ecology of sponges on Caribbean reefs: natural products shapes natural systems. *BioScience*, 61: 888–898. DOI:10.1525/bio.2011.61.11.8

Pawlik, J.R., S.E. McMurray, P. Erwin and S. Zea. 2015. A review of evidence for food limitation of sponges on Caribbean reefs. *Marine Ecology Progress Series*, 519: 265–283. DOI:10.3354/meps11093

Pearl, J. and D. MacKenzie. 2018. *The Book of Why: the New Science of Cause and Effect*. Basic Books, New York, 419 p.

Pearre, S. 1991. Growth and reproduction, p. 61–75 *In:* Q. Bone, H. Kapp and A.C. Pierrot-Bults (eds) *The Biology of Chaetognaths*. Oxford University Press, Oxford.

Pearson, R.G. and T.P. Dawson. 2003. Predicting the impact of climate change on the distribution of species: are bioclimate envelope models useful? *Global Ecolology and Biogeography,* 12: 361–371.

Pelster, B., C.M. Wood, B.M. Susana and A.L. Val. 2020. Gills and air-breathing organs in O_2 uptake, CO_2 excretion, N-waste excretion, and ionoregulation in small and large pirarucu (*Arapaima gigas*). *Journal of Comparative Physiology, B* 190(5): 569–583.

Perales-Raya, C., A. Bartolomé, E. Hernández-Rodríguez, M. Carrillo, V. Martín, E. Fraile-Nuez. 2020. How old are giant squids? First approach to aging *Architeuthis* beaks. *Bulletin of Marine Science,* 96(2): 357–374.

Peralta-Maraver, I. and E.L. Rezende. 2021. Heat tolerance in ectotherms scales predictably with body size. *Nature Climate Change,* 11(1): 58-63.

Perrone, Jr, M. and T.M. Zaret. 1979. Parental care patterns of fishes. *American Naturalist,* 113(3): 351–361.

Perry, W.B. 2024. James and the giant perch: a fresh framework for understanding gigantism in fish. *Journal of Fish Biology,* 6: 1653-1653. DOI:10.1111/jfb.15838

Perry, A.L, P.J. Ellis and J.D. Reynolds. 2005. Climate change and distribution shifts in marine fishes. *Science,* 308: 1912–1915.

Persaud, D.I., I.W. Ramnarine and J.B. Agard. 2006. Trade-off between digestion and respiration in two airbreathing callichthyid catfishes *Holposternum littorale* (Hancock) and *Corydoras aeneus* (Gill). *Environmental Biology of Fishes,* 76(2-4): 159–165.

Pezner, A.K., T.A. Courtney, H.C. Barkley, W.C. Chou, H.C. Chu, S.M. Clements, T. Cyronak, M.D. DeGrandpre, S.A. Kekuewa, D.I. Kline and Y.B. Liang. 2023. Increasing hypoxia on global coral reefs under ocean warming. *Nature Climate Change,* 13(4): 403–409.

Phan, L.T., K. Masagounder, J. Mas-Muñoz and J. Schrama. 2021. Differences in energy utilization efficiency of digested protein, fat and carbohydrates in snakehead (*Channa striata*). *Aquaculture,* 32: 736066. DOI:10.1016/j.aquaculture.2020.736066

Pholprasit, S. 1994. *The story of Mekong giant catfish. Proceedings of the fourth Indo-Pacific Fish Conference, Bangkok,* 1993, pp. 23-26.

Pholprasith, S. and P. Tavarutmaneegul. 1997. Biology and culture of Mekong giant catfish *Pangasianodon gigas* (Chevey 1930). Extension Paper No. 31. National Inland Fisheries Institute, Bangkok, 79 p.

Pickford, G.E. and F.B. Grant. 1967. Serum Osmolality in the Coelacanth, *Latimeria chalumnae*: Urea Retention and Ion Regulation. *Science,* 155: 568–570.

Pietsch, TW. and J.M. Orr. 2010. *Fishes of the Salish Sea – Puget Sound and the Straits of Georgia and Juan de Fuca.* Vol. 2, University of California Press, Seattle, p. 217–646.

Pillans, R.D. and C.E. Franklin. 2004. Plasma osmolyte concentrations and rectal gland mass of bull sharks *Carcharhinus leucas*, captured along a salinity gradient. *Comparative Biochemistry and Physiology Part A,* 138: 363–371. DOI:10.1016/j.cbpb.2004.05.006

Pimiento, C., J.L. Cantalapiedra, K. Shimada, D.J. Field and J.B. Smaers. 2019. Evolutionary pathways toward gigantism in sharks and rays. *Evolution,* 73(3): 588–599.

Pitt, K. A. and M. J. Kingsford. 2000. Reproductive biology of the edible jellyfish *Catostylus mosaicus* (Rhizostomeae). *Marine Biology,* 137: 791–799.

Poloczanska, E.S., C.J. Brown, W.J. Sydeman, W. Kiessling, D.S. Schoeman, P.J. Moore, K. Brander, J.F. Bruno, L.B. Buckley, M.T. Burrows, C.M. Duarte, B.S. Halpern, J. Holding, J., Kappel, C.V., O'Connor, M.I., Pandolfi, J.M., Parmesan, C., Schwing, F., Thompson,

S.A. and A.J. Richardson. 2013. Global imprint of climate change on marine life. *Nature Climate Change,* 3(10): 919–925.

Polymeropoulos, E.T., N.G. Elliott and P.B. Frappell. 2016. The maternal effect of differences in egg size influence metabolic rate and hypoxia induced hatching in Atlantic salmon eggs: implications for respiratory gas exchange across the egg capsule. *Canadian Journal of Fisheries and Aquatic Sciences,* 73(8): 1173–1181.

Pope, E.C., G.C., Hays, T.M. Thys, T.K. Doyle, D.W. Sims, N. Queiroz, V.J. Hobson, L. Kubicek and J.D. Houghton. 2010. The biology and ecology of the ocean sunfish *Mola mola*: a review of current knowledge and future research perspectives. *Reviews in Fish Biology and Fisheries,* 20: 471–487.

Porter, R.K. and M.D. Brand. 1995. Cellular oxygen consumption depends on body mass. *American Journal of Physiology - Regulatory, Integrative and Comparative Physiology* 269(1): R226-R228.

Pörtner, H.O. 2001. Climate change and temperature-dependent biogeography: oxygen limitation of thermal tolerance in animals. *Naturwissenschaften,* 88: 137–146.

Pörtner, H.O. and R. Knust. 2007. Climate change affects marine fishes through the oxygen limitation of thermal tolerance. *Science,* 315: 95–97. DOI:10.1126/science.1135471

Pörtner, H.O. 2010. Oxygen-and capacity-limitation of thermal tolerance: a matrix for integrating climate-related stressor effects in marine ecosystems. *Journal of Experimental Biology,* 213(6): 881–893.

Pörtner, H.O. 2002. Climate variations and the physiological basis of temperature dependent biogeography: systemic to molecular hierarchy of thermal tolerance in animals. *Comparative Biochemistry and Physiology A,* 132(4): 739–761.

Pörtner, H.O., D.C. Roberts, E.S. Poloczanska, K.Mintenbeck, M. Tignor, A. Alegría, M. Craig, S. Langsdorf, S. Löschke, V. Möller and A. Okem. 2022. IPCC 2022: summary for policymakers, p. 3-33 *In: Climate change 2022: Impacts, Adaptation, and Vulnerability: Contribution of Working Group II to the sixth assessment report of the Intergovernmental Panel on Climate change.* Cambridge, UK, and New York, NY.

Post, J.R. and J.A. Lee. 1996. Metabolic ontogeny of teleost fishes. *Canadian Journal of Fisheries and Aquatic Sciences,* 53: 910–923.

Prasad, M.S. and P. Prasad. 1984. Blood-water diffusion barrier at the secondary gill lamellae in *Anabas testudineus* (Bloch) during early ontogenesis. *Current Science,* 53(3): 156–158.

Priede, I.G. 1985. Metabolic scope in fishes, p. 33–64 *In:* P. Tyler and P. Calow (eds) *Fish Energetics: New Perspectives.* Springer, London/Dordrecht.

Prince, E.E. 1904. The maximum size of fishes and its causes. 36th Annual Report of the Department of Marine and Fisheries for 1903, Sessional Paper 22, pages lix-lxvii. https://publications.gc.ca/collections/collection_2015/mpo-dfo/MA1-1-1903-eng.pdf

Prince, J., A. Hordyk, S.R. Valencia, N. Loneragan and K. Sainsbury. 2015. Revisiting the concept of Beverton–Holt life-history invariants with the aim of informing data-poor fisheries assessment. *ICES Journal of Marine Science,* 72: 194–203. DOI:10.1093/icesjms/fsu011

Prinzing, T.S., J.S. Bigman, Z.R. Skelton, N.K. Dulvy and N.C. Wegner. 2023. The allometric scaling of oxygen supply and demand in the California horn shark, *Heterodontus francisci. Journal of Experimental Biology* 226. DOI:10.1242/jeb.246054

Privalov, P.L. 1990. Cold Denaturation of Protein. *Critical Reviews in Biochemistry and Molecular Biology,* 25(4): 281–306. doi/org/10.3109/10409239009090612

Prokop, J., K. Rosová, A. Leipner and P. Sroka. 2023. Thoracic and abdominal outgrowths in early pterygotes: a clue to the common ancestor of winged insects? *Communications Biology,* 6(1): 1262.

Pütter, A. 1909. *Die Ernährung der Wassertiere und der Stoffhaushalt der Gewässer.* G. Fischer, Jena, 168 p.

Pütter, A. 1911. *Vergleichende Physiologie.* G. Fischer, Jena, 721 p.

Pütter, A. 1914. Der Stoffwechsel der Kieselschwämme. *Zeitschrift für Allgemeine Physiologie,* 16: 65–114.

Pütter, A. 1920. Studien über physiologische Ähnlichkeit VI. Wachstumsähnlichkeiten. *Pflüger's Archiv für die gesamte Physiologie des Menschen und der Tiere,* 180: 298–340. DOI:10.1007/BF01755094

Quince, C., P.A. Abrams, B.J. Shuter and N.P. Lester. 2008. Biphasic growth in fish I: Theoretical foundations. *Journal of Theoretical Biology,* 254(2): 197–206.

Queiroz, H.L. 2000. Natural history and conservation of pirarucu, *Arapaima gigas,* at the Amazonian várzea: red giants in muddy waters. Doctoral dissertation. University of St. Andrews, Scotland, 190 p.

Rabalais, N.N., W.J. Cai, J. Carstensen, D.J. Conley, B. Fry, X. Hu, Z. Quinones-Rivera, R. Rosenberg, C.P. Slomp, R.E. Turner and M. Voss. 2014. Eutrophication-driven deoxygenation in the coastal ocean. *Oceanography,* 27(1): 172–183.

Radtke, L.D. 1966. Distribution of smelt, juvenile sturgeon and starry flounder in the Sacramento-San Joaquin Delta with observations on food of sturgeon, p. 115-129 *In:* J. L. Turner and D. W. Kelly (ed.) *Ecological Studies of the Sacramento-San Joaquin Delta. Part II Fishes of the Delta. Fish Bulletin,* 136. https://escholarship.org/uc/item/4z31776k

Ragonese, S. and P. Jereb. 1991. Length-weight relationships and growth of the pink and elegant cuttlefish *Sepia orbignyana* and *Sepia elegans* in the Sicilan channel, p. 31-47 *In*: E. Boucaud-Camou (ed.) *La Seiche – The Cuttlefish: 1st International Syposium on the Cuttlefish Sepia,* Caen, June 1-3 1989. Centre de Publication de l'Université de Caen, France.

Rainboth, W.J. 1996. *Fishes of the Cambodian Mekong.* FAO species identification field guide for fishery purposes. FAO, Rome. 265 p.

Raja, V. and M.L. Anderson. 2021. Behavior considered as an enabling constraint, p. 209–232 *In:* F. Calzavarini and M. Viola (eds) *Neural Mechanisms - Studies in Brain and Mind,* Vol 17. Springer, Cham, Switzerland. DOI:10.1007/978-3-030-54092-0_10

Randall, D.J., J.N. Cameron, C. Daxboeck and N. Smatresk. 1981. Aspects of bimodal gas exchange in the bowfin: *Amia calva* L. (Actinopterygii: Amiiformes). *Respiration Physiology,* 43(3): 339–348.

Ray, C. 1960. The application of Bergmann's and Allen's rules to the poikilotherms. *Journal of Morphology,* 106(1): 85–108.

Read, K.R.H. 1962. Respiration of the bivalved molluscs *Mytilus edulis* L. and *Brachidontes demissus plicatulus* Lamarck as a function of size and temperature. *Comparative Biochemistry and Physiology,* 7: 89–101.

Ready, J., K. Kaschner, A.B. South, P.D. Eastwood, T. Rees, J. Rius, E. Agbayani, S. Kullander and R. Froese. 2010. Predicting the distributions of marine organisms at the global scale. *Ecological Modelling,* 221(3): 467–478.

Reeve, M.R. 1970. The biology of Chaetognatha I. Quantitative aspects of growth and egg production in *Sagitta hispida*, p.168–189 *In:* J.H Steele. (ed.) *Marine Food Chains*. University of California Press, Berkeley.

Reeve, M.R. and L.D. Baker. 1975. Production of Two Planktonic Carnivores (Chaetognatha and Ctenophore) in South Florida Inshore Waters. *Fishery Bulletin*, 73(2): 238–248.

Reeve, M.R. and M.A. Walter. 1972. Conditions of Culture, Food-Size Selection, and the Effects of Temperature and Salinity on Growth Rate and Generation Time in *Sagitta hispida* Conant. *Journal of Experimental Marine Biology and Ecology.* 9(2): 191–200. DOI:10.1016/0022-0981(72)90048-2

Regier, H., J.A. Holmes and D. Pauly. 1990. Influence of temperature changes on aquatic ecosystems: an interpretation of empirical data *In:* H.A. Regier, J.J. Magnuson and C.C. Coutant (eds). Proceedings of the Symposium on Effects of Climate Change on Fish. *Transactions of the American Fisheries Society*, 119: 374–389.

Reygondeau, G., Y., Egorova, K. Boerder, D.P. Tittensor, K. Kaschner, K. Kesner-Reyes, N. Bailly and W.W.L. Cheung. 2025. AquaX: An enhanced and revised AquaMaps methodology to model marine species distributions and biodiversity. *PLOS One* 21(2):10.1371. DOI: 10.1371/journal.pone.0335823

Renn, J. 2005. Einstein's invention of Brownian motion. *Annalen der Physik*, 14 Supplement: 23–37. DOI:10.1002/andp.200410131

Rensch, B. 1948. Histological changes correlated with evolutionary changes of body size. *Evolution*, 218–230.

Revell, L.J. 2010. Phylogenetic signal and linear regression on species data. *Methods in Ecology and Evolution*, 1: 319–329. DOI:10.1111/j.2041-210X.2010.00044.x

Reznick, D., T. Hrbek, S. Caura, J. De Greef and D. Roff, D. 2007. Life history of *Xenodexia ctenolepis*: implications for life history evolution in the family Poeciliidae. *Biological Journal of the Linnean Society*, 92(1): 77–85.

Rhumbler, L. 1909. *Die Foraminiferen (Thalamophoren) der Plankton-Expedition - zugleich Entwurf eines natürlichen Systems der Foraminiferen auf Grund selektionistischer und mechanisch-physiologischer Faktoren*, Vol. 3. *Ergebnisse der Plankton-Expedition der Humboldt-Stiftung*. Lipsius and Tischer, Kiel, 144 p.

Richards, R.J. 2017. Did Goethe and Schelling endorse species evolution? *Marking Time: Romanticism and Evolution*, 219–238.

Ricker, W.E. 1975. Computation and interpretation of biological statistics of fish populations. *Bulletin of the Fisheries Research Board of Canada*, 191, 382 p.

Ricker, W.E. 1958. Handbook of computations for biological statistics of fish populations. *Bulletin of the Fisheries Research Board of Canada*, 119, 300 p.

Ricklefs, R.E., J.M. and Starck. 1996. Applications of phylogenetically independent contrasts: a mixed progress report. *Oikos*, 77:167-172. DOI:10.2307/3545598

Rideout, R.M., G.A. Rose and M. Burton. 2005. Skipped spawning in female iteroparous fish. *Fish and Fisheries*, 6: 50–62. DOI:10.1111/j.1467-2679.2005.00174.x

Ridley, M. 1978. Paternal care. *Animal Behaviour*, 26: 904–932.

Rien, T.A. and R.C. Beamesderfer. 1994. Accuracy and precision of white sturgeon age estimates from pectoral fin rays. *Transactions of the American Fisheries Society*, 123(2): 255–265.

Rijnsdorp, A.D. 1989. Maturation of male and female North Sea plaice (*Pleuronectes platessa* L.). *Journal du Conseil international pour l'Exploration de la Mer,* 46: 35–51. DOI:10.1093/icesjms/46.1.35

Riley, W.D., M.J. Ives, Pawson, M.G. and D.L. Maxwell. 2006. Seasonal variation in habitat use by salmon, *Salmo salar*, trout, *Salmo trutta* and grayling, *Thymallus thymallus*, in a chalk stream. *Fisheries Management and Ecology,* 13(4): 221–236.

Robb, T. and M.V. Abrahams. 2003. Variation in tolerance to hypoxia in a predator and prey species: an ecological advantage of being small? *Journal of Fish Biology,* 62(5): 1067–1081. DOI:10.1046/j.1095-8649.2003.00097.x

Roberts, T.R. and C. Vidthayanon. 1991. Systematic revision of the Asian catfish family Pangasiidae, with biological observations and descriptions of three new species. *Proceedings of the Academy of Natural Sciences of Philadelphia,* 143: 97–143.

Robischon, M. 2017. Surface-area-to-volume ratios, fluid dynamics & gas diffusion: Four frogs & their oxygen flux. *The American Biology Teacher,* 79(1): 64-67.

Robinson D.P., M.Y. Jaidah, S. Bach, K. Lee, R.W. Jabado, C.A. Rohner, A. March, S. Caprodossi, A.C. Henderson, J.M. Mair, R. Ormon and S.J. Pierce. 2016. Population structure, abundance and movement of whale sharks in the Arabian Gulf and the Gulf of Oman. *PLoS ONE,* 11(6): e0158593. DOI:10.1371/journal.pone.0158593

Rodhouse, P.G. 1998. Physiological progenesis in cephalopod molluscs. *Biological Bulletin,* 195: 17–20.

Rogers, L.A. and A.B. Dougherty. 2019. Effects of climate and demography on reproductive phenology of a harvested marine fish population. *Global Change Biology,* 25(2): 708–720. DOI:10.1111/gcb.14483

Rohde, S., S. Nietzer and P.J. Schupp. 2015. Prevalence and mechanisms of dynamic chemical defenses in tropical sponges. *PLoS ONE,* 10(7): e0132236. DOI:10.1371/journal.pone.0132236

Rohle, F.J. 2006. A comment on phylogenetic correction. *Evolution,* 60: 1509–1515. DOI:10.1111/j.0014-3820.2006.tb01229.x.

Rollinson, N. and L. Rowe. 2018a. Temperature-dependent oxygen limitation and the rise of Bergmann's rule in species with aquatic respiration. *Evolution,* 72(4): 977–988. DOI:10.1111/evo.13458

Rollinson, N. and L. Rowe. 2018b. Oxygen limitation at the larval stage and the evolution of maternal investment per offspring in aquatic environments. *American Naturalist,* 191(5): 604–619.

Romanes, G.J. 1892. *Darwin and After Darwin: The Darwinian theory,* Vol. 1. Longmans, Green and Company, London. 460 p.

Rombough, P. 2007. The functional ontogeny of the teleost gill: which comes first, gas or ion exchange? *Comparative Biochemistry and Physiology A,* 148(4): 732–742.

Rombough, P. and B. Moroz. 1990. The scaling and potential importance of cutaneous and branchial surfaces in respiratory gas exchange in young chinook salmon (*Oncorhynchus tshawytscha*). *Journal of Experimental Biology,* 154: 1–12.

Roper, C.F.E., M.J. Sweeney and C.E. Nauen. 1984. *Cephalopods of the World: an annotated and illustrated catalogue of species of interest to fisheries.* FAO Species Catalogue, Vol. 3. FAO, Rome. 351 p. + 12 plates.

Rosa, R. and Seibel, B.A. 2008. Synergistic effects of climate-related variables suggest future physiological impairment in a top oceanic predator. *Proceedings of the National Academy of Sciences,* 105: 20776–20780.
Rosa, R., V.M. Lopes, M. Guerreiro, K. Bolstad and J.C. Xavier. 2017. Biology and ecology of the world's largest invertebrate, the colossal squid (*Mesonychoteuthis hamiltoni*): a short review. *Polar Biology,* 40: 1871–1883.
Rösch, R. 2000. Gonadosomatic Index (GSI) of female whitefish (*Coregonus lavaretus*) in Lake Constance. *Limnologica,* 30(2): 193–196.
Rosenfeld, J., T. Van Leeuwen, J. Richards and D. Allen. 2015. Relationship between growth and standard metabolic rate: measurement artifacts and implications for habitat use and life-history adaptation in salmonids. *Journal of Animal Ecology,* 84(1): 4–20.
Ross, L.N. 2023. The explanatory nature of constraints: law-based, mathematical, and causal. *Synthese,* 202(2): 56. DOI:10.1007/s11229-023-04281-5
Rothschild, A. and M. Rothschild. 1939. Some observations on the growth of *Peringia ulvae* (Pennant, 1777) in the laboratory. *Novitates Zoologicae,* 41: 240–247.
Rowe V.L. and K. Mangold. 1975. Effect of starvation on sexual maturation in *Illex illecebrosus* (Lesueur) (Cephalopoda-Teuthoidea). *Journal of Experimental Marine Biology and Ecology,* 17: 157–163.
Rubalcaba, J.G., W.C. Verberk, A.J. Hendriks, B. Saris and H.A. Woods. 2020. Oxygen limitation may affect the temperature and size dependence of metabolism in aquatic ectotherms. *Proceedings of the National Academy of Sciences,* 117(50): 31963–31968.
Rubner, M. 1894. Die Quelle der thierischen Wärme. *Zeitschrift für Biologie,* 30: 73–142.
Rummer, J.L., C.S. Couturier, J.A. Stecyk, N.M. Gardiner, J.P Kinch, G.E. Nilsson and P.L. Munday. 2014. Life on the edge: Thermal optima for aerobic scope of equatorial reef fishes are close to current day temperatures. *Global Change Biology,* 20: 1055–1066.
Rummer, J.L. and C.J. Brauner. 2020. Gas exchange, p. 33–46 *In:* S. Currie and D.H. Evans (eds) *The Physiology of Fishes,* 5th Edition. CRC Press, Boca Raton.
Russell, E. S. 1921. *Form and function: A contribution to the history of animal morphology.* University of Glasgow, 360.
Russell, F.S. 1932. On the Biology of *Sagitta* I. The Breeding and Growth of *Sagitta elegans* Verrill in the Plymouth Area, 1930-1931. *Journal of the Marine Biological Association of the United Kingdom,* 18: 131–146.
Rutjes, H.A. 2006. *Phenotypic responses to lifelong hypoxia in cichlids.* Doctoral dissertation, Leiden University, 156 p.https://hdl.handle.net/1887/4925
Rutjes, H.A., M.P. De Zeeuw, G.E. Van Den Thillart and F. Witte. 2009. Changes in ventral head width, a discriminating shape factor among African cichlids, can be induced by chronic hypoxia. *Biological Journal of the Linnean Society,* 98(3): 608–619.
Saba, G.K., A.B. Burd, J.P. Dunne, S. Hernández-León, A.H. Martin, K.A. Rose, J. Salisbury, D.K. Steinberg, C.N. Trueman, R.D. Wilson and S.E. Wilson. 2021. Toward a better understanding of fish-based contribution to ocean carbon flux. *Limnology and Oceanography,* 66(5): 1639–1664.
Sabatié, R., M. Potier, C. Broudin, B. Seret, F. Ménard and F. Marsac. 2003. Preliminary analysis of some pelagic fish diet in the eastern Central Atlantic. *Collective Volume of Scientific Papers ICCAT,* 55(1): 292–302.

Salihoglu, B.A.R.I.Ş., B.A. Fach and T. Oguz. 2011. Control mechanisms on the ctenophore *Mnemiopsis* population dynamics: A modeling study. *Journal of Marine Systems,* 87(1): 55–65.

Saikkonen, A., J. Kekäläinen and J. Piironen. 2011. Rapid growth of Atlantic salmon juveniles in captivity may indicate poor performance in nature. *Biological Conservation,* 144(9): 2320–2327.

Sakyi-Djan P. 2017. *Aspects of biology and status of fisheries of the giant African cuttlefish* (Sepia hierredda, *Rang, 1835*) *in Ghanaian waters.* MSc. thesis, Cape Coast University, 135 p.

Sakyi-Djan P., Fynn, J.A. and Lazar, N. 2022. Habitat distribution, growth and mortality of giant African cuttlefish *Sepia hierredda* (Rang, 1835) in Ghana waters. *Journal of Fisheries and Coastal Management,* 3: 1–9.

Sallan L. and A.K. Galimberti. 2015. Body-size reduction in vertebrates following the end-Devonian mass extinction. *Science* 350: 812–815.

Samb, B. 1988. Seasonal growth, mortality and recruitment pattern of *Sardinella maderensis* off Senegal, p. 257–271 *In:* S. Venema, J. Möller-Christensen and D. Pauly (eds) *Contributions to tropical fisheries biology: papers by the participants of FAO/DANIDA follow-up training courses.* FAO Fisheries Report 389.

Samb, B. 1990. News from Senegal. *Fishbyte, Newsletter of the Network of Tropical Fisheries Scientists,* 8(2): 32.

Samb, B. and D. Pauly. 2000. On 'variability' as a sampling artifact: the case of Sardinella in North-western Africa. *Fish and Fisheries,* 1: 206–210.

Samoilys, M.A. 1997. Periodicity of spawning aggregations of coral trout *Plectropomus leopardus* (Pisces: Serranidae) on the northern Great Barrier Reef. *Marine Ecology Progress Series,* 160: 149–159.

Samuel (n.d.) First Book of Samuel 17: 4 *In: The Bible.* Various Editions.

Sanfelicea, D. and P.A. Temussia. 2016. Cold denaturation as a tool to measure protein stability. *Biophysical Chemistry,* 208: 4–8. DOI:10.1016/j.bpc.2015.05.007

Sankey, J., J. Biewer, J. Basuga, F. Palacios, H. Wagner and D. Garber. 2016. The giant, spike-toothed salmon, *Oncorhynchus rastrosus* and the "Proto-Tuolumne River." *PaleoBios,* 33: 1-16.

Santana, T.M., H. A. Elias, F.A.L. da Fonseca, O.R. Freitas, J.T. Kojima and L.U. Gonçalves. 2020. Stocking density for arapaima larviculture. *Aquaculture,* 528: 735565.

Sargent, R.C., P.D. Taylor and M.R. Gross. 1987. Parental care and the evolution of egg size in fishes. *American Naturalist,* 129(1): 32–46.

Sarrus, F. and J.F. Rameaux. 1839. Application des sciences accessoires et principalement des mathématiques à la physiologie générale (Rapport sur une mémoire adressé à l'Académic royale de Médecine, séance du 23 juillet 1839). *Bulletin de l'Académie Royale de Médecine.(Paris),* 3: 1094–1100.

Saunders, W.B. 1983. Natural rates of growth and longevity in *Nautilus belauensis. Paleobiology,* 9: 280–288.

Savitz, J. 1969. Effects of temperature and body weight on endogenous nitrogen excretion in the bluegill sunfish (*Lepomis macrochirus*). *Journal of the Fisheries Board of Canada,* 26: 1813–1821.

Savitz, J. 1971. Effects of starvation on body protein utilization of bluegill sunfish (*Lepomis macrochirus* Rafinesque) with a calculation of caloric requirements. *Transactions of the American Fisheries Society*, 100(1): 18–21.

Sawai, E., Y. Yamanoue, M. Nyegaard and Y. Sakai. 2018. Redescription of the bump-head sunfish *Mola alexandrini* (Ranzani 1839), senior synonym of *Mola ramsayi* (Giglioli 1883), with designation of a neotype for *Mola mola* (Linnaeus 1758) (Tetraodontiformes: Molidae). *Ichthyological Research,* 65: 142–160. DOI: 0.1007/s10228-017-0603-6

Schaack, S. and L.J. Chapman. 2004. Interdemic variation in the foraging ecology of the African cyprinid, *Barbus neumayeri. Environmental Biology of Fishes,* 70: 95–105.

Shapin S. and S. Schaffer. 1985. *Leviathan and the Air-Pump: Hobbes, Boyle, and the Experimental Life*. Princeton University Press, Princeton, US, 440 p.

Scheuffele, H., F. Jutfelt and T.D. Clark. 2021. Investigating the gill-oxygen limitation hypothesis in fishes: intraspecific scaling relationships of metabolic rate and gill surface area. *Conservation Physiology,* 9(1): coab040.

Schloss, J.V. 2002. Oxygen toxicity from plants to people. *Planta* 216(1): 38-43.

Schlosser, I.J. 1987. The role of predation in age-and size-related habitat use by stream fishes. *Ecology*. 68(3): 651–659.

Schmidt, M., E.E. Philipp and D. Abele, D. 2008. Size and age-dependent changes of escape response to predator attack in the Queen scallop *Aequipecten opercularis. Marine Biology Research,* 4(6): 442–450.

Schmidt-Nielsen, K. 1984. *Scaling: Why is Animal Size so Important?* Cambridge University Press, Cambridge. 241 p.

Schmidt-Rhaesa, A. and V. Vieler. 2018. *Spadella kappae*, a new small centhic chaetognath (Spadellidae) from Roscoff, France. *Cahier de Biologie Marine,* 59: 257–265. DOI: 10.21411/CBM.A.801C4E6C

Schofield, C.B. and C.R. White. 2025. The effect of hypoxia and hyperoxia on the growth and metabolic rate of tadpoles *Rhinella marina. Journal of Experimental Biology* 228(18): jeb250327. DOI: 10.1242/jeb.250327

Schuck-Paim, C., W.J. Alonso, P.A. Pereira, J.L. Saraiva, M. Cerqueira, C. Chiang and L.U. Sneddon. 2025. Quantifying the welfare impact of air asphyxia in rainbow trout slaughter for policy and practice. *Scientific Reports,* 15(1), p.19850.

Schulz, H.N. and B.B. Jørgensen. (2001). Big bacteria. *Annual Reviews in Microbiology,* 55(1): 105-137.

Schuman, D. and J. Piiper. 1966. Der Sauerstoffbedarf der Atmung bei Fischen nach Messungen and der narkotisierten Schlei (*Tinca tinca*). *Pflüger's Archiv für die gesamte Physiologie des Menschen und der Tiere,* 288: 15–25.

Scott, K. 2005. Allometry of gill weights, gill surface areas, and foot biomass δ^{13}C values of the chemoautotroph–bivalve symbiosis *Solemya velum. Marine Biology,* 147: 935–941.

Sebens, K.P. 1987. The ecology of indeterminate growth in animals. *Annual Reviews of Ecology and Systematics,* 18: 371–407.

Secord, J.A. 1981. Nature's fancy: Charles Darwin and the breeding of pigeons. *Isis,* 72(2): 163–186.

Seebacher, F., G.C. Grigg and L.A. Beard. 1999. Crocodiles as dinosaurs: behavioural thermoregulation in very large ectotherms leads to high and stable body temperatures. *Journal of Experimental Biology,* 202(1): 77–86.

Seibel, B.A. and C. Deutsch. 2020. Oxygen supply capacity in animals evolves to meet maximum demand at the current oxygen partial pressure regardless of size or temperature. *Journal of Experimental Biology,* 223(12): jeb210492.

Seibel, B.A. 2007. On the depth and scale of metabolic rate variation: scaling of oxygen consumption rate and enzymatic activity in the Class Cephalopoda (Mollusca). *Journal of Experimental Biology,* 210: 1–11.

Sekiguchi, K., H. Seshimo and H. Sugita. 1988. Post-embryonic development p. 181-195 *In:* K. Sekiguchi (ed.) *Biology of Horseshoe Crabs*. Science House Press, Tokyo.

Sella, M. 1929. Migrazioni e habitat del tonno (*Thunnus thynnus*) studiati col metodo degli ami, con osservazioni su l'accrecimento, etc. *Memoria Reale Comitato Talasssografico Italiano,* 156: 511–542.

Semakula, S.N. and P.A. Larkin. 1968. Age, growth, food and yield of the white sturgeon (*Acipenser transmontanus*) of the Fraser River, British Columbia. *Journal of the Fisheries Board of Canada,* 25: 2589–2602. DOI:10.1139/f68-229

Seppänen, E., J. Piironen and H. Huuskonen. 2010. Consistency of standard metabolic rate in relation to life history strategy of juvenile Atlantic salmon *Salmo salar*. *Comparative Biochemistry and Physiology A,* 156: 278–284.

Seth, H., A. Gräns, E. Sandblom, C. Olsson, K. Wiklander, J.L. Johnsson and M. Axelsson, M. 2013. Metabolic scope and interspecific competition in sculpins of Greenland are influenced by increased temperatures due to climate change. *PLoS ONE,* 8: e62859.

Severinghaus, J.W. 2016. Eight sages over five centuries share oxygen's discovery. *Advances in Physiology Education,* 40(3): 370–376.

Sheaves, M. 2001. Are there really few piscivorous fishes in shallow estuarine habitats? *Marine Ecology Progress Series,* 222: 279–290.

Sheridan, J.A. and D. Bickford. 2011. Shrinking body size as an ecological response to climate change. *Nature Climate Change,* 1(8): 401–406.

Shi, D., K. Zhang, Y. Cai Geng, M. Xu, M. Sun and Z. Chen. 2020. Population structure of *Trichiurus japonicus* in northern South China Sea and parameters of its growth, mortality and maturity. *South China Fisheries Science,* 16(5): 51–59.

Shimada, K., M.A. Becker and M.L. Griffiths. 2021. Body, jaw, and dentition lengths of macrophagous lamniform sharks, and body size evolution in Lamniformes with special reference to 'off-the-scale' gigantism of the megatooth shark, *Otodus megalodon*. *Historical Biology,* 33(11): 2543–2559.

Shimada, K., H.M. I.V. Maisch, V.J. Perez, M.A. Becker and M.L. Griffiths. 2023. Revisiting body size trends and nursery areas of the Neogene megatooth shark, *Otodus megalodon* (Lamniformes: Otodontidae), reveals Bergmann's rule possibly enhanced its gigantism in cooler waters. *Historical Biology,* 35(2): 208–217.

Shul'man, G.E. 1974. *Life Cycles of Fish: Physiology and Biochemistry*. Israel Program of Scientific Translations & Wiley & Sons, New York, 288 p.

Shulman, M.J. 1985. Coral reef fish assemblages: intra-and interspecific competition for shelter sites. *Environmental Biology of Fishes,* 13: 81–92.

Shuster, C.N. and K. Sekiguchi. 2003. Growing up takes about ten years and eighteen stages, p. 103–130 *In:* C.N. Shuster, R.B. Barlow and H.J. Brockman (eds) *The American Horseshoe Crab*. Harvard University Press, Cambridge.

Siegel, V. 1987. Age and growth of Antarctic Euphausiacea (Crustacea) under natural conditions. *Marine Biology,* 96(4): 483–495. DOI:10.1007/BF00397966

Silvert, W. and D. Pauly. 1987. On the compatibility of a new expression for gross conversion efficiency with the von Bertalanffy growth equation. *Fishery Bulletin*, 85(1): 139–140.

Singh, G.G., Z. Sajid and C. Charles Mather. 2024. Quantitative analysis of mass mortality events in salmon aquaculture shows increasing scale of fish loss events around the world. *Scientific Reports*, 14: 3763. DOI:10.1038/s41598-024-54033-9

Skeeles, M.R. and T.D. Clark. 2023. Evidence for energy reallocation, not oxygen limitation, driving the deceleration in growth of adult fish. *Journal of Experimental Biology*, 10.1242/jeb.246012.

Skeeles, M.R., H. Scheuffele and T.D. Clark. 2023. Supplemental oxygen does not improve growth but can enhance reproductive capacity of fish. *Proceedings of the Royal Society B*, 290(2010): 20231779.

Skov, P.V., J.F. Steffensen, T.F. Sørensen and K. Qvortrup. 2010. Embryonic suckling and maternal specializations in the live-bearing teleost *Zoarces viviparus*. *Journal of Experimental Marine Biology and Ecology*, 395(1–2): 120–127.

Slobodkin, L. B. 1964. The strategy of evolution. *American Scientist*, 52(3): 342–357.

Sloman, K.A., R.D. Sloman, G. De Boeck, G.R. Scott, F.I. Iftikar, C.M. Wood, V.M. Almeida-Val and A.L. Val. 2009. The role of size in synchronous air breathing of *Hoplosternum littorale*. *Physiological and Biochemical Zoology*, 82(6): 625–634.

Smale, D.A., T. Wernberg, E.C. Oliver, M. Thomsen, B.P. Harvey, S.C. Straub, M.T. Burrows, L.V. Alexander, J.A. Benthuysen, M.G. Donat and M. Feng. 2019. Marine heatwaves threaten global biodiversity and the provision of ecosystem services. *Nature Climate Change*, 9(4): 306–312.

Smith, C.L., C.S. Rand, B. Schaeffer and J.W. Atz. 1975. *Latimeria*, the living coelacanth, is ovoviviparous. *Science*, 190: 1105–1106. DOI:10.1126/science.190.4219.1105

Smith, J.W. and C.C. Lamborn. 2024. The influence of population growth and weather on the value of recreational angling trips within Utah (USA). *Fisheries Research*, 74: 106976.

Smith, L.J., K.M. Fiebig, H. Schwalbe and C.M. Dobson. 1996. The concept of a random coil: Residual structure in peptides and denatured proteins. *Folding and Design*, 1(5): R95–R106.

Smith, K.E., P.J. Moore, N.G. King and D.A. Smale. 2022. Examining the influence of regional-scale variability in temperature and light availability on the depth distribution of subtidal kelp forests. *Limnology and Oceanography*, 67(2): 314-328.

Smith, R.S. and D.L. Kramer. 1986. The effect of apparent predation risk on the respiratory behavior of the Florida gar (*Lepisosteus platyrhincus*). *Canadian Journal of Zoology*, 64(10): 2133–2136.

Smith, W.L. 2021. Introduction to Ichthyology and Herpetology. *Ichthyology & Herpetology*, 109(1): 1–2.

Sokamte, T., P. Mbougueng, B. Mohammadou, N. Tatsadjieu and N. Sachindra. 2020. Proximal composition and fatty acid profile of fresh and smoked fillets of *Pangasius hypophthalmus*. *Scientific African*, 9: e00534.

Sokolova, I M. 2023. Ectotherm mitochondrial economy and responses to global warming. *Acta Physiologica*, 237(4): e13950.

Sollid, J. and G.E. Nilsson. 2006. Plasticity of respiratory structures—adaptive remodeling of fish gills induced by ambient oxygen and temperature. *Respiratory Physiology and Neurobiology,* 154(1-2): 241-251.

Somero, G.N. and J.J. Childress. 1980. A violation of the metabolism-size scaling paradigm: activities of glycolytic enzymes in muscle increase in larger-size fish. *Physiological Zoology,* 53: 322–337. DOI:10.1086/physzool.53.3.30155794

Somers, I.F. 1988. On a seasonally oscillating growth function. *Fishbyte, Newsletter of the Network of Tropical Fisheries Scientists,* 6: 8–11.

Soriano, M., J. Moreau, J.M. Hoenig and D. Pauly. 1992. New functions for the analysis of two-phase growth of juvenile and adult fishes, with application to Nile perch. *Transactions of the American Fisheries Society,* 121: 486–493.

Spallanzani, L. 1803. *Mémoires sur la respiration, traduit en français par J. Senebier d'après un manuscrit inédit.* J.J. Paschoud, Genève, 373 p.

Spencer, H. 1866. *The Principles of Biology,* Vol. 1. D. Appleton, New York and London, 475 p.

Skomal, G.B., S.I. Zeeman, J.H. Chisholm, E.L. Summers, H.J. Walsh, K.W. McMahon and S.R. Thorrold. 2009. Transequatorial migrations by basking sharks in the western Atlantic Ocean. *Current Biology,* 19(12): 1019–1022.

Staden, H. von. 1989. *Herophilus: The Art of Medicine in Early Alexandria: Edition, Translation and Essays.* Cambridge University Press, Cambridge. 712 p.

Staehr, P.A. and T. Wernberg. 2009. Physiological responses of *Ecklonia radiata* (Laminariales) to a latitudinal gradient in ocean temperature. *Journal of Phycology,* 45(1): 91–99.

Stang, A. 2023. A historical note about the sufficient cause model with special consideration of the work of Max Verworn 1912. *Annals of Epidemiology,* 85, pp.1-2. DOI:10.1016/j.annepidem.2023.06.005

Stearley, R.F. and G.R. Smith. 2016. Fishes of the Mio-Pleiocene Western Snake River Plain and Vicinity. I. Salmonid Fishes from Mio-Pleiocene Lake Sediments in the Western Snake River and the Great Basin. *Miscellaneous Publications, Museum of Zoology, University of Michigan,* 43 p.

Steine, G.R.O., F. Alfnes and M.B. Rørå. 2005. The effect of color on consumer WTP for farmed salmon. *Marine Resource Economics,* 20(2): 211–219.

Steneck, R.S., M.H. Graham, B.J. Bourque, D. Corbett, J.M. Erlandson, J.A. Estes and M.J. Tegner. 2002. Kelp forest ecosystems: biodiversity, stability, resilience and future. *Environmental Conservation,* 29(4): 436–459.

Stéquert, B., J.N. Rodriguez, B. Cuisset and F. Le Menn. 2001. Gonadosomatic index and seasonal variations of plasma sex steroids in skipjack tuna (*Katsuwonus pelamis*) and yellowfin tuna (*Thunnus albacares*) from the western Indian Ocean. *Aquatic Living Resources,* 14: 313–318.

Stevely, J.M., J.C. Thompson and R.E. Warner. 1978. The biology and utilization of Florida's commercial sponges. *Florida Sea Grant Program, Technical Paper 8,* 45 p. http://ufdc.ufl.edu/UF00072259/00001

Stokes, G.L., L. Castello, T.A. Petersen, S.J. Cooke, M. Power, J. Zuanon and E.G. Martins. 2021. Air-breathing ecology of *Arapaima* sp.: Conservation implications for an imperiled fish. *Aquatic Conservation: Marine and Freshwater Ecosystems,* 31(6): 1257–1268.

Storr, J.F. 1964. Ecology of the Gulf of Mexico commercial sponges and its relation to the fishery. *U.S. Fish and Wildlife Service, Special Scientific Report – Fisheries* 466. 72 p.

Subramoniam, T. 2016. *Sexual Biology and Reproduction in Crustaceans*. Academic Press, London. 726 p.

Suskiewicz, T.S., J.E. Byrnes, R.S. Steneck, R. Russell, C.J. Wilson and D.B. Rasher. 2024. Ocean warming undermines the recovery resilience of New England kelp forests following a fishery-induced trophic cascade. *Ecology*, 105(7), p.e4334.

Suzuki, Y., A. Kondo and J. Bergström. 2008. Morphological requirements in limulid and decapod gills: A case study in deducing the function of lamellipedian exopod lamellae. *Acta Palaeontologica Polonica*, 53(2): 275–283.

Sweatt, A.J. and R.B. Forward. 1985. Diel vertical migration and photoresponses of the chaetognath *Sagitta hispida* Conant. *Biological Bulletin*, 168: 18–31.

Sweetman, A.K., A.J. Smith, D.S.W. de Jonge, T. Hahn, P. Schroedl, M. Silverstein, C. Andrade, R.L. Edwards, A.J.M. Lough, C. Woulds, W.B. Homoky, A. Koschinsky, S. Fuchs, T. Kuhn, F. Geiger and J.J. Marlow. 2024. Evidence of dark oxygen production at the abyssal seafloor. *Nature Geoscience*, 17(8): 737–739. DOI:10.1038/s41561-024-01480-8

Swift, J. 1726. *Travels into Several Remote Nations of the World. In Four Parts. By Lemuel Gulliver, First a Surgeon, and then a Captain of Several Ships* (Book II). Benjamin Motte, London.

Sytova, M.V., H.N. Harenko, V.A. Belyaev and A.P. Shmigirilov. 2004. The analysis of quantitative and weight parameters of sturgeon spawning groups at the Amur River basin. *Voprosy Rybolovstva / Problems of Fisheries*, 5(3): 470–481.

Takakuwa, Y. 2021. The first fossil record of giant spike-toothed salmon, *Oncorhynchus rastrosus* from the Northwestern Pacific region. *Bull.Gunma Museum of Natural History*, (25):41–48 [In Japanese with English abstract]

Tao, J., D. He, M.J. Kennard, C. Ding, S.E. Bunn, C. Liu, Y. Jia, R. Che and Y. Chen. 2018. Strong evidence for changing fish reproductive phenology under climate warming on the Tibetan Plateau. *Global Change Biology*, 24(5): 2093–2104. DOI:10.1111/gcb.14050.

Taranger, G.L., M. Carrillo, R.W. Schulz, P. Fontaine, S. Zanuy, A. Felip, F.A. Weltzien, S. Dufour, O. Karlsen, B. Norberg, E. Andersson and T. Hansen. 2010. Control of puberty in farmed fish. *General and Comparative Endocrinology*, 165(3): 483–515.

Taylor, C.R. and E.R. Weibel. 1981. Design of the mammalian respiratory system. I. Problem and strategy. *Respiration Physiology*, 44(1): 1–10.

Te Winkel, L.E. 1935. A study of *Mistichthys luzonensis* with special reference to conditions correlated with reduced size. *Journal of Morphology*, 58: 463-535.

Teletchea, F. and D. Pauly. 2024. Why do fish larvae hatch when they do? *Environmental Biology of Fishes*, 107: 583–591. DOI:10.1007/s10641-024-01553-y

Thagard, P.R. 1978. The best explanation: criteria for theory choice. *The Journal of Philosophy*, 75(2): 76–92. DOI:10.2307/2025685

Thom, R. 1975. *Structural Stability and Morphogenesis: An Outline of a General Theory of Models*. Translated by D.H. Fowler. W.A. Benjamin, Reading, MA, 392 p.

Thompson, D.W. 1917. *On Growth and Form*. Cambridge University Press, Cambridge, 793 p.

Thornton, S.A., S.E. Dudin, S.E. Page, C. Upton and M.E. Harrison. 2018. Peatland fish of Sebangau, Borneo: diversity, monitoring and conservation. *Mires and Peat,* 22(04): 1–25.

Thorpe, J.E. 1986. Age at first maturity in Atlantic salmon, *Salmo salar*: freshwater period influences and conflicts with smolting. *Canadian Special Publication of Fisheries and Aquatic Sciences,* 89: 7–14.

Thorpe, J.E. 1990. Variation in life-history strategy in salmonids. *Polish Archive for Hydrobiology,* 37: 3–12.

Thuesen, E.V., L.D. Rutherford, Jr., P.L. Brommer, K. Garrison, M.A. Gutowska and T. Towanda. 2005. Intragel oxygen promotes hypoxia tolerance of scyphomedusae. *Journal of Experimental Biology,* 208: 2475–2482. DOI:10.1242/jeb.01655.

Tiews, K. 1963. Synopsis of biological data on bluefin tuna *Thunnus thynnus* Linnaeus 1758 (Atlantic and Mediterranean), p. 422–481 *In*: *Proceedings of the World Scientific Meeting on the biology of tunas and related species, La Jolla, California, USA, 2-4 July 1962.* FAO Report 6, Food and Agriculture Organization of the United Nations, Rome.

Timmerman, C.M. and L.J. Chapman. 2003. The effect of gestational state on oxygen consumption and response to hypoxia in the sailfin molly, *Poecilia latipinna. Environmental Biology of Fishes,* 68(3): 293–299.

Timofeev, S.F. 2001. Bergmann's Principle and Deep-Water Gigantism in Marine Custaceans. *Biology Bulletin,* 28(6): 646–650.

Tirsgaard B., J.W. Behrens and J.F. Steffensen. 2015. The effect of temperature and body size on metabolic scope of activity in juvenile Atlantic cod *Gadus morhua* L. *Comparative Biochemistry and Physiology A,* 179: 89–94. DOI:10.1016/j.cbpa.2014.09.033.

Todd, E.V., Liu, H., Muncaster, S. and Gemmell, N. J. 2016. Bending genders: the biology of natural sex change in fish. *Sexual Development,* 10(5-6): 223–241.

Tomita, T., T.F. Cotton and M. Toda. 2016. Ultrasound and physical models shed light on the respiratory system of embryonic dogfishes. *Zoology,* 119:36–41. DOI:10.1016/j.zool.2015.09.002

Tomita, T., R. Nozu, M. Nakamura, S. Matsuzaki, K. Miyamoto and K. Sato. 2017. Live-bearing without placenta: Physical estimation indicates the high oxygen-supplying ability of white shark uterus to the embryo. *Scientific Reports,* 7(1): 1–7.

Tomita, T., M. Toda, M., K. Uchida and K. Nakaya. 2012. Live-bearing manta ray: How the embryo acquires oxygen without placenta and umbilical cord. *Biology Letters,* 23: 721–724.

Torres, J.J. and G.N. Somero. 1988. Metabolism, enzymic activities and cold adaptation in Antarctic mesopelagic fishes. *Marine Biology,* 98: 169–180.

Tran-Duy, A., A.A. van Dam and J.W. Schrama. 2012. Feed intake, growth and metabolism of Nile tilapia (*Oreochromis niloticus*) in relation to dissolved oxygen concentration. *Aquaculture Research,* 43(5): 730–744.

Trippel, E.A., O.S. Kjesbu and P. Solemdial. 1997. Effects of adult age and size structure on reproductive output in marine fishes, p. 31–61 *In:* R.C. Chambers and E.A. Trippel (eds) *Early life history in fish populations.* Chapman & Hall–Kluwer, Dordrecht.

Trippel, E.A., I.A. Butts, A. Babin, S.R. Neil, N.J. Feindel and T.J. Benfey. 2014. Effects of reproduction on growth and survival in Atlantic cod, *Gadus morhua,* assessed by comparison to triploids. *Journal of Experimental Marine Biology and Ecology,* 451: 35-43.

Troyer, E.M., R. Betancur-R, L.C. Hughes, M. Westneat, G. Carnevale, W.T. White, J.J. Pogonoski, J.C. Tyler, C.C. Baldwin, G. Ortí and A. Brinkworth. 2022. The impact of paleoclimatic changes on body size evolution in marine fishes. *Proceedings of the National Academy of Sciences,* 119(29): p.e2122486119.

Tsikliras, A.C. and K.I. Stergiou. 2015. Age at maturity of Mediterranean marine fishes. *Mediterranean Marine Science,* 16(1): 5–20. DOI:10.12681/mms.659

Tsikliras, A.C. and K.I. Stergiou. 2014. Mean temperature of the catch increases quickly in the Mediterranean Sea. *Marine Ecology Progress Series,* 515: 281–284.

Tsikliras, A.C., Peristeraki, P., Tserpes, G. and K.I. Stergiou. 2015. Mean temperature of the catch (MTC) in the Greek Seas based on landings and survey data. *Frontiers in Marine Science,* 2: 23. DOI:10.3389/fmars.2015.00023

Tu, Z., L. Li, X.L. Yuan, Y. Huang and D. Johnson. 2012. Aerobic swimming performance of juvenile largemouth bronze gudgeon (*Coreius guichenoti*) in the Yangtze River. *Journal of Experimental Zoology Part A,* 317: 294–302.

Tunnicliffe, V., R. Gasbarro, F. Juanes, J. Qualley, N. Soderberg and J.W. Chu. 2020. An hypoxia-tolerant flatfish: consequences of sustained stress on the slender sole *Lyopsetta exilis* (Pleuronectidae) in the context of a changing ocean. *Journal of Fish Biology,* 96(2): 394–407.

Tuponogov, V.N., A.M. Orlov and L.S. Kodolov. 2008. The most abundant grenadiers of the Russian Far East EEZ: distribution and basic biological patterns, p. 285–316 *In:* E.M. Orlov and T. Iwamoto (eds) *Grenadiers of the World Oceans: Biology, Stock Assessment and Fisheries.* American Fisheries Society Symposium 63, Bethesda, MD.

Turay, I., J.M. Vakily, M.L.D. Palomares and D. Pauly. 2006. Growth, food and reproduction of the mudskipper, *Periophthalmus barbarus* on mudflats of Freetown, Sierra Leone, p. 49-54 *In:* M.L.D. Palomares, K.I. Stergiou and D. Pauly (eds) *Fishes in Databases and Ecosystems.* Fisheries Centre Research Reports 14(4), University of British Columbia, Vancouver.

Turko, A.J., C.A. Cooper and P.A. Wright. 2012. Gill remodelling during terrestrial acclimation reduces aquatic respiratory function of the amphibious fish *Kryptolebias marmoratus. Journal of Experimental Biology,* 215(22): 3973–3980.

Turner, E.C. 2021. Possible poriferan body fossils in early Neoproterozoic microbial reefs. *Nature,* 596: 87–97. DOI:10.1038/s41586-021-03773-z

Turner, R.E., 2017. Smaller size-at-age menhaden with coastal warming and fishing intensity. *Geo - Geography and Environment,* 4(2): e00044.

Ultsch, G.R. 1976. Respiratory surface area as a factor controlling the standard rate of O_2 consumption of aquatic salamanders. *Respiration Physiology,* 6(3): 357–369.

Ulman, A., S. Kalogirou and D. Pauly. 2022. The dynamics of maximum lengths for the invasive silver-cheeked toadfish (*Lagocephalus sceleratus*) in which Eastern Mediterranean. *Journal of Marine Science and Engineering,* 10(3): 387. DOI:10.3390/jmse10030387

Urbina, M.A. and C.N. Glover. 2013. Relationship between fish size and metabolic rate in the oxyconforming inanga *Galaxias maculatus* reveals size-dependent strategies to withstand hypoxia. *Physiological Biochememical Zoology,* 86: 740–749.

Urbina, M. A., C.N. Glover and M.E. Forster. 2012. A novel oxyconforming response in the freshwater fish *Galaxias maculatus. Comparative and Biochemical Physiology A,* 161: 301–306.

Ursin, E. 1967. A mathematical model of some aspects of fish growth, respiration and mortality. *Journal of the Fisheries Research Board of Canada*, 24(11): 2355–2453.
Ursin, E. 1979. Principles of growth in fishes, p. 63–87 *In:* P.J. Miller (ed.) *Fish Phenology: Anabolic Adaptiveness in Teleost.* Academic Press, New York.
U.S. Environmental Protection Agency. 1972. *Investigation of fish kills.* Oklahoma Cooperative Fishery Unit, Stillwater, USA, 262 p.
Vagner, M., T. Lacoue-Labarthe, J.L. Zambonino Infante, D. Mazurais, E. Dubillot, H. Le Delliou, P. Quazuguel and C. Lefrancois. 2015. Depletion of essential fatty acids in the food source affects aerobic capacities of the golden grey mullet *Liza aurata* in a warming seawater context. *PLoS ONE*, 10(6): e0126489.
Vahl, O. 1978. Seasonal changes in oxygen consumption of the Iceland scallop (*Chlamys islandica* (O.F. Müller)) from 70°N. *Ophelia*, 17(1): 143–154.
Vahl, O. 1984. The relationship between specific dynamic action (SDA) and growth in the common starfish, *Asterias rubens* L. *Oecologia*, 61: 122–125.
Vakily, J.M. 1992. Determination and comparison of bivalve growth, with emphasis on Thailand and other tropical areas. *ICLARM Technical Reports* 36, 125 p. https://www.worldfishcenter.org/publication/determination-and-comparison-bivalve-growth-emphasis-thailand-and-other-tropical-areas
Valentine, M.M. 2019. *Sponge community biocomplexity, competition, and functional significance in hard-bottom habitats of the Florida Keys, FL (USA).* Doctoral dissertation, Old Dominion University, Norfolk, VA, 150 p. DOI:10.25777/chcq-f487
Van der Have, T.M. and G. De Jong. 1996. Adult size in ectotherms: temperature effects on growth and differentiation. *Journal of Theoretical Biology*, 183(3): 329-340.
Van der Meer, D.L., G.E. van den Thillart, F. Witte, M.A. de Bakker, J. Besser, M.K. Richardson, H.P. Spaink, J.T.D. Leito and C.P. Bagowski. 2005. Gene expression profiling of the long-term adaptive response to hypoxia in the gills of adult zebrafish. *American Journal of Physiology - Regulatory, Integrative and Comparative Physiology*, 289(5): R1512–R1519.
Van Hal, R., T. van Kooten and A.D. Rijnsdorp. 2016. Temperature induced changes in size dependent distributions of two boreal and three Lusitanian flatfish species: A comparative study. *Journal of Sea Research*, 107: 14–22.
Van Oosten, J. 1923. The whitefishes (*Coregonus clupeaformis*). A study of the scales of whitefishes of known ages. *Zoologica - Scientific Contributions of the New York Zoological Society*, 2: 381–412.
Van Poorten, B.T. and C.J. Walters. 2016. How can bioenergetics help us predict changes in fish growth patterns? *Fisheries Research*, 180: 23-30.
Van Roy, P., D.E. Briggs,and R.R. Gaines. 2015. The Fezouata fossils of Morocco: an extraordinary record of marine life in the Early Ordovician. *Journal of the Geological Society*, 172(5): 541–549.
Van Soest, R.W.M. 1978. Marine sponges from Curaçao and other Caribbean localities. Part I. Keratosa. *Studies on the Fauna of Curaçao and Other Caribbean Islands*, 56(179): 1–94.
Van Voorhies, W.A. 1996. Bergman size clines: a simple explanation for their occurrence in ectotherms. *Evolution*, 50: 1259–1264

Vehanen, T., A. Mäki-Petäys, J. Aspi and T. Muotka. 1999. Intercohort competition causes spatial segregation in brown trout in artificial streams. *Journal of Fish Biology*, 55(1): 35–46.

Verberk, W.C. and D. Atkinson. 2013. Why polar gigantism and Palaeozoic gigantism are not equivalent: effects of oxygen and temperature on the body size of ectotherms. *Functional Ecology*, 27(6): 1275–1285.

Verberk, W.C. and D.T. Bilton. 2011. Can oxygen set thermal limits in an insect and drive gigantism? *PLoS ONE*, 6(7): e22610.

Verberk, W.C., D. Atkinson, K.N. Hoefnagel, A.G. Hirst, C.R. Horne and H. Siepel. 2021. Shrinking body sizes in response to warming: explanations for the temperature–size rule with special emphasis on the role of oxygen. *Biological Reviews*, 96(1): 247–268.

Verberk, W.C., J. Overgaard, R. Ern, M. Bayley, T. Wang, L. Boardman and J.S. Terblanche. 2016. Does oxygen limit thermal tolerance in arthropods? A critical review of current evidence. *Comparative Biochemistry and Physiology A*, 192: 64–78.

Verberk, W.C., J.F. Sandker, I.L. van de Pol, M.A. Urbina, R.W. Wilson, D.J. McKenzie and F.P. Leiva. 2022. Body mass and cell size shape the tolerance of fishes to low oxygen in a temperature-dependent manner. *Global Change Biology*, 28(19): 5695–5707.

Vermeer, G.K. 1986. Effects of air exposure on desiccation rate, hemolymph chemistry, and escape behavior of the spiny lobster, *Panulirus argus*. *Fishery Bulletin*, 85: 45–51.

Vermeij, G.J. 2016. Gigantism and its implications for the history of life. *PLoS ONE*, 11(1): e0146092.

Versteeg, M.A., C. MacDonald, M.F. Bennett-Smith, P. Buston and T. Burger. 2025. Individual clown anemonefish shrink to survive heat stress and social conflict. *Science Advances*, 11(21): eadt7079.

Verworn, M. 1901. *Allgemeine Physiologie: Ein Grundriss der Lehre vom Leben* (8th edition). G. Fischer Verlag, Jena. 631 p.

Verworn, M. 1909. Letter to Max Wolff (A 9452; December 12). Ernst Haeckel-Archiv Jena. https://haeckel-briefwechsel-projekt.uni-jena.de/de/document/b_9452

Vincent, R.E. 1960. Some influences of domestication upon three stocks of brook trout (*Salvelinus fontinalis* Mitchill). *Transactions of the American Fisheries Society*, 89(1): 35–52. DOI:10.1577/1548-8659(1960)89(35:SIODUT)2.0.CO;2

Viravong. S, S. Phounsavath. C. Photitay, P. Solyda. C. Sokheng. J. Kolding. J.V. Jørgensen and K. Phoutavong. 2006. Hydro-acoustic surveys of deep pools in Southern Lao PDR and Northern Cambodia. MRC Technical Paper No.11, Mekong River Commission, Vientiane. 70 p.

Vogt, G. 2012. Ageing and longevity in the Decapoda (Crustacea): A review. *Zoologischer Anzeiger*, 251: 1–25.

von Bertalanffy, L.: *see* Bertalanffy, L. von.

Vosseler, J. 1901. *Die Amphipoden der Plankton-Expedition: I. Theil. Hyperiidea 1.* Lipsius and Tischer, Kiel and Leipzig. 129 p. + xii plates.

Walters, C.J. 2000. Natural selection for predation avoidance tactics. *Marine Ecology Progress Series*, 208: 309–313.

Walters, C.J., B.T. van Poorten and L.G. Coggins. 2012. Bioenergetics and population dynamics of Flannelmouth Sucker and Bluehead Sucker in Grand Canyon as evidenced by tag recapture observations. *Transactions of the American Fisheries Society*, 141(1): 158–173.

Wang, C., H. Li, Z. Xu, S. Zheng, Q. Hao, Y. Dong, L. Zhao, W. Zhang, Y. Zhao, G. Grégori and T. Xiao. 2020. Difference of planktonic ciliate communities of the tropical West Pacific, the Bering Sea and the Arctic Ocean. *Acta Oceanologica Sinica,* 39(4): 9–17.

Wang, K., B.P.V. Hunt, C. Liang, D. Pauly and E.A. Pakhomov. 2017. Reassessment of the life cycle of the pteropod *Limacina helicina* from a high resolution interannual time series in the temperate North Pacific. *ICES Journal of Marine Science,* 74(7): 1906–1920. DOI:10.1093/icesjms/fsx014

Wang, T., X. Gao, J. Wang, I. Jakovlić, S.G. Dan and H.Z. Liu. 2015. Life history traits and implications for conservation of rock carp *Procypris rabaudi* Tchang, an endemic fish in the upper Yangtze River, China. *Fisheries Science,* 81(3): 515–523.

Wang, X., P. Liang, C. Li and F. Wu. 2014. Analysis of regional temperature variation characteristics in the Lancang River Basin in southwestern China. *Quaternary International,* 333: 198–206.

Warburg, O.H. 1930. *The Metabolism of Tumors*. Constable, London. 327 p.

Ward, P.D. 1987. *The Natural History of the Nautilus*. Allen & Unwin, Winchester, MA, 267 p.

Ward, P.D. 2006. *Out of Thin Air: Dinosaurs, Birds and the Earth Ancient History.* National Academies Press, Washington, D.C. 296 p.

Ward, P.D. 2018. *Lamarck's Revenge: How Epigenetics is Revolutionizing Our Understanding of Evolution's Past and Present*. Bloomsbury Publishing, New York, 288 p.

Warkentin, K.M. 2007. Oxygen, gills, and embryo behavior: mechanisms of adaptive plasticity in hatching. *Comparative Biochemistry and Physiology A,* 148(4): 720–731.

Warlen, S.M. and J.S. Burke. 1990. Immigration of larvae of fall/winter spawning marine fishes into a North Carolina estuary. *Estuaries,* 13(4): 453–461.

Warne, C., M.M. Guzzo, K. Cazelles, C. Chu, N. Rooney and K.S. McCann. 2024. Thermal tolerance and habitat preferences mediate how freshwater fish body sizes respond to warming. *Canadian Journal of Fisheries and Aquatic Sciences,* 81(4): 488–496. DOI:10.1139/cjfas-2023-0233

Warren, M.A. 2023. *Urea (ka)! The role of urea and oxygen in the relationship between length at first maturity and maximum length in ureotelic and non-ureotelic cartilaginous fishes.* Msc Thesis, University of British Columbia. 279 p. https://open.library.ubc.ca/collections/ubctheses/24/items/1.0431066

Warren, M.A. and D. Pauly. 2024. The likely role of urea in delaying the size at first maturity of ureosmotic Chondrichthyes. *Environmental Biology of Fishes* 107: 107(5): 523–536.

Webb, P.W. and J.R. Brett. 1972. Respiratory adaptations of prenatal young in the ovary of two species of viviparous seaperch, *Rhacochilus vacca* and *Embiotoca lateralis. Journal of the Fisheries Board of Canad,a* 29(11): 1525-1542.

Webb, P.W. 1978. Partitioning of energy into metabolism and growth, p. 184–214 *In:* S.D. Gerking (ed.) *Ecology of Freshwater Fish Production*. Blackwell Scientific Publications, Oxford.

Weber, K T. and J.S. Janicki. 1979. The metabolic demand and oxygen supply of the heart: physiologic and clinical considerations. *American Journal of Cardiology,* 44:722–729. DOI:10.1016/0002-9149(79)90294–7

Weeks, S.C. 1996. The hidden cost of reproduction: reduced food intake caused by spatial constraints in the body cavity. *Oikos,* 75(2): 345–349.

Wegner, N.C. 2015. Elasmobranch gill structure, p. 102–15 *In:* R.E. Shadwick, A.P. Farrell and C.J. Brauner (eds) *Physiology of Elasmobranch Fishes: Structure and Interaction with Environment*. Academic Press, London.

Weisz, J. 2006. Measuring impacts of associated microbial communities on Caribbean reef sponges: searching for symbiosis. Doctoral dissertation, University of North Carolina, Chapel Hill, NC, 165 p. https://cdr.lib.unc.edu/concern/dissertations/1r66j1662

Welch, D.W., Y. Ishida and K. Nagasawa. 1998. Thermal limits and ocean migrations of sockeye salmon (*Oncorhynchus nerka*): long-term consequences of global warming. *Canadian Journal of Fisheries and Aquatic Science,* 55: 937–948.

Welch, H.E., T.D. Siferd and P. Bruecker. 1996. Population Densities, Growth, and Respiration of the Chaetognath *Parasagitta elegans* in the Canadian High Arctic. *Canadian Journal of Fisheries and Aquatic Science,* 53: 520–527. DOI:10.1139/f95-227

Wels, P. 1925. Der Einfluss der Tiergrösse auf die Oxydationsgeschwindigkeit im überlebenden Gewebe. *Pflüger's Archiv der Gesamten Physiologie der Menschen und Tiere,* 209: 32–48.

West, G.B., J.H. Brown and B.J. Enquist. 1997. A general model for the origin of allometric scaling laws in biology. *Science,* 276 (7): 122–126.

West, G.B., W.H. Woodruff and J.H Brown. 2002. Allometric scaling of metabolic rate from molecules and mitochondria to cells and mammals. *Proceedings of the National Academy of Sciences,* 99: 2473–2478.

West, J.B. 2013a. Torricelli and the ocean of air: the first measurement of barometric pressure. *Physiology*, 28(2): 66–73.

West, J.B. 2013b. The collaboration of Antoine and Marie-Anne Lavoisier and the first measurements of human oxygen consumption. *American Journal of Physiology -Lung, Cellular and Molecular Physiology,* 305(11): L775–L785.

West, J.B. 2015. *Essays on the history of respiratory physiology*. Springer, London and Dordrecht, 342 p.

West-Eberhard, M.J. 2003. *Developmental Plasticity and Evolution*. Oxford University Press, New York. 816 p.

White, C.R. and D.J. Marshall. 2023a. Optimisation and constraint: explaining metabolic patterns in biology. *Journal of Experimental Biology,* 226(11): jeb245426.

White, C.R. and D.J. Marshall. 2023b. How and why does metabolism scale with body mass? *Physiology,* 38(6): 266–274.

White, C.R., L.A. Alton, C.L. Bywater, E.J. Lombardi and D.J. Marshall. 2022. Metabolic scaling is the product of life-history optimization. *Science,* 377(6608): 834–839.

White, C.R. and D.J. Marshall. 2019. Should we care if models are phenomenological or mechanistic? *Trends in Ecology and Evolution,* 34(4): 276–278.

White, C.R., L.A. Alton and P.B. Frappell. 2012. Metabolic cold adaptation in fishes occurs at the level of whole animal, mitochondria and enzyme. *Proceedings of the Royal Society B: Biological Sciences,* 279(1734): 1740–1747.

White, C.R., L.A. Alton, C.L. Bywater, E.J. Lombardi and D.J. Marshall. 2023. Response to Comments on "Metabolic scaling is the product of life-history optimization." *Science*, 380(6643): eadf5188.

White, C.R., L.A. Alton, C.L. Bywater, E.J. Lombardi and D.J. Marshall. 2022. Metabolic scaling is the product of life-history optimization. *Science,* 377: 834–839.

Whitehead, P.J.P. 1985. *Clupeoid Fishes of the World. An Annotated and Illustrated Catalogue of the Herrings, Sardines, Pilchards, Sprats, Anchovies and Wolf Herrings. Part 1 - Chirocentridae, Clupeidae and Pristigasteridae*. FAO species catalogue, Vol. 7 - FAO Fisheries Synopses 125. 303 p.

Whitledge, G.W., R.S. Hayward and C.F. Rabeni. 2002. Effects of temperature on specific daily metabolic demand and growth scope of sub-adult and adult smallmouth bass. *Journal of Freshwater Ecology,* 17(3): 353–361.

Whitney, R.J. 1942. The relation of animal size to oxygen consumption in some fresh-water turbellarian worms. *Journal of Experimental Biology,* 19: 168–175.

Wiener, N. 1948. *Cybernetics*. MIT Press, Cambridge, MA. 212 p.

Wieser, W. 1995. Energetics of fish larvae, the smallest vertebrates. *Acta Physiologica Scandinavica*, 154(3): 279–290.

Wiesner, J. 1892. *Die Elementarstructur und das Wachstum der lebenden Substanz*. A. Hölder, Vienna. 269 p.

Williamson, P., H.-O. Pörtner, S. Widdicombe and J.P. Gattuso. 2021. When ocean acidification experiments are not the same, repeatability is not tested. *Biogeosciences,* 18: 1787–1792.

Wilson, L.G. 1960. The transformation of ancient concepts of respiration in the seventeenth century. *Isis,* 51(2): 161–172.

Winberg, G.G. 1960. *Rate of Metabolism and Food Requirements of Fishes*, Minsk, USSR. Fisheries Research Board of Canada, Translation Series No. 194. 239 p.

Windsland, K., Hvingel, C., Nilssen, E.M. and Sundet, J.H. 2013. Evaluation of von Bertalanffy growth curves for the introduced red king crab (*Paralithodes* *camtschaticus*). *Fisheries Research,* 145: 15–21. DOI:10.1016/j.fishres.2013.03.003

Withers, P.C. 1998. Urea: diverse functions of a "waste" product. Clinical and *Experimental Pharmacology and Physiology*, 25: 722–727. DOI:10.1111/j.1440-1681.1998.tb02284.x

Wittgenstein, L. 1958. *The Blue and Brown Books*. Blackwell, Oxford. 208 p.

Witzell, W.N. 1998. The origin of the Florida sponge fishery. *Marine Fisheries Review,* 60(1): 27–32.

Wöhlisch, E. 1929. August Pütter: Nachruf. *Ergebnisse der Physiologie*, 28: 690–692.

Wohlschlag, D. 1964. Respiratory metabolism and ecological characteristics of some fishes in McMurdo Sound, Antarctica, p. 33–62 *In:* M.O. Lee (ed.) *Biology of the Antarctic Seas.* Antarctic Research Series, Vol. I. American Geophysical Union, Baltimore.

Wolfe J.M., J.W. Breinholt, K.A. Crandall, A.R. Lemmon, E.M. Lemmon, L.E. Timm, M.E. Siddall and H.D. Bracken-Grissom. 2019. Phylogenomic framework, evolutionary timeline and genomic resources for comparative studies of decapod crustaceans. *Proceedings of the Royal Society B,* 286: 20190079. DOI:10.1098/rspb.2019.0079

Won, E.T. and R.J. Borski. 2013. Endocrine regulation of compensatory growth in fish. *Frontiers in Endocrinology* 4:74. DOI:10.3389/fendo.2013.00074

Wood, C. M. and J. Eom. 2021. The osmorespiratory compromise in the fish gill. *Comparative Biochemistry and Physiology Part A*, 254: 110895.

Woodcock, A. and M. Davis. 1978. *Catastrophe Theory: A New Way of Understanding how Things Change.* E.P. Dutton, New York, 152 p.

Woods, H.A., A.L. Moran, D. Atkinson, A. Audzijonyte, M. Berenbrink, F.O. Borges, K.G. Burnett, L.E. Burnett, C.J. Coates, R. Collin, E.M. Costa-Paiva, M.I. Duncan, R. Ern, E.M.J. Laetz, L.A. Levin, M. Lindmark, N.M. Lucey, L.R. McCormick, J.J. Pierson, R. Rosa, M. R. Roman, E. Sampaio, P.M. Schulte, E.A. Sperling, A. Walczyńska and W.C. Verberk. 2022. Integrative approaches to understanding organismal responses to aquatic deoxygenation. *Biological Bulletin,* 243(2): 85–103.

Wootton, T.P. 2011. *Gill morphometrics of thresher sharks (genus* Alopias*): An investigation of the evolutionary pressures influencing gill morphology*. Doctoral dissertation, University of California, San Diego, 30 p.

Wootton, H.F., J.R. Morrongiello, T. Schmitt and A. Audzijonyte. 2022. Smaller adult fish size in warmer water is not explained by elevated metabolism. *Ecology Letters,* 25(5): 1177–1188.

Wootton, T. P., Sepulveda, C. A. and Wegner, N. C. 2015. Gill morphometrics of the thresher sharks (Genus *Alopias*): Correlation of gill dimensions with aerobic demand and environmental oxygen. *Journal of Morphology,* 276(5): 589–600.

Wosnitza, C. 1984. The growth of *Arapaima gigas* (Cuvier) after stocking in a Peruvian lake. *Archiv* für *Fischereiwissenschaft,* 35: 1-5.

Wourms, J.P. 1991. Reproduction and development of *Sebastes* in the context of the evolution of piscine viviparity. *Environmental Biology of Fishes,* 30(1): 111–126.

Wouters, A.G. 2007. Design explanation: determining the constraints on what can be alive. *Erkenntnis,* 67(1): 65–80.

Wright, P.A. and C.M. Wood. 2015. Regulation of Ions, Acid-Base, and Nitrogenous Wastes in Elasmobranchs. *Fish Physiology,* 34: 279–345. DOI:10.1016/B978-0-12-801286-4.00005-8

Wu, R.S. and Y.Y. Or. 2005. Bioenergetics, growth and reproduction of amphipods are affected by moderately low oxygen regimes. *Marine Ecology Progress Series,* 297: 215–223. DOI:10.3354/meps5297215

Wu, Y., S. Pates, D. Pauly, X. Zhang and D. Fu. 2024. Rapid growth in a large Cambrian apex predator. *National Science Review* 11: nwad284. DOI:10.1093/nsr/nwad284

Wulff, J.L. 2006. Ecological interactions of marine sponges. *Canadian Journal of Zoology,* 84: 146–166. DOI:10.1139/z06-019.

Xu, Z. and Y.J. Xu. 2015. Determination of trophic state changes with diel dissolved oxygen: a case study in a shallow lake. *Water Environment Research,* 87(11): 1970–1979

Yan, R., P. De Los Rios, A. Pastore and P.A. Temussi. 2018. The cold denaturation of IscU highlights structure-function dualism in marginally stable proteins. *Communications Chemistry,* 1:13. DOI:10.1038/s42004-018-0015-1

Yang, W., H. Quan, M.A. Meyers and R.O. Ritchie. 2019. *Arapaima* Fish Scale: One of the toughest Flexible Biological Materials. *Matter,* 1: 1557–1566.

Yesilyurt, İ.N., C. Türeli and Ö Duysak. 2019. Growth parameters of the cuttlefish, *Sepia officinalis* Linnaeus, 1758 from the Karataş coasts of the Mediterranean Sea, Turkey, p. 25–36. Third International Congress on Advances in Bioscience and Biotechnology, July 10-14, 2019. Kiev. https://cabidigitallibrary.org

Yokouchi, K., F. Daverat, M.J. Miller, N. Fukuda, R. Sudo, K. Tsukamoto, P., Elie and W.R. Poole. 2018. Growth potential can affect timing of maturity in a long-lived semelparous fish. *Biological Letters,* 14: 20180269. DOI:10.1098/rsbl.2018.0269

You, X. 2015. The effect of temperature on sex differentiation, sex steroid hormones and growth in yellow catfish, *Pseudobagrus fulvidraco* Richardson. Master's thesis, Huazhong Agriculture University, Wuhan, 47 p. [In Chinese; English abstract at https://kns.cnki.net/KCMS/detail/detail.aspx?dbname=CMFD201601&filename=1015380778.nh].

Zeller, D. and D. Pauly. 2001. Visualisation of standardized life history patterns. *Fish and Fisheries,* 2(4): 344–355. DOI:10.1046/j.1467-2960.2001.00051.x

Zeng, Y. and X. Wan. 2000. A new mathematical model and its application to the growth of crustaceans. *Crustaceana,* 3(5): 565–573. DOI:10.1163/156854000504660

Zhang Y., H.W. Sun, X.Y. Liu, Q.Z. Qu and D.J. Sun. 2012. Histological observation of gonadal differentiation and effect of rearing temperature on sex differentiation in Amur sturgeon *Acipenser schrenckii. Journal of Fishery Sciences of China,* 19(6): 1008–1017. DOI:10.3724/SP.J.1118.2012.01008

Zhang, W., Z.D. Cao, J.L. Peng, B.J. Chen and S.J. Fu. 2010. The effects of dissolved oxygen level on the metabolic interaction between digestion and locomotion in juvenile southern catfish (*Silurus meridionalis* Chen). *Comparative Biochemistry and Physiology Part A,* 157(3): 212–219.

Zhang, Y., F.U. Gao, D. He and S. Li. 2007. Comparison of spatial-temporal distribution characteristics of water temperatures between Lancang River and Mekong River. *Chinese Science Bulletin,* 52(2): 141–147.

Zhivotovsky, L.A. 2015. Genetic history of salmonid fishes of the genus *Oncorhynchus. Russian Journal of Genetics,* 51: 491–505.

Zhou, P. 2004. Determining Protein Half-Lives, p. 67-77 *In:* Dickson, R.C., Mendenhall, M.D. (eds) *Signal Transduction Protocols. Methods in Molecular Biology*, Vol 284. Humana Press. DOI:10.1385/1-59259-816-1:067

Zo, Z. 1973. Breeding and growth of the chaetognath *Sagitta elegans* in Bedford Basin1. *Limnology and Oceanography,* 18: 750–756. DOI:10.4319/lo.1973.18.5.0750

Zworykin, D., J. Müller, H. Grundmeijer and E. Pavlov. 2024. Paternal mouthbrooding in the chocolate gourami *Sphaerichthys osphromenoides* (Osphronemidae). *Environmental Biology of Fishes,* 107(4):1–9.

About the authors

Daniel Pauly studied fisheries science in Germany but spent much of his career in the tropics, notably in the Philippines. Since 1994, he is a Professor of Fisheries at the University of British Columbia, in Vancouver, Canada, where he directs the *Sea Around Us*. Its research is devoted to studying, documenting and mitigating the impact of fisheries, and increasingly, the effects of global warming, on the world's fish and marine ecosystems. The concepts, methods and software Daniel Pauly (co-)developed, are documented in over 1000 widely-cited publications, and have led to his receiving multiple scientific awards.

Johannes Müller is an environmental historian specializing in environmental change and its human perceptions as well as the history of knowledge and science. He is an assistant professor at the Leiden University Centre for the Arts in Society and a guest researcher at the Naturalis Biodiversity Center. He currently leads and participates in several interdisciplinary research projects that focus on the history of biodiversity and environmental change in the Netherlands and its former colonies. Beyond his academic work, he has long experience in fish breeding and actively participates in various local biodiversity initiatives.

Daniel Pauly

Johannes Müller

Endnotes

Chapter 1

1. One example of a thoroughly theoretical – albeit brief – engagement with the biological consequences of the physical properties of air and water as different media was presented by Denny (1993). Historically, one of the few biologists who highlighted the need to understand the physical properties of different respiratory media and their impact on animal physiology was Étienne Geoffroy Saint-Hilaire (1772 – 1844), who put particular emphasis on the differences between air- and water-breathing as two fundamentally different respiratory modes. According to E.S. Russell (1921), Saint-Hilaire's interest in the physical properties of different respiratory media and "*the stress which Geoffroy lays upon respiration and the respiratory milieu*" informed his pre-Darwinian notions of evolution to the point that it became a "*constant obsession.*"
2. Despite the seriousness of this topic and the need to find societal and political solutions for the problems discussed, we are committed to avoiding alarmist tropes and using a tone of urgency only as far as strictly necessary. This, however, is admittedly a rather low bar at a time when it has become clear that even the most half-hearted global climate goals will not be met.
3. We would like to add here that genuine "*paradigm shifts,*" as defined by Kuhn (1962), are rather rare, and the concept of such shifts is nowadays trivialized. Even in cases of major scientific advances, instances of epistemic incommensurability are quite rare, and science keeps building on (and transforming) the body of knowledge established earlier. While earlier findings and insights require continuous reinterpretation, they usually remain relevant.
4. The term 'climate crisis' seems inadequate since 'crisis' implies a notion of reversibility, i.e., to imply the possibility of returning to a state of 'normalcy;' we are skeptical about this notion.
5. When Charles Darwin joined the *Beagle* voyage in late December 1831, Captain Fitzroy gave him a copy of the first volume of Lyell's *Principle of Geology* (1830), which enabled him to conceive his (Darwin's) theory of coral reef formation before he had really studied coral reefs and later provided him with the deep time that his nascent theory of slow evolution through natural selection required (Pauly 2004).
6. Various authors have referred to the theory presented here as the "*Gill-Oxygen Limitation Hypothesis*" (see, e.g., Audzijonyte *et al.* 2019; Scheuffele *et al.* 2021). We use the term 'theory' rather than 'hypothesis' not because of grandiosity, but because the GOLT is not based on a single idea (or hypothesis) but presents a larger explanatory framework that incorporates several mutually compatible, testable hypotheses articulated within a mechanistic framework that allows for their interpretation.
7. Incorporating relevant forms of traditional and practical knowledge and expertise from communities outside science can be understood as a question of "*epistemic justice,*" as defined by Michela Massimi, who stresses the importance of applied "*craftmanship*" for scientific findings (Massimi 2022). Another powerful example of the interconnection between scientific (and theoretical) knowledge and the applied expertise of non-academic communities is presented in the first chapter of *The Origin of Species*, where Darwin reflects on his experience with pigeon breeding and his connections with pigeon fanciers (see also Secord 1981). As we have argued elsewhere (Müller *et al.* 2023), the practical expertise of fieldworkers in freshwater fisheries management is of great value

for understanding the physiological effects on fish in heatwaves or hypoxic zones. Indeed, such expertise may be more valuable than one based on the result of short-term experiments conducted under laboratory conditions that do not reflect real-life circumstances.

8. Ross (2023) describes 3 kinds of constraints: *"Law-based, mathematical, and causal,"* all of which are discussed in this book.

9. The knowledge and awareness of surface-volume relationships appear to have first emerged in the context of architectural practices. One of the many examples of a pre-modern solution to surface-volume constraints was an early Egyptian public works project (the 'Bent Pyramid' at Dashur), which started with an acute (55°) angle (implying that a huge mass of stone would have to be supported by a relatively small cross-section), which but was completed with a top built at an obtuse (43°) angle to prevent a collapse. While such measures may have been taken *ad hoc*, ancient Egypt (like China and India) did develop highly theorized knowledge of surface-mass problems (Bardi 2017).

10. The authors also refer to organisms as 'Kantian wholes,' based on Kant (1790), who explored the internal organization of living and mechanical systems in his critique of teleology in art and nature.

11. Charles Darwin also used this argument to explain why more complex brain functions depended on homeothermy and, therefore, on air-breathing (Darwin 1872). He used this observation in different contexts and for different purposes. We will return to this in Chapter 11.

12. The great ichthyologist Georges Cuvier also wrote on this; we translated his remarks from French and commented on his views in Pauly and Müller (2022); see also Fitting (2005).

13. Hesse also stressed the possibility of simplifying and reducing complicated morphological structures after the transition from water- to air-breathing. This applies for example in gastropods, which maintain filamentous and vulnerable gill structures with large surface areas in water, but whose terrestrial species rely on comparably small respiratory surfaces.

14. The exception is the phylum Onychophora, or velvet worms, which evolved in water, but now occur only in terrestrial ecosystems.

15. We thank our colleague Mark Butler (pers. comm. to D.P.) for this apt comparison.

16. Schuman and Piiper (1966) reported an estimate of 1/3 for the tench *Tinca tinca*. This high value, noted by Pauly (1979), strongly influenced the GOLT's development.

17. In this respect, our theory differs from other theories on growth and metabolic scaling, e.g., *Ontogenetic Growth Model* or the *Metabolic Theory of Ecology* developed by West, Brown, and Enquist (1997). While these authors claim that internal fractal network structures are a constraining factor for aquatic animals, we argue that the growth of WBE will first be constrained by other factors (see also our discussion of the *Metabolic Theory of Ecology* further below in the text, in the section on *Constraints and Optimizations*.

18. This is due to their diminutive sizes; see Chapters 3 for the guppies and Chapter 12 for the brine shrimp (*Artemia*).

19. The reason is that if the dimensions of gills were the same as that of a volume (i.e., $d_G = 1$), water could not flow through them. This is difficult enough for tuna, which, by forcing water across their gills through 'ram ventilation' over extremely thin lamellae, can reach $d_G = 0.9$. However, the distance between their gill lamellae is so small, and the lamellae are so thin that their gills accommodate up to 160 lamellae per mm (Muir and Hughes 1969). This is one reason why tuna do not tolerate silty or turbid water, as the owner(s) of a tuna-farming operation in Port Lincoln, Australia, discovered when, in 1996, a storm lifted sediment from the sea floor and asphyxiated 2/3 of the southern bluefin tuna (*Thunnus maccoyii*) in their pens, worth an estimated 14 million Australian dollars (Dawes 1998).

20. This work, which involved numerous studies that included very small (young) specimens of larger species, i.e., stages at which *d* values are still transiting from the high values ($d >> 1$) in larvae toward the values of $d < 1$ in late juveniles and adults (see **Figure 6.2** for an example) was bound to over-estimate the 'average' *d* value in fish, and it does. Note, however, that the precise value of such 'average' will depend on the mix of species considered. If many species of guppies and/or tiny gobies are thrown into the mix, the average will be lower than that of a sample consisting mainly of large, well-studied commercial fish species (see Pauly 1981, and Chapter 7).

21. These slopes, which are indeed proportional to surface-volume relationships in spherical objects, may be typical for small endotherms with high metabolic demands, which is in line with Bergmann's argument Bergmann (1847).

22. The scaling factor *d* is usually estimated as the *slope* of a (double-logarithmic) linear regression, for which the letter *b* is normally used. However, given its importance in this book, this slope is here labelled differently, i.e., *d* in all cases where gill surface area (*S* or *GSA*) or absolute respiratory rate (*R*) and body mass or weight (*W*) are related, as in $GSA \propto W^d$, or $R \propto W^d$. The choice of the letter '*d*' was inspired by Ursin (1967), who used it for "[t]*he power of weight with which the rate of oxygen consumption is proportional, disregarding the feeding condition of the fish.*" Note also that, when this difference matters, we use d_Q for the metabolism-related power and d_G for the power linking gill surface area and weight.
23. The same is true for plants; see Glazier (2010).
24. Strangely, this observation has been used as an argument against the GOLT by Scheuffele *et al.* (2021), who assume that the GOLT requires lower scaling exponents for gill surface area than for metabolic rate. However, the GOLT is built on the fact, established by De Jager and Dekkers (1975) and other authors, that these two exponents take the same or similar values when properly estimated (see also Pauly and Müller 2025).
25. The debate around the evolvability of morphological novelties within given geometrical constraints between George McGhee, Wim Hordijk, Stuart Kauffman, and others (see Hordijk 2016) is illustrative in this respect, and we will return to this topic in Chapter 15.
26. See De Silva (1974) for gill development in larval plaice and herring, and Oikawa and Itazawa (1985) for the same in common carp (*Cyprinus carpio*; **Figure 6.2**). However, these larvae at first rely exclusively on cutaneous respiration, which is magnified by their primordial fin. Also, Post and Lee (1996) report that early juveniles have weight-proportional metabolic rates, i.e., their $d = 1$.
27. Here, 'gill surface area' could be replaced by 'respiratory surface area,' as many WBE lack gills (see Chapter 12). However, in this book, we will generally skip over this fact for simplicity's sake.
28. Few authors make explicit that oxygen demand is distinct from oxygen supply. One of them is Brett (1972), who noted that various "*technical advances made it possible to establish the power-performance relation for the gill-respiring aquatic vertebrate, and to determine the likely demand for oxygen associated with activities which exceeded the capacity to supply oxygen, resulting in fatigue.*"
29. Von Bertalanffy's theory of growth will play a central role in this book, as we will see in the next chapter. However, his conceptual approach to organisms as open systems is also relevant in the context of constraints as traits that enable operative functions. While organisms function as thermodynamically open systems, which allows them to exchange energy and resources with their environment, they also rely on closure, and on constraints to maintain their functional identity.
30. Epigenetics, where species can turn on and off different genes to help their survival in different environmental conditions may be mentioned here, although we can't elaborate (but see Ward 2018).
31. As Gould (1989) has argued, "*positive*" meanings of the term "*constraint*" were gradually outweighed by the negative ones. "*Constraints*" had a far more positive meaning in the King James Version of the Old Testament and used it in the sense of 'urge,' 'compel' or 'make,' indicating its constitutive role in decision making or similar processes: "*Constraint also has a set of positive meanings, well and long established in our language. The King James Bible, for example, uses the word 10 times, nine in the positive sense of directing or forcing an action in a particular way. The most famous passage illustrates the confusion that results when only negative definitions are entertained. When Elihu the Buzite gives his unsolicited advice to Job, he justifies his decision to speak, despite his youth (Job 32:18-19): "For I am full of matter; the spirit within me constraineth me. Behold my belly is as wine which hath no vent; it is ready to burst." I used to interpret "constraineth" negatively, and the passage made no sense. Why should Elihu be full to bursting yet still holding it all in. Of course, he means quite the opposite – that the spirit within is compelling him to belch forth his thoughts.*"
32. Many of these limitations may not even be experienced. As Raja and Anderson (2020) put it in the context of constraint-based models of scientific explanation: "*Humans cannot fly. This is a little bit disappointing but we have largely made our peace with it (recurring dream motif notwithstanding) and have invented planes. Planes allow us to fly, albeit within several mechanical, temporal, and legal restrictions.*" Not even planes function without limitations, even though they enable the very activity that would otherwise be inaccessible to humans.

33. This practice is strangely common among colleagues studying snappers, Family Lutjanidae (see, e.g., Jurado-Molina *et al.* 2018; Moncrief *et al.* 2018), but can also be found among researchers studying cephalopods (see Chapter 13).
34. The term *"constrained generality"* has also been used in other contexts, for example, by Butler (2002), who employs it to develop a generalizing definition of *"critique."* Our use of the term is slightly different as it allows us to bridge different models of explanation. Some authors have suggested separating mechanistic and constraint-based accounts of explanations from each other (see, e.g., Green and Jones 2016). Such seemingly different explanatory models can be connected if we acknowledge that mechanisms are necessarily constrained and also act *as* constraints on other processes. An interesting and potentially fruitful model in this context is Wouters' (2007) *"design explanation,"* which proposes to determine the *"constraints on what can be alive"* and base its explanatory model on these constraining factors. This approach is compatible with the GOLT, but leaves open questions about which specific options (within the range of possibilities) are realized.
35. This point may be conceived simply as a call for 'Occam's Razor,' named after a medieval monk, William of Ockham, who suggested that *"entities should not be multiplied beyond necessity"* (Pauly 1994b). However, this point is also relevant within the context of the so-called *"bottoming out-problem"* (Felline 2016; Glennan 1996), which refers to the level of causality at which mechanistic explanations should operate. Since it lies in the nature of mechanistic explanations that causation must be assumed at several levels, valid explanations do not necessarily need to delve into the deepest possible of causation. *"Bottoming out"* the various levels of causation too deeply may also lead to constant revision: if one specific link in the explanatory chain is later found inviable, the theory, as a whole, will need reformulation. More importantly, if causation always takes place at multiple levels, the most fundamental levels of causality are often less consequential and immediate for the actual phenomena that are eventually described. Choosing a higher (or more 'superficial') level can add to the explanatory power and, simultaneously, be a simplification because deeper levels of causations can be neglected (i.e., infinite regress is avoided; Pauly 2019, p. 4). For example, Bergmann's model explains why, e.g., in cold climates, homeotherms benefit from being large (which reduces their surface area relative to the mass of their body, and potential heat loss; Bergmann 1847). However, there is no need to explain why heat loss of a body is proportional to its surface (though it was once a legitimate research question; Boltzmann 1884), or going even deeper, why heat flows from warm bodies to cold ones, etc.
36. These books, and particularly the more recent one, still contain a large amount of evidence for the GOLT that is aligned with the new material in this present volume, but which was not incorporated into it.
37. Both FishBase and SeaLifeBase coordinate their taxonomy with the World Register of Marine Species (WoRMS; www.marinespecies.org), which covers all marine (and many non-marine) species, and provides the taxonomic nomenclature to SeaLifeBase.
38. Indeed, asymptotic growth may also have occurred in Ediacaran biota such as *Parvancorina minchami* (Ivantsov *et al.* 2025). The Ediacaran, by the way, was the period before predation was invented. Ediacaran organisms had no appendages to grab or mouths to eat each other. Later, in the Cambrian, metazoans learned not to make their newly acquired appendages easily available to newly evolved predators, which could see them through their newly acquired eyes (Parker 2003). Leafy plants, which evolved much later, don't mind being 'out there' with their leaves; that's why herbivores have such great time. However, the GOLT doesn't care about appendages, mouths, or eyes; all that is needed for it to apply is to have a solid body that must be supplied via a surface with a gas extracted from a liquid.
39. There is ample evidence of octopuses' intelligence, as illustrated by the film *"My Octopus Teacher."* One aspect of this, pertinent to the theme of this book, is the observation reported by Learn (2024) that a scientist at the Marine Biological Station in Naples, Piero Amodio, said *"that he has seen fights between octopuses where one will fill the gill hole other the other with an arm, asphyxiating its foe."*
40. A second edition (Pauly 2024) of this work is available. In addition to its original content (in better layout), it has several footnotes with updates and reflections.

Chapter 2

41. Cited from Ibn Al-Nafis (1968). Other translations read "*a continuous state of dissolution and nourishment*" (see also Fernie and Pichersky 2015).
42. Two-dimensional worlds may be conceived. Dewdney (1984) proposed and illustrated one, to which Pauly (1989) added considerations relative to fishing and 'fish' respiration – obviously!
43. It is likely that this form of growth represents the ancestral state of biological organisms (see Hariharan *et al.* 2016).
44. The paradoxical nature of fish growth ceasing (at W_{max}) despite being surrounded by potential prey in the wild or being given food *ad libitum* in captivity is not appreciated as such by colleagues who reject the GOLT's explanation of why growth ceases. In fact, they double down and, in the absence of the 'internal' limit to growth and size that is presented in this book, they evoke either unknown mechanisms (i.e., "more research is needed") or postulate the existence of specific and, by necessity, different prey limitations for each population of each species in each ecosystem, with some additional *ad hoc* assertions thrown in for captive fish.
45. Note that the term "*intussusception*" did not yet have its contemporary meaning of an intestinal defect.
46. In another publication, Kassowitz (1905) used the example of cell plasma disintegration and repair to solve one of the major philosophical problems in biology at the time. He argued against the teleological notions of contemporary vitalism and the idea that underlying aims and purposes caused physiological processes.
47. The Gompertz growth function (Gompertz 1825) and the logistic growth function developed by Pierre-François Verhulst (1804 – 1849) were already being used in biology in the nineteenth century, but primarily to model population growth and not the organic growth curves of individual organisms.
48. Both authors of this book speak German fluently, which was a definite advantage, given that most of the foundational work considered here was initially published in that language. A small fraction of this work also exists in English translations, of which some are inaccurate.
49. See Kearney (2021) and Müller and Koster (2024) for the dissemination and reception of Pütter's growth model in twentieth-century physiology.
50. This theory was dismissed by many researchers at the time, notably by August Krogh, but continues to be a topic of investigation today. Max Verworn, Pütter's mentor, was one of the few colleagues who supported his idea about dissolved organic matter as an important food source for aquatic animals (Verworn 1909).
51. Verworn defined "*assimilation*" in a very broad sense, and as "*the entirety of processes that constitute the formation of new living substances*" with the synthesis of proteins as its "*climax*" (Verworn 1901, p. 517). "*Dissimilation*" was then understood as the opposite, i.e., the breakdown of cellular materials. In adult organisms, a state of "*metabolic equilibrium*" was reached in which new building materials were only acquired at a rate enabling assimilation to equal dissimilation.
52. While Spencer wrote extensively on "*anabolism*" and "*catabolism*" in various contexts, his definition of "*expenditures*" remained open and included all the energetic costs that organisms had to carry. Most attention was devoted to the energetic costs of large terrestrial animals, whose growth was thought to be limited by the high costs of simply having a heavy body.
53. Geddes and Thomson (1889) further elaborate on this as follows: "*The early growth of the cell, the increasing bulk of contained protoplasm, the accumulation of nutritive material, correspond to a predominance of protoplasmic processes which are constructive or anabolic. The growing disproportion between mass and surface must however imply a relative decrease of anabolism. Yet the life, or general metabolism, continues, and this entails a gradually increasing preponderance of destructive processes, or katabolism. As long as growth continues, the algebraic sum of the protoplasmic processes must of course be plus on the side of anabolism, and growth may be now more precisely defined as the outcome of the preponderance of an anabolic tendency, rhythm, or bias.* [...] *What is true for the cell, is true for cell-aggregates. Organisms in their entirety have very definite limits of growth.*"
54. Note, however, that d is always smaller than 1, except in larvae and other very young individuals; we will return to this point repeatedly.
55. See, e.g., von Bertalanffy (1934): "*Pütter (1920) deserves the credit for first providing dynamic and physiologically based ideas to have developed growth. Of course, his work is subject to some points of justified criticism.*"

56. Von Bertalanffy's growth "*type 2*" mainly applied to insects, whose respiration involved tracheae that occur throughout their body. Hence, he assumed that their respiration would be weight-proportional. However, this is wrong because the sum of their tracheal openings is a surface (as is also the case for the pores of sponges). Therefore, insects are oxygen limited, as indirectly demonstrated by the huge dragonflies of the Jurassic, when atmospheric O_2 were much higher than today (Atkinson 2005; Ward 2006), and by the fact that compact insects cannot grow beyond the 17 cm of *Titanus giganteus*, the largest beetle (Kaiser *et al.* 2007). Von Bertalanffy growth "*Type 3*" was limited to water pulmonates, i.e., lung-breathing snails, whose growth curve resembled the logistic growth function (see von Bertalanffy 1951).

57. Here, for simplicity's sake, Pütter's symbols, and L were replaced by their equivalents, L_t and L_∞, respectively. Also note that this equation assumes that $W \propto L^3$, as is also the case for the integration of the standard VBGF (see further in main text).

58. The value of t_0 should always be negative because the different versions of the von Bertalanffy equations all underestimate the growth of larvae and very young WBE, which are too small to be limited by their oxygen supply.

59. In addition to his growth equation, Pütter stated that final size (L_∞) could be expressed as the division of the coefficients of buildup and breakdown metabolism, i.e., as $L_\infty = k/k'$. In addition, he proposed that the time in which L_∞ was reached could be written as $k - k' = c/L$ (where c is the growth coefficient). As von Bertalanffy pointed out, this assumption was based on a different interpretation of the growth coefficients c and K). One should add that it also implies a different concept of L_∞, as it is never 'reached' in the VBGF.

60. See Pauly (2024) for the details of two ways to integrate the general VBGF: the longer original version, kindly adapted from Pauly (1979) by Cui 'Elsa' Liang, and a shorter, more elegant version provided by the mathematician Ivar Ekeland (see Wikipedia).

61. The special VBGF fitted to the age-at-length data of Sella (1929) produces estimates of $K = 0.043$ year^{-1} and $L_\infty =$ 505 cm (FL), which is >> L_{max} (330 cm FL; Tiews 1963). The generalized VBGF, with $D = 0.3$ (corresponding to $d = 0.9$), produces an estimate of $K = 0.41$ year^{-1} and $L_\infty = 332$ cm (FL), with $L_\infty \approx L_{max}$ (Pauly 1981, 2021a).

62. This was also the argument of von Bertalanffy (1951), who lamented that the increased focus on statistics blurred actual "*analysis,*" and precluded scientific explanation. We share the view that despite the indispensability of statistical models, they cannot replace actual mechanistic explanations of biological phenomena.

63. As von Bertalanffy (1951, p. 279) further argued, smaller deviations from a direct mass proportionality of breakdown metabolism were possible but did not make a more than negligible difference for growth calculations.

64. Note that in German scientific discourse, the term "*Katabolismus*" was not used in the modern sense before the second half of the 20th century.

65. Many physiologists, however, do so, for example Hochachka (1969); see his definition of 'catabolism' in Chapter 4.

66. 'Catabolic term' is what we shall use throughout this book when referring to the 'breakdown' of Pütter and von Bertalanffy.

67. See also Beverton and Holt (1957, p. 32) and Ricker (1958, p. 196), whose interpretation of von Bertalanffy confused generations of fisheries scientists.

68. Pütter was among the first to estimate the surface area of the lamellae of fish gills, for example, in *Hippocampus* sp. (Pütter 1911, p. 315).

69. Ricker's critique of the hypothesis that gut surface limitation could play role in food assimilation remains valid nonetheless: "*This could seem reasonable if food were always available in excess, so that absorptive surface could actually be a factor limiting growth; and in the guppy experiments which are quoted in support of this relationship, food was actually provided in excess. In nature, fish are usually less fortunate; this is shown by the small average volume of food commonly found in their stomachs, and also by the great variability of their observed growth rates, both when we compare individual fish in the same environment and when we compare populations from different (but physically similar) waters.*" (Ricker 1958).

70. See, e.g., the introduction to Vol. 1 of von Bertalanffy's *Theoretical Biology*: "*It may be a commonly accepted view that biology in its current form finds itself in a state of crisis. Until roughly the turn of the century, it was widely believed that Darwinism possessed the solution to all biological puzzles.*" According to von Bertalanffy, Darwinism lacked real explanatory potential, since explanatory accounts based on natural selection did not provide answers for the

morphological and physical features and traits that could be selected. His critique was not uncommon at the time and reflected a wider discontent with the state of evolutionary theory before the Modern Synthesis. As Drack (2015) puts it: "*Bertalanffy's early critique on the theory of evolution, Darwinism in the early 20th century as it were, is in line with the dominant critique at the time. At that point, Darwinism as a theory was no longer to be taken seriously. Only later, when mutations were investigated in more detail, did Darwin become important again.*"

71. Karpouzi and Pauly (2008) found that the fit of the VBGF (for weight) to the growth of seabirds was not particularly tight (Gompertz curves would probably have done a better job). Their VBGF's growth parameters were retained only because they enabled comparisons (in SeaLifeBase) with other groups, notably with WBE, for which the VBGF generally provides a good fit.

72. The special VBGF has also been used to model the growth of cancerous tumors, which represents a kind of growth which is not controlled by an anatomical body plan but only by the balance between supply and demand (see, e.g., Kühleitner *et al.* 2019). In this case, the VBGF may be better suited to describe growth than when applied to air-breathing vertebrates, whose growth patterns are shaped by different factors.

73. A surface limitation to the acquisition of oxygen occurs in fish eggs (and, by extension, in the eggs of other WBE), however (Teletchea and Pauly 2024; see also Chapter 17).

74. The proportionality between the metabolic rate and gill surface area is one of the key assumptions of the GOLT's growth model. Surprisingly, the empirical confirmation of this assumption has sometimes led colleagues to criticize the theory. Scheuffele *et al.* (2021) suggest that the GOLT depends on the assumption that the slope of the gill surface is lower than the slope of the metabolic rate. This would contradict Fick's Law (Pauly and Müller 2025) and, as well, result in the paradoxical situation that organisms would consume energy which is, in fact, unavailable (or would have been produced anaerobically, which is only possible during short periods). The error here is the notion that measurements of metabolic rates in fish below W_{max} and L_{max} (or W_∞ and L_∞) do not include costs associated with growth. This is a misunderstanding of the methodologies on which such measurements are based (see, e.g., Parry 1983; Rosenfeld *et al.* 2015); see also Chapter 6.

75. This is also the reason why Prinzing *et al.* (2023), in their eagerness to demonstrate the invalidity of the GOLT, broke a regression showing that log(*GSA*) vs. log(*W*) in 19 horn shark had a slope of d = 0.877 (i.e., d <1, as expected) into two parts, one with 3 data points for small individuals, yielding '*d*' = 0.564, the other, with 16 large individuals, producing a '*d*' value of 1.012. This questionable procedure is examined in more detail in Chapter 6.

76. Besides the 3 primary criteria for the GOLT validity, the following predictions of the GOLT for WBE can also be seen as additional criteria for its applicability:

    1. Growth is exponential in larvae and very early juveniles, for which $d < 1$;
    2. Within the tolerated range of temperature, increasing temperatures should result in smaller maximum and asymptotic size (except for very cold environments where cold denaturation can cause maximum sizes to remain low);
    3. Growth and attained size should decline under sub-lethal hypoxia;
    4. In populations where maximum size is low, size at first maturity must be low as well;
    5. Food conversion efficiency declines with size, and approaches zero near the maximum sizes reached in a population;
    6. If a taxon includes both water- and obligate air-breathing species, as do crabs or bony fishes, the air-breathers should be larger and/or be more active, and/or tolerant of higher temperatures than their closest water-breathing relatives;
    7. Larger-size individuals should be more sensitive to hypoxic or high temperature, and in hypoxic conditions, the larger individuals should die first.

    The various chapters of this book deal, mostly tacitly, with these various criteria, and also with seeming exceptions, e.g., the asymptotic growth of chironomid larvae.

Chapter 3

77. Hence the title of Gandhi, the founder of modern India was the (Sanskrit) word 'mahatma,' consisting of 'maha' (= great) and 'atman' (soul), the latter root being, in several Indo-European languages, related to vapour or air, and thus to breathing. See, e.g., Greek 'ατμός' = vapour (Beekes 2010), or German: 'Atmung' = respiration (https://www.dwds.de/wb/atmen; consulted August 28, 2024).

78. Note that respiration in terrestrial animals can indeed have this (additional and secondary) function, especially in species that use their tongue and the inner wall of their mouth to cool themselves. This is why the San people of Namibia can run after antelopes until they overheat and have to stop running (Liebenberg 2006).

79. An exception is the treatise *De Spiritu*, which was long ascribed to Aristotle but is now considered Pseudo-Aristotelian. According to the author of this text, fish do not respire because water contains no air; see Bos and Ferwerda (2008).

80. It is unclear how Malpighi conceived of the role of respiration in the transformation and digestion of food. However, his writings suggest that he assumed that respiration was essential in these processes. See also Nordenskiöld (1946, p. 162), and Pauly (2019, p. 19).

81. It was only later established that plants consumed oxygen, too. The botanists Julius Sachs (1823 – 1897) and Wilhelm Pfeffer (1845 – 1920) demonstrated that plants could only release oxygen when exposed to light, a crucial observation that laid the basis for our modern understanding of photosynthesis.

82. It is astonishing that nearly 2½ centuries later, 28 physiologists would author a paper in which oxygen is described as "*only a permissive factor compared with other constraints (e.g., food availability)*" (Jutfelt *et al.* 2018), and to which we shall return in Chapter 6. The paper by Jutfelt *et al.* (2018) is also a powerful illustration of the point made by Franks (2002) that "*the common practice of uncritically equating 'recent literature' with 'progress' is of dubious value.*" It also highlights the pertinence of the advice that any newcomer to a field "*should consider whether a recently published article actually constitutes and advance of our understanding, or whether, at best, it rehashes earlier work or, at worst, actually obscures older insights into complex problems.*"

83. There is, high up in the Bolivian Andes, a species of amphibians, the Lake Titicaca water frog (*Telmatobius culeus*), which reaches up to 1 kg, and whose adults (and not only their tadpoles) breathe water, and no air. They live at depths between 10 and 120 m in Lake Titicaca and water bodies in the surrounding basin, which is possible not only because these water bodies are very cold, but mainly because their cutaneous respiration occurs through a hugely distended skin which hangs in folds from their bodies, with their 'bagginess' increasing with size (Robischon 2017).

84. An additional reason why cutaneous respiration in adult fish is rare is that it is precluded by the mineralization of their skin (i.e., their coverage by scales), which protects them from mechanical damage by some predators and parasites (see Rombough 2007).

85. Early methods for studying the gill surface areas of fish and other WBE were refined and further developed by George M. Hughes (1925 – 2011). The following quotes may give an idea of his approach and the considerable work involved: "*The most important features to be analysed are the area and thickness of the respiratory surfaces. Gill areas have been measured by a number of authors* [...] *All such measurements depend on sampling, particularly with respect to the area of the ultimate exchange unit, the secondary lamella. Measurements of filament length and frequency of secondary lamellae are fairly certain, but the errors involved in obtaining a representative figure for the area of a single secondary lamella may be quite considerable. The use of weighted averages (Muir and Hughes, 1969) greatly reduces the errors.*" (Hughes 1972).

    In one of the following publications, Hughes and Morgan (1973) further refined their approach: "*Estimates of total surface area are made from basic measurements of the length of the gill filaments, their total number, the surface area of individual secondary lamellae and their spacing along the filaments. The relationship between these various parameters and the total gill area in fish (Hughes, 1966) is: total area of all secondary lamellae = [2(L/d')] bl, where L is the total length (mm) of all gill filaments, I/d' is the number of secondary lamellae/mm on one side of a filament, and bl is the surface area of both sides of an average secondary lamella. The very large number of secondary lamellae obviously precludes measurement of all of them in a given fish and it is therefore necessary to select samples for actual measurement from which the total gill area can then be estimated. Fixed material is usually used for comparison with fresh gills and in most cases the two show only slight differences.*"

86. In fish larvae, the surface of the body is magnified by a primordial fin (which extends on top from the head to the tail and down forward to the edge of the yolk sack), and which also functions as a respiratory organ (Pauly 2019, 162-163).

87. There are some small inconsistencies in table 1 of Kunwar and Munshi (1988), but they do not affect the stark difference of relative gill surface area between the 17 g and the 17 kg specimens of *C. catla* that are shown in **Figure 3.4**.

88. This logic – demand generating supply – is actually taught in Economics classes at UBC, in Vancouver, though the students then have problems understanding why their demand for cheap housing is not met by an increasing supply.

89. This very question puzzled Somero and Childress (1980), as can be inferred from the title of their contribution. They suggested that the solution to this puzzle would be found in "*surface-volume relationships,*" which is precisely what we argue is the case.

90. The entire structure of Wouters' "*design explanation*" for the existence of fish that breathe through their gills is as follows:
"*(1a) They take their oxygen from water. (1b) They are moderately active and their size is within certain limits. (1c) They use oxygen as an electron acceptor to generate organic energy. (2) If the respiratory medium is water and the fishes' level of activity and size are as mentioned in 1b, the biological role of respiration is better performed with gills than with lungs. (3a) Given the activity and size mentioned in 1b the organism needs a certain supply of oxygen (laws of physics and the way that aerobic organisms get their energy). (3b) Due to water's low oxygen content and very small diffusion coefficient, the organism must ventilate huge amounts of water in order to get enough oxygen (arithmetic + Fick's law). (3c) Due to water's density and viscosity, much power is needed to move this amount of water (physics). (3d) Because of their smaller respiratory area, their larger distance between the medium and the blood, the dead air in their alveoli and the tidal character of their air flow, lungs are less efficient than gills (Fick's law). This means that if fishes were to use lungs to take up oxygen from water, they would need even more power than they do now. The lungs' ventilation mechanism is less powerful than that of gills (the way lungs and gills are built) and does not have enough power to move the required amount of water. (3e) Due to their large respiratory area, their small distance between the medium and the blood, and their continuous, unidirectional flow over the respiratory surface, gills are more efficient than lungs (Fick's law). In addition, they have much more power (the way lungs and gills are built). As a result, they are able to provide the required amount of oxygen at the required rate.*" (Wouters 2007; see also Fick 1855).

91. A pelagic fish must have a fusiform body with a smooth pointed head to swim in an energy-efficient manner, especially if it uses ram ventilation. However, sleek ram-ventilator, such as the grey reef shark (*Carcharhinus amblyrhynchos*) can exhibit "*resting behavior,*" and use buccal pumping to meet a reduced O_2 demand (Bullock *et al.* 2023). Weird departures from fusiform shapes, such as in seadragons (Syngnathidae) or the red handfish (*Thymichthys politus*), are possible. However, these fishes live on or near the sea floor, and metabolic and growth performance are much reduced (Pauly *et al.* 2022c, 2024b).

92. In fact, ablation of gill arches was previously an accepted method of demonstrating the importance of a large gill surface area for fish (see Duthie and Hughes 1982).

93. The complex tradeoff between iono- and osmoregulation and gas exchange in gills in the context of evolutionary optimization has puzzled many biologists – Rombough (2007) even calls the functional adaptation of fish gills "*irrational*": "*A large surface area is necessary for efficient gas exchange. However, neither freshwater nor marine fish are in ionic equilibrium with their environments and a large surface area compounds the problems of trying to maintain ionic homeostasis under such circumstances (e.g., Gonzalez and McDonald, 1992). In terms of development, the "rational" approach (recognizing that evolution does not act rationally in the human sense) would be to have the ionoregulatory capacity to compensate for increased rates of passive ion flux in place before lamellar expansion occurs.*"

94. On the final point, see also Schaack and Chapman (2004). Note also that the largest fishes, the whale shark (*Rhincodon typus*), the basking shark (*Cetorhinus maximus*), and the (now extinct) giant bony fish *Leedsichthys problematicus* are (or were) plankton feeders, which 'solved' the conflict between feeding and breathing by evolving gill structures which combine both.

95. A more general point relevant here is that maximization of oxygen uptake capacity is not always an optimization, especially in air-breathing organisms for which oxygen is available in concentrations that can be toxic. Oxygen limitation through regulating uptake can also be an adaptive feature. Many organisms require a mechanism that protects them from lethal overdoses, such as insects, many of which breathe discontinuously to avoid oxygen toxicity (Hetz and Bradley 2005; Schloss 2002). A limitation can also be a safety mechanism that later became stabilized in the *Bauplan* of the organism (Kutschera and Niklas 2013). Limitations also could result in phenotypical features that might not initially be beneficial but, at the same time, protect animals from entering a phenotypical trap in the form of gigantism that would bring the organism into a situation where changes in temperature result in dangerous states of oxygen limitation (Kaiser *et al.* 2007; Verberk and Bilton 2011; Verberk *et al.* 2016, 2021; see also Chapter 10).

96. Von Bertalanffy's ambiguity regarding the question of which exact surface limited metabolic processes, and hence growth, was possibly inspired by his quest to develop explanatory models with a maximum degree of general applicability, as is evident from his seminal text (von Bertalanffy 1951). Some of his biological work on growth and metabolism already foreshadowed his later systems theory, which claimed to be applicable across many disciplines and even across the natural and social sciences. Such a theory depended on a maximal degree of generality – which could have easily been undermined by too many details that only applied to a limited number of taxa. Having said this, we note however, that he wrote on p. 246 of his text that "*the respiratory surface will be not be, in general, the outer surface of the body, but rather any surface within the body, which obviously implies that one must think first of the respiratory surfaces,*" while similar statements may be found on p. 251, among others.

97. In fact, there seems to be an inverse relationship between cell size and fish swimming performance (Cavalier-Smith 1991; Pauly *et al.* 2000).

98. We return to this theme in Chapter 16.

99. Ultsch (1976) makes a similar point: "*it is the oxygen transport capacity* of [salamanders breathing water] *that at least partially regulates the standard rate of O_2 consumption rather than all such control being exerted at the cellular level.*"

100. The extent to which gill remodeling is important to tropical marine species is disputed. As Bowden *et al.* (2014) suggest, the gills of many tropical marine fish already operate at their physiological and morphological limit, and further remodeling is difficult. An exception in this study was the clownfish *Pomacentrus moluccensis,* which can apparently undergo extensive morphological modifications in response to increased temperature (Bowden *et al.* 2014). However, the orange clownfish *Amphiprion percula* has been reported to *shrink* during a heatwave by authors who considered the GOLT as a possible explanation (see Versteeg *et al.* 2025).

101. Rutjes (2006) investigated the increase in gill surface in hypoxia-raised cichlids and concluded the following: "*Measurements on the third gill arch of* [Haplochromis] *pyrrhocephalus showed that the cross-sectional area of individual respiratory channels was increased in [hypoxia-raised] fish as well as their number and secondary lamellae were longer. However, a ~26.9% increase of the number of respiratory channels, a 9.2% increase of the height of the channels and a 37.7% increase of the length of the channels, increases the retention time of water in the gills by less than a factor 2. Thus, at an eightfold increased water flow, the retention time of water in the gills was more than 4 times shorter, resulting in lower efficiency. Additionally, an 80% increase in gill surface does not compensate for the eight-fold decrease in O_2 flux caused by the lower partial gas pressure under 10%* [air saturation]." However, even major morphological modifications cannot fully overcome the geometrical constraints on effective oxygen uptake, and gill remodeling is only effective up to a specific limit (aside from the osmoregulatory problems that would result from an ever-larger gill surface area).

102. Hypoxia can induce plastic responses in cell structures elsewhere in the body; however, this does not typically alter the cells' surface-volume ratio but rather their biochemical composition (Farhat *et al.* 2019).

103. In a related issue, Holeton (1976) write that "[t]*he heart of [*Chaenocephalus*]* aceratus *is 2 – 3 times larger than that of the other fish examined. Its spleen and many other parts of its body are comparatively large and it is suggested that this may be a consequence of reduced axial musculature.*" A better explanation may be that high viscosity of very cold water puts a higher demand on the circulatory system (Verberk and Atlinson 2013).

Chapter 4

104. This does not mean that thermoregulation in ectotherms has no implications for body size. Large reptiles, for example, occur mostly in the tropics, but only in habitats where either water, or shade are available where they can cool their bodies down when required (see, e.g., Seebacher *et al.* 1999).

105. This pattern had a major influence on the development of what became the GOLT, particularly because it is very well documented for Atlantic menhaden (*Brevoortia tyrannus*), whose stocks range along the East coast of the United States from the cold waters of Maine in the North, where individuals can reach 8-10 years and an maximum weight around 800 g, to the warmer waters of the Carolinas, where they reach only 4-5 years and a weight around 200 g (Henry 1971). Also, *Brevoortia tyrannus* is replaced in Florida and the North Gulf of Mexico by Gulf menhaden (*B. patronus*), which an even shorter lifespan and smaller maximum weight (see FishBase). This exemplary demonstration of the TSR was at the time disputed by various colleagues, who suggested that it was the result of more intense fishing in the south of the distribution of *B. tyrannus*, while in fact, its exploitation was strong throughout its range.

    And still, given that Atlantic menhaden have mostly recovered from overfishing, and their TSR patterns, now intensified by ocean warming, are even more pronounced, there are colleagues who contest the conclusion of an analysis of 65 years of size records for Atlantic and Gulf menhaden, that the "*most plausible explanation for these size changes is that they are a consequence of recent coastal and oceanic warming*" (Turner 2017). This is the reason why, in this book, we emphasize the mechanisms of 'temperature'-induced changes, which in reality, are changes in the balance between the O_2 supply and demand of WBE.

106. A major impulse for the awareness of this pattern was the German '*Plankton-Expedition*' of 1889 which stimulated continued research on Arctic plankton species and renewed surveys in polar waters (see, e.g., Nyhart 2012). The growing interest in 'polar gigantism' was probably the main factor for the discovery of correlations between temperature and final size in ectotherms. Many nineteenth-century textbooks still mention not only faster growth, but also *larger sizes at higher temperatures* in fish and other ectotherms (see, e.g., Jäger 1884).

107. At the highest Arctic and Antarctic latitudes, the size-latitude pattern reverses itself due to 'cold denaturation.' However, this was discovered only well after the Second World War (see further down in Chapter 4).

108. An example of a recent publication on planktonic species being larger at higher latitudes is Wang *et al.* (2020), who reported on the size of tintinnids (*Phylum Ciliophora*) being smallest in the tropical South China Sea, larger in the Bering Sea and largest in the Arctic Ocean (see their figure 7).

109. Note that this 'explanation' for large sizes proposed a 'reason' – delayed maturation – which itself requires an explanation. Such non-explanation is what motivates us and other researchers to only accept explanations that include (or describe) a mechanism.

110. An early attempt to explain differences in the maximum sizes reached by the individuals of various populations within species may be found in Prince (1904), who listed the following factors as affecting the maximum size of fish: "*heredity, food, physical environment or physical surroundings, age, congenital variation or inherent strength or weakness, adaptability, and security from pernicious influences*" and documented each with multiple anecdotes. Temperature is mentioned in a few of them, but the only general statement about its role is that "*a low temperature appears to be favourable to the welfare of the most valuable food fishes and it is in the colder waters of the northern hemisphere that these prized economic fishes reach their largest dimensions.*"

111. As it turns out, when temperatures decline below 4 °C, 'cold denaturation' causes an increase in metabolic rate, due to the need to re-synthesize 'cold-denatured' body proteins – see further down in the main text.

112. The 'explanation' for the larger size reached by nematodes at lower temperatures provided by Van Voorhies (1996), who published the data for this figure, is that this "*may simply be a consequence of developmental processes that cause cells to grow larger at lower temperature. This would provide a general explanation for the increases size of ectotherms at lower temperatures independent of species-specific ecology.*" Actually, what this illustrates is the expression 'kicking the can down the street' because the question now becomes why "*developmental processes*" cause cells to grow larger at lower temperature. It should be noted that this author didn't cite any paper demonstrating that nematodes have larger cells at lower temperatures. Also note that the VBGF fits size-at-age data of at least one species of

nematodes reasonably well, despite their growth proceeding via successive moults (generally 4; Bird 1971). This is similar in crustaceans (see Chapter 12).

113. This implies, conversely, that shell formation based on precipitating calcium carbonate is easier in warmer water, a trend that has its apogee in the (tropical) giant clams of the genus *Tridacna*, e.g., in *T. gigas*, which get an additional boost of oxygen during daytime from their symbiotic zooxanthellae (Klumpp *et al.* 1992; see also Chapter 13).

114. This point explains that, for example *"with warming conditions, cold-adapted species (Lake Whitefish) decreased in adult body size, while cool and warm-adapted species (Walleye and Smallmouth Bass) increased in adult body size"* (Warne *et al.* 2024), which is in line with Pütter's argument as the observed mean sizes can be much larger in "*warm*" populations where animals grow faster and reach adulthood earlier while their "*colder*" counterparts remain small for much longer, but eventually grow bigger (Pütter 1920). In addition, sudden growth spurts of warm-adapted species at higher temperatures can easily be explained by Pütter's model which predicts an initially sharper increase of the anabolic term at higher temperature. However, this factor is not always incorporated in studies on temperature-induced size changes. An example is the article of Audzijonyte *et al.* (2020), which has a title that is easy to misunderstand, i.e., "*Fish body sizes change with temperature but not all species shrink with warming*" and is therefore now widely cited as implying that elevated temperature may or may not reduce the *maximum* sizes that fish can reach, although it dealt with their *mean sizes*. This last point becomes clear only when carefully reading the article in question, which unfortunately, if often not done.

115. A similar view is echoed in some more recent publications, where metabolic scaling is explained as a result of decreasing maintenance costs at larger size (instead of declining growth rates). See, e.g., White and Marshall (2023b) whose *"model assumes that the scaling exponents of total metabolic rate and maintenance metabolic rate* [...] *are identical, such that f is independent of size."* In this model it is "*the fraction of energy expenditure that is allocated to the metabolic work of production (growth + reproduction).*" We will discuss the problems with this view in more detail in Chapter 6.

116. Costello *et al.* (2023) make a strong case that most life processes on Earth are optimized at temperatures around 20 °C. This could be tested, in the context of the GOLT, by checking if the rate of spontaneous denaturation of (many or most?) enzymes involved in protein synthesis (e.g., alkaline phosphatase) has a minimum near 20 °C. This would neatly fit with the GOLT, because alkaline phosphatase, which works only in a O_2-rich, high pH environment, is ubiquitous in small (young) fish, but absent in large (old) ones (see, e.g., Shul'man 1974, p. 188-194, and Somero and Childress. 1980).

117. These amino-acid pools are labelled I and II, respectively in **Figure 2.1**.

118. Note that the kinetic energy of Brownian motion was used by Einstein (1905) to demonstrate the existence of molecules, via the ability of aggregates of water molecules to move larger particles around (Renn 2005). Here, this kinetic energy explains why larger particles, such as proteins, can be denatured as a result of shocks with similar aggregates of water molecules. Note also that it is the kinetic energy increase with temperature, which provides the mechanism for increased denaturation at higher temperature (see further in the main text for more on this point).

119. Correcting an error in Pauly and Lam (2022), where this species was called *Anarhichas 'lupus.'*

120. Note that Brownian motion does not occur in desiccated substances, such as the resting eggs ('Dauereier') of *Artemia* spp. and some fish. As a result, they can last without respiring which nicely illustrates the intrinsic connection between metabolism and Brownian motion.

121. In our experience with journalists and other non-biologists (but also, alas, with some biologists), we noted that fact that warmer water contains less oxygen is rather well known, while the fact that fish and other WBE also require more oxygen at higher temperatures, and that the latter effect is much stronger than the former, regularly comes as a surprise. On the other hand, anglers and aquarists tend to be aware of the second point, as well (see Chapter 5).

122. Using FishBase data, Lin and Costello (2024) found (among other patterns predicted by the GOLT), a strong pattern of maximum size reduction in Antarctic fish with decreasing temperature (below 4-5 °C) which they attributed to food scarcity (as occurs in very deep water). We think, however, that the observed size reduction is consistent with what one would expect from cold denaturation. This suggests that their findings do not only support the GOLT but also reinforce the theory precisely in a context that initially seemed anomalous but is, in fact, a predicted outcome (and Mark Costello agrees; pers. comm. to DP, July 2024).

Chapter 5

123. As Lindmark *et al.* (2022) in the first instance state, Pütter's equation would predict the temperature-size effects, but then they dismiss this idea because *"the assumptions underlying the VBGE were recently questioned because of the lack of empirical basis for the scaling exponents and the effects of those on the predicted effects of temperature on asymptotic size* [...]. *Specifically, the allometric exponent of energy gains (a) is assumed to be smaller than that of energetic costs (b). This is based on the assumption that anabolism scales with the same power as surfaces to volumes, and catabolism, or maintenance metabolism, is proportional to body mass (a = 2/3). In contrast, maintenance costs are commonly thought to instead be proportional to standard metabolic rate, which in turn often is proportional to intake rates at the interspecific level* [...]. *This leads to a ≈ b, resulting in unrealistic growth trajectories and temperature dependencies of growth dynamics in Pütter models."*

This self-induced conundrum can easily be solved if we confront the erroneous assumptions that (*i*) that the anabolic term in Pütter's equation universally scale with an exponents of 2/3 (see also Chapters 1 and 3), and (*ii*) that maintenance metabolism (i.e., excluding costs of growth) is proportional to standard metabolic rate. As a reference for these assumptions, the authors cite Lefevre *et al.* (2018) and Marshall and White (2019), whose work we discuss in more detail in the following chapter (Chapter 6). However, as already mentioned, it is one of the central tenets of the GOLT that the scaling exponents of anabolism are not the same in all species of WBE but are always <1 in their adults. Furthermore, the idea that maintenance metabolism is proportional to standard metabolic rate relies on the assumption that respirometry measurements on growing WBE allow for establishing realistic values of maintenance costs *excluding* growth (which they don't; see Rosenfeld *et al.* 2015). Since intraspecific metabolic scaling is mainly a result of declining energetic investment in growth, we believe that the rejection of Pütter's equation and its mechanistic foundations by Lindmark *et al.* (2022) is based on faulty premises.

124. The obvious relationship documented by Heincke (1913) was part of the introductory fisheries science course at the Institute of Marine Science in Kiel, Germany, where the first author studied. The version of this relationship in a bay in the Philippines (see **Figure 5.4**) and another version in a textbook on the ecology of tropical fishes (Longhurst and Pauly 1987, p. 280, here modified as **Figure 5.2**).

125. Indeed, the fact that the studied species – now *Eubleekeria splendens* – and the ecosystem in which it was abundant at the time were then unexploited, jointly with the size-depth trend of that species, allowed the estimation of natural mortality in this species (Pauly 1980b).

126. In addition to **Figure 5.6**, this is loss of thermal habitat and the possible responses by fish are nicely illustrated by the following extract from the abstract in Humphries *et al.* (2024), which states that *"yellowfin tuna respond to low* [dissolved oxygen] *at depth by spending more time in shallower, more oxygenated waters* [...while] *bigeye tuna increased the frequency of brief upward vertical excursions they performed by four times when* [dissolved oxygen] *at depth was lower, but with no concomitant significant difference in temperature, suggesting that this behaviour is driven in part by the need to re-oxygenate following time spent in hypoxic waters. These findings suggest that increasing* [oxygen minimum zones] *will impact the behaviour of these commercially important species, and it is therefore likely that other water-breathing predators with higher metabolic rates will face similar pressures."*

Another such response to the spread of hypoxic waters is highlighted in Bianucci *et al.* (2016); it pertains to the (demersal) Atlantic wolffish (*Anarhichas lupus*) of the Scotian Shelf in Canada, whose population is predicted to experience a *"contraction."*

127. 'Thermal niche' should be understood here as the temperature prevailing in the region that a population inhabits within a given habitat. This definition is compatible with Gvoždík (2018), even though this author uses the term for slightly different purposes. For ours, it is sufficient to distinguish between habitat and niche as the latter also refers to the specific space and time zone within a given physical locality, e.g., daytime in a deep cool pool vs. nighttime at the water's surface.

128. Interestingly, it is in such challenging habitats that the only fishes (except 3 species of marine seadragon, Family Syngnathidae) to have evolved to attach their maturing eggs to their bodies, are the 5 species of 'pelvic brooders' in the Family Adrianichthyidae (see also Chapter 14, mainly devoted to crustaceans). What unites the fish of the families Syngnathidae and Adrianichthyidae with the Crustacea is that they all have gills with small surface areas and/or inefficient gills, at least when compared with pelagic fishes.

129. The point here is that large freshwater fish species are all riverine, because flowing water has – other things being equal – a higher O_2 concentration than the water of stagnant pools or lakes. Potamodromous species ascend rivers not because they provide abundant food, but rather because they provide an oxygen-rich environment.

130. A similar mechanism was described by Amitai *et al.* (2024) who linked hypoxia and heat-induced fish kills to biblical descriptions of miraculous catches in the Sea of Galilee. It may require a leap of faith to project modern observations of fish kills into a Bible story, but the occurrence of such hypoxic events is well-documented.

131. For more examples in fish kill manuals, see Müller *et al.* (2023).

132. Indeed, Norwegian-style intensive salmon aquaculture is experiencing increasing problems with temperature-induced mass-mortalities even in Norway (Singh *et al.* 2023), which is not surprising given that *Salmo salar* is a psychrophilic species whose members, additionally, are exposed to horrendous conditions in their underwater cages.

133. This is similar to the tail-flipping escape behaviour in small (young) vs large (old) lobsters and the lactate-limited fighting among hermit crabs mentioned near the end of Chapter 14.

134. Similarly, after fighting back when hooked on a line, yellowfin tuna, quickly go anaerobic, with the pH of their tissue dropping precipitously so that their muscle cells go into apoptosis and lysis. The result is that their flesh, *"which seems paler and softer than normal is characterized* [...] *as being 'burnt.' Burnt tuna, because of its poor texture, color and slightly sour taste, while edible, is undesirable for raw consumption (as sashimi)"* (Cramer *et al.* 1981).

135. Marine heatwaves, which are becoming increasingly frequent, are not analogous to heatwaves on land, however challenging the latter may be for many terrestrial animals. Instead, marine heatwaves are analogous to forest fires, which not only kill the most sensitive or exposed members of various species (including older adults and infants) but kill (via respiratory stress) most members of most species in an affected ecosystem (Pauly 2023).

136. AquaMaps 2.0 is a next-generation marine species distribution modeling framework that builds upon the foundational AquaMaps approach by integrating cutting-edge data sources, machine learning techniques, and ecological theory, and which currently covers over 30,000 marine species of WBE. Designed to address the limitations of earlier models – such as coarse resolution, simplified habitat representations, and limited validation – AquaMaps 2.0 employs an ensemble of ten algorithms, high-resolution (5 km) environmental predictors from Bio-Oracle v. 3.0, and expert-informed biogeographical constraints to produce accurate and ecologically realistic habitat suitability projections. It standardizes taxonomy via the World Register of Marine Species (WoRMS), cleans and validates occurrence data from AquaMaps, OBIS, and GBIF, and incorporates species-specific ecological profiles through the novel Distance-3 Ocean System (D3OS). AquaMaps 2.0 enables climate change scenario forecasting (SSP1–2.6 to SSP5–8.5) for both mid- and end-century timeframes, providing fine-scale, maps of current and projected future distributions (Reygondeau *et al.* 2025).

137. This discussion does not cover the sea – to – freshwater or freshwater – to – sea of diadromous fish, which are not *driven* by temperature. However, they can be strongly affected by it. One example is the spawning migration of several 'runs' of sockeye salmon (*Oncorhynchus nerka*) in British Columbia, Canada, which requires them to enter the Fraser River. However, as sockeye salmon now try to return weeks earlier to their native stream (which is itself likely to be a temperature and/or oxygen effect; see Welch *et al.* 1998), the fall temperatures of the Fraser River are still too high for the aerobic scope of sockeye salmon, which end up (*i*) milling around in its estuary until the temperature drops, while falling prey to marine mammals and parasites, or (*ii*) initiating a river ascent that many of them won't be able to complete (Martins *et al.* 2011).

138. This is also the case for the second-largest Recent fish, the basking shark *Cetorhinus maximus*, which spends its summers in food-rich temperate and boreal surface waters and its winter in deep mesopelagic waters at tropical latitudes (Skomal *et al.* 2009).

139. Lake Peipsi (in Estonian) is famous because of the 'Battle on the Ice' during which the ragtag army of Alexander Nevsky, Prince of Novgorod, inflicted a crushing defeat on the Teutonic Knights. The army's disastrous plunge through the ice is portrayed in the 1938 classic movie by Sergei Eisenstein. When I (D.P.) read about shifts in the fish communities of this lake, I could not resist and contacted the Estonian colleagues who had written about these shifts and told them about the MTC as a way to interpret them; the paper (Kangur *et al.* 2021) then nearly wrote itself.

140. This hockey-stick pattern justified, in this case, the use of a segmented regression, in contrast to the abuse of this method described in Chapter 6.

Chapter 6

141. See Pauly (2024) for a 2nd edition of this thesis, which has its original content but with new footnotes providing corrections and/or extensions of this content. It includes, in a new appendix, a much shorter and elegant integration of the generalized VBGF (by Ivar Ekeland, a kind professor of mathematics) than initially presented.

142. The abstract of the contribution by Pörtner (2010), which summarizes his '*Oxygen- and Capacity-limited Thermal Tolerance* theory (OCLTT), reads, in part, as follows: "*The concept of oxygen- and capacity-dependent thermal tolerance in aquatic ectotherms has successfully explained climate-induced effects of rising temperatures on species abundance in the field. Oxygen supply to tissues and the resulting aerobic performance characters thus form a primary link between organismal fitness and its role and functioning at the ecosystem level. The thermal window of performance in water breathers matches their window of aerobic scope. Loss of performance reflects the earliest level of thermal stress, caused by hypoxemia and the progressive mismatch of oxygen supply and demand at the borders of the thermal envelope. Oxygen deficiency elicits the transition to passive tolerance and associated systemic and cellular stress signals like hormonal responses or oxidative stress as well as the use of protection mechanisms like heat shock proteins at thermal extremes. Thermal acclimatization between seasons or adaptation to a climate regime involves shifting thermal windows and adjusting window widths.* [...] *The conceptual analysis suggests that the relationships between energy turnover, the capacities of activity and other functions and the width of thermal windows may lead to an integrative understanding of specialization on climate and, as a thermal matrix, of sensitivity to climate change and the factors involved. Such functional relationships might also relate to climate-induced changes in species interactions and, thus, community responses at the ecosystem level.*"

As such, the OCLTT may be more ambitious than the GOLT; however, these two theories have in common that (*i*) they both focus on the disparity between the oxygen that is available through the respiratory organs of water-breathers and the oxygen demand of their cells – mainly at extreme temperature in the OCLTT, and mainly in large (old) WBE in the GOLT – and (*ii*) the OLCTT is rejected by the group of physiologists who also diss the GOLT (see Jutfelt *et al.* 2018), and of which some are mentioned in this chapter.

It is evident from Jutfelt *et al.* (2018) that its 28 authors are unable or unwilling to distinguish between the oxygen *supply* a fish receives through its gills and the oxygen requirements or *demand* of its tissues, which is the reason these physiologists attacked one of their own, Hans-Otto Pörtner, who dared to report that fish exposed to excessive temperatures suffer from internal hypoxia, clearly the result of oxygen supply not meeting demand (Pörtner *et al.* 2001; Pörtner and Knust 2007; Pörtner 2010).

However, all 28 authors found it "*hard to imagine why animals would allow tissue hypoxia to become severe enough to inflict performance decline at moderate levels of activity when possessing the functional capacity to significantly increase oxygen delivery to tissue.*" This is a classical argument from ignorance: as if the inability of a group of people to imagine a given process was evidence against the existence of that process! The funny thing is that it is straightforward to imagine why the respiration of fish cannot correspond to peak performance all the time: such performance has high costs in terms of fitness (Priede 1985; Pauly and Cheung 2017). Yes, Usain Bolt can run very fast, but when he runs, he can't eat, socialize, reproduce, etc. In other words, the peak respiratory performance that physiologists force laboratory fish to exhibit in their respirometers does not translate into a performance that fish usually perform in nature. Peak performance is costly and is used only under extreme circumstances, such as to escape predators.

Finally, the title of the 28-author paper, "*Oxygen- and capacity-limited thermal tolerance: blurring ecology and physiology,*" implies that inferences from one specialized discipline to another should be avoided. But it is usually consilience, i.e., cross-linkages between different disciplines, that provide the evidence that 'clinches' advances in science, be it evolution through natural selection or plate tectonics (Pauly 2002b). That physiology should be the sole avenue to insights about fish is as preposterous as claiming that theology is.

143. Powerful examples of the explanatory potential of seemingly contradictory phenomena to the formulation of biological theories are provided in the 6th edition (1872) of Charles Darwin's *Origin of Species*; see Chapter 6, "*Difficulties on Theory,*" p. 133-167, and Chapter 7, "*Miscellaneous Objections to the Theory of Natural Selection,*" p. 168-204. Darwin's examples show how a fruitful theory can benefit from addressing challenges, which can eventually broaden its explanatory scope.

144. The contribution by Pauly (2021a) and its expended version (Pauly 2021c) dealt mainly with critiques published before those reviewed in this chapter. These earlier critiques do not need to be dealt with again here, as they were inconsequential and had been refuted earlier, notably by Pauly and Cheung (2017). Incidentally, the reviews of our initial submission of the latter article and the responses we provided are available online (as Pauly 2018b), which allows for assessing the changes that were demanded; see also Pauly (2018a, 2020) and Pauly and Cheung 2018).

145. It took us some time to understand how anyone could interpret the GOLT as implying that gills behave like a sphere, but we believe we have now grasped the reasoning — see the main text below.

146. Scheuffele *et al.* (2021) state that the GOLT *"proposes that there is a mismatch between the body mass scaling exponents of gill surface area (GSA) and aerobic metabolic rate (MR), as the former is considered to be a 2D surface and the latter is considered to be a product of (3D) body volume."* However, the GOLT proposes nothing of the sort. Rather, it explicitly states that *both* gill surface area *and* metabolic rate act as surfaces (because the former *is* one, and the latter is proportional to one – due to Fick's Law). What is proportional to body volume (i.e., its mass or weight) is the oxygen *requirement* for maintenance and growth, with the former being met even in large (old) individuals. The requirements for additional growth are no longer met.

147. Lefevre *et al.* (2021) argue in this context that aerobic scope (AS) *"does not decline with mass."* However, the GOLT's point is not that growing fish lose their aerobic scope when they get bigger but that they invest less energy (i.e., oxygen) in growth. Curiously, the latter view seems to be expressed in the very same paper (Lefevre *et al.* 2020), when they state that *"if AS is lower at high temperatures, fishes may limit some oxygen-demanding activities, like foraging and growth, if these would constrain their ability to face transient demanding conditions such as disease, episodes of hypoxia or predator attack."* This sentence captures the essence of the GOLT, and we wonder if the entire controversy is not based on some simple misunderstanding.

    This also seems to be the case with the paper by Johansen *et al.* (2024), who compared the maximum size of coral reef fish in the Persian Gulf (with temperatures above 35 °C) with conspecifics in cooler areas outside the Gulf. They found, as expected, that the fish in the Persian Gulf were smaller. Nevertheless, they suggested that their *"findings challenge traditional oxygen-limitation theories as the primary driver of size reductions. Our data support a modified resource-acquisition theory to explain how ocean warming leads to species-specific size reductions and why smaller individuals are evolutionarily favored under elevated temperatures."* We suggest they found nothing of the sort and misinterpreted their data. All they have done is clutter the literature with yet another set of *ad hoc*, species-specific hypotheses (notably that the Gulf fish eat less because they are in hot environments... as if this were not precisely what fish do – and need to do – when they are oxygen limited...). We hope that this book will help declutter things.

148. Another version of the P-diagram is figure 9.14 in Longhurst and Pauly (1987) where the height of the 'maintenance box' declines smoothly from left to right, to reflect the reduction of maintenance metabolism due to larger individuals living in deeper, cooler waters. This differs from the one-step decline illustrated in **Figure 6.1B** (see also Chapter 10), and implies a complication that may have to be considered in future elaborations of the GOLT.

149. McKenzie (2023), in what was supposed to be a review of Pauly (2019), asserted that *"the 'maintenance' variable is never properly defined,"* while on p. 159 of the book in question, which this 'reviewer' manifestly did not read, it was proposed that *"it would have been appropriate, throughout this book, to replace the term 'maintenance metabolism' by the awkward term 'backward projected terminal metabolic rate' because that would be the correct definition. Thus, by maintenance metabolism, I mean the metabolic rate that maintains the weight of fishes that have stopped growing (in the wild or in captivity), hence the word 'terminal.'"*

    The belief that standard metabolic rates (*SMR*), as commonly measured, reflect only maintenance costs appears to be the central problem of McKenzie's (2023) interpretation of the GOLT. The main difference between *SMR* and maintenance metabolism, as represented in a P-diagram (**Figure 6.1A**), is that maintenance metabolism (when growth is indeterminate, as in fish) is a primarily theoretical value that can only be estimated retroactively from the metabolic rate of individuals that have stopped growing.

    This definition, whose very existence was denied by McKenzie (2023), implies – and this should have been stated explicitly – that maintenance metabolism in the wild *includes* the cost of food acquisition and digestion, of escaping predators, and even of reproduction (via the mechanism presented in Chapter 9), i.e., all the things that large (old) fish do except growing. While a parameter that can be empirically measured at a given size would be preferable, how data on metabolic rates are fitted into models immediately shows the parameterization problem. If we follow McKenzie's suggestion to use *SMR* instead of maintenance metabolism to think about growth, everything depends on the way in which the parameters of the *SMR* are estimated.

150. That aerobic scope can be maintained at the cost of growth is neatly illustrated by the study of Gräns *et al.* (2024), in which juvenile halibut (*Hippoglossus hippoglossus*) grew much less when kept at 16-18 °Cthan at 10-14 °C, which was supposed to document that the tension between oxygen supply and demand does not impact growth, but failed to do so. Figure 3 in Gräns *et al.* (2024), which also shows that juvenile halibut grow more slowly in low temperature (4-6 °C), doesn't change that, and neither does the obvious finding that growth is impacted by low pH. The point here is that the aerobic scope concept is far less useful than its frequent uncritical use suggests.
151. The logic here is structurally similar to a claim that vitamin C is not effective against scurvy (which it is if your diet doesn't include fruits and/or other sources of vitamin C) because people who consume 100 pills of vitamin C per day do not fare better than people who take 1 such pill per day. Skeeles and Clarke (2023) then piled on their flawed argument that oxygen is not a limiting factor to growth the claim that energy allocation to reproduction is the limiting factor for fish growth (i.e., the 'Reproductive Drain Hypothesis'). This, however, is a zombie hypothesis (see Chapter 7), refuted by every lone goldfish in its fishbowl which never reproduced but ceased to grow despite ample food being provided. Yes, we mention this lone goldfish several times in this book, but it is because the argument holds that the growth of fish that never reproduced (a lone goldfish or sterile triploid fish) also gradually slows down and then stops. A reasonable theory of growth must be able to deal with this fact.
152. The choice of the species used for the experiment by Skeeles and Clarke (2023), the inanga (*Galaxia maculatus*), is also bizarre given that this species is known to be an oxyconformer, which is rare among bony fishes. Even more curiously, they mention none of the recent papers that reported on the lack of a capacity for oxyregulation in *G. maculatus* (see, e.g., Urbina *et al.* 2012; Urbina and Glover 2013, and the many other studies of these authors on inanga). Instead of trying to refute the GOLT using exceptional species, Skeeles and Clarke (2023) could have used it to identify its limits of applicability, which still have to be defined.
153. Similar results were presented in a study on the tadpoles of the cane toad (*Rhinella marina*), whose growth was reduced under hypoxic conditions (Schofield and White 2025). The same study also found that hypoxia led to an increase of the size of external gills (via developmental plasticity), which allowed the tadpoles which grew under hypoxia to ultimately reach the same size as those that grew in normoxia.
154. The irony is that these *lower* values of *d* for *large* (old) shark and ray species were originally published by no other than Wegner (2015), one of the co-authors of the Prinzing *et al.* (2023) paper and who, therefore, should have known better.
155. Lonthair *et al.* (2024b) cited 2 single-species studies in support of their contention that there are "*robust datasets in the literature showing the GSA can scale close 2 or even above* [d] = *1*." One is Prinzing *et al.* (2023), and the other is Holeton (1976) which pertained to young individuals still in the transition from the high values of *d* observed in larvae and early juveniles (see **Figure 6.2**). 'Robust' evidence for *d* <1 in adult fish (and breathing only water, not air), which would have to come from a serious review or meta-analysis, would indeed challenge the GOLT. However, the fact that Lonthair *et al.* (2024b) had to rely on the shady data of Prinzing *et al.* (2023) for 50 % of their "*robust*" evidence is indicative of the depth of the hole they have dug themselves in.
156. It is difficult to believe this was due to a good-faith error, as a vast literature on fish and bivalve gill surface area and respiration documents that they both scale with weight raised to a power <1.
157. Scheuffele *et al.* (2020) state that the GOLT's "*model also disregards universal allometric scaling of basal metabolic rate with mass in animals.*" This, again, is a fundamental misunderstanding of intraspecific (not interspecific) metabolic scaling since the decline in metabolic rates that occurs during individual growth trajectories is mainly a result of less energy being invested in growth.
158. The 'cooling dives' of whale sharks are described in Meekan *et al.* (2015); however, these authors consider these dives to be 'feeding dives,' with the sharks' returns to the surface being necessary to re-acquire the high temperature that they supposedly prefer. This is the inverse of what we think occurs, and it contradicts multiple observations of whale sharks feeding predominantly in surface waters. The note below elaborates on this theme.

159. This last point brings us to the question of how a comprehensive theory such as the GOLT may be tested in practice, an issue it shares with other generalizations (see Williamson *et al.* 2021; Cheung *et al.* 2013a). One approach would be to continue with the present approach, which argues that "It may work for the 1,399 species you mention, but species X doesn't fit the pattern, and therefore the GOLT is wrong." To make their argument against the GOLT, Lefevre *et al.* (2017) made whale shark (*Rhincodon typus*) their species X, because occurs in the tropics and is the world's biggest fish. However, when the GOLT predicts that fish remain smaller at a higher temperature, it is meant to apply *within*, not *between* species (see main text above).

In any case, it seems that a more fruitful procedure would have been to look at what makes *R. typus* the exception that it seemingly is. They would have quickly found it: *R. typus* swims in yo-yo fashion, with near-surface feeding periods followed by 'cooling dives,' which may reach 1000 m and beyond (Brunnschweiler *et al.* 2009). And, to top it up, large adult *R. typus* (10 – 14 m) whose juveniles (< 8 m) are abundant in the Red Sea (Cochran *et al.* 2019) and the Persian Gulf (Robinson *et al.* 2016) do not occur in these two water bodies, as the deeper waters of the Red Sea are warm, and the Persian Gulf is too shallow.

Pauly (2019, p. 84-89) devoted an entire section to this theme ("*An apparent anomaly: very big fish in warm water*"), where *R. typus* and other large 'tropical' fish are shown to (*i*) spend much if not most of their time in deep, cold water (to the extent that some of them evolved a heating system for their eyes and brain, as in billfishes), or (*ii*) undertake regular cooling dives, or (*iii*) they were not very big.

For example, 'big' *Mola mola* consists of up to 65% fatty jelly, which is inert and does not require oxygen for its maintenance. Also, *M. mola*, which reach greater weights than *Mola alexandrini*, prefer cooler temperatures in Japanese waters, i.e., 11.5 – 25.6 °C, average 17.7 °C *vs* 16.8 – 25.6 °C, average 19.9 °C for *M. alexandrini* (Sawai *et al.* 2018). Indeed, while Pope *et al.* (2010) describe the distribution of *Mola mola* as "*tropical*," their distribution map shows that this fish is reported from subtropic temperate locations far more than from the Tropics.

Clearly, such a whac-a-mole approach gets us and the field nowhere. Further below in the main text, we propose an alternative, potentially more fruitful approach.

Chapter 7

160. This statement applies only to iteroparous bony fishes whose maximum length exceeds 10 cm, for reasons explained further in this chapter. A possible origin for the semelparity of Pacific salmon species is presented at the end of Chapter 10, while the semelparity of cephalopods is a topic in Chapter 13.

161. The Reproductive Drain Hypothesis is also contradicted by the fact that, despite their higher reproductive investment, the females of most bony fishes or elasmobranch species such as sharks tend to reach larger sizes than the males (Pauly 2018a, 2020).

162. The observation that reproducing and non-reproducing fishes do not differ in final size is also informative in the context of the debate around the concept of 'equifinality' which describes equal outcomes under different circumstances. The concept of equifinality was used by vitalists like Hans Driesch to argue for a teleological principle that shaped the growth trajectories of organisms. As von Bertalanffy argued from a systems-oriented perspective, vitalist principles were unnecessary to explain equal outcomes under different growth conditions (von Bertalanffy 1950). As von Bertalanffy argued, it was the capacity of a biological system to take up energy that set the limit to growth and size rather than an inherent 'program.' While growth under unfavorable conditions could indeed be slower, animals and plants could later 'catch up' in growth with their conspecifics and reach identical final sizes (see Jobling 2010 for the concept of 'catch-up' growth).

163. The research by Yokouchi *et al.* (2018) is particularly enlightening, as it underscores the close relationship between the initiation of maturity and the subsequent migration in the eel *Anguilla anguilla*, which is closely linked to their somatic growth rate.

164. However, some authors view maturation and spawning as processes so separate from growth that they publish lengths at first maturity (L_m) without the corresponding values of L_{max} or L_∞ (see, e.g., Duponchelle and Panfili 1998).

165. That the internal hypoxia in fish may be the reason for the higher protein content in the otoliths of small (young) individuals compared to large (old) ones (Morales-Nin 1986a, 1986b) is another surprising corollary of the GOLT.

166. As Seibel and Deutsch (2020) argue, "*as whole-animal metabolic rate increases with size, the oxygen supply capacity increases to match it.*" As we outline in Chapter 6, this is correct for maintenance (including activity) but only at the

cost of growth. Animals do not grow into asphyxiation (which seems to be how these authors interpret the GOLT) but preserve their aerobic scope by investing less in growth.

167. The 56 'cases' are due, in the 1984 study and the subsequent ones, to $L_{max}D$ and L_mD data pairs being available for several populations of the same species and/or separate $L_{max}D$ and L_mD data pairs being available for females and male fish.

168. We chose 'A' because it signifies a 'beginning' or 'start' in German ('*Anfang*'), as the letters of the English alphabet are exhausted.

169. The extension of the conceptualization presented here would require further elaboration to account for fall spawning in temperate fish (Warlen and Burke 1990), monsoonal spawning in fishes of the Indo-Pacific (Longhurst and Pauly 1987), flood-related spawning in fish of floodplains (Pauly 1994a), etc.

170. As pointed out in Chapter 1, mechanistic explanations are possible (and often more powerful) without involving all levels of causations that play a role in various processes. Indeed, as Felline (2016) outlined, it is not always the deepest layer of causation that may be invoked in explanations since "*mechanistic causality*" is typically not located at the most fundamental level of causation.

171. The reason why zooplankton are reported not to have switched their phenology may be because the dynamics of their populations are closely tied to that of their food, i.e., phytoplankton, and that the phenology of the latter (outside of the tropics) is tied to seasonal variation of sunlight, which are not affected by climate change.

172. The reason for this limitation of the model is because the arcos(X) function of Equation (7.15) is only defined for X values ranging from -1 to 1. Therefore, $-2B/Amp$ must satisfy the constraint $-1 \leq (-2B/Amp) \leq 1$. This does not apply to low amplitudes of the seasonal temperature oscillations (Amp) and increasing mean annual temperature differences (B), which results in the triangular 'no solution' space on the right of **Figure** 7.7**B**.

173. Internal hypoxia, in fish, triggers pain (Schuck-Paim *et al.* 2015) and having to supply oxygen to a growing mass of gametes can be conceived as causing a state of (slight) hypoxia in the other organs of maturing fish. Spawning, which leads to a sudden increase of the oxygen available for the organ systems other than the gonads (see Chapter 9) may be experienced by fish as relief from (dull?) pain (see the section on 'pain' in Chapter 17).

174. Dusk is also the time when larger predators, at least in freshwater, where hypoxia is more frequent, may experience oxygen limitation and are more likely to be sluggish, which does, of course offer additional advantages to dusk-spawners. In this sense, the predator avoidance hypothesis might also be related to oxygen limitation.

Chapter 8

175. The information assembled for this is presented in the appendices of the MSc thesis upon which this chapter is based. It is and will remain available via *https://open.library.ubc.ca/collections/ubctheses/24/items/1.0431066.*

176. The basic information for each species was complemented by other life-history characteristics, including differences in length with sex, average number of pups per litter, and reproductive mode. Comprehensive compendia for sharks and rays, *Sharks of the World* (Ebert *et al.* 2021) and *Rays of the World* (Last *et al.* 2016), respectively, were used as a baseline source of data for marine species and *Rayas de agua dulce (Potamotrygonidae) de Suramérica Parte I* and *Parte II* (Lasso *et al.* 2013, 2016) were used for freshwater rays. Stenohaline freshwater rays, which have lost the ability to retain urea, are represented mainly by a single family, Potamotrygonidae, confined to South America. Several species belonging to the Dasyatidae are also confined to freshwater environments in Asia (Last *et al.* 2016), and were also considered stenohaline freshwater species. No single compendium exists for the life-history of chimaeras, and they were covered by assembling information from scattered references.

All euryhaline rays belonged to species that are known to reside in brackish or freshwater environments, typically estuaries and river mouths, either temporarily or for extended periods of time (Constance *et al.* 2023). This criterion was chosen because even temporary habitation of brackish environments reduces urea retention (Wright and Wood 2016). A species was determined euryhaline based on information from Last *et al.* (2016) or individual IUCN assessments under 'habitat and ecology' (www.iucnredlist.org).

Contrary to the Batoidea, which include marine, euryhaline, and freshwater rays, all species of Holocephali (chimaeras) are stenohaline marine, never entering brackish or freshwater environments (Ballantyne and Fraser 2012). Also, while some shark species, such as the bull shark *Carcharhinus leucas* and 'river sharks' (*Glyphis* spp.), are known to occur frequently in brackish and freshwater environments, these species still retain high levels of urea in their tissues, as do stenohaline marine species (Cramp *et al.* 2015; Dwyer *et al.* 2020). In bull shark, this is achieved through a rectal grand (Pillans and Franklin 2004) which "*can adjust the amount of internal solutes by releasing a compound called urea to balance with the solutes of the water outside*" as stated on p. 99 in the neat book by Graham (2024). Therefore, they are considered 'marine' species.

177. Indeed, it would have been better if Equation (8.1) were based on Chondrichthyes. However, this equation produces estimates of d similar to available empirical estimates in sharks; also, the plots presented in this chapter would lead to similar conclusions, had, for example, an average value of $d = 0.85$ been used for all Chondrichthyes species.

178. Some colleagues have suggested that forcing the intercept of plots of $L_{max}D$ and L_mD through zero forces a correlation to appear between $L_{max}D$ and L_mD when none may occur. However, these colleagues overlook the fact that the purpose of plots of $L_{max}D$ vs. L_mD is not to assess if a correlation exists between these variables but to estimate the mean ratio between them and the standard error of this ratio (see also Pauly and Müller (2025). A normal linear regression would not be appropriate for these purposes because some of the variance in this ratio would be 'explained' by its (non-zero) intercept, which, moreover, would have no biological interpretation. If more sophisticated analyses are required, we suggest that they should be performed with the residuals of a plot of $L_{max}D$ vs L_mD with a zero intercept. For example, it can be hypothesized that since the $L_{max}D/L_mD$ ratio can be described as a heuristic threshold that individual fish use to decide when to mature and spawn, the residuals should be larger in species or populations occurring in highly variable habitats.

Chapter 9

179. Note that, with Jobling (2010), we distinguish 'compensatory growth' from 'catch-up growth.'

180. Strangely, the reason why *FCE* declines with body weight (Equations 9.1 and 9.2) does not have a clear explanation–outside of the explanatory context provided by the GOLT. Gerking (1952, 1971) presented the following list of hypotheses:

    (1) The reduction of growth efficiency may result from 'aging.' Gerking (1952) considered this a pseudo-explanation, which it is;
    (2) Stomach and gut surface may increase proportionally to a power of weight lower than unity (this is dealt with in Chapter 2);
    (3) The digestive enzymes may not supply the same amount of nutrient material per unit of body weight in fishes of different sizes;
    (4) The decreased protein utilization is associated with a change in metabolism or with some bodily process that controls metabolism;
    (5) The thyroid hormones may directly influence the conversion of nutrient protein to body substance;
    (6) The differential rate of protein synthesis may explain the decrease in protein conversion efficiency.

    Pauly (2019, p. 69-70) discusses these and other hypotheses (some quite fanciful) proposed by other authors. He also singled out Gerking's hypothesis # 4 as likely correct but too vague to be of much use. Reformulated in terms of the GOLT, this hypothesis states that metabolic rate "*is associated with a change in metabolism*" because $d < 1$. As fish grow, their gills supply them with less O_2 per unit body weight, which results in less O_2 being available to convert food into growth. As W_∞ is approached, all the O_2 available is used for maintenance (in the broadest sense, i.e., to support various activities (see Chapters 2, 4, and 7), with no oxygen available for growth (thus, $FCE \rightarrow 0$).

181. It shouldn't be necessary to mention that for *FCE* estimates to be reasonable, they must be based on growth increments and food intake pairs expressed in comparable units, eg., g wet weight a animal tissue. Yet, the Norwegian-style salmon aquaculture industry claims thermodynamically impossible 'food conversion efficiencies' near (or even 'better than) 1/1, based on the salmon weight gains expressed as *wet* weight divided by food consumption (consisting of a large fraction of fishmeal) expressed as *dry* weight (Majluf *et al.* 2024). Unfortunately, it is possible that this tomfoolery misleads some people as to the role of an industry which *consumes* approximately 4 kg of perfectly edible fish such as sardine, anchovies or mackerel, much imported from food-deficient countries (Cashion *et al.* 2017) to produce 1 kg of salmon, then feed them coloring agents so their flesh turns from grey to pink (Steine *et al.* 2005).

182. Equation (9.2) was proposed as a replacement for the conventional relationship between food conversion efficiency (*FCE*) and body weight (*W*) of the form $FCE = a \cdot W^{\beta 1}$ (see, e.g., Jones 1976) because, as briefly mentioned in the main

text, it has two major deficiencies. One is that, once fitted to a data set, it predicts *FCE* >1 for very small fish (e.g., larvae). However, even if the *FCE* of fish larvae can be very high when they 'feed' on their yolk sac, e.g., 0.93 in rainbow trout *Oncorhynchus mykiss* (From and Rasmussen 1984) or 0.85 in the sole *Solea solea* (Flüchter and Pandian 1968), it cannot be ≥1. The other deficiency of traditional models is that they predict *FCE* never to reach zero, even if it must be zero at W_∞, per definition. The new equation resolved both of these deficiencies.

183. Note that Edward *et al.* (1972) also estimated oxygen consumption (*R*) in cod, which is related to weight according to $R \propto W^{0.82}$. This implies, with *S* as the gill surface area, that $S \propto W^{0.82}$; 0.82 is a good estimate of d_Q for cod (see De Jager and Dekkers 1974).

184. The increase of *FCE* occurs because spawning does not affect gill surface area. A sudden weight decrease (due to eggs or sperm being shed) will immediately increase the gill surface area/body weight ratio, i.e., in the body being suddenly suffused with higher levels of oxygen, resulting in a higher *FCE*.

185. Pauly *et al.* (2023, table 4) cite, in addition to Trippel *et al.* (2014), seven references stating that post-spawning cod feed at elevated rates, while citing (in their table 5) 11 references in which the same was reported from other species of fish, ranging from China to the North Atlantic, including the Mediterranean and Black Seas.

186. Note that nest 'guarding' (and importantly aerating the eggs therein), when it occurs, is performed by male fish in the majority of fish species, while the females may get themselves back to being able to produce a new batch of eggs within the same spawning season by continuing to feed (see, e.g., Blumer 1986). The sex-specific differences in parental care may be even stronger in mouthbrooding species (see, e.g., Balshine-Earn 1995).

187. The seasonal dynamic of fat, which is metabolically inert explains how storing energy in the form of fat allows fish, in cold temperate latitudes, to feed and gain weight during the warm season (see Pauly 2019, p. 79 – 81).

188. This is line with the seminal contribution by Ginther *et al.* (2024), who demonstrated that female fish spend far less energy for reproduction than other vertebrates.

189. This point about the seasonal variations of *GSI* being small in low latitude fish species is also made by López *et al.* (2025), who distinguish between 'income breeders' and 'capital breeders.' The former live (in marine waters) in the tropics, where conditions tends to be more even throughout the year, thus enabling smaller and spread-out spawning events. The latter live in temperate and cold temperate regions, where intensive feeding during the warm season leads to an accumulation of 'energy' during the warm season (in the form of fat stored, e.g., in the liver), which is used several months later for the gonad production. Indeed, the example of this dynamics in Patagonian grouper (*Acanthistius patachonicus*) in López et al. (2025) corresponds precisely to the sequence of events presented in this chapter.

Chapter 10

190. This chapter is based on the pre-submission version of a paper (Pauly *et al.* 2024) which was selected to be featured "*Between the JFB Covers*" by the editors of the *Journal of Fish Biology* (see Perry 2024).

191. This observation applies to fish (e.g., whale sharks), but not necessarily to cephalopods (see Chapter 13).

192. Still, large bony fish appear to have been rare, which contrasts to the abundant fossil record of large sharks. A tentative GOLT-related hypothesis to explain this difference may be that, in the ancestors of Chondrichthyes, bones were replaced by cartilage to reduce maintenance costs, which then contributed to some sharks becoming really big and outcompeting big bony fish (see also Freedman and Noakes 2002 who mention the energetic costs of bone resorption as a potentially limiting factor on the size of bony fish).

193. Planktivory was probably also the reason why the Cambrian radiodont, *Aegirocassis benmoulai*, was able to attain the (then) huge length of 2.1 m (see Chapter 12).

194. This notion is thoroughly refuted in Chapters 7 and 9.

195. The following account of this extraordinary fish was published by a Senegalese scientist knowledgeable on the biology of sardinella species (see Samb 1988; Samb and Pauly 2000), and it reads as follows: "*On 15 May 1990 at Mbour, in Senegal, a canoe equipped with a purse seine caught an extremely large* Sardinella maderensis, *with the following characteristics: Fork length = 37.3 cm; Total length = 45.5 cm; Total weight = 927 g. Sex determination indicated that it was a male. This fish was brought to the Centre de Recherches Océanographiques de Dakar-Thiaroye by Mr. Mor Sylla*" (Samb 1990). This account implies a length increase of only 16% over the reported maximum length of 32.1 cm FL, and an increase of weight of 57% from 591 to 927 g, as can be calculated assuming $b = 3$. Yet,

such cases are rare, which may be explained by the fact that the reduction of metabolic demand is more difficult to accomplish for species where *d* is low, and where Q_{maint2} needs to be much lower than Q_{maint1}.

196. The term 'Cope's rule' was first coined by the German biologist Bernhard Rensch (1948), in honor of the US-American zoologist Edward Drinker Cope (1840-1897). Aside from the debates on the empirical support for this 'rule,' we do not consider it unlikely that its name will be changed in the near future. The journal *Copeia*, published by the American Society of Ichthyologists and Herpetologists (ASIH) has recently changed its name to *Ichthyology & Herpetology* (see Smith 2021), the reason for this being the racist and sexist views of Cope, which were extreme even for his time and social *milieu*. We welcome this name change but refer to the term 'Cope's rule' pending a more suitable name, which we hope will become available for future editions of this book (which we also hope for...).
197. One of these 'boundary conditions' is that the ecosystems in which the species of interest are reasonably stable over long periods, which is not necessarily the case in lakes (Kolding and van Zwieten 2012; Gownaris *et al.* 2018), which may be the reason why the largest fish and other WBE are marine.
198. As outlined in the introduction to this book (Chapter 1), the GOLT is a theory built on mechanisms rather than general 'rules' or 'laws.' While we do not deny that biological rules exist, we argue that their statistical existence does not yet constitute a biological explanation. For rules to be incorporated in a general theory, it is necessary to formulate (and test) hypotheses on their underlying mechanisms. We agree with Glennan (2017), that "*a great variety of scientific generalizations, some of which we call laws, are in fact explained by mechanisms*" (p. 45).
199. This is different for larval and early juvenile fish, whose gill surface area grows with $d_G \geq 1$ (see Chapter 3). For these young forms, however, slower growth due to food scarcity results in higher mortality, which has the effect that the relationship between food availability and growth is (at least partly) masked (Fiksen and Jørgensen 2011; Le Pape and Bonhommeau 2015).

Chapter 11

200. This point, which is valid for fish, does not apply to marine mammals, which are highly active compared to fish, obviously due to their breathing air. It is therefore surprising to read in the 3rd edition of '*On the Origin of Species*' (Darwin 1861) that "*physiologists believe that the brain must be bathed by warm blood to be highly active, and this requires aërial respiration; so that warm-blooded mammals when inhabiting the water live under some disadvantages compared with fishes.*" Darwin's perception of this putative disadvantage was "*that warm-blooded mammals when inhabiting the water lie under a disadvantage in having to come continually to the surface to breathe,*" as stated in the 6th and last edition of '*Origin*' (Darwin 1972).
201. Another manner of adapting to tropical heat is to remain covered in moist leaf mold, which appears to be the main 'trick' of terrestrial amphipod of the Family Talitridae, another group of crustaceans which managed the transition to air breathing (Hurley 2015).
202. The increasing popularity of air-breathing fish, and the increase of their production, in aquaculture may, in part, be due to the lean nature of their flesh, which may be explained by their respiratory behavior. Many air-breathing fishes that are commercially important, such as *Channa* spp., *Clarias* spp., *Heteropneustes fossilis*, or *Pangasius* and *Pangasianodon* spp., are known for the lower fat content of their tissues (Graham 1997; Sokamte *et al.* 2020). Studies on *Channa striata* and other *Channa* species show that fat percentage remains low regardless of their diet (Phan *et al.* 2021). This is not the case in all air-breathers: the lungfishes (*Lepidosiren paradoxa* and *Protopterus* spp.) may be exceptions. However, data on most other air-breathing species underlines the importance of oxygen in converting food into body tissues. An example of food conversion efficiency declining linearly with (log) body weight in *Channa striata* is given in Pandian (1967; see also Pauly 2019, p. 71).

    Fishes with large fat reserves are typically temperate species, notably freshwater inhabitants with large seasonal temperature fluctuations, such as carp or other cyprinids. The explanation for their high fat percentage is that it serves as a buffer to mitigate the effect of large seasonal temperature fluctuations in their habitat. During the warm seasons, they can gain weight quickly, but experience reduced oxygen levels while the oxygen demand is high. They convert their food energy into metabolically inert weight, i.e., fat, and not "*oxygen-hungry proteins*;" later, when temperatures decline, they can convert the fat into gonad tissue without encountering excessive maintenance cost (Pauly 2019). Similar processes seem to play a role when fish are well-fed but face other sub-optimal conditions, for example, in overstocked and poorly aerated ponds where carp and other fishes build up large fat reserves that do not have any metabolic costs (Shul'man 1974). Air-breathing species with the ability to minimize branchial oxygen loss and fully exploit their capacity of aerial respiration can process energy into metabolically effective body mass without running the risk of creating unaffordable maintenance costs, which results in high protein, but lower fat content of their tissue.

203. As Lefevre *et al.* (2013) argue (based on unpublished data), hypoxia reduced growth in *Pangasianodon hypophthalmus* even though it is difficult to measure the exact impact of other factors that coincided with poor aeration (increase in ammonia, nitrate, or nitrite levels).
204. Some studies report even higher growth rates of air-breathing fish in small containers, even though this effect is more typical for larval stages, when the air-breathing organs are not fully developed; in this case, it may be due to easier access to food, which can there be reached without crossing larger distances (Santana *et al.* 2020). Either way, this observation suggests a reduced impact of water-dissolved oxygen on the growth of air-beathing fish growth in small and heavily stocked containers.
205. See https://secure.petaasia.com/page/51327/action/1?locale=en-US; accessed 29 February 2024.
206. Lorenzen *et al.* (2006) did not specify the length type used. There are 92 mentions of the word 'length' (or 'lengths') in their report, but only one of 'total length' (TL). Given this single mention, and because assuming that fork length (FL) or standard length (SL) was used in the 91 remaining instances would generate unrealistically high maximum weights in conjunction with their length-weight relationship, we assume that Lorenzen *et al.* (2006) used TL throughout.
207. Air-breathing in fishes was first described by Erman (1808), who examined the gut as an accessory breathing organ in weatherfish (*Cobitis fossilis*). Later research on other air-breathing organs showed that bi-modal respiration in fishes could take highly complex forms, even though a better understanding of the physiological mechanisms remained highly fragmentary until the early twentieth century (see, e.g., Müller 2021).
208. According to Randall *et al.* (1981), *Amia calva* suffers ca. 7% of branchial oxygen loss in hypoxia.
209. Recall that the scaling factor d is estimated as the slope of a regression of (log) gill surface area (*GSA*) against (log) body weight (W). This regression also yields an intercept ('a'); if this is low, the *GSA* will be low even if the estimate of d is high, as long as W remains small. This explains why, e.g., in mudskippers, *GSA* grows with values of $d > 1$, while they will drown (i.e., asphyxiate) when kept underwater. This, and other issues must be considered when considering the statement by Palzenberger and Pohla (1992) that in fish, "[t]*he scaling exponent of gill area ranges from 0.36 to 1.13"* (Pauly 2019, p. 157-158).
210. Note that arapaima surfacing rates decrease with greater lengths which could imply lower oxygen demand but also be a result of hyper-allometric scaling of the growth of their air-breathing organ which can grow with a slope of >1 relative to body mass (Stokes *et al.* 2021). A similar conclusion is drawn in a study on *Hoplosternum littorale* by Sloman *et al.* (2009) who suggest that bigger individuals might be more efficient air-breathers, perhaps due to an increased storage capacity of their air-breathing organ.

Chapter 12

211. However, we will mention here that the abundance of oxygen in the air, as opposed to water, that probably enabled the insect to develop a strong but light exoskeleton. Asano *et al.* (2023) write that a *"potential advantage of* [the formation of their cuticle] *is the utilization of molecular oxygen abundant in the atmosphere, which is different from the case in crustaceans (close relatives of insects) that utilize calcium ions. Accumulation of calcium ions leads to an increase in weight, but the lightweight cuticle without calcification might have been a critical factor enabling insects to evolve flight first in the history of Metazoa. Our theory also provides a simple explanation to a long-standing question of why insects are so rare in marine environments."*
212. Jónasson and Kristiansen (1967) show that the larvae of *Chironomus anthracinus,* which spent nearly two years at the bottom of a lake in Denmark grew well (and asymptotically; see **Table 14.4**) when *both* their food (sedimented phytoplankton) and oxygen were abundant, but did not grow (and even shrunk) when either food or O_2 were scarce, particularly when temperature was high. The GOLT applies to them.
213. Here the term 'crustaceans' is used in the traditional sense, i.e., excluding insects, although they have recently been included in a group called 'Pancrustacea.'

214. In Chapter 11, which deals with air-breathing fish, some passing references are made to air-breathing crustaceans. We leave it to specialized colleagues to follow up on the implications of their physiology for the GOLT.

215. Hughes (1983) presented data on the gill surface area obtained *"using the method of logarithmic transformation with data for 16 species of North American crabs and seven species of British decapod crustaceans. It has been shown that a wide range of relationships exist each of which is typical for a given species. The slope b of the log/log regression lines varied from 0.5 to 1.0, the lower values being most commonly found in the Macrura. For the North American species, the average slope is about 0.8 whereas for the British brachyurans, the relationship is closer to a linear one (b = about 0.97)."* Given that the latter average (0.97) is based on a mix of water- and air-breathing brachyuran species, we believe we can maintain that in water-breathing crabs, $d < 1$ (see Chapter 11 on the gills of air-breathing fishes, e.g., mudskippers, for which d is also very high). Also, Zeuthen (1953) demonstrated that in the large (old) crustaceans, d_G is lower than in small (young) ones, a likely case of ontogenetic change in scaling, or OCS (see also **Figure 3.8**).

216. For more than 2000 amphipod species, Chapelle and Peck (2004) demonstrated a strong relationship between the maximum sizes that they reach and ambient oxygen concentrations.

217. In tropical ecosystems with strong seasonal variations of water properties and levels (e.g., in the Amazon), the 'Winter Point' occurs (or should be set) at the time when the environmental conditions are most inimical to the growth of fish and other WBE.

218. See figure 8 in Pauly *et al.* (2022).

219. As well, Gutiérrez-Marco *et al.* (2009) wrote about the trilobite fossils they studied that their *"giant"* size in *"a region at high southern latitudes on the margin of Gondwana, close to the Ordovician south pole (Fortey and Cocks, 2003), may be a cold-water adaptation, as suggested for some Ibero-Armorican trilobites in the Middle Ordovician (Henry, 1989)."* This is also in line with the GOLT.

220. Interestingly, it has been recently proposed that such side flaps, which may have supported gills, may also be associated with the eventual development of insects' wings (Prokop *et al.* 2023).

221. Sekiguchi *et al.* (1998) write that *"several* Limulus *juveniles were grown to the fourteen-instar stage in the sixth year. They lived for another year or two without molting and then died."* Also, they wrote that "T. tridentatus *grows much slower than* L. polyphemus, *as shown in Figure [12.11A]; further,* T. tridentatus *is very slow in feeding action."* In other words, the horseshoe crabs in these experiments did not feed and grow well and died prematurely, as most animals do when kept under stressful conditions.

222. Oddly, the fact that a *'Deus ex Machina'* – here an additional physiological process – needs to be evoked to stop (at some point) the growth of horseshoe crabs as described by Shuster and Sekiguch (2003), is similar to issues raised by the growth concept of Scheuffele *et al.* (2021; see **Figure 6.4**), which also requires an external agent to stop the growth of fishes. Asymptotic growth does not have this problem.

223. Radical changes in the feeding habits of horseshoe crabs, such that would suddenly reduce their cost of food acquisition (as in the case of fish transiting from zooplankton to feeding on other fish, as discussed in Chapter 10) and accelerate their growth, have never been demonstrated, despite multiple studies on the feeding habits of horseshoe crabs (see, e.g., Botton 1984; Botton and Shuster 2003; Chatterji *et al.* 1992). It is, therefore, more parsimonious to assume that the grafting of the growth patterns of late juvenile adult horseshoe crabs onto the earlier growth patterns of individuals held in captivity misled the authors, who had no theoretical conception of growth to constrain their assumptions.

224. As may also be expected, the level of spontaneous activity in crustaceans is reduced when the ambient pH level is low (see, e.g., Jarrett *et al.* 2024).

Chapter 13

225. This probably also applies to abalones (*Haliotis* spp.), as the growth of their juveniles has been shown to depend on ambient oxygen levels (Harris 1999).

226. Note that large adults are typically more vulnerable to external hypoxia despite their superior capacity to generate ATP through anaerobic metabolism. While their higher tolerance to the waste products of anaerobic metabolism gives them an advantage during short-term hypoxia, this resource is not endless. Once it is exhausted, their higher maintenance costs make them more vulnerable (see Chapter 4 and Müller *et al.* 2023). Another study, on the bivalve *Aequipecten opercularis*, suggested that larger individuals might have lower anaerobic capacities (Schmidt *et al.*

2008). However, as the authors showed, the larger individuals opened their shells for longer periods than smaller conspecifics after being stressed, which reflects longer regeneration time.

227. These asymptotic shell lengths of *Limacina helicina* correspond to 20 and 30 mg (wet weight; WW), respectively, based on the length-weight relationship DW = $0.257 \cdot L^{2.141}$ and (dry weight) DW = WW·0.28 in Bednarsek *et al.* (2012); see **Figure 13.3**.

228. Lavin *et al.* (2022) confirmed that this pattern (TSR) occurs not only for Wellington squid along a South-North gradient around New Zealand, but for the 4 fish and 1 crustacean species that also they studied.

229. The largest (i.e., longest) invertebrate is the giant squid (*Architeuthis dux*), whose females grow to 13 m; however, most of this length consists of tentacles, while the mantle contributes only 2.25 m; the males are smaller (see Wikipedia).

230. The heaviest invertebrate appears to be the colossal squid (*Mesonychoteuthis hamiltoni*), which reaches 500 kg (Rosa *et al.* 2017) and may even reach 600-700 kg (see Wikipedia).

231. Still, in the 21st century, one can read that, in giant squid, the "*age estimation remains controversial with many estimates of maximum age ranging from 1 to 14 years*" (Perales-Raya *et al.* 2020). However, the 5 untenable tenets, which have been applied to squid and, to a somewhat lesser extent, to octopuses, do not seem to have been applied to cuttlefish, whose length-frequency data are routinely studied with ELEFAN, an approach and software which assume asymptotic growth (Pauly and David 1981; Gayanilo *et al.* 2005). One result is that numerous estimates of cuttlefish VBGF parameters are available for several species from multiple locations, as can be ascertained with the search term 'sepia+bertalanffy' in Google Scholar (see also Pauly *et al.* 2024b). At least in *Sepia* spp., asymptotic growth and longevities of 3 to 4 years are non-controversial.

232. See Pauly (1998c) and Pauly (2019, p. 113-117) for the mechanism that links hypoxia and large sizes to the non-formation of the 'daily' markings on squid statoliths and fish otoliths.

233. Hypoxic waters, common in the Eastern Pacific between 200 and 1000 m, are where myctophids and other mesopelagic fishes 'hide' during the day and where *D. gigas* often feed on them. At dusk, these fishes migrate toward the sea surface, where they feed and where tuna can feed on them. Then, at dawn, they migrate back to their hypoxic refuge (Pauly *et al.* 2021b).

Chapter 14

234. Amphipods (such as *Leucothoe spinicarpa*) occurring in sponges may be seen as parasitic: while they do not consume parts of their host, they do consume bacteria and detritus which their sponge host would assimilate.

235. Note that sponges, over their long evolutionary history, have turned anoxia from a liability into an asset (Hentschel *et al.* 2006).

236. Seven of the 8 datasets in this this graph were presented by Hirst and Forster (2013) in their paper titled "*When Growth Models Are Not Universal: Evidence from Marine Invertebrates.*" These authors do not identify why "*some growth models*" fail. Also, contrary to what they stated, the VBGF did not fail in the 7 cases of chaetognaths that they presented. Finally, it can be expected that many of their other cases were based on subjective analysis of length-frequency data as illustrated in **Figure 13.5**. The point here is that while the VBGF is not universal, it has a wide range of applicability which should not be ignored because some people don't know how to use objective methods to fit growth curves to length-frequency data.

Chapter 15

237. We note here that there are numerous studies on metabolic rate (i.e., O_2 consumption) of chaetognaths (see Pauly *et al.* 2021), very little thought has been devoted to how they breathe, i.e., how they transfer oxygen dissolved in the water surrounding them into their bodies. For example, there is no mention of breathing, respiration, or oxygen in the otherwise wonderfully detailed 12 chapters in the book '*The Biology of Chaetognaths*' (Bone *et al.* 1991).

238. Critiques of simplistic views on adaptationism are now often associated with Gould and Lewontin's 1979 paper on "*the Spandrels of San Marco and the Panglossian Paradigm,*" named after the optimistic Leibnitzian doctor Pangloss in Voltaire's *Candide*. According to Dr. Pangloss, "*everything is made for the best purpose,*" and every trait or feature of an organism has a function that is perfectly and optimally capable of fulfilling it (Gould and Lewontin 1979). This view, as Gould and Lewontin asserted, was still very much alive in the world of Anglophone evolutionary biologists,

who tried to interpret every feature of an organism as an adaptation selected for in the course of evolutionary processes. Using several examples, from adaptationist explanations of alleged Aztec cannibalism to physiognomic speculations about the form of human faces in different climates, Gould and Lewontin's argument centered around an architectural metaphor: the spandrels between the pillars of medieval cathedrals whose inlay was highly decorated with ornamental figures depicting the Christian history of salvation. Adaptationists like the successors of the pedantic Dr. Pangloss, they asserted, could not help but see a deeper purpose, meaning, or adaptive advantage in these decorated spandrels, while their existence is explained far more prosaically: where there are pillars, there must also be space between the pillars (since pillars without space in between would be nothing but a wall). The spandrels were not an adaptation in themselves but rather a necessary by-product of the construction of pillars. While this critique of adaptationism was undoubtedly important, the problem of assuming infinite optimization is more fundamental. None of the morphological features of fish gills can be considered genuine "*spandrels,*" and constraints to their development do not arise as mere "*by-products*" of their evolution but are indistinguishable from their functional optimization. In the context of Darwinian theory, it is not necessary to distinguish between optimized features and their *non*-optimal by-products; their optimization can constrain their functionality.

239. Similar ideas and observations entered anthropological debates in the late eighteenth and early nineteenth century, where human opportunities were also explained as consequences of "*deficiencies.*" The characterization of humans as "*deficient beings*" ("*Mängelwesen;*" Herder 1772) is still pre-Darwinian, but it intriguingly describes how *Homo sapiens* have never fully adapted to specific ecological niches: compared to other species, humans perform very poorly as swimmers, climbers, runners, etc. It would be more apt to state that the human body has evolved *from* all previous niches, and its crucial physical deficiencies might be the only legitimate foundation for any ideas of human exceptionalism. In short, because of these deficiencies or *limitations*, humans do what they do – build computers, write books, play music, or study fish.

240. For global biogeographical patterns, see Chapter 5 and Blanchet *et al.* (2010); for paleontological data that show correlates between temperature and body size in fishes, see, e.g., Troyer *et al.* (2022), Shimada *et al.* (2023), and Adhikari (2015).

241. An earlier formulation of this idea was "*Evolution is like a game, but the only payoff is to stay in the game*" (Slobotkin 1964)

242. Many manifestations may be better understood regarding negative selections, i.e., the selection of non-viable traits *from* the gene pool. As Holly Dunsworth put it, evolutionary adaptation is mostly not about the "*survival of the fittest*" but rather the "*survival of that which does not suck too badly*" (Holly Dunsworth, pers. comm. to J.M., 29 December 2021). Mark Costello (pers. comm. June 2025) commented on this "*I note that all molluscs suck, because they cannot get large items of solid food through their throat because if the ganglia encircling it. Another odd feature is their waste excreta discharge near the gills. Every adaptation has a cost and limitation, as do evolutionary constraints.*"

Chapter 16

243. Wootton *et al.* (2022) also measured gonadal weight and concluded that investment in reproductive activities might be responsible for reduced growth, implying a mechanism similar to the 'Reproductive Drain Hypothesis' as discussed in several chapters. However, the results of the experiment fail to support this hypothesis. Two points are important here:

As their figure 3 clearly shows, gonad weights in mature F6 females are significantly lower in warmer than in colder temperatures. This allocation of energy to reproduction (i.e., gonad weight) implies that energy is available at lower temperatures that would be needed for other purposes at higher temperatures (e.g., replacing proteins denatured by the higher temperature; see Chapter 4). This observation supports the limitation hypothesis: when this limitation is reduced, more energy (i.e., oxygen) is available for other purposes.

The data presentation leaves two points unclear: the scaling of *SMR* and juvenile growth. Both parameters are extensively discussed in the text and the supplementary document, but only for mature and post-spawn adults. In a short and evasive remark, the authors claim that "*juvenile SMR remained similar between temperatures yet by the sixth-generation juveniles were ~30% larger (in weight).* However, it is unclear how this statement relates to their figure 2a and 2b. These figures suggest a direct correlation between (temporarily) increasing *SMR* and juvenile growth in the F6 generation. Why are the data on juveniles not included in the supplementary document?

As long as this point remains unclear, it would be reasonable to conclude that the short early growth spurt that is needed to reach (earlier) maturation is 'paid for' by the short spike in *SMR* (note that juvenile *SMR* is higher in the F6-generation than in F1, which seems to contradict the paper's argument). The organism can only lower its *SMR* at a stage where the organism has virtually stopped growing.

The data on gonad weight (1) and juvenile growth (2) support the limitation hypothesis. The only way to challenge this view would be to present more comprehensive data on juvenile growth, and it is not clear why they are lacking. If figures 2a and 2b are correct, the answer should be clear: fish adapt to elevated temperatures by reducing their *SMR*, but this reduction comes at the cost of growth. The limitation hypothesis is also supported by the fact that *MMR* remains high and adaptation to lower rates is not possible over six generations: this implies that some energetic costs within the *SMR* 'budget' need to be reduced. This is also incompatible with the idea that higher temperatures would redistribute costs of growth reproduction – where would that energy come from if it is invisible in the (lower) SMR measurements? What is new and important about the paper is the insight that adaptation to higher temperatures occurs via the reduction of *both* growth and *SMR*. However, this generates even more support for a hypothesis the paper seeks to refute. The title might be turned around: it is not a *higher* but (adaptation to) lower standard metabolic rates that explains growth reduction.

244. In a recent review of different metabolic scaling theories, Glazier (2022) proposed a shift from nomothetic "*Newtonian to Darwinian approaches*" to scaling in living systems. According to him, earlier theories focused too much on universally valid patterns by proposing power laws of either 2/3 or 3/4. Such theories ignored the fact that natural selection acted upon scaling relationships in optimizing ways and allowed adaptation to specific environments or lifestyles. Glazier argues that a "*paradigm shift*" was needed to overcome outdated "*Newtonian*" models and acknowledges the importance of context-dependent adaptive processes. While this criticism may apply indeed to some theories of scaling, for example, to the models proposed by Pütter (1920), Kleiber (1934), or von Bertalanffy (1951), the GOLT shows that such a "*paradigm shift*" is not needed and that this conundrum can be overcome by applying a mechanistic model that explains the factors upon which natural selection can act. In its earliest versions (Pauly 1979, 1981), the GOLT provided a clearly defined mechanism for the respective metabolic slopes in different water-breathing animal species. Depending on the lifestyle and environment of an animal, selection favors specific metabolic slopes, and it can clearly act towards optimization, both on the sides of demand and supply.
245. Note that these authors focused on interspecific scaling patterns for which reductions in maintenance demand as a strategy to enable allometric scaling are far more plausible than changes in ontogenetic scaling.
246. Strangely enough, the focus on *causal* relationships is recent in various scientific disciplines. Throughout much of the 20th century, papers presenting only correlative associations were happily accepted. Fortunately, this has changed, and there is now much more emphasis on mechanistic explanations (see Pearl and MacKenzie 2018). A foundational text on causal inferences in biology is Mayr (1961).
247. In addition, it may be relevant to note that the cited study by Elliot (1976) states that fishes kept on maintenance rations "*met their requirements for maintenance metabolism and if these requirements were not fulfilled by the daily energy intake, then the extra energy was obtained from body tissues.*" The conclusion of White and Marshall (2023b) that maintenance metabolism (understood as excluding the costs of growth and reproduction) declined with size is therefore premature in relation to this study. If larger animals received smaller food rations, the study suggests that they compensated for the lack of food by consuming energy stored in their body tissues. The capacity to obtain energy for maintenance from body reserves usually depends on body mass and is greater in larger animals.
248. In this respect, our argument is similar to and partly compatible with Banavar *et al.* (2002). While these authors focus on fractal-shaped transportation networks in the vein of the *Metabolic Theory of Ecology* (with scaling exponents of 3/4), their model relies on adjusting demand to supply. Banavar *et al.* (2002) did not detail "*the physiological mechanisms that result in the maintenance of approximate equality of the scaling exponents of r_1 and r_2 [i.e., supply and demand]*" and state that "*the actual mechanisms may differ among different types of organisms,*" we believe that we were able to define the most critical mechanisms (for at least aquatic ectotherms) on the right side of **Table 16.1**.
249. It is telling that demand-based models do not present mechanisms for diminishing maintenance metabolic rate at larger body size (White and Marshall 2023b; Kozłowski *et al.* 2020; see also Harrison 2017, who, however, only

discusses interspecific scaling patterns). The only hypothesis that would justify such a model would be changes in tissue composition and a lower percentage of tissues with a high maintenance demand (an assumption that seems to underlie the reasoning of Lefevre *et al.* 2017). While this may indeed be the case in many animals, and perhaps more so in long-lived species, it is unclear how important these reductions are for maintenance energy consumption. Also, it is unclear if they genuinely represent a decline in demand or rather illustrate that supply capacities only allow for the addition of such economic tissues.

250. Plasticity itself has a cost, because "*everything in life has a cost, so expanding gills in real time to cope with hypoxia, which is questionable unless it is a very slow depletion, if even possible, will have other costs to the individual (energetic, osmotic, etc.)*" (Mark J. Costello, pers. comm. June 2025).
251. Although Kozłowski's model of metabolic scaling seems to fundamentally differ from the basic assumptions of the GOLT, we agree with his account of evolutionary optimization regarding scaling patterns on an interspecific level. In addition, we believe that his argument that "*cells never sleep*" aligns with the GOLT's interpretation of the catabolic term of Pütter's equation. Rather than representing this term as an oxygen-consuming process that produces energy for maintenance, we interpret it as the breakdown of cellular material that needs to be either repaired or renewed.
252. For example, the model of Scheufele *et al.* (2021; see **Figure 6.3**) requires external factors to stop the growth of fish at some point, presumably a different one for each population, which would require identifying and quantifying before any prediction, e.g., of the effect of warming on maximum size (and William of Ockham turns in his grave).
253. In this respect, we agree with Kozłowski *et al.* (2020).

Chapter 17

254. While similar accounts of causality have been developed for neural control mechanisms (see, e.g., Bechtel 2017), they are also applicable to metabolic supply and demand questions, even though more theoretical engagements with this issue would be useful.
255. This is mainly true in terms of biodiversity and species richness. Regarding total biomass, plants dominate, but are followed by bacteria (Bar-On *et al.* 2018). Animals comprise 2 gigatons of carbon (out of 546 gigatons of the entire biosphere), half of which consists of arthropods.
256. On land, the challenges are gravity (see Chapter 1) and others (desiccation, heat, etc.) that we don't need to address here.
257. The article by Regier *et al.* (1990) contains numerous bivariate graphs documenting that numerous temperature-related biological phenomena, and not only fish egg development time, can be represented by 'Arrhenius plots,' which are (jointly with the underlying equations) related to various aspects of the GOLT. This theme is not followed up on here, but it would probably open a fruitful line of research.
258. More precisely, the estimated mean ratio was 1:1.52, with a 95% confidence interval of 1.39 to 1.63. Also note that an earlier account suggested that oxygen availability provided "*a signal for hatching*" (Latham and Just 1989).
259. Despite what the title of their article promises, Gajbhiye *et al.* (2024) did not present the mechanism that "controls" (i.e., *causes*) hatching in fish. They assume that "[t]*he decision to hatch is often carefully timed to coincide with favorable conditions that will improve survival though through early life stages*" which is certainly not the case. If it were, it wouldn't be possible to predict precisely egg incubation times only from their diameter and the temperature to which they are exposed using, e.g., Equation (17.2).

    What these authors did, rather, is identify "*how the relevant cues are relayed to trigger hatching,*" which turned out to be thyrotropin-releasing hormone (TRH). But they missed out on the real "*relevant cues,*" which we show are sent by growing embryos running out of oxygen.
260. In this respect, salamanders are an interesting exception, as many species hatch with fully developed gills or even possess gills while still inside the egg. Whether this morphological trait represents an adaptation to high predatory pressure and for quick growth (which would be enabled by an adequate oxygen uptake capacity) is a hypothesis that could be tested by comparing the development times of amphibians with and without gills at the immediate post-hatching stage. For more information on oxygen as a size-limiting factor in larval amphibians; see Rollinson and Rowe (2018a, 2018b).
261. While Haeckel's recapitulation theory (see Haeckel 1877) can no longer be regarded as valid, the notion that "*phylogeny recapitulates ontogeny*" may hold at least metaphorically when it comes to the evolution of body sizes

since individual organisms (and entire taxa) need to find ways to overcome surface constraints if they are to grow beyond larval size.

262. Unfortunately, we couldn't cover *all* metazoan phyla in this book. Still, a cursory examination of the literature on the growth and respiration of species in groups such as the nematodes (Bird 1971; see also **Figure 4.2**), flatworms (Whitney 1942), annelids (Davies *et al.* 1992; Forbes and Lopez 1990), echinoderms (Ebert 1975; Nauen 1978), cephalochordates (Pauly and Chu 2021) and others suggest that we could have straightforwardly written accounts documenting that the GOLT applies to them, similar to those included here on arthropods (Chapter 12), molluscs (Chapter 13) and sponges, cnidarians and chaetognaths (Chapter 14).

263. ... although all the animals depending on these bacteria still need O_2 from the water. On the other hand, a newly discovered (but still-controversial) process involving seafloor nodules that produces 'dark oxygen' without photosynthesis (Sweetman *et al.* 2024), may also occur in the oceans of alien moons.

264. The regulative role of oxygen constraints for maintaining functioning ecosystems is also illustrated by the invasive potential of species that can overcome these constraints, such as air-breathing fishes that quickly take over new habitats. The massive invasion of Asian snakeheads (*Channa* spp.) and South American loricariid catfishes in the United States and elsewhere are illustrative examples (for more examples, see Knight 2010; Mäkinen *et al.* 2013; Hossain *et al.* 2018).

265. It even seems that the GOLT, at least in part, applies to kelp, which also 'breathe' water. Robert Steneck, proposed a 'KOLT,' and the following argument:

"Anatomical limitations constrain an organism's ability to obtain oxygen. This holds both for oxygen consumers such as fishes and crustaceans (GOLT) and for oxygen producers such as kelp (KOLT). Warming oceans lead to oxygen limitation for fish and water-breathing invertebrates. For kelp, warming creates internal hypoxia by affecting respiration relative to photosynthesis.

Kelp are the world's largest algae. These large brown algae evolved convergently in two orders (Laminariales and Fucales) during the Cretaceous (Bringloe et al. *2020), but they only evolved large size after the poles cooled in the last 15 million years. Kelp uniquely has the lowest ratio of photosynthetic area to biomass and this cause a nighttime reduction of O_2 in the thick 'tissues' of kelp. This effect leads to two overarching patterns of the KOLT: (i) kelp can only live in cold-water settings; as water temperatures warm, kelp biomass declines and fronds becomes smaller and more dissected, which leads to higher surface area to volume ratios (González* et al. *2012), and (ii) with warmer temperatures, photosynthesis first increases, but eventually levels off and declines whereas respiration keeps increasing (Davison* et al. *1991; Staehr and Wernberg 2009).*

Item (i) was effectively tested in the natural experiments provided by marine heatwaves, whose . acute warming results in kelp loss (Steneck et al. *2002; Smale* et al. *2019). Item (i) has also been documented in cases of gradually increasing sea temperatures (Suskiewicz* et al. *2024). Usually there is a threshold temperature between 15 and 20 oC where kelp become pale and scarce at depth due to low light and reduced primary productivity relative to respiration (e.g., Smith* et al. *2022).*

The phenomenon described here is rather simple. The low photosynthetic surface area to volume ratio of kelp requires lower rates of respiration than the increasingly warmer at high latitudes can sustain. As a result, former kelp forests are being replaced by diminutive filamentous 'turf' algae that have an large high surface area to volume ratio (Filbee-Dexter and Wernberg 2018). It it can be argued that both GOLT and KOLT link reduced sizes with increasing ocean temperatures." (R.S. Steneck, University of Maine, pers. comm. March 10, 2025).

266. The applicability of the GOLT to the echinoderms, annelids such as the polychaetes, the ribbon worms (Nemertea) and other taxa with hydrostatic skeleton has not been demonstrated. But Bahtiar *et al.* (2024), who cited positively several GOLT-related articles, demonstrated that in the peanut worm (*Siphonosoma australe-australe*), a sipunculid consumed in various Asian and Pacific countries, the size at first maturity increases with the observed maximum size.

267. This does not apply to the tilapia (Cichlidae), notably Nile tilapia (*Oreochromis niloticus*) which is extremely important in aquaculture, and where it is the females that care for their young by keeping them in their mouths. Here, it is male mouth-brooding, as in backchin tilapia (*Sarotherodon melanotheron*; Pauly 1976) that is the exception.

268. These levels of pain were assessed by various objective methods, the most important being the quantification of chemicals being produced (e.g., when a fish is pulled out of the water) that directly impact on the fish' nociceptors.

We should also note, for 'sport' fishers who do not know the origin of the work 'excruciating' that it refers to the intensity of pain one experiences when crucified. Happy angling!

269. Some of the papers and books on the bioenergetics of WBE (and predominately fish) are Brody (1945), Hartman and Brandt (1995), Hansen *et al.* (1993), Jobling (1993), Niimi and Beamish (1974), van Poorten and Walters (2016), Walters *et al.* (2012), Wu and Or (2005). Several of these references contradict each other or propose solutions (i.e., models) with so many parameters that they can only describe things but can't make interesting predictions. Still, it should still be possible to identify some basic principles that are common to several of these models and compatible with the GOLT.

270. Several dozen of recent articles on bioenergetics could be cited here but they do not even mention oxygen. There are also books that review 'all' aspects of the biology of various WBE, except they fail to mention that they respire, as well.

271. This index omits the names of broad grouping such as 'metazoans,' 'animals,' 'vertebrates' or 'fish,' and most cognates and plural forms of the names that are included. The currently valid scientific (and English common) names of freshwater and marine fish were checked against FishBase (Froese and Pauly 2024, http://www.fishbase.org); while those of marine invertebrates were checked against SeaLifeBase (Palomares and Pauly 2024, http://www.sealifebase.org); therefore, they may differ from those used in the references cited here.

272. Since the focus of this is on the commonalities that are shared between different water-breathing taxa and not so much on the particularities of individual species or local ecosystems or regions, the subject does not geographical entries.

Taxonomic index

Subject index